AGROBIODIVERSITY
Characterization, Utilization and Management

Dedicated to Mirrie Lenné; for her support

Agrobiodiversity
Characterization, Utilization and Management

Edited by

D. Wood
Agrobiodiversity International, Milnthorpe, UK

and

J.M. Lenné
Agrobiodiversity International, Milnthorpe, UK

CABI *Publishing*

CABI *Publishing* is a division of CAB *International*

CABI Publishing
CAB International
Wallingford
Oxon OX10 8DE
UK

CABI Publishing
10 E. 40th Street
Suite 3203
New York, NY 10016
USA

Tel: +44 (0)1491 832111
Fax: +44 (0)1491 833508
Email: cabi@cabi.org

Tel: +1 212 481 7018
Fax: +1 212 686 7993
Email: cabi-nao@cabi.org

A catalogue record for this book is available from the British Library, London, UK.

Library of Congress Cataloging-in-Publication Data
Wood, D. (David)
    Agrobiodiversity : characterization, utilization, and management /
edited by D. Wood and J.M. Lenné.
        p.   cm.
    Includes bibliographical references and index.
    ISBN 0-85199-337-0 (alk. paper)
    1. Agrobiodiversity.   I. Lenné, Jillian M.   II. Title.
S494.5.A43W66   1999
333.95′34--dc21   98–52952
                                                                       CIP

ISBN 0 85199 337 0

Typeset by Solidus (Bristol) Limited
Printed and bound in the UK by Biddles Ltd, Guildford and King's Lynn

Contents

v

The colour plate section can be found between pages 34 and 35

Acknowledgements

The editors and publishers would like to thank the UK Department for International Development (DFID) Crop Protection Programme (CPP) for their financial contribution towards the colour plate section and the colour cover. Further information about the CPP can be found via the Internet at http://www.nrinternational.co.uk.

Contributors

C. Abivardi, Geobotanisches Institut ETH, Zürichbergstrasse 38, 8044 Zürich, Switzerland

D.J. Allen, Higher Quantock, Stockland, Honiton, Devon EX14 9DX, UK

G.M. Barker, Landcare Research, Private Bag 3127, Hamilton, New Zealand

M.R. Bellon, CIMMYT, Apartado Postal 6-641, 06600, Mexico DF, Mexico

T.S. Cox, B-128 Sainikpuri, Secunderabad 500094, India

P.J. Edwards, Geobotanisches Institut ETH, Zürichbergstrasse 38, 8044 Zürich, Switzerland

J.M.M. Engels, IPGRI, Via delle Sette Chiese 142, 00145 Rome, Italy

K.E. Giller, Department of Biological Sciences, Wye College, University of London, Wye, Ashford, Kent TN25 5AH, UK; currently at Department of Soil Science and Agricultural Engineering, University of Zimbabwe, Box MP 167, Mount Pleasant, Harare, Zimbabwe

D. Gisselquist, International Economics Department, The World Bank, 1818 H Street NW, Washington, DC 20433, USA

J. Kollmann, Geobotanisches Institut ETH, Zürichbergstrasse 38, 8044 Zürich, Switzerland

J. LaSalle, Unit of Parasitoid Systematics, CABI Bioscience UK Centre, Department of Biology, Imperial College at Silwood Park, Ascot, Berkshire SL5 7PY, UK

J.M. Lenné, Agrobiodiversity International, 13 Herons Quay, Sandside, Milnthorpe, Cumbria LA7 7HN, UK

R. McDowell, Department of Animal Science, North Carolina State University, Raleigh, NC 27695, USA

J.-L. Pham, Genetic Resources Centre, International Rice Research Institute, PO Box 933, 1099 Manila, Philippines

A. Polaszek, CABI Bioscience UK Centre, Silwood Park, Ascot, Berkshire SL5 7PY, UK

C. Riches, Natural Resources Institute, University of Greenwich, Central Avenue, Chatham Maritime, Kent ME4 4TB, UK

W. Richner, Institute of Plant Sciences, ETH-Zentrum, 8092 Zürich, Switzerland

J. Salick, Department of Botany, Porter Hall, Ohio University, Athens, OH 45701, USA

M. Smale, CIMMYT, Apartado Postal 6-641, 06600, Mexico DF, Mexico; currently Visiting Research Fellow, International Food Policy Research Institute (IFPRI), 2033 K Street N.W., Washington, DC 20006, USA

M.E. Smith, Department of Plant Breeding, Cornell University, Ithaca, NY 14850, USA

D. Steane, Conservation and Use of Animal Genetic Resources in Asia and the Pacific, FAO Regional Office, Maliwan Mansion, 39 Phra Atit Road, Bangkok 10200, Thailand

H.D. Thurston, Department of Plant Pathology, Cornell University, Ithaca, NY 14850, USA

P. Trutmann, Institut für Agrarwirtschaft, ETH-Zentrum, 8092 Zürich, Switzerland

M. Turner, Seed Unit, ICARDA, PO Box 5466, Aleppo, Syria

J.M. Waller, CABI Bioscience UK Centre, Bakeham Lane, Egham, Surrey TW20 9TY, UK

D.A. Wardle, Landcare Research, PO Box 69, Lincoln 8152, New Zealand

J.R. Witcombe, Centre for Arid Zone Studies, University of Wales, Bangor, Gwynedd LL57 2UW, UK

D. Wood, Agrobiodiversity International, 13 Herons Quay, Sandside, Milnthorpe, Cumbria LA7 7HN, UK

M. Wright, VSO Programme Office, PO Box 137, Belmopan, Belize

Acronyms and Abbreviations

AFLP	amplified fragment length polymorphism
AI	artificial insemination
ai	active ingredient
BLUP	best linear unbiased prediction
CATIE	Centro Agronómico Tropical de Investigacíon y Enseñanza
CBD	Convention on Biological Diversity
CGIAR	Consultative Group on International Agricultural Research
CGRFA	Commission on Genetic Resources for Food and Agriculture
CIMMYT	Centro International de Mejoramiento de Maíz y Trigo (International Maize and Wheat Improvement Centre)
COGENT	Coconut Genetic Resources Network
COP	Conference of the Parties
CP	coefficient of parentage
CT	conventional tillage
CTIC	Conservation Tillage Information Centre
DAD-IS	Domestic Animal Diversity Information System
DNA	deoxyribonucleic acid
DUS	distinct, uniform, stable
EBV	estimated breeding value
ECP/GR	European Cooperative Programme for Genetic Resources
EPA	Environmental Protection Agency
ET	embryo transfer
FAO	Food and Agriculture Organization of the United Nations
FIS	International Seed Trade Federation
FLB	fingerprint linkage blocks
GEF	Global Environment Facility

GM	genetically modified
GPA	Global Plan of Action
HRS	highly repeated sequences
IARC	International Agricultural Research Centre
IBPGR	International Board for Plant Genetic Resources
ICARDA	International Centre for Agricultural Research in Dry Areas
*i*DAD	Initiative for Domestic Animal Diversity
IDRC	International Development Research Centre
ILCA	International Livestock Centre for Africa (now ILRI International Livestock Research Institute)
INIBAP	International Network for the Improvement of Banana and Plantain
IPGRI	International Plant Genetic Resources Institute
IPM	integrated pest management
IPR	intellectual property rights
IRRI	International Rice Research Institute
ISTA	International Seed Testing Association
IUCN	International Union for the Conservation of Nature and Natural Resources
IVF	*in vitro* fertilization
IVM	*in vitro* maturation
LAMP	Latin American Maize Programme
MAS	marker-assisted selection
MV	modern varieties
NFI	net feed intake
NGO	non-governmental organization
NPP	net primary productivity
NT	non-tillage
OECD	Organization for Economic Cooperation and Development
ONBS	open nucleus breeding system
PCR	polymerase chain reaction
PGRFA	Plant Genetic Resources for Food and Agriculture
PLFA	phospholipid fatty acids
PPB	participatory plant breeding
PVP	plant varietal protection
PVS	participatory varietal selection
QTL	quantitative trait loci
RAPD	random amplified polymorphic DNA
RFI	residual feed intake
RFLP	restriction fragment length polymorphism
SADC	Southern African Development Council
SBSTTA	Subsidiary Body for Scientific, Technical and Technology Advice
SGRP	System-wide Genetic Resources Programme
SINGER	System-wide Information Network for Genetic Resources

SSR	simple sequence repeats
TIC	transgenic insecticidal cultivar
TRIPS	Trade Related Intellectual Property Rights Agreement
UNDP	United Nations Development Programme
UNEP	United Nations Environment Programme
UNESCO	United Nations Educational, Scientific and Cultural Organization
UPOV	Union for the Protection of New Plant Varieties
USDA	United States Department of Agriculture
VCU	value for cultivation and use
WCGALP	World Congress on Genetics Applied to Livestock Production
WTO	World Trade Organization
WWF	World-wide Fund for Nature
WWLDAD	World Watch List for Domestic Animal Biodiversity

Why Agrobiodiversity?

D. Wood and J.M. Lenné

Agrobiodiversity International, 13 Herons Quay, Sandside, Milnthorpe, Cumbria LA7 7HN, UK

> the dynamism of agrodiversity, a constantly changing patchwork of relations between people, plants, and the environment, always coping with new problems, always finding new ways
>
> (Brookfield, 1998)

Within the past decade there has been an explosion in the use of the word 'biodiversity'. Biodiversity has its own International Convention. Biodiversity is widely referred to in the titles of books, scientific and popular articles, and grant proposals. Biodiversity has now become a 'good thing' which all agree should be preserved for the future, the more the better.

'Biodiversity' refers to all living things and the interactions between them: a vast array of organisms with an almost infinite complexity of relationships. Given this complexity, and given the range of disciplines – ecology, systematics, conservation biology, economics and many others – that predated, and now contribute to, our concepts of biodiversity, there are many and contrasting views on how to value, manage and conserve biodiversity. Repeatedly we are being told that time is running out for biodiversity – forests are shrinking, climate is changing, pollution is spreading, species are winking out – and that there is now a biodiversity crisis (Wilson, 1985).

Agriculture and Biodiversity

Agricultural biodiversity – the 'agrobiodiversity' of our title – is a subset of biodiversity. It is, however, an exceptionally important subset. 'Agrobiodiversity' has been defined by Qualset *et al.* (1995) as including all crops and livestock and their wild relatives, and all interacting species of pollinators, symbionts, pests, parasites, predators and competitors. Under this definition, the immediate importance of agrobiodiversity is clear: our interaction with crops and domestic animals directly provides most of our food. Food from the wild (of which fish is the most important) can meet less than 3% of the food needs of a developed country (our estimate for the USA from Prescott-Allen and Prescott-Allen, 1986). We exclude wild plants and animals of food value outside the agroecosystem. Although such wild food is often of critical importance to farm families (Scoones *et al.*, 1992), it is not part of the farm and is not agrobiodiversity. We use a broader concept of 'agroecosystem' than Conway (1987): 'an ecological and socio-economic system, comprising domesticated plants and/or animals and the people who husband them, intended for the purpose of producing food, fibre, or other agricultural products.' Agrobiodiversity includes far more than the husbandry of domesticated plants and animals.

The interactions of 'food agrobiodiversity' with other agrobiodiversity within agroecosystems – including pests, pathogens, pollinators and much else – are also important, determining whether there will be more food or less food. These interactions may be direct, as with insects eating crops, or indirect, through the modification of agroecosystem function. There are also complex interactions with the wild, including the transfer into the agro-ecosystem of pollen and seed from wild relatives, and the migration of pests, pathogens, predators and weeds: these too impact directly on food produc-tion and on agroecosystem function.

Our usage of 'agrobiodiversity' therefore closely follows that of Qualset *et al.* (1995). Crops and domestic animals are firmly included in 'agro-biodiversity' (Fig. 1.1). Also included are the great range of organisms above

Distribution of agrobiodiversity

Fields	Pastures/rangeland	Wild
Crop *agrobiodiversity*	Not applicable	*Agrobiodiversity* of wild relatives
Domestic animal *agrobiodiversity*		
Agrobiodiversity of pasture species		
Harmful *agrobiodiversity*: weeds/pests/deleterious microorganisms		
Useful *agrobiodiversity*: pollinators/IPM agents/useful microorganisms		

Impacts of agriculture on wild biodiversity

'Incidental' biodiversity in agroecosystems	Not applicable
Wild foods (economic complementarity)	
Wild biodiversity (by 'land-saving')	

Fig. 1.1. The relation of components of agrobiodiversity to both agriculture and wild biodiversity.

and below ground that can harm or help agriculture, such as pests, diseases and weeds, pollinators and biological control organisms, and the many organisms controlling nutrient cycling. The importance of the functional role of agrobiodiversity has been stressed by Swift and Anderson (1994) in their division of the biotic components of agroecosystems into three types: productive, resource (beneficial) and destructive. *Productive* biota include crop plants and livestock, producing food, fibre or other products for consumption. *Resource* biota contribute positively to the productivity of the system: pollinators, plants of fallows, and much soil biota controlling nutrient cycling are included here. *Destructive* biota include weeds, pests and pathogens. As stressed by Swift and Anderson, this classification underpins management – the role of farmers and agricultural scientists in increasing crop and animal production by encouraging resource biota and discouraging destructive biota.

Significant agrobiodiversity may be found well outside agroecosystems. For example, wild potatoes (*Solanum demissum*) in Mexico represent agrobiodiversity of value for breeding resistance to potato blight – *Phytophthora infestans*. Yet potatoes were domesticated several thousand kilometres to the south, in the Andean highlands. The wild rice, *Oryza nivara*, is an important source of resistance to grassy stunt virus in rice, but is found beyond the limits of rice fields, as a weed of waste places. Agrobiodiversity represented by pests and diseases may move intercontinentally, and between agroecosystems and the wild. Desert locusts, multiplying in the natural vegetation of northeast Africa and the Arabian peninsula, were until recently a significant part of the seasonal agrobiodiversity of the fields of eastern Africa.

In contrast with the existence of agrobiodiversity outside agroecosystems, not all biodiversity regularly present within agroecosystems should be regarded as 'agrobiodiversity'. Function will determine this, rather than just presence: agrobiodiversity has an actual or potential impact on agricultural production. Much biodiversity in agroecosystems may have no specific role in agriculture: indeed, it has been estimated that a greater part of terrestrial biodiversity persists in agricultural landscapes. Such landscapes could therefore be a key to global biodiversity conservation: 'The struggle to maintain biodiversity is going to be won or lost in agricultural systems. Management of agricultural landscapes will be the litmus test of our ability to conserve species . . . most terrestrial biota will eventually have to coexist with human agriculture' (McIntyre *et al.*, 1992). However important, this 'incidental' biodiversity falls outside our definition of 'agrobiodiversity', yet its conservation could be a bonus of sound agroecosystem management.

Even with an agreed definition, the physical and temporal boundaries of agrobiodiversity may be blurred. Bennett (1992) illustrates the problem of delimiting tropical agroecosystems spatially: 'Unlike temperate ecosystems, there is often very little distinction between cultivated species and wild species. Neither is there a clear boundary between fields and fallows or between fallows and forests.' Similarly, for Alcorn (1989): 'The traditional agroecosystem is a fluid complex of planted fields, fallows, savannas, dooryards, forests, river, and river banks – the entire range of resource zones that are open for human exploitation.' There may also be a large, often unrecognized, relationship over time. For example, shifting cultivation in tropical forest involves 500 million people on 8.3% of the world's tropical land area (Lanly, 1982) and produces a mosaic of successional stages as forest regrows in transient fields. The effect of past shifting cultivation can be profound; in central Africa: 'all present-day forests areas are really a patchwork of various successional stages of growth created by people, and no areas are what proposals and reports refer to as "pristine," "untouched," "primary," or "mature" forest. In short, these forests are human cultural artifacts' (Bailey *et al.*, 1992; further reviewed by Wood, 1993).

The Policy Setting: Agrobiodiversity within the International Biodiversity Agenda

Given the massive economic value of agrobiodiversity, the topic has been relatively neglected in the global biodiversity debate.

The *World Heritage Convention* is an international convention for protected sites. It was adopted with the title 'The Convention Concerning the Protection of the World Cultural and Natural Heritage' in Paris in 1972. World Heritage Sites can include 'cultural landscapes' for the 'combined works of nature and man' including 'organically evolving landscapes resulting from successive social and economic imperatives and in response to the natural environment' and a 'social, traditional way of life in which the evolutionary process is still in progress' (Anon., 1993). Despite its mandate, the World Heritage Convention has neglected agrobiodiversity.

Also, the system of biosphere reserves established under the UNESCO *Man and the Biosphere Programme* has paid little attention to agrobiodiversity. Reserves can include 'examples of harmonious landscapes resulting from traditional patterns of land use' (UNESCO, 1984). 'Buffer zones' around reserves may contain a variety of agricultural activities. One reserve, the Sierra de Manatlán Biosphere Reserve in Mexico, was established to protect *Zea diploperennis* (Guzman and Iltis, 1991), a wild grass of potential value in maize breeding.

The 1992 Convention on Biological Diversity (CBD) was a watershed for agrobiodiversity, although the word was not used in the text of the Convention (UNEP, 1992). In early drafts, the CBD was a conservationists' charter, with emphasis on endangered species and habitats, and with no positive references to agriculture. This was a result of its parentage – out of the UN Environment Programme (UNEP) with inputs from the two major transnational conservation agencies, IUCN and WWF. However, more emphasis was given to agriculture during negotiation and the CBD now includes key elements recognizing domesticated or cultivated species, the need for scientific research on genetic resources, and *in situ* and *ex situ* conservation. Significantly, the term 'equitable sharing of benefits' from the use of biodiversity was included (CBD Article 1), and emphasis given to the concept of 'sustainable development' and the sustainable use of biodiversity in meeting food needs (and agriculture is by far the most important mechanism for this). The agreed and extensive 'Agenda 21' gave a fuller coverage to agriculture.

More specifically for agrobiodiversity, the UN Food and Agriculture Organization (FAO) held an international conference in 1996 on what the FAO call 'Plant Genetic Resources for Food and Agriculture' (PGRFA). In his introductory address, the Director General of FAO used the word 'agro-biodiversity' both as a unifying theme, and also as a convenient synonym for PGRFA. While we accept the conceptual power of the word 'agro-biodiversity', we argue that 'PGRFA' as more formally used by FAO is not congruent with agrobiodiversity. PGRFA (i) includes only plants; (ii) covers only useful (mainly edible) plants, while 'agrobiodiversity' includes harmful organisms such as pests and diseases; (iii) includes wild plant foods that are not part of agricultural biodiversity. The FAO Leipzig Conference, which did not have the full weight of an international convention such as the CBD, produced a 'Global Plan of Action' (FAO, 1996) for the conservation and sustainable utilization of PGRFA. One expected output of the Leipzig Conference – new funding for the conservation and utilization of PGRFA – was notably lacking.

The Global Environment Facility (GEF) of the World Bank, UNDP and UNEP is important as the main project implementing mechanism for the CBD (GEF has other mandates too, such as ozone depletion and climatic change). Emphasis in the GEF has been on biodiversity conservation, often in protected areas. Of the $773 million currently allocated to biodiversity projects within the GEF, only $19 million (2.46%) are specifically for agrobiodiversity (with projects mainly in crop centres of diversity such as Ethiopia, Turkey and Peru). A similar neglect of agrobiodiversity is evident from the survey by Abramovitz (1989) of US bioconservation funding: of a total of 873 projects, there are only four projects (0.46%) dealing with genetic resources for agriculture, compared with multiple projects on elephants, birds and pandas. These figures are remarkable, given that 97% of US food comes from agrobiodiversity, and not from wild food sources. And perhaps worse than neglect, protected areas, still the predominant global mechanism for conservation, almost entirely exclude human use of the area, and therefore preclude agriculture, associated agrobiodiversity (and its continued evolution) and the local knowledge of how to manage it.

Agriculture, Biodiversity and the Environment

The milestone publication *Biodiversity* (Wilson and Peter, 1988) contained 57 chapters, of which only three dealt with agrobiodiversity. By 1997, the updated *Biodiversity II* (Reaka-Kudla *et al.*, 1997), with a total of 33 chapters, had only one chapter (with parts of two others) on agriculture. Some of the continued neglect of agrobiodiversity is due to the long-standing antipathy of environmentalists to agriculture. The negative view correctly identifies agriculture as the predominant land-use system on Earth, but then sees 'man the farmer' as always perverse and destructive to the environment and its biodiversity, turning natural ecosystems into biological wastelands and, eventually, through mismanagement, degrading fields and pastures into unproductive biodeserts. At its most strident, the debate over 'feeding people versus saving nature' finds in favour of nature. The shortcoming of this 'new environmental ethic' – allowing people to starve to save nature – has been discussed by Brennan (1998).

A damaging polarity now exists in the debate over agriculture and the environment. A move to the 'middle ground' is needed, soundly based on fact, rather than misconception. The reality may be that:

> It is possible to increase food production without damaging the resource base

and that

> the protection of the environment and the development of food production are closely linked. Not only are they compatible, but we cannot have one without the other. If we do not protect the environment, we cannot continue to produce food. If we do not develop sustainable food production systems we cannot protect the environment

> (Mba, 1989)

The productive management of agrobiodiversity will be a key to developing the middle ground between agriculture and the environment.

In our preparations for compiling this book, it became obvious that the concept of agriculture in conflict with 'nature' was at its worst wrong and, at its best, simplistic and self-defeating. We return to this theme in Chapter 17, but emphasize here that 'nature' will certainly be defeated if agrobiodiversity is mismanaged and its potential squandered. Global food production will need to double over the next half century. Attempting to double agricultural production by increasing the land in production would create 'a veritable environmental disaster' (Thompson, 1998). There is repeated emphasis to this danger in the agricultural literature; for example, Abdulai and Hazell (1995) predict that agricultural intensification will be essential for relieving the pressure on natural resources and containing further degradation. They argue that a move to extensive farming systems would lead to mismanagement of resources and damage environmentally fragile areas.

The alternative to yet more marginal farming is to intensify production on existing land, but this will need considerably more research (not less) to understand and exploit the potential of agrobiodiversity without damaging nature. And as part of nature, agrobiodiversity will remain a sensitive indicator of the health of agroecosystems.

The Importance of Agrobiodiversity

This book will attempt to repair the misconceptions, neglect and ignorance over agrobiodiversity, its potential and its management. A dismissive view of agrobiodiversity is partly a result of poor presentation. The biodiversity debate has brought together, in a readily understandable way, ideas from an existing range of disciplines, including ecology and conservation biology. There has been ample funding and promotion for this synthesis into 'biodiversity'; for example, the $3.5 million to prepare the single volume *Global Biodiversity Assessment* (Heywood, 1995), the billion dollars that will be spent on biodiversity under the Global Environment Facility, and much more.

In contrast with the expanding debate over biodiversity, there has been no synthesis of agricultural knowledge and practice into an 'agro-biodiversity' agenda, and no conceptual advances comparable to the bio-diversity debate (and certainly no new funding; rather the reverse). Agricultural researchers have had to concentrate on the immediate job of feeding an expanding global population. The results of this understandable single-mindedness have been lost opportunities to promote the importance of agrobiodiversity and agriculture itself in the global biodiversity debate. As a consequence, international development policy has favoured 'bio-diversity'.

> The US Agency for International Development, for example, appears to have decreased its funding of agricultural development and research and increased its funding of tropical biodiversity programs. Philanthropic foundations have clearly begun to abandon agriculture and devote their resources to rainforests and other nonagricultural biodiversity issues.
>
> (Buttel, 1992)

Yet agriculture is the largest global user of biodiversity. Agriculture has selected and added value to wild biodiversity over more than 10,000 years of managing agrobiodiversity. Agriculture has conserved biodiversity on the hoof and as seed and planting materials over this long period. Agriculture extracts value from biodiversity at each harvest or cull, but nurtures the productive and renewable base. Indeed, it is certain that the most immediately valuable part of global biodiversity is the agrobiodiversity on which farming, and in turn, global food security depends.

This book is premised on the fact that agrobiodiversity is irreplaceably important in its own right, as providing most of our food. In the past, agrobiodiversity management has underpinned our civilization, based as this is on the transfer of the surplus production of the agricultural landscape to cities. The management of agrobiodiversity will determine our future, both in cities and the countryside, with particular impact on the unequal distri-bution of resources between the rich and the poor. While negative views of agriculture have permeated the environmental debate, more positive approaches to agrobiodiversity – as presented here – are urgently needed.

The new approach to conservation in the wild has moved from an emphasis on rare and endangered species to embrace ecosystem function and the importance of ecosystem services. The knowledge base exists for a comparable synthesis in agriculture. Agroecosystems – mediated through agrobiodiversity – have always provided the essential ecosystem service of food production, and can be designed to deliver a further range of ecosystem services as needs and knowledge change. Present knowledge extends from a greater appreciation of traditional agriculture and the needs of farmers, through classical agricultural research in animal husbandry, genetics, statistics, replicated experiments, plant breeding, agronomy, crop protection, rural sociology, information management and many more, through to biotechnology. Contributors to this book will review our practical knowledge of agrobiodiversity management and attempt to place it in greater prominence in the global debate over biodiversity and sustainable development.

Outline of Chapters

This book provides a broad view of the current thinking on agrobiodiversity: what it is made up of, how it is managed, how it is conserved, and how it can be better utilized through sustainable farming practices for productive agriculture. It brings together contributors from a wide geographical and disciplinary background. This synthesis is especially timely in the light of the challenges facing global food production during the next 20–30 years, especially the need to produce more food from less land using technologies which have minimal adverse environmental impacts.

To achieve our main purpose – to promote a more rational, technically sound and functional view of agrobiodiversity – this book needs to cover three themes. The first theme is the current status of the concept and usage of the word agrobiodiversity and its relation to biodiversity (attempted above).

The second theme is wider: a review of the components of agrobiodiversity, how they relate together functionally, how they impact on agricultural production, and how agrobiodiversity can best be managed for sustained food production. This second theme represents the bulk of the chapters by the varied contributors, and these chapters illustrate and amplify our definition. There is some emphasis on tropical agrobiodiversity – for sound reasons: there is more of it; its management is more important for local food security; and interactions with the 'wild', however defined, are closer. The knowledge base is nothing less than that generated by 10,000 years of crop and animal production, with humans attempting to understand and benefit from the multitude of interacting organisms in a wide range of terrestrial environments. This environmental management on the widest scale through agriculture is now supported by a detailed formal research process (still dependent on a tool initially developed for agricultural research and now of the widest use – the statistical analysis of replicated experiments).

The third theme depends directly on the second: can this extensive knowledge of the management of agrobiodiversity provide models and practices for the wider management of biodiversity? Note that this emphasis is quite the opposite of that normally linking agriculture and biodiversity management – of ecological and biodiversity practice attempting to provide models for agriculture (Conway, 1997). However, agricultural knowledge and practice, both traditional and formal, have been seriously neglected by ecologists and conservation biologists. This neglect has even extended to topics on which agricultural knowledge could provide essential guidance to the biodiversity agenda, including the behaviour of introduced species (most crop and animal production is from such species); genetic mapping; mating systems; plant and animal physiology and stress tolerance; plant and animal health; pest management strategies (including quarantine); the rhizosphere; *ex situ* conservation; and, not least, indigenous knowledge as a key to environmental management.

This introductory chapter sets the scene by providing an overview of technical and political issues and key definitions. It discusses the importance of agrobiodiversity and the problematical relationships between biodiversity, agriculture and the environment. The second chapter reviews the historical dimensions of agrobiodiversity with particular emphasis on crop domestication. It considers the management of diversity before agriculture; the transition to agriculture; the domestication process; the human impact on diversity; and introduces the beneficial and harmful components of agrobiodiversity which are covered in more detail in the following chapters.

The next five chapters examine the nature, role and function of various important components of agrobiodiversity. Chapter 3 is concerned primarily with genetic diversity between and within species of crop plants, the varieties of which are the functional units of food production. It examines the ways in which crop diversity is distributed, assessed and organized into agroecosystems. It effectively uses two case studies on common wheat and sorghum to illustrate the varied ways in which interaction between natural and human-directed evolutionary forces help to set the genetic boundaries for crop diversity. Chapter 4 considers the biodiversity of domesticated animals used for food and other agricultural services. It reviews the nature, extent, erosion, importance and utilization of domesticated animal biodiversity. It stresses the need to improve support for the characterization of animal biodiversity to ensure that appropriate conservation strategies are developed and implemented. Particular attention is given to the FAO Global Strategy for Animal Genetic Resources and the development of the Domestic Animal Diversity Information System (DAD-IS) which is helping to ensure the successful management of domesticated animal biodiversity to enable the world to feed itself. The regulation and functional significance of soil biodiversity are considered in Chapter 5. It examines in some detail how agricultural practices, and in particular intensification, impact upon the soil biota (such as soil-associated food webs, microorganisms involved in nutrient transformations, and soil-associated herbivores) and the biodiversity of the below-ground system. It also considers what effects shifts in soil biodiversity have on the processes governed by soil organisms, the overall functioning of the below-ground system, and, ultimately, crop productivity. Chapters 6 and 7 provide interesting contrasts between the largely harmful nature of pathogen diversity and the more beneficial character of insect biodiversity in agroecosystems. Chapter 6 traces the evolution of disease in plants and gives emphasis to the mechanisms by which pathogen diversity arises. Functional diversity in pathogens is discussed in terms of fitness, survival, and spatial and temporal dynamics in agroecosystems, focusing on the physiological race as the principal unit of diversity. This chapter then appraises the consequences of pathogen diversity for effective management of diseases in agroecosystems. Chapter 7 looks at the function of beneficial insect biodiversity and theoretical and practical considerations for optimizing insect biodiversity in agroecosystems. It presents a model project for studying and efficiently utilizing insect biodiversity. Some critical issues in biological control and conservation biology are then examined using recent work on the citrus leafminer as a case study.

The management of agrobiodiversity in the broader environment is introduced through a consideration of the agroecosystem in the landscape in Chapter 8. The differences between agroecosystems and other ecosystems from an ecological context are examined under the concept of the habitat templet, the disturbance axis, the adversity axis and the landscape structure. Two contrasting agroecosystems, a traditional agropastoral system in Spain and a paddy rice system in Thailand, are presented as case studies to assess how much diversity an agroecosystem can support and to provide a foundation for a discussion of the effects of intensification in agroecosystems. Finally, the relationship between biodiversity and ecosystem function is briefly discussed. Management issues are then considered in more detail in the next five chapters.

Traditional management of agrobiodiversity is reviewed in Chapter 9. Some of the generic characteristics of the many different, traditional farming systems are firstly addressed. This is followed by a series of case studies on the management of diversity by farmers in specific crops (cassava, maize, common bean and rice) and of domesticated animals. The concluding section highlights the remarkable parallels across crops, cultures and continents and stresses the continuing need for farmer management of agrobiodiversity. Chapter 10 critically reviews the effects of plant breeding on genetic diversity in crops. It examines how crop biodiversity is measured and how genetic diversity is affected by genetic improvement with emphasis on the role of farmer participation in the breeding process and the breeding methods employed. It then considers the potential impact of modern plant breeding techniques on agrobiodiversity. The effects of pest management on biodiversity in agroecosystems is addressed by Chapter 11. After a brief consideration of the importance of insect pests, weeds and pathogens in agroecosystems, much of the chapter is devoted to a review of how management practices based on chemical, biological, cultural and genetic techniques affect various components of agrobiodiversity. Some attention is paid to the effects of management practices on the interrelationships between pests, crops and the surrounding vegetation. Chapter 12 looks at the effects of using alternative tillage systems on biodiversity in agroecosystems through a comparative analysis of wheat with maize and rice systems. In each case, the perceived problems of conventional tillage are examined before considering the consequences of these practices on agrobiodiversity. Some general conclusions on the effects of alternative methods of tillage on the functioning of agroecosystems are then drawn. The relationships between seed management systems and genetic diversity are addressed in Chapter 13. A comparison is made between traditional farmer-managed systems and modern commercial systems within a framework of food production and sustainability. The need to bring the two seed-supply systems together under a concept of integrated seed systems is highlighted.

The next two chapters consider various aspects of the conservation of agrobiodiversity. Chapter 14 addresses, firstly, the evolution of approaches to conservation and, secondly, recent arguments for conservation of agrobiodiversity. Emphasis is given to crops. It then considers the policy and technology of conservation using the role of IPGRI as an example of an international approach to conservation of plant genetic resources. The final section discusses the complementarity of conservation methods and places this in the context of agroecosystem conservation. A conceptual framework for valuing crop genetic resources on-farm is presented in Chapter 15. The framework is designed to support strategic decisions about which crop populations in a predetermined area are suitable candidates for conservation. The framework is presented as a tool for discussion and testing. The importance of farmers' preferences and the opportunity costs of maintaining specific varieties are highlighted.

Chapter 16 looks at the effects of regulatory issues on agrobiodiversity. It considers the standard practices used by governments to regulate inputs to agricultural production with emphasis on seed and pesticides; it identifies the issues where input regulation can have a significant impact on agrobiodiversity; and examines possible changes in common regulatory practices that may be considered to more effectively promote or protect agrobiodiversity. Chapter 17 attempts to relate the evolving debate over diversity, stability and ecosystem functioning to agrobiodiversity. Is it reasonable to apply analogues from natural ecosystems to agriculture? Can knowledge of the traditional and modern management of agrobiodiversity perhaps throw light on some of the problems of natural resource conservation? The final chapter brings together key issues that have arisen throughout the book. Ways of optimizing agrobiodiversity for productive agricultural development are suggested. We recognize that study, increased understanding, and the sustainable management of agrobiodiversity may well be critical not just to agricultural production, but also to the future of biodiversity globally. Agrobiodiversity depends on people, and people must be continually – and knowledgeably – involved in its management.

We hope that this book will redress the past neglect and demonstrate that the long and productive history of the human management of agro-biodiversity can provide a corpus of knowledge and practice which is both of supreme value in its own right, and also of the greatest value as a model for wider biodiversity conservation and utilization.

References

Abdulai, A. and Hazell, P. (1995) The role of agriculture in sustainable economic development in Africa. *Journal of Sustainable Agriculture* 7, 101–119.

Abramovitz, J.N. (1989) *A Survey of U.S.-based Efforts to Research and Conserve Biological Diversity in Developing Countries.* World Resources Institute, Washington, DC.

Alcorn, J.B. (1989) Process as resource: the traditional agricultural ideology of Bora and Huastec resource management and its implications for research. In: Posey, D.A. and Balee, W. (eds) *Resource Management in Amazonia: Indigenous and Folk Strategies. Advances in Economic Botany* 7, 63–77.

Anon. (1993) *World Heritage Newsletter* No. 1, Feb. 1993, p. 15.

Bailey, R.C., Bahuchet, S. and Hewlett, B. (1992) Development in the Central African rainforest: concern for forest peoples. In: *Conservation of West and Central African Rainforests.* World Bank, Environment Paper No. 1, pp. 202–211.

Bennett, B.C. (1992) Plants and people of the Amazonian rainforests. *BioScience* 42, 599–607.

Brennan, A. (1998) Poverty, puritanism and environmental conflict. *Environmental Values* 7, 305–331.

Brookfield, H. (1998) Review of Zimmerer, K.S. (1996) Changing fortunes, bio-diversity and peasant livelihood in the Peruvian Andes. *Annals of the American Association of Geographers* 88, 180–182.

Buttel, F.H. (1992) The 'environmentalism' of plant genetic resources: possible benefits, possible risks. *Diversity* 8, 36–39.

Conway, G.R. (1987) The properties of agroecosystems. *Agricultural Systems* 24, 95–117.

Conway, G. (1997) *The Doubly Green Revolution: Food for All in the Twenty-first Century.* Penguin Books, London.

FAO (1996) *Global Plan of Action for the Conservation and Sustainable Utilization of Plant Genetic Resources for Food and Agriculture.* FAO, Rome.

Guzman, M.R. and Iltis, H.H. (1991) Biosphere reserve established in Mexico to protect rare maize relative. *Laboratory for Information Science in Agriculture* 7(1/2), 82–84.

Heywood, V.H. (1995) *Global Biodiversity Assessment.* UNEP and Cambridge University Press, Cambridge.

Lanly, J.P. (1982) *Tropical Forest Resources.* FAO Forestry Paper, No. 30. FAO, Rome.

Mba, D.A. (1989) Foreword. In: Delleré, R. (ed.) *Land and Food.* Technical Centre for Agricultural and Rural Cooperation (CTA), Wageningen, p. 7.

McIntyre, S., Barrett, G.W., Kitching, R.L. and Recher, H.F. (1992) Species triage-seeing beyond wounded rhinos. *Conservation Biology* 6, 604–606.

Prescott-Allen, C. and Prescott-Allen, R. (1986) *The First Resource: Wild Species in the North American Economy.* Yale University Press, New Haven, Connecticut.

Qualset, C.O., McGuire, P.E. and Warburton, M.L. (1995) 'Agrobiodiversity': key to agricultural productivity. *California Agriculture* 49(6), 45–49.

Reaka-Kudla, M.L., Wilson, D.E. and Wilson, E.O. (eds) (1997) *Biodiversity II: Understanding and Protecting our Biological Resources.* John Henry Press, Washington, DC.

Scoones, I., Melnyk, M. and Pretty, J.N. (1992) *The Hidden Harvest: Wild Foods and Agricultural Systems.* IIED/WWF, London.

Swift, M.J. and Anderson, J.M. (1994) Biodiversity and ecosystem function in agricultural systems. In: Schulze, E.-D. and Mooney, H.A. (eds) *Biodiversity and Ecosystem Function.* Springer-Verlag, Berlin, pp. 15–41.

Thompson, R.L. (1998) Public policy for sustainable agriculture and rural equity. *Food Policy* 23, 1–7.

UNEP (1992) *Convention on Biological Diversity.* UNEP, Geneva.

UNESCO (1984) *Programme on Man and the Biosphere (MAB) – Action Plan for Biosphere Reserves.* UNESCO, Paris.

Wilson, E.O. (1985) The biological diversity crisis. *BioScience* 35, 700–706.

Wilson, E.O. and Peter, F.M. (eds) (1988) *Biodiversity.* National Academy Press, Washington, DC.

Wood, D. (1993) Forests to fields: restoring tropical lands to agriculture. *Land Use Policy* 10, 91–107.

The Origins of Agrobiodiversity in Agriculture

2

D. Wood and J.M. Lenné

Agrobiodiversity International, 13 Herons Quay, Sandside, Milnthorpe, Cumbria LA7 7HN, UK

In the sweat of thy face shalt thou eat bread . . .

(Genesis 3:19)

It is not the purpose of this book to try to solve the sometimes 'chicken and egg' questions about the complex reasons for agricultural origins which have been raised by a wealth of texts (Sauer, 1969; Cohen, 1977; Rindos, 1984; Smith, 1995). Indeed, a recent account (Harlan, 1995, p. 239) considers that humans everywhere post-Pleistocene had to change their way of living and that trends pushed on under their own momentum towards the cultivation of plants and the domestication of animals. For our purposes, to begin to understand the complex interrelationships of agrobiodiversity, it is more fruitful to look at the *processes* of agricultural origins. This chapter will outline two parallel processes: the increasing environmental management that was crucial to domestication, and the process of domestication itself, based on the raw materials of wild species. These two related processes gave rise to our crops and domestic animals in the domesticated habitat. Emphasis here is given to crops, but we also look briefly at crop-associated biodiversity (see also Steane, Chapter 4, and Thurston *et al.*, Chapter 9, this volume, for domesticated animals).

Agrobiodiversity and History

Our high dependence on food from farming is relatively recent in human history. The transition from a total dependence on wild biodiversity for food began over 10,000 years ago, when there was no farming. The evolution of farming depended on the independent domestication of many plants and animals in separate parts of the world. There was an associated 'domestication of the environment'. Intensification of food production, for example by irrigation, allowed human cultural evolution (and social stratification) to the complex degree described by Wittfogel (1957) for the 'hydraulic civilizations' of Asia. Some results of our dependence on farming are obvious: a massive increase in human population, vast cities, and elaborate intercontinental trade to feed these cities. Mannion (1997, p. 75) notes that 'Food security has always conferred power, and it continues to do so today'.

It is useful to step back a little more than 10,000 years. Then, all food came from the wild – fishing, hunting, and gathering wild food, mainly plants. Now only rarely do these three non-farming food sources by themselves supply enough food for the survival of human communities. If we had never progressed beyond hunting and food-gathering, the world's surface could only sustain an estimated 20 to 30 million people (Dimbleby, 1967). At a rough estimate, for every one person supportable by hunter–gathering, ten could be fed from extensive agriculture, and possibly more than 100 from intensive (including irrigated) agriculture. For Iran, in one of the regions of domestication, Flannery (1969) estimated human dependence on wild and domesticated ecosystems. Sixty-five per cent of Iran would have been marginal for hunter–gatherers, who survived on the remaining 35%. However, the change from hunter–gathering to dry-farming was accompanied by a 20-fold human population increase, but this was based on the 10% of the land surface considered to be arable. There was a further intensification of food production with irrigation – possible on only 1% of the land surface of Iran, but producing 30% of crops.

Flannery's estimates were based on two key ideas: firstly, the increase in carrying capacity based on the *technology* of acquiring food through domestication (the transition from hunter–gathering to agriculture); and, secondly, the *quality of the land* – or perhaps more accurately, the suitability of land for food production.

In a monumental survey of the carrying capacity of land in Africa, Allan (1965) emphasized the importance of this second idea of land quality, and noted that: 'the traditional land-use systems of Africa are adapted to the limitations of their environments, as any viable system of agriculture must be'. Allan related land use to 'unwritten knowledge' of land quality by those involved in 'hunting and food-gathering', 'herding and shepherding' and 'hoe cultivation'. In many cases this knowledge was precise and remarkably complete, and without it no community could have survived for long (Allan, 1965, p. 2). Thus the technology of acquiring food by gathering, hunting, herding or growing was closely related to the quality of the land.

However, Allan also recognized two further factors that determined land use, both illustrated by his description of a visit made to a Gishu farmer high on Mt Elgon in Uganda: the upper farms, with banana groves, mulching, vegetable production and hand-fed livestock, were more intensive and perennial than the lower farms. The farmer explained the intensification of cultivation higher on the mountain as a response to past cattle-raiding by pastoralists of the plains. The upper slopes were readily defensible and difficult to access by the raiders. However, as population increased on the upper slopes, more and more intensive practices had to be adopted. As cattle-raiding declined, the Gishu colonized the lower slopes, and abandoned the intensive practices of the uplands. The Gishu farmer thought that this was sensible, as no farmer loves labour for its own sake (Allan, 1965, p. 168). The factors of *community security* balanced against the *labour* needed for intensive farming have always been major features in the use and management of agrobiodiversity.

It is the complex relation between the four dynamic factors of developing *agricultural technology*, land *quality*, returns to invested *labour* and community *security* that mainly determine the origin, use and early management of agrobiodiversity in less or more intensive systems of producing food. In addition, very early in the history of agriculture a fifth factor – *trade in farm produce* – became important. In Bronze Age Levant – one of the areas of agricultural origins – agricultural surpluses allowed early urbanization, and farmers shifted towards intensified management to exploit growing market opportunities, mainly 'surplus-oriented tree cropping' of olive, grape, fig and date (Fall *et al.*, 1998).

In turn, these factors of early agriculture define much of subsequent human history, not least the repeated invasive spread of pastoralists from the centre to the agricultural fringes of Eurasia (and from the drier to the wetter regions of Africa), and the general retreat of hunter–gatherers to the margins of land and history.

Management of Diversity Before Agriculture: the Role of Hunter–Gatherers

The process of domestication was not a random one either taxonomically or geographically: it focused on a limited range of species in nuclear areas of domestication. There were two prerequisites – the presence of hunter–gatherers (at least seasonally) and the presence of wild relatives suitable for domestication.

While the economic results of domestication – cities and trade – may be obvious, the reasons for the relatively rapid change from hunter–gathering to farming remain subject to debate (see Harlan, 1995, for a review). This debate draws on a complex interdependency of issues involving the evolution of human culture, agrobiodiversity and the management of agricultural systems. As yet unresolved are the questions: did climatic change after the last glacial period provide the conditions that stimulated and permitted farming to produce food (with the food from farming allowing a population increase)? – or did increasing human population (and therefore pressure on wild food resources) demand a more reliable food source from farming? Indeed, did farming (and the sedentary life necessitated by farming) initially provide a more secure food supply than contemporary hunter–gathering?

In contrast with the conjecture over agricultural origins, there is considerable information from archaeology and the study of present-day hunter–gatherers on their *environmental manipulation* and their *knowledge of biological resources* for food (Harris and Hillman, 1989). Before agriculture, human communities met their entire food needs through hunter–gathering and fishing: there was no agrobiodiversity. However, there was often intense human management of the ecosystem, and detailed human knowledge of food plants and animals.

Environmental manipulation

Environmental manipulation – the 'domestication of the environment' in Yen's terminology (Yen, 1989, p. 57) – is shared between hunter–gatherer and agricultural subsistence production. Although hunter–gatherers tapped the seasonal food production of diverse environments, they were not just passive exploiters of this seasonal bounty. There could be a 'manipulation of the environmental settings of the exploited species . . .' (Yen, 1989, p. 57) with the objective of greater food security. There will also have been environmental changes incidental to hunter–gathering, such as soil enrichment through kitchen and human waste around temporary settlements. However, the main human impact on the hunter–gatherer environment would have been through control of flooding in valley bottoms, and, more importantly, the use of fire to clear vegetation. There is a significant continuity (see Wood and Lenné, Chapter 17, this volume) between natural fire and flood; the use of fire and flood by hunter–gatherers; and the value of land clearing by fire and of flood irrigation to early agriculture.

Flooding
Harlan (1995, pp. 11–13) reports two separate examples of recent hunter–gatherers using water diversion to promote access to food. Australian Aborigines diverted water to flood forests in the dry season, to attract birds, and to increase the growth of food plants. The Great Basin Indians of North America also diverted water to flood extensive tracts (up to 13 km²) to ensure increased production of wild food plants.

Fire
Notwithstanding these examples of flooding, fire was the most important tool for environmental management available to hunter–gatherers. Burning of dry-season vegetation would have a major impact on the environment of hunters. Burning was used to clear cover, to drive game, and to provide more nutritious young growth to attract game (Hallam, 1989, p. 143). Sauer

considered that man's greatest power to disturb the balance of nature lay in his employment of fire (Sauer, 1969). The use of burning by Australian aboriginal hunter–gatherers is thought to account for the spread of fire resistant eucalypts as a component of the vegetation (Clark, 1983). There is evidence from pollen records of forest interference in New Guinea as early as 30,000 years ago (Groube, 1989).

The use of burning associated with plant food gathering has a close relevance to agrobiodiversity. Judicious burning can both *reduce competition* with unwanted plants and also *release nutrients* which encourages plant growth. Such burning was used extensively as part of the process of gathering food, and has been called 'firestick farming' (Jones, 1969) in Australia, where aboriginals never developed agriculture. Hallam (1989, p. 147), for the non-agricultural environment of Australia, goes as far as arguing that aboriginal husbandry and management practices, including firing, encouraged the maintenance and improvement of tuber and pasture resources. Inevitably, there would be associated effects on other (non-target) biodiversity in the soil and the above-ground parts of the gatherer ecosystem.

Knowledge of biological resources: seasonality

Harlan (1981, p. 4) noted that: 'Hunter–gatherers seem to regard plants similarly in Africa, Australia, Asia, Europe or the Americas. They are attracted to similar species and develop similar methods of modifying vegetation, detoxifying poisonous plants and processing edible plant foods'. There is archaeological evidence for a wide range of animals and plants used for food, and also for changes in human food sources.

Commonly, hunter–gatherers rely on seasonally available food, for example changing their resource base between wet and dry seasons and between animals and plants, as available: this may not entail migration. For example, the environment of Australia, with sharp rainfall gradients between the arid interior and the coast, and otherwise seasonal climate – mediterranean or monsoonal – encourages what Yen (1989, p. 57) calls 'the cyclic methods of foraging by hunter–gatherers, regulated in time and space ... conditioned by the distribution and seasonal production of important species'. Exploitation ranged from grass-seed gathering in the arid hinterland, to gathering *Dioscorea* yams and tuberous *Ipomoea* species. Seasonal reliance on the harvest of *Dioscorea* yams (*D. hastifolia*) in southwest Australia was almost year-round, and involved a considerable degree of sedentism (Hallam, 1989, p. 139). Concentrations of yam-digging areas were often adjacent to lagoons, with edible reed rhizomes, fish and wildfowl. In northern Australia (Arnhemland) there was seasonal resource exploitation from swamps in the wet season (stalks and seed of waterlilies (*Nymphaea* spp.)), early dry season (yams), and in the later dry season (spike rush;

Eleocharis corms) geese, cycad nuts (Jones and Meehan, 1989, p. 125). The cycle of exploitation involved moving the home base seasonally.

Food value and processing

To this knowledge of the seasonality of production of wild resources was added a knowledge of the value of different species as food.

Many plant species now grown by farmers, and many more that were and are gathered as wild food, are toxic to humans. This includes a full range of food types – fruits, nuts, tubers, grain legumes and, to a much lesser extent, cereals. Considerable effort is needed to detoxify such wild foods. The presumed advantage to human gatherers is that potential competition from animals for the same toxic food would be low, as animals would not have the ability to detoxify.

Other important species of grains and grain legumes produced small seeds. Effort was needed, first to collect, and then to process, these grains as food, usually by grinding (Smith, 1989). Exploiting this wild resource for food depended on the development of the seed-processing technology that was an essential precursor to grain agriculture.

There is a remarkable similarity between favoured foods of plant gatherers and domesticated crops. In Australia alone, many genera which have been domesticated in other regions were gathered as food. These include tubers from *Dioscorea*, *Ipomoea* and *Colocasia*, and a wide range of grass seed, including the wild rice *Oryza meridionalis* (Jones and Meehan, 1989). Harlan (1989) also reports wild rice harvesting (*Oryza barthii* and *O. longistaminata*) by foragers in Africa: other wild foods related to crops include *Brachiaria*, *Eragrostis* and *Panicum*.

The Transition to Agriculture: Favouring Food Species

Hunter–gatherers had discovered the food value of the wild relatives of most of our present crops and domestic animals. Hunter–gatherers had also developed the ability of manipulating the ecosystem to encourage the growth of these favoured food species. However, plant food reserves, particularly if in reproductive structures such as fruits, seeds and tubers, are an essential part of the competitive strategy of wild plant species.

Competition has been defined by Harper (1961) as the response of plants to density-induced shortages, and by Keddy (1989, p. 2) as 'the negative effects which one organism has on another by consuming, or controlling access to, a resource that is limited in availability'.

Harvesting of plant parts – particularly the reproductive parts – for human food would decrease the competitive ability of targeted plants. Unless there was some compensation for this, the population of the food plant would decline in competition with non-food plants: there would be less food for gatherers the following season. Simply, if we eat the plant's reproductive strategy, it will not be able to compete with less useful species. To maintain the food supply, we need to compensate the plant in some way. With compensation, the food species could expand at the expense of non-food plants, giving us a more assured food supply.

Compensation is the key to the coevolution of food plants and human exploitation. Compensation can be applied in several ways.

- Through *environmental manipulation*, by deliberately or accidentally reducing interspecific competition through burning or habitat disturbance; and the hunting of grazing animals or fruit- and seed-eating birds.
- Through *accidental seed dispersal*: for example, gathering and eating fruits, and dispersing seeds to soil possibly enriched with the excreta of humans and domestic animals or nutrient-rich domestic refuse.
- Through *deliberate sowing* of the food plant after control of competition by slashing, burning and tilling, with later thinning and weeding. For example (reviewed by Yen, 1989), yams may be replanted following harvest by food gatherers in Australia. This has been called plant husbandry (see Shipek, 1989).

Two of these compensation mechanisms are the essential characteristics of agriculture: environmental manipulation by the *tilling of fields* (to reduce competition from non-food plants); and the *sowing of seed* (to enhance the competitive ability of the food plant). Agriculture is therefore a combination of ways in which we help food-providing species to compete: making more and more plants and animals dependent on our 'sufferance or favor' (Sauer, 1969, p. 13).

Compensation implies increased human effort for the future returns of a more concentrated and reliable food supply. However, the effort of compensation would only be worth while if the benefits could be captured by those investing this effort (and not other groups of food gatherers or wild and domestic grazing animals). Protection of plant stands resulting from 'compensation' could be enhanced by semi-permanent camps and even defined and defended fields. It is surely no coincidence that the earliest domestic animal, the dog, excels at this defensive role. Another early domesticate, the cat, can protect seasonal surplus of agriculture (stored grain) against rats and mice.

Agricultural Diversity and the Domestication Process

Agriculture depends on environmental management, a knowledge of the food value and processing needs of crops, and the replanting of propagules – all features of hunter–gathering. Precise definitions of 'agriculture' and 'domestication' to distinguish them from hunter–gathering are a problem, with immediate dangers of circular reasoning. Agriculture was defined by Harris (1969) as the manipulation of the natural ecosystem by substituting domesticated species for wild species in appropriate ecological niches. However, Rindos (1984, p. 101) argued that the process of domestication is the evolution of plants used in agricultural systems (and then further defined 'incidental', 'specialized' and 'agricultural' domestication).

Harris's definition of *agriculture* could include brewing (or any controlled fermentation), and Rindos's definition of *domestication* – as he himself noted – could apply to fungi in the 'gardens' of termites, and many other animal–plant relationships. Rather than attempt an all-embracing definition, we here consider agriculture as managing (as far as practicable) the conditions of production, and domestication as genetically changing the species which produces the food or other economic product to do so more effectively. The agricultural management of fields will tend to be more productive with domesticated species, but could be worth while with wild species (as with many of our present-day horticultural systems growing undomesticated ornamentals and medicinal plants). Domestication could have occurred more rapidly and to a greater extent with the controlled habitat of managed fields, but domestication could have been possible in predominantly natural ecosystems.

The attractiveness of agriculture based on the control of competition and on compensation mechanisms would be strengthened by domestication, where genetic change of the domesticate increases the reward to farmers for their work in increasing the level of compensation (more tilling, weeding, nutrient enrichment, and eventually sophisticated irrigation and cropping systems).

Darlington (1969, p. 69) noted that, following domestication, 'Men, crops and stock had become overwhelmingly and continuously dependent on one another for their survival. Their fitness was therefore an integrated property and their relationship reciprocal.' The future competitive success of domesticated plants and animals would depend largely on human management. Human management, rewarded by crop surplus, would largely determine the form of agricultural systems, and their content of agro-biodiversity.

Features of domestication

Crops

Domestication ranks with toolmaking and speech as a key stage in human evolution. Domesticated plants (and, to a lesser extent, animals) are certainly the most valuable part of biodiversity. Yet domestication over space, time, the range of organisms domesticated, and the genetic mechanisms involved appears to have been strongly selective, and to have emphasized the quality of diversity, rather than the quantity of diversity itself. However, fail-safe mechanisms have often been maintained, to allow a restitution of diversity as a raw material for further selection.

Domestication did not take place worldwide, but in *limited areas*. Vavilov (1926) had recognized the concentration of crop diversity and the presence of wild ancestors in defining his centres of origin of crops. These have often been redefined: they include the Near Eastern, North Chinese, and Mesoamerican centres, and, as redefined by Harlan (1971), more diffuse 'non-centres' in Africa, Southeast Asia and South America. Harlan (1995) now queries the validity of specific centres of origin, and prefers to treat each crop separately, on an ecological basis.

Domestication occurred globally over a *relatively short time span*, beginning a little over 10,000 years ago. For each crop, domestication could have been rapid. Even without conscious selection, it has been estimated that wild einkorn and emmer wheats and wild barley could have been domesticated within two centuries (Hillman, 1990). Even with the technology of plant breeding, there have been few recent additions to the crop portfolio of early farmers (perhaps rubber, triticale and sugar beet, and the industrial crops *Vernonia* and *Cuphea*).

Domestication was confined to a *limited list of plant families* (and parts of families). Intriguingly, these families included many of the 'natural', easily recognized families whose older names did not end in 'aceae', such as *Gramineae, Leguminosae, Umbelliferae, Cruciferae* and *Palmae* (Wood and Lenné, 1993). There are obvious concentrations of major crops within the *Gramineae* and *Leguminosae*, and, within these families, in certain tribes. These include *Triticeae* in the grass family, with *Hordeum, Secale* and *Triticum*; the *Phaseoleae* in the legume family with *Dolichos, Glycine, Phaseolus* and *Vigna*; and the tribe *Viciëae*, also in the legume family, with *Cicer, Lathyrus, Lens, Pisum* and *Vicia*. Other concentrations of crops are to be found, for example, in the *Solanaceae* (tobacco, potato, tomato, chilli pepper, aubergine) and *Cucurbitaceae* (*Citrullus, Cucumis, Cucurbita, Sechium*). Genera within families may also show a concentration of cultivated species: *Allium, Brassica, Citrus, Coffea, Dioscorea, Diospyros, Gossypium, Passiflora, Phaseolus* and several other genera in different families contain many domesticated species (Harlan, 1992; Wood and Lenné, 1993; Plate 1).

Domestication often tapped only a *limited part of the diversity* of each ancestral species, perhaps through a single domestication event. Riley (1965) noted for tetraploid wheat: 'the extreme probability that all the tetraploid forms of *Triticum* are of monophyletic origin.' The significance of the domestication bottleneck for one of our most important crops – bread wheat, which is a hexaploid – is treated in detail by Cox and Wood, Chapter 3, this volume. Many other crops appear to be monophyletic, that is, resulting from a single origin. For these, domestication took place in only part of the range of the ancestral species, so that genetic variation related to ecogeographical adaptation of the ancestor was not captured in the domesticate. However, for other crops – notably common bean (*Phaseolus vulgaris*; Evans, 1976) and chilli peppers (*Capsicum* spp.; Pickersgill, 1989) – separate domestication events in different parts of the range of the wild ancestor resulted in different suites of genetic variation passing from the ancestor into the crop.

During and subsequent to domestication, *barriers to gene flow* within and between populations of domesticates may have evolved. Such barriers include:

- the transition from diploid to polyploid state at the time of domestication;
- the predominance of self-fertilization in annual crops (common bean, barley and rice), over the more prevalent cross-fertilization of wild relatives; and
- the predominance of vegetative propagation in roots, tubers and fruit crops, to the extent that many varieties can no longer produce seed.

Subsequent to domestication, there may have been strong selection by farmers for *varietal purity* within the crop, based on morphological markers (often controlled by single genes, such as seed colour variants). There is still a widespread impression that traditional varieties are all genetic soups, with rich within-variety variation. In reality, even in outcrossing crops, farmers may maintain pure varieties. A remarkable passage by Anderson (1954, p. 186) indicates the strength of farmer-selection of maize in Guatemala:

> One conclusion was particularly interesting because it was directly opposed to what I had been led to believe. It concerned the purity of Indian varieties of corn, which are ordinarily described as very mixed compared with modern ones. . . . I found to my surprise that their cornfields had been more rigidly selected for type than those of their Latin-speaking neighbours. Their fields were quite as true to type as had been prize-winning American cornfields in the great corn-show era when the American farmer was paying exquisite attention to such fancy show points such as uniformity. This fact was amazing, considering the great variability of Guatemalan maize as a whole, and the fact that corn crosses so easily. . . . Only the most finicky selection of seed ears and the pulling out of plants which are off type could keep a variety pure under such conditions . . . wherever the old Indian cultures have survived most completely the corn is the least variable within the variety . . . Only a fanatical adherence to an ideal type could have kept these varieties so pure when they were being traded from family to family and from tribe to tribe. It is apparently not true, as has so frequently been stated, that the most primitive people have the most variable varieties. Quite the opposite. It is rather those natives most frequently seen by travellers, the ones who live along modern highways and near big cities, the ones whose ancient cultures have most completely broken down, who have given rise to the impression that primitive people are careless plant breeders.

The value of varietal purity in managing crops is further confirmed by the prevalence of clonal propagation in many roots and tubers (potato, taro, sweet potato), some of which may also have lost the ability to produce seed (as with sterile triploid bananas). For many perennials (e.g. grape, citrus, date, tea, but with the exception of arabica coffee) clonal propagation, and therefore varietal fixity, is the rule. This 'quality over quantity' became most refined with wine grape, where desired traditional varieties such as Chardonnay and Cabernet Sauvignon are maintained clonally and varietally pure for hundreds of years. It seems that for many crops, well-adapted and distinct varieties, with assured characteristics, have been a major result of domestication, and not maximum within-variety variation. However, whether clonal or selfing, each type retains some ability to recombine genetic variation, either by seed cycles in clonal (often heterozygous) crops, or low levels of outcrossing in selfing (homozygous) seed crops. We will return to this theme in Chapter 3.

This great range of origin and diversity of crop varieties has been deployed in farmers fields. Frankel *et al.* (1995) have argued that there is an evolutionary continuum linking wild pre-domesticates with present-day cultivars. They also identify an ecological continuum: just as crop diversity can be influenced by natural selection and farmer management of within and between varietal diversity, so too can field type be first selected from the landscape and then managed at various levels of intensity.

Fields

During the transition from hunter–gathering to agriculture, human management continued to change ecosystems, with an impact both on the crops and also on associated agrobiodiversity (see Edwards *et al.*, Chapter 8, this volume, for impacts at the landscape scale). However, some of the profound impacts of hunter–gatherers on their physical and biotic environment was carried through to field management, notably through fire and managed flooding. While it is debatable whether the pressure of human population increase gave rise to agriculture, later population increases certainly induced intensification, with early examples of irrigation, raised beds and terracing. During fieldwork in Yemen, one of the authors came across the deeply impressive remains of the Mârib dam, destroyed some 1500 years ago, but once a key to the irrigated agriculture of the kingdom of Saba' (Bowen and Albright, 1958).

It is likely that the present spectrum of agriculture, from intensively managed quality land through to the barely productive margins of agriculture, has been with us for several thousand years. The wide spectrum of land quality would influence which crops could be grown (and how) and all associated agrobiodiversity. In turn, the chosen crops and their management would impact on the field and its agrobiodiversity. One example, of coffee plantations in central America, is now receiving attention, as the crop management affects insects in the plantation, and in turn the success of bird migrations to North America.

It is important to recognize that many, if not most, agricultural communities maintain a distinction between fields and gardens, both in the type of crops grown, and in the intensity of management. Particularly in wetter tropical areas, systems may be predominantly horticultural (described by Weightman (1989, p. 31) as 'the cunning nurturing of individual plants') often with a strong component of perennials, and the use of the hoe, rather than the plough.

Some types of deliberate human management are shown in Box 2.1. Impacts could be complex; for example, weeding the crop could have many impacts: (i) reducing the potential for gene flow through crop–weed introgression; (ii) reducing the alternative hosts for crop pests and diseases; (iii) reducing competition, thereby allowing the plant to grow in a wider range of ecogeographical niches than hitherto.

This last impact alone (noted by Purseglove, 1968, p. 10) is of great general importance for agrobiodiversity. Niche expansion will have a direct impact in enhancing the productive agrobiodiversity of farms, allowing crops from many different ecogeographical origins to be brought together and to add diversity to any one agricultural system. Farmers have exploited this greatly. For example, a farmer in upland Kenya may grow maize, common beans, potatoes, cassava and sweet potatoes, from the Americas. At lower altitudes another farm may grow sugar cane, bananas, rice and mango from Asia, with African sorghum, cowpea and millets. There will be indirect impacts, as new evolutionary pathways will emerge with local pests and

---

Box 2.1. Human management of agricultural systems.

Control of direct plant competition	• burning
	• tilling
	• weeding
	• mulching
	• removal of trees
	• thinning
	• transplanting (rice)
	• off-season grazing by stock/domestic fowl
Land modification	• irrigation
	• drainage
	• terracing
	• erosion-stimulation
Control of pests/predators and feeding animals	• animal scaring (scarecrows)
	• hunting on site
	• traditional pest management/IPM
	• plant introduction away from co-evolved pests
Soil modification	• nutrient enrichment (burning, household waste, green manure, termite soil)
	• increasing water percolation

pollinators adapting to newly encountered crops (see Allen *et al.*, Chapter 6, LaSalle, Chapter 7, and Polaszek *et al.*, Chapter 11, this volume).

There would have been further secondary impacts of field management on agrobiodiversity. For example, the practice of burning – almost universal for shifting cultivation – can destroy weeds in the soil seed bank, remove harmful pests from the soil, and modify the impact of beneficial soil organisms. Indeed, it is probable that the management of fields, to increase or stabilize crop production, had more profound impacts on crop-associated agrobiodiversity than the diversity of the actual crop itself. Settle *et al.* (1996) give an example of the indirect effect of rice production on the insect agrobiodiversity of the irrigated rice field. As the rice crop develops, detritus from previous crops in the flooded field allows the build-up of detritus feeders; these are then preyed upon by predators, whose populations build up to a level that permits the early and effective control of rice pests (see Edwards *et al.*, Chapter 12, this volume, for more on the agrobiodiversity of rice systems).

Crop-associated agrobiodiversity: weeds as examples

Associated with the great variation of crop and field there are a multitude of other organisms. Much of this book deals with our knowledge and management of this associated agrobiodiversity – both beneficial and harmful – but little is known of the history of these associations. We will return to this in Chapter 18.

This chapter concludes with a brief look at weeds as an example of crop-associated agrobiodiversity. Evidence from archaeology and genetics indicates a long, important, complex and still-evolving association between farmers, crops and associated weeds.

Weeds as crop ancestors

Weeds have been a continuous part of the human association with land since long before domestication: 'The history of weeds is the history of man ...' (Anderson, 1954, p. 131). Indeed, it has been regularly postulated that the disturbance associated with early human settlements encouraged weeds, and that these then became domesticated. In the words of Sauer (1969, p. 71), such plants 'usable by man, were first tolerated, then protected, and finally planted'. The progenitors of crops are often weedy. Daehler (1998) showed that the grass family (*Poaceae*) were 'spectacularly over-represented' among weeds (see Wood and Lenné, Chapter 17 this volume, for a more ample discussion of the significance of this). Indeed, early sites of crop domestication could have been on sites of natural disturbance that favoured weedy progenitors of crops. For example, some of the areas of origin of domesticated plants in Central America, the Fertile Crescent and Southeast Asia were immediately downstream of mountain ranges, with seasonal soil enrichment on alluvial fans, floodplains, and great, seasonally flooded, deltas. Some plant species (perhaps more weedy than others) adapting over time to this natural seasonal disturbance and soil enrichment would be preadapted to the managed disturbance of cultivation and the disturbed or enriched soils associated with human camps.

Crop–weed introgression

Marshall (1990) has noted that for crops with associated cross-compatible and coevolving weedy associates, the associated weeds will provide ample opportunities for future evolution. Examples of wild relatives in genetic contact with the crop include oats (*Avena sativa* ssp. *fatua*); rice (*Oryza sativa* ssp. *spontanea*); barley (*Hordeum vulgare* ssp. *spontaneum*); sorghum (*Sorghum arundinaceum*); and maize (*Zea mays* ssp. *mexicana*). Marshall (1990) believes that these wild species provide large evolving gene pools for the respective crops (see also Cox and Wood, Chapter 3, and Thurston *et al.*, Chapter 9, this volume). However, introgression (and strong selection associated with cultivation and harvesting) may lead to aggressive weeds that so mimic the crop in phenology and seed size that they cannot be eradicated (Harlan, 1975; see Edwards *et al.*, Chapter 12, this volume). Furthermore, the introgression of weedy characters, such as shattering seed dispersal, can render the crop worthless as seed. Another, still emerging, problem with this facility for genetic introgression is the potential for the escape of genes from genetically engineered crops into weeds and then into the wider wild flora (see Polaszek *et al.*, Chapter 11, this volume).

Weeds and pests

Weeds are direct competitors with crops, and the defining feature of agriculture, the 'tilling of fields', is a major method of weed control (and before tilling, burning to reduce the soil seed bank). But weeds can also harbour pathogens and pests of crops and be an important refuge of natural enemies (see Allen *et al.*, Chapter 6, LaSalle, Chapter 7, and Polaszek *et al.*, Chapter 11, this volume).

Weeds as invasive aliens

Agriculture has been a major determinant influencing the composition of agrestal floras (Altieri, 1988). The clearing of indigenous vegetation for either temporary arable fields within a swidden system or for sedentary agriculture has provided habitats which are vulnerable to colonization by both opportunist native species and introduced exotics. Along with proliferation and spread of plants which become noxious weeds after either intentional or unintentional introduction, agriculture and weed management play a key role in the resulting diversity of vegetation at both the landscape and field scale. Conservationists are becoming increasingly aware of the threat to biodiversity from invasive, alien plants which may attain complete dominance of native vegetation. This may lead to extinctions, as has happened with invasion by *Psidium cattleianum* in Mauritius, threaten native fauna, or lead to altered soil chemistry, geomorphological processes, fire regimes and hydrology (Cronk and Fuller, 1995).

For the agriculturalist, invasion is not a recent phenomenon and weed floras in many parts of the world include a sizeable proportion of immigrants (Turner, 1988) which have spread as impurities in crop seeds, with fodder, livestock and trade goods (Ridley, 1930; Wild, 1968). Early in the 20th century the weed flora of Transvaal in South Africa comprised 17% species of American origin, 40% from Mediterranean and Asiatic and 8% from Europe (Burtt Davy, 1904). Europe has contributed 60% of the weed species found in Canada and 80% of those in New Zealand (Salisbury, 1961). *Euphorbia heterophylla,* from tropical America, which is now considered to be one of the most serious weeds of upland rice in the moist savannah of West Africa, and the Asian grass *Rottboellia cochinchinensis,* which has become widespread in Central America during the past 30 years, have both been spread as contaminants in rice seed (Wilson, 1981; Rojas *et al.,* 1993). The rhizomatous perennials *Cynodon dactylon* and *Sorghum halepense* (Johnsongrass), introduced from the Old World as fodder grasses, have become intractable problems in field crops, particularly the subtropics (Holm *et al.,* 1977). These are examples of deliberate introductions which have become noxious weeds and demonstrate human influence on the dynamic and changing nature of the weed flora.

Agrobiodiversity and change

The almost limitless combinations of more or less intensive management, the varied local biotic and abiotic environments, and the human ability to introduce crops and livestock (and, unfortunately, their pests and diseases) from elsewhere, and then to select within and between these varieties, resulted in a great diversity of planned agrobiodiversity and a yet greater diversity of associated, unplanned, and even unmanageable agrobiodiversity. As the relationship between crop and weeds shows, agrobiodiversity is complex and dynamic, both in the crop and the associated component. Agrobiodiversity has been determined largely by human management: knowledge of past changes could help with future evolutionary management; past constraints could be overcome with new approaches.

Acknowledgement

We thank Charlie Riches for a significant contribution on weeds.

References

Allan, W. (1965) *The African Husbandman.* Oliver & Boyd, Edinburgh.

Altieri, M.A. (1988) The impact, uses and ecological role of weeds in agroeco-systems. In: Altieri, M.A. and Liebman, M. (eds) *Weed Management in Agro-ecosystems: Ecological Approaches.* CRC Press, Boca Raton, Florida, pp. 1–6.

Anderson, E. (1954) *Plants, Man and Life.* Melrose, London.

Bowen, R.L. and Albright, F.P. (1958) *Archaeological Discoveries in South Arabia.* Johns Hopkins University Press, Baltimore, Maryland.

Bulmer, S. (1989) Gardens in the south: diversity and change in prehistoric Maori agriculture. In: Harris, D.R. and Hillman, G.C. (eds) *Foraging and Farming: the Evolution of Plant Exploitation.* Unwin Hyman, London, pp. 688–705.

Burtt-Davy, J. (1904) Alien plants spontaneous in the Transvaal. *Republic of South Africa Association for the Advancement of Science, Johannesburg Meeting,* pp. 252–299.

Clark, R. (1983) Pollen and charcoal evidence for the effects of Aboriginal burning on the vegetation of Australia. *Archaeology in Oceania* 18, 32–37.

Cohen, M.N. (1977) *The Food Crisis in Prehistory: Overpopulation and the Origins of Agriculture.* Yale University Press, New Haven, Connecticut.

Conway, G.R. (1987) The properties of agroecosystems. *Agricultural Systems* 24, 95–117.

Cronk, C.B. and Fuller, J.L. (1995) *Plant Invaders.* Chapman & Hall, London.

Daehler, C.C. (1998) The taxonomic distribution of invasive angiosperm plants: ecological insights and comparison to agricultural weeds. *Biological Con-servation* 84, 167–180.

Darlington, C.D. (1969) The silent millennia in the origin of agriculture. In: Ucko, P.J. and Dimbleby, G.W. (eds) *The Domestication and Exploitation of Plants and Animals.* Duckworth, London, pp. 67–72.

Dimbleby, G.W. (1967) *Plants and Archaeology.* John Baker, London.

Evans, A.M. (1976) Beans: *Phaseolus* spp. (Leguminosae – Papilionatae). In: Simmonds, N.W. (ed.) *Evolution of Crop Plants.* Longman, Harlow, pp. 168–172.

Fall, P.L., Lines, L. and Falconer, S.E. (1998) Seeds of civilization: Bronze Age rural economy and ecology in southern Levant. *Annals of the American Association of Geographers* 88, 107–125.

Flannery, K.V. (1969) Origins and ecological effects of early domestication in Iran and Near East. In: Ucko, P.J. and Dimbleby, G.W. (eds) *The Domestication and exploitation of Plants and Animals.* Duckworth, London, pp. 73–100.

Frankel, O.H., Brown, A.H.D. and Burdon, J.J. (1995) *The Conservation of Plant Biodiversity.* Cambridge University Press, Cambridge.

Groube, L. (1989) The taming of the rain forests: a model for Late Pleistocene forest exploitation in New Guinea. In: Harris, D.R. and Hillman, G.C. (eds) *Foraging and Farming: The Evolution of Plant Exploitation.* Unwin Hyman, London, pp. 292–304.

Hallam, S.J. (1989) Plant usage and management in Southwest Australian Aboriginal societies. In: Harris, D.R. and Hillman, G.C. (eds) *Foraging and Farming: the Evolution of Plant Exploitation.* Unwin Hyman, London, pp. 136–151.

Harlan, J.R. (1971) Agricultural origins: centers and noncenters. *Science* 174, 468–474.

Harlan, J.R. (1975) *Crops and Man.* American Society of Agronomy, Madison, Wisconsin.

Harlan, J.R. (1981) Ecological settings for the emergence of agriculture. In: Thresh, J.M. (ed.) *Pests, Pathogens and Vegetation*. Pitman, London, pp. 3–22.

Harlan, J.R. (1989) Wild-grass harvesting in the Sahara and sub-Sahara of Africa. In: Harris, D.R. and Hillman, G.C. (eds) *Foraging and Farming: the Evolution of Plant Exploitation*. Unwin Hyman, London, pp. 79–98.

Harlan, J.R. (1992) *Crops and Man*, 2nd edn. American Society of Agronomy, Madison, Wisconsin.

Harlan, J.R. (1995) *The Living Fields: Our Agricultural Heritage*. Cambridge University Press, Cambridge.

Harper, J.L. (1961) Approaches to the study of plant competition. In: Milnthorpe, F.L. (ed.) *Mechanisms in Biological Competition*. Symposia of the Society of Experimental Biology no. 15, pp. 1–39.

Harris, D.R. (1969) Agricultural systems, ecosystems and the origin of agriculture. In: Ucko, P.J. and Dimbleby, G.W. (eds) *The Domestication and Exploitation of Plants and Animals*. Duckworth, London, pp. 3–15.

Harris, D.R. and Hillman, G.C. (eds) (1989) *Foraging and Farming: the Evolution of Plant Exploitation*. Unwin Hyman, London.

Hawtin, G. (1996) Safeguarding and sharing plant genetic resources. *Outlook on Agriculture* 25, 81–87.

Hillman, G.C. (1990) Domestication rates of wild-type wheats and barley under primitive cultivation. *Biological Journal of the Linnean Society* 39, 39–78.

Holm, L., Plunknett, D., Pancho, J. and Herberger, J. (1977) *The World's Worst Weeds: Distribution and Biology*. University of Hawaii Press, Honolulu, Hawaii.

Jones, R. (1969) Firestick farming. *Australian Natural History* 16, 224–228.

Jones, R. and Meehan, B. (1989) Plant foods of the Gidjingali: ethnographic and archaeological perspectives from northern Australia on tuber and seed exploitation. In: Harris, D.R. and Hillman, G.C. (eds) *Foraging and Farming: the Evolution of Plant Exploitation*. Unwin Hyman, London, pp. 120–135.

Keddy, P.A. (1989) *Competition*. Chapman & Hall, London.

Lenné, J.M. and Wood, D. (1991) Plant diseases and the use of wild germplasm. *Annual Review of Phytopathology* 29, 35–68.

Mannion, A.M. (1997) Agriculture and land transformation. 1. Temporal and spatial dimensions. *Outlook on Agriculture* 26, 71–78.

Marshall, D.R. (1990) Crop genetic resources: current and emerging issues. In: Brown, A.H.D., Clegg, M.T., Kahler, A.L. and Weir, B.S. (eds) *Plant Population Genetics, Breeding, and Genetic Resources*. Sinauer, Sunderland, Massachusetts, pp. 367–388.

Martin, P.S. (1966) Africa and Pleistocene overkill. *Nature* 212, 339–342.

Pickersgill, B. (1989) Cytological and genetic evidence on the domestication and diffusion of crops within the Americas. In: Harris, D.R. and Hillman, G.C. (eds) *Foraging and Farming: the Evolution of Plant Exploitation*. Unwin Hyman, London, pp. 426–439.

Purseglove, J.W. (1968) *Tropical Crops, Dicotyledons*. Longman, London.

Ridley, H.N. (1930) *The Dispersal of Plants throughout the World*. L. Reeve, Ashford, UK.

Riley, R. (1965) Cytogenetics and the origin of wheat. In: Hutchinson, J. (ed.) *Essays on Crop Plant Evolution*. Cambridge University Press, Cambridge, pp. 103–122.

Rindos, D. (1984) *The Origins of Agriculture: an Evolutionary Perspective*. Academic Press, London.

Rojas, C.E., de la Cruz, R., Shannon, P.J. and Merayo, A. (1993) Study and management of itchgrass (*Rottboellia cochinchinensis*) in the Pacific region of Central America. In: *Proceedings of the Brighton Crop Protection Conference – Weeds*, pp. 1183–1188.

Salisbury, E. (1961) *Weeds and Aliens*. Collins, London.

Sauer, C.O. (1969) *Seeds, Spades, Hearths, and Herbs: the Domestication of Animals and Foodstuffs*, 2nd edn. MIT Press, Cambridge, Massachusetts.

Scoones, I., Melnyk, M. and Pretty, J.N. (1992) *The Hidden Harvest: Wild Foods and Agricultural Systems*. IIED/WWF, London.

Settle, W.H., Ariawan, H., Astuti, E.T., Cahayana, W., Hakim, A.L., Hindayana, D., Sri Lestari, A. and Pajarningsih (1996) Managing tropical rice pests through conservation of generalist natural enemies and alternative prey. *Ecology* 77, 1975–1988.

Shipek, F.C. (1989) An example of intensive plant husbandry: the Kumeyaay of southern California. In: Harris, D.R. and Hillman, G.C. (eds) *Foraging and Farming: The Evolution of Plant Exploitation*. Unwin Hyman, London, pp. 159–170.

Simmonds, N.W. (1995) Food crops: 500 years of travels. In: *International Germplasm Transfer: Past and Present*. Crop Science Society of America, Madison, Wisconsin, pp. 31–45.

Smith, B.D. (1995) *The Emergence of Agriculture*. Scientific American Library, New York.

Smith, M.A. (1989) Seed gathering in inland Australia: current evidence from seed-grinders on the antiquity of the ethnohistorical pattern of exploitation. In: Harris, D.R. and Hillman, G.C. (eds) *Foraging and Farming: The Evolution of Plant Exploitation*. Unwin Hyman, London, pp. 305–317.

Sprecher, S.L. (1988) Isozyme genotype differences between large seeded and small seeded gene pools in *Phaseolus vulgaris* L. *Bean Improvement Cooperative* 31, 36–37.

Turner, C.E. (1988) Ecology of invasions by weeds. In: Altieri, M.A. and Liebman, M. (eds) *Weed Management in Agroecosystems: Ecological Approaches*. CRC Press, Boca Raton, Florida, pp. 41–55.

Vavilov, N.I. (1926) *Studies on the Origin of Cultivated Plants*. Institute of Applied Botany and Plant Breeding, Leningrad.

Weightman, B. (1989) *Agriculture in Vanuatu: a Historical Review*. British Friends of Vanuatu, Cheam.

Wild, H. (1968) *Weeds and Aliens in Africa: the American Immigrant*. Inaugural Lecture, University College of Rhodesia, Salisbury.

Williams, M.D.H.M., Peña, R.J. and Mujeeb-Kazi, A. (1993) Seed protein and isozyme variations in *Triticum tauschii* (*Aegilops squarrosa*). *Theoretical and Applied Genetics* 87, 257–263.

Wilson, A.K. (1981) *Euphorbia heterophylla*: a review of distribution, importance and control. *Tropical Pest Management* 27, 32–38.

Wittfogel, K. (1957) *Oriental Despotism: a Comparative Study of Total Power*. Yale University Press, New Haven, Connecticut.

Wood, D. (1988) Introduced crops in developing countries: a sustainable agriculture? *Food Policy* 3, 167–172.

Wood, D. and Lenné, J. (1993) Dynamic management of domesticated biodiversity by farming communities. *Proceedings of the Norway/UNEP Conference on*

Biodiversity, Trondheim, May, 1993, pp. 84–98.

Wood, D. and Lenné, J. (1997) The conservation of agrobiodiversity on-farm: questioning the emerging paradigm. *Biodiversity and Conservation* 6, 109–129.

Yen, D.E. (1989) The domestication of environment. In: Harris, D.R. and Hillman, G.C. (eds) *Foraging and Farming: the Evolution of Plant Exploitation*. Unwin Hyman, London, pp. 55–75.

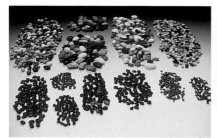

Plate 1

Plate 2

Plate 3

Plate 4

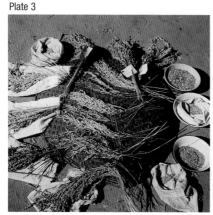

Plate 1. Seed of four ancestral and domesticated *Phaseolus* species (common, tepary, lima and scarlet runner beans) (courtesy of CIAT).

Plate 2. Morphological diversity in sorghum landraces (courtesy of ICRISAT).

Plate 3. Morphological diversity of rice from a village in Lao PDR (courtesy of IRRI).

Plate 4. Morphological diversity in the maize landrace 'Bolita' from the Central Valleys region of Oaxaca, Mexico (courtesy of Mike Listman, CIMMYT).

CROP PROTECTION PROGRAMME

The editors and publishers would like to thank the UK Department for International Development (DFID) Crop Protection Programme (CPP) for their financial contribution towards the colour plate section and colour cover. Further information about CPP can be found via the Internet at www.nrinternational.co.uk.

Plate 5. A Thüringer Waldziege goat from Germany (courtesy of Hans Hinrich and FAO).

Plate 5

Plate 6. A swamp buffalo from Vietnam (courtesy of FAO).

Plate 6

Plate 7

Plate 7. A meishan sow and piglet from Taiwan (courtesy of Hans Hinrich and FAO).

Plate 8. A Öländskt Dvärghöns hen from Sweden (courtesy of FAO).

Plate 8

Plate 9

Plate 9. A dehesa with *Quercus ilex*, the traditional agropastoral system near Cáceres in southwestern Spain (courtesy of K. Ullrich).

Plate 10

Plate 10. Terraced cereal landscape in the Himalayan foothills, Uttar Pradesh, India (courtesy of D. Wood).

Plate 11

Plate 11. A mosaic landscape in the Andes of Peru (courtesy of J.M. Lenné).

Plate 12

Plate 12. A home garden with banana, maize, sweet potato, common bean, traditional vegetables and multipurpose trees in the highlands of Kenya (courtesy of J.M. Lenné).

Plate 13

Plate 13. Parasitic weed *Striga hermonthica* on pearl millet in Niger (courtesy of ICRISAT).

Plate 14. No-till maize sown into frost-killed residues of white mustard (courtesy of W. Richner).

Plate 14

Plate 15

Plate 15. Traditional granaries for seed and grain storage in Nigeria (courtesy of M. Turner).

Plate 16. *Ex situ* conservation of genetic resources at -18°C (courtesy of CENARGEN).

Plate 16

The Nature and Role of Crop Biodiversity

3

T.S. Cox[1] and D. Wood[2]

[1]B-218 Sainikpuri, Secunderabad 500094, India;
[2]Agrobiodiversity International, 13 Herons Quay, Sandside,
Milnthorpe, Cumbria LA7 7HN, UK

Introduction

Biodiversity exists on several levels at once: ecological, organismal, genetic and cultural (Heywood and Baste, 1995). Although the relationship between genetics and culture is a matter for debate, we can safely accept that genetic diversity forms the foundation of ecological and organismal diversity. When scientists and policymakers discuss the destruction, manipulation, protection, or assessment of 'biodiversity' – especially in the agricultural context – they usually have in mind the *genetic* diversity that underlies it. This chapter will be concerned primarily with genetic diversity between and within species of crop plants.

Outside of agriculture, much of the discussion of plant biodiversity is concerned with species: their numbers, distribution, relationships and loss (Groombridge, 1992; Reaka-Kudla *et al.*, 1996). There, the focus remains on species, despite a growing awareness over the past three decades – thanks to the development of protein and DNA markers – that extensive intraspecific genetic variation also exists. It has been estimated that wild species have, on average, about 220 genetically different populations (Hughes *et al.*, 1997). In contrast, the number of distinct varieties in major crops will number many thousands. There is a pronounced human ability to promote, partition and enhance this genetic diversity through selection. Varieties, whether landraces or modern varieties, are a result of the long-standing human interest in the genetic diversity that we find *within* species. This interest is, above all, practical: crop varieties are the functional units of food production.

While formal taxonomy is the key to defining diversity through species relationships and numbers, it has been less useful for crops (and pasture species). There has been a notable and general lack of interest in the taxonomy of cultivated plants (Zohary in Bennett, 1968; Li, 1974). This may be understandable: crops and pasture species are often 'taxonomic nightmares' (Clayton, 1983). However, the problems of taxonomists with agricultural species may be biologically significant. Many of the species chosen for domestication were already in an active state of evolution, and often weedy, with introgression and polyploidization (Hawkes, 1986). It seems that this evolutionary dynamism was a positive feature of early domesticates (see Wood and Lenné, Chapter 2, this volume). Furthermore, humans have meddled with the evolution of crop species over several millennia since domestication. Crop biodiversity has become too complex, and we have become too familiar with its complexity, for it to be described or partitioned by traditional taxonomic methods (de Wet *et al.*, 1986).

Methods of partitioning crop diversity vary from species to species and from scientist to scientist. We will now attempt to steer clear of detailed taxonomic issues and concentrate on two issues: the ways in which crop diversity – with emphasis on genetic diversity – is distributed and assessed, and the ways in which crop diversity is organized into agroecosystems. The interrelationships between crop biodiversity and pests, pathogens and soil biodiversity are dealt with in detail in Chapters 5, 6, 7, 11 and 12, this volume. In this chapter, we will quite naturally take the human point of view, which sees crop biodiversity as serving two purposes: providing the genetic raw material needed to breed new genotypes for future agroecosystems (see Witcombe, Chapter 10, this volume), and helping current agroecosystems to function properly (i.e. to produce food as efficiently and sustainably as possible, see Lenné and Wood, Chapter 18, this volume).

Diversity in Crop-based Agroecosystems

Agroecosystems, like other types of ecosystems, vary widely in the amount of biodiversity they contain and how that biodiversity is organized among species, among varieties or cultivars within species, and within cultivars. Crop ecosystems are well known for having low species diversity when compared with most non-agricultural ecosystems; indeed, the farmer's intention is to eliminate all but one or a few species from a field if possible. Discrimination based on species is also applied by breeders when they separate the species in an ecosystem into two groups: the ones that they are trying to improve (comprising 'genotypes') and all others (which are part of the 'environment'). In general, knowledge of crop diversity at the species level is good, but at the cultivar level it is variable: for improved varieties it is patchy but adequate for some crops (rice, maize, wheat), while for landraces it is poor (Boyle and Lenné, 1997). Firstly, we look at the distribution and uses of this diversity in crop-based agroecosystems.

Among species

Diversity among species – in space, in time, or in both – is important because of the complementary contributions that crop species make to the human or animal diet, as well as to the functioning and productivity of the agroecosystems of which they are a part. A large part of farmers' skills in managing crops depends on the ability first to recognize species, and then to develop a knowledge of the properties of species to allow their deployment, whether promotion or removal, in the production system (see Thurston *et al.*, Chapter 9, and Witcombe, Chapter 10, this volume).

In general, the more important the role of the species – whether productive or harmful – the more detailed the farmers' knowledge and management skills. However, this will only extend to the more obvious species – the above-ground crops and more obvious pests. For less obvious, but still important productive and harmful organisms, formal taxonomic science is strongly complementary to farmers' ability to recognize species. The expansion of formal agricultural research into new, and especially tropical, environments, created a need for taxonomic identification to allow the promotion of useful species and the control of species harmful to agroecosystems. For productive species, botanic gardens and associated herbaria (Kew, 1887; Brockway, 1979) met the need. For harmful organisms, a good example of taxonomic identification and the development of methods of control is the series of international institutes (now CABI Bioscience) with disciplines of entomology, mycology, nematology and biological control (Hammond, 1998).

Among cultivars within species

Diversity among cultivars within a species is often not easily described, and its effects are more difficult to evaluate. Most cultivars have names, but the number of different names representing the cultivars in a given field, farm or region is not usually a good indicator of genetic diversity (Morris and Heisey, 1997; Wood and Lenné, 1997). Both the International Maize and Wheat Improvement Centre (CIMMYT) and the International Rice Research Institute (IRRI) have parentage tracking systems (Hargrove *et al.*, 1988; Skovmand *et al.*, 1992). The complete genetic parentage, a measure of diversity, of all bred lines is stored in databases and the contribution of any variety or landrace to a bred line can be analysed (see Witcombe, Chapter 10, this volume).

The relative degree of genetic relationship among cultivars must also be considered. For example, as the total number of wheat (*Triticum aestivum*) cultivars grown in the eastern USA increased steadily through the 1970s, overall genetic uniformity rose dramatically. This occurred because many cultivars were grown over small areas, while the popular winter wheat

'Arthur' and related cultivars dominated the landscape (Cox *et al.*, 1986). This a good example of the problems of contrasting ways of measuring diversity: by *richness*, which looks at the total number of variants, or *evenness*, which measures the frequency of different variants (Frankel *et al.*, 1995). It is now routine to estimate genetic diversity between wheat cultivars at both the national (Yang and Smale, 1996) and international level (Rejesus *et al.*, 1996; and see Witcombe, Chapter 10, this volume).

Duvick (1984) has pointed out that we should not concentrate on a particular farm or region, but consider all intraspecific diversity, including that in unreleased breeding material, to be directly relevant (see Witcombe, Chapter 10, this volume). Current cultivars can be, and usually are, replaced rapidly by new ones, and unreleased breeding material can be a ready source of needed genes. For winter wheat in the USA, as for most crops, the rate of cultivar turnover has increased steadily during the 20th century (Cox *et al.*, 1986).

The 'gene pool' concept (Harlan and de Wet, 1971) allows an extension of Duvick's argument. We can visualize the cultivars of a species that are actually in production as being backed up by gene pools of currently unused cultivars, experimental lines, ancestral taxa, wild relatives – any genotype which can be hybridized to produce new cultivars, either in the short term (within the primary gene pool) or the long term (within the secondary and tertiary gene pools – see Witcombe, Chapter 10, this volume). In the past, the lack of definition of gene pools for many crops may have limited the use of species in the primary and secondary gene pools in breeding programmes (Wood and Lenné, 1997). In recent years, the techniques of biotechnology have made these gene pools much wider but only a little deeper; that is, many more species now can contribute to a particular gene pool via genetic transformation, but their contributions are limited to individual genes with high profit potential.

Many crops have multiple uses (food, fibre, fodder, oil) or are used in different ways by different tribes, cultures and countries (Boyle and Lenné, 1997; Wood and Lenné, 1997; see Thurston *et al.*, Chapter 9, this volume). For example, cassava is processed in many ways in Latin America which are unknown in Africa where the crop is also important. There is a need for a global inventory of crops, their direct and processed uses, and their negative characters (weediness, toxins) which can also be disseminated when crops are introduced to new areas (Boyle and Lenné, 1997).

Among plants within cultivars

The amount of genetic diversity within a cultivar depends on the relative frequencies of self- and cross-pollination in the species to which it belongs, the methodology used by breeders or farmers to develop the cultivar, and the magnitude of natural gene flow (Frankel *et al.*, 1995). In general, within-cultivar diversity is greater in predominantly cross-pollinated species than in predominantly selfing species, greater in landraces (traditional cultivars handed down through generations of farmers) than in cultivars developed by modern breeding programmes (see Witcombe, Chapter 10, this volume), and greater in open-pollinated cultivars than in ones produced by controlled pollination (e.g. F_1 hybrids), *but there are many exceptions to these generalizations.*

Maize (*Zea mays* L.) is a cross-pollinated species, but a single-cross hybrid has, by design, very little variation within it. A so-called 'pure-line' cultivar of wheat (*Triticum aestivum* L.) – a self-pollinated species – may be just that if it was developed in a European country with licensing requirements that require strict genetic uniformity. Conversely, a wheat cultivar developed for the central or northern Great Plains of the USA – where uniformity requirements are looser and expression of genetic variation is diminished by the harsher environment – may represent not a pure line but the bulked progeny of a plant from an early generation of self-pollination (e.g. F_3 or F_4), leaving considerable variation among plants within the cultivar. A modern Great Plains cultivar may harbour even more genetic diversity than a wheat landrace that is descended from a single, highly homozygous plant selected and propagated by a farmer.

Landraces tend to be the most genetically diverse group of crop cultivars (Frankel *et al.*, 1995; see Thurston *et al.*, Chapter 9, this volume). They evolved under both environmental and farmer selection pressures and remain the predominant form of crop diversity in less-favoured and marginal environments. Landraces generally have a population genetic structure which accommodates a wide range of genetic diversity probably due to the weakness and lack of consistency of the selection pressures to which they have been exposed (Frankel *et al.*, 1995). The lack of a generally usable classification and nomenclature system for landraces is a serious constraint to describing this important component of agrobiodiversity (Wood and Lenné, 1997).

Measuring Crop Diversity

Types of data

Ecological techniques exist for describing crop biodiversity at the level of the production system and farm but have rarely been applied to agroecosystems (Boyle and Lenné, 1997). This is in great contrast to the extensive knowledge base generated for wild biodiversity, resulting in the discipline of conservation biology (Wood and Lenné, 1997).

Quantifying genetic diversity between and within populations or cultivars has always been problematic. Methods of data generation generally fall into three categories: pedigree-based, phenotype-based, or genetic marker-based (Table 3.1). Each method has strengths and weaknesses.

Table 3.1. Quantifying genetic diversity.

Type of data	Example	Genome coverage	Proximity to DNA diversity	Problems
Pedigree	Coefficients of parentage; ancestral composition	Complete	Low	Inaccuracy of breeders' records; selection bias; intraculvar and intraline variation
Phenotype	Polygenic traits	Low to high (depending on traits)	Low	Expression of similar phenotypes by dissimilar genotypes; genotype × environment interaction
	Genotypes at morphological loci	Low	High	Usually very few loci available; loci may be subject to strong selection; dominance
Genetic marker	Protein level	Low to moderate	High	Number of loci limited; some loci subject to selection; dominance
	DNA level	Moderate to high	Highest	'Identical' alleles not necessarily identical by descent; relationship, if any, between markers and phenotypes usually unknown; lack of polymorphism

Taxonomic techniques for classification of crop genus and species are available; however, few major crop plants have been reclassified at the genus and species levels since first being named (Wood and Lenné, 1997). Computation of pairwise relationships (e.g. 'coefficients of parentage') from cultivar pedigree records has low precision for any one data point, but averages and patterns computed from those data may be useful for describing relative diversity in different regions or time periods (Cox *et al.*, 1986; Murphy *et al.*, 1986). Mean coefficients of parentage show the reduction in diversity that organized breeding (which, for most crops, started in the early 20th century) has *superimposed* on the genetic similarity that *already* existed among ancestral genotypes.

Agreed lists of descriptors are available for many crops but their compilation tends to be very time-consuming (Boyle and Lenné, 1997). Indices of diversity for polygenic traits (observed phenotypically) are useful when applied to traits of immediate interest, but not as indicators of overall genetic diversity (Table 3.1). Some phenotypic traits are determined by well-defined (usually single-locus) genotypes and are useful as genetic markers, but there are rarely, if ever, enough of them in a given population to provide even minimal coverage of the genome.

It is useful to consider farmers' ability to measure diversity. While plant breeders can measure and deploy genetic distinctness, this cannot be done directly by farmers. Traditional farmers usually have a detailed ability to discriminate between their own varieties, but this is based on phenotypic characters (see Boster, 1985, for cassava; Louette *et al.*, 1997, for maize; see Thurston *et al.*, Chapter 9, this volume; Plates 1–4) and not necessarily on genetic distinctness. Strong selection by a farmer of an obvious variant, such as seed colour based on a single gene, may actually reduce overall genetic variation. For example, common bean (*Phaseolus vulgaris*) has a great range of distinct seed colours which are used as a guide to composing mixtures each planting season to meet agronomic and production needs (Voss, 1992; see Thurston *et al.*, Chapter 9, this volume; Plate 1). However, bean seed colour is controlled by very few genes, and different colours are not a measure of genetic distinctness (Briand *et al.*, 1998).

The situation for beans is complicated by the existence of two distinct gene pools: Central American and South American (Evans, 1976). Because of parallel evolution of seed colours in these two gene pools, seeds of the same colour may be from quite distinct origins (Sprecher 1988; Briand *et al.*, 1998): here farmers cannot use phenotype to select against genetic diversity. If, for example, farmers select for black beans, they will include a range of genetic variation from Central and South America. In contrast, seed size in beans can be a good proxy for genetic distinctness: small-seeded beans evolved from their wild ancestors in Central America, and large-seeded beans evolved in South America. If farmers select for seed size (as they do), they will favour one gene pool over the other and reduce genetic diversity.

Farmers in eastern Africa do not know the geographical origin of their bean varieties. When farmers can determine origin, there is some evidence from seed flows that traditional farmers worldwide may be using geographical distance (in source of seed) as a proxy for genetic distance. Farmers often go to some trouble to obtain 'distant' varieties. This is profoundly sensible, as the technologies for establishing 'core collections' in genebanks have shown that origin is a simple and effective way to partition variation in germplasm collections (Crossa *et al.*, 1995).

In contrast with the continuing assessment by farmers of diversity based on phenotypic characters, in the past few decades the evaluation of diversity by breeders has moved to the use of genetic markers, which are directly indicators of DNA-level diversity and, as technology improves, provide better and better genome coverage (Brown, 1978; Frankel *et al.*, 1995; see Witcombe, Chapter 10, this volume). But genetic distance or diversity estimated from markers is still a matter of relative probabilities. Two cultivars may have the same allele of a particular marker simply by chance (i.e. they are 'identical in state'), or they may have inherited their alleles from a recent common parent (i.e. they are 'identical by descent'). But with a large enough array of markers, one can conclude that the relative genetic distance between any two cultivars of any pair is indicated by the relative number of markers for which they carry contrasting alleles.

Experimental methods

Methods for collecting and analysing diversity data vary even more widely. The research questions to be answered are, of course, the main determinants of methodology (Table 3.2). The questions and the analyses that usually apply to crop species differ from those encountered when evaluating diversity in natural systems (Bisby, 1995; Hawksworth and Kalin-Arroyo, 1995).

Plant breeders and germplasm curators often want to identify geographical regions where crops or related wild taxa most often express a desired trait, exhibit the greatest genetic diversity, or harbour rare alleles. The methods applied will depend on whether the interest is in overall diversity or in a specific trait, for example disease resistance (Table 3.2; see Burdon 1987; Frankel *et al.*, 1995).

Policymakers and agriculturalists may be interested in 'genetic vulnerability' of a species or region – that is, its relative lack of genetic diversity, especially regarding susceptibility to pests or environmental stresses that could trigger widespread crop failures. The range of diversity within a crop – the genetic base – is a measure of the evolutionary resilience of the crop (Simmonds, 1962; and see Witcombe, Chapter 10, this volume). We have argued above that morphological diversity may be no indication of genetic diversity (e.g. for common bean varieties in Tanzania; Briand *et al.*, 1998). Also, morphological uniformity may be no indication of a lack of genetic

Table 3.2. Locating genetic diversity.

Question to be answered	Sources of data	Types of data: examples	Types of statistics: examples
Where is desirable expression of a trait found?	Germplasm collections; genotypes from breeding programmes	Phenotypic	Means and ranges
Where are regions of high (and low) genetic diversity?	Random samples among and within fields of a crop or stands of a wild species	Genetic markers are best; phenotypic data useful only for specific traits of interest; pedigree data usually not complete enough	Polymorphic indices; genetic variances
Where are globally rare genes or genotypes found?	As above, but targeting regions or fields, based on knowledge or experience, may be done	Genetic markers are best; disease or pest resistance genes if they are of interest	Simple allele frequencies
Is a particular agroecosystem or region 'genetically vulnerable'?	Cultivars in production, weighted by land area occupied; representatives of the 'available' gene pool	Pedigrees combined with production data if complete or nearly so; resistance genes; genetic markers	Weighted mean coefficient of parentage; numbers of alleles; polymorphic indices; frequencies of resistance alleles deployed
What are the patterns of relationships among genotypes from different regions (or breeding programmes, etc.) for a particular crop?	Representative cultivars	Genetic markers if polymorphism is sufficient; pedigree data	Cluster analysis (hierarchical or non-hierarchical) or principal component or coordinate analysis, based on genetic distances
What are relative genetic contributions of different ancestral genotypes to the current gene pool?	Cultivars of interest plus all available ancestors	Pedigrees; genetic markers if polymorphism is sufficient	Summed coefficients of parentage or similarity indices between an ancestor and all cultivars
Which genotypes should be selected to make up a 'core' collection?	Accessions from the germplasm collection of interest	'Passport' data; genetic markers	Random sampling; stratified sampling based on geography or allele frequencies

diversity: for example, the pedigree of the rice variety 'IR66' includes 42 landraces with multiple disease and pest resistances, drought tolerance and earliness (Hargrove *et al.*, 1988). A mythology may have arisen through an overemphasis on the value of traditional 'micro-varieties' based on morphological diversity (Wood and Lenné, 1997).

Agricultural historians and others may want to investigate relationships among cultivars from different regions or breeding programmes, or the relative genetic contributions of ancestral landraces. And efficiency-minded germplasm curators may want to identify subsamples from their genebanks – 'core collections' – that can be used to represent as accurately as possible the entire collection (Brown, 1995).

Chakraborty and Rao (1991) have shown that almost all measures of genetic diversity based on discrete loci belong to the same class of mathematical formulae, and that there are advantages and disadvantages associated with any single method of evaluating diversity. The most commonly used measures are based simply on the total number of alleles in a population or on the probability that any two alleles drawn at random are different. Choice of diversity measure often rests on the relative weight to be given to rare alleles (Brown, 1978; Chakraborty and Rao, 1991). A myriad of computer packages now exist for computed statistics appropriate for analysing both allelic and quantitative data, but this is not the place for a comprehensive review of them. Suffice it to say that the simplest sorts of statistics (allele frequencies, variances, polymorphic indices) often can provide the most useful information.

Development of Crop Diversity

The arrays of species that inhabit crop ecosystems are determined largely by farmers, who base their decisions on experience, influenced by many forces: economic, social, cultural, natural, historical. But their choices are limited, ultimately, by the range of plant species and cultivars available to them. That range, in turn, depends on the magnitude of genetic diversity within the crop species of interest and the capacity of breeders and farmers to manipulate them. As we saw in Chapter 2, every crop species has an evolutionary history that has shaped its gene pool. The stories of two cereal grains – common wheat and sorghum (*Sorghum bicolor*) – illustrate the varied ways in which interaction between natural and human-directed evolutionary forces help to set the genetic boundaries for crop diversity.

Common wheat

Wheat, especially common wheat (also known as hexaploid or bread wheat), always has teetered on a precariously narrow genetic base. Whereas genetic uniformity in many other crop species is attributable to modern breeding and farming practices (Raeburn, 1995), wheat lacks diversity because it evolved through a *natural* genetic bottleneck (Cox, 1998).

Common wheat is one of the youngest plant species on the planet. It appeared abruptly about 7000 years ago in southwest Asia, when plant(s) of emmer wheat (*Triticum turgidum*), growing on an early Neolithic farm, were fertilized by windborne pollen from plant(s) of a weedy diploid goatgrass, *Aegilops tauschii* (Zohary and Hopf, 1988). The resulting hybrid seeds produced partially fertile plants the following season, and progeny of those plants were incorporated by farmers into the domestication process.

This simple event with monumental impact was graphically described by Harlan (1981):

> Some time during the neolithic of the Near East, the genomes of tetraploid wheat combined with that of *Aegilops squarrosa* [now = *Ae. tauschii*]. This little weedy goatgrass is the only member of the genus with a continental distribution and the only one extending into the Central Asian steppes. It transformed a rather ordinary cereal into the most widely grown food crop on earth.

Since Kihara (1944) and McFadden and Sears (1946) first determined that *T. turgidum* and *Ae. tauschii* are the parents of common wheat, wheat geneticists have reproduced such crosses many times. But they have been able to obtain viable seeds only by rescuing the hybrid embryo on an artificial culture medium. Therefore, the spontaneous formation of viable *T. turgidum* × *Ae. tauschii* hybrid seed in Neolithic fields must have been extremely rare. And if the hybrid seed managed to germinate and survive to reproductive stage, the hybrid plant would have been largely sterile. To produce any seed for the next generation, it had to form 'unreduced' male and female gametes, that is, eggs and sperm that each had twice the normal chromosome number (Xu and Joppa, 1995).

The rarity of hybrid formation was not the only genetic bottleneck. In the eyes of a Stone Age farmer, the many wild-type traits exhibited by the new species would have rendered it undesirable, even downright ugly, compared with neighbouring emmer wheat plants. Only when spontaneous mutants with alternative traits arose were farmers likely to have intentionally sown the new species.

For early farmers who developed common wheat as a crop, the process of breeding via selection of spontaneous mutants would have been long and slow. Once more desirable wheats were developed, farmers would have rejected any newly arising hybrids – which were just as undesirable as the original hybrid had been – thereby closing off opportunities to widen the genetic bottleneck (Ladizinsky, 1984). Because no wild hexaploid ancestor existed, and new hybrids were probably being weeded out of fields, enrichment of the gene pool via natural hybridization, as has occurred in other crops (Harlan, 1965), would have been insignificant.

Although we do not know precisely how many times *T. turgidum* × *Ae.*

tauschii hybrids appeared and were domesticated, the number of events must have been small. It is even possible that all common wheat is descended from a single hybrid. If there were more than a few such ancestral plants, subsequent genetic drift must have eliminated much of the variation. Common wheat has extremely low levels of polymorphism for all genetic markers, compared with related species. Nishikawa *et al.* (1980) found seven alleles of an α-amylase gene in a collection of 60 *Ae. tauschii* strains. Only one allele of the gene occurred in common wheat. Among 79 *Ae. tauschii* strains, Lagudah and Halloran (1988) identified no fewer than 72 alleles coding for one seed-storage protein gene, 57 alleles of another, and 14 of a third. In contrast, among 29 wheat cultivars of diverse origin, only one allele existed for the first gene, one for the second, and two for the third.

Lubbers *et al.* (1991) found very high levels of diversity among 102 *Ae. tauschii* strains for molecular markers, but when A.K. Fritz *et al.* (Kansas State University, USDA-ARS, unpublished) surveyed 21 common-wheat cultivars from around the world using 33 molecular markers (representing almost 100 locations on wheat's chromosomes), only ten showed any diversity at all. Over all markers, *Ae. tauschii* strains had a mean of 3.4 alleles per marker (compared with 1.1 for wheat) and a mean diversity index of 0.41 (compared with 0.04 for wheat, on a scale of 0.0–1.0).

We have cited extensive experimental evidence supporting the hypothesis that all common wheat is descended from a small number of interspecific hybrid plants. Yet, despite this dearth of detectable genetic diversity, we do see great phenotypic variation among wheats from around the world – a result of farmers' and breeders' efforts to take advantage of the trickle of mutation and gene flow that have introduced new alleles over the millennia. Today, wheat breeders in most regions of the world expend much of their effort on attempts to protect this genetically vulnerable species from diseases and insects. Fortunately, common wheat is somewhat of a 'genetic sponge', capable of soaking up variation from many different species. Many disease and insect resistance genes now deployed were obtained from other species (Cox, 1990). Indeed, over the past few decades, humans have used interspecific crossing to widen wheat's genetic base at an ever-increasing rate. For example, only two of the first 20 genes to receive symbols (*Lr1–Lr20*) for resistance to leaf rust (caused by *Puccinia recondita* Rob. ex Desm.) were introduced from outside the species *T. aestivum*, whereas 18 more recently named genes in the range *Lr21–Lr44* were introgressed from other species (McIntosh, 1991; McIntosh *et al.*, 1995).

In wheat, as in most species, transfer of resistance genes always has been the most prominent objective of interspecific crossing. But why should related species not be regarded as sources of useful genetic variation for all economic traits? Humans would have been fortunate indeed if the rare hybrid(s) that gave rise to common wheat carried the ideal allele of every gene. If the genetic diversity of wheat is to be increased, humans will have to continue introducing new alleles from other species (Cox, 1998).

Sorghum

Sorghum belongs to the grass family, as does wheat, and, like wheat, it is usually grown as a cereal grain. But the genetic structure of sorghum differs greatly from that of wheat, because of its very different evolutionary history. Sorghum was domesticated in Africa at least 5000 years ago in the Ethiopia–Sudan region (Doggett, 1988), and may have been cultivated as early as 7000 years ago along what is now the Egypt–Sudan border. The cultivated sorghums were developed gradually by direct selection of non-shattering, large-grained variants of wild sorghum (Harlan and Stemler, 1976; Doggett, 1988), and the crop and ancestor (both belonging to *Sorghum bicolor*) continued to grow in close proximity through much of Africa (Doggett, 1988), with ample gene flow between them (Aldrich *et al.*, 1992; Cui *et al.*, 1995). Therefore, virtually the entire natural gene pool of the species has been available to farmers or breeders who develop sorghum cultivars.

The wild and cultivated sorghums are assigned to different subspecies within one biological species. Patterns in the vast morphological variation in the species (Plate 2) allow identification of sorghum 'races' within the wild and cultivated subspecies. The races are interfertile and do not differ enough from one another to merit formal taxonomic status. Similar races have been defined for maize (e.g. 'flint' and 'dent') and other crops (see Smartt and Simmonds, 1995). The natural enhancement of biodiversity in sorghum has occurred through two routes that were severely restricted in common wheat: multiple domestication and gene flow between subspecies and races. Of the four races of wild sorghum, three have been identified as potential ancestors (Harlan and Stemler, 1976; Doggett, 1988), each perhaps giving rise to a different cultivated race. However, variation of genetic markers in wild and cultivated sorghums probably results more from gene flow between the crop and the wild races than from multiple domestication events (Aldrich and Doebly, 1992).

A founder effect can be seen in the evolution of the sorghum crop, but the bottleneck was relatively wide. The magnitude of genetic variation detectable with genetic markers is much greater in sorghum than would be expected with a tight bottleneck such as the one through which common wheat evolved. Examining 53 marker genes, Aldrich and Doebly (1992) found an average of 3.45 alleles per gene in wild sorghums and 2.28 in cultivated types; 87% of wild markers and 75% of cultivated markers showed variation (i.e. at least two alleles). Whereas this indicates that much variation survived the domestication bottleneck, a significant 44% of all marker alleles found in the wild sorghums were not found in the cultivated gene pool. In contrast, de Oliveira *et al.* (1996) found very few marker alleles that were unique to the wild subspecies. Morden *et al.* (1989,1990) obtained results strikingly similar to those of Aldrich and Doebly (1992): 3.2 versus 1.8 alleles per gene and 83 versus 59% variable markers in the wild and cultivated sorghums, respectively.

We have contrasted the evolutionary histories of sorghum and wheat –

two annual grasses grown for their grain – to emphasize the differences in patterns of crop biodiversity that can be found even among species occupying similar niches in agriculture. Obviously, the highly restricted diversity in wheat and the much greater diversity in sorghum have differing implications for breeders. One example: hybrid sorghum has been highly successful, whereas hybrid wheat has never succeeded on a large scale. Germplasm specialists also must view the crops differently. Whereas conservation of landraces (either *in situ* or *ex situ*) may be a high priority in sorghum, wheat's future may depend more on maintenance of the Middle Eastern ecosystems that harbour its wild relatives.

It is important to remember that evolution of these two species has not stopped. The processes of introgression and selection in the crop–wild sorghum complexes of Africa described by Doggett (1988) will continue as long as the crop is grown there. And although the evolution of common wheat remains solely in the hands of humans, the many wild relatives of wheat – which form a critically important gene pool – continue to evolve in southwest Asia.

Partitioning of Crop Diversity in Agroecosystems

There are many different crop-based agroecosystems and it is difficult to generalize about them (see Thurston *et al.*, Chapter 9, this volume). They vary from monocultures of single varieties (including plantations), crop multilines, landraces, varietal mixtures, intercrops, crop rotations, multiple crops, and polycultures (including home gardens) (see Plates 9–12). Let us consider some examples of crop-based agroecosystems and compare their biodiversity, in particular, the partitioning of their genetic diversity.

Eight examples of agroecosystems (most of them wheat-based) are arranged in Table 3.3, from the simplest to the most complex. We have partitioned the total genetic diversity in any single field into those portions among species, among cultivars within crop species, and within cultivars. (Although the species diversity includes not only crops but also weeds, pathogens, insect pests, pollinators, other animals, soil organisms, and parasites or predators that attack pest species, we will consider only the crop here, as most of the others are considered elsewhere in this book.) The amount of diversity shown for each level (low, medium or high) is relative to other types of cropping systems.

The first four examples (Table 3.3) are continuous monocultures of wheat. In the first, a monoculture of a single, modern wheat cultivar, genetic diversity is low at all levels in any one growing season. If the cultivar was derived from an early generation of selfing, there will be some residual genetic diversity among plants within the cultivar. But, at other levels, uniformity reigns. The second example – a monoculture of a multiline cultivar – is also highly uniform. A multiline cultivar, in the strictest sense, is a blend of different

Table 3.3. Examples of genetic diversity in agroecosystems.

Example of agroecosystem	Relative diversity			
	Among species		Among cultivars within species	Among individuals within cultivars
	In space	In time		
Modern wheat cultivar: continuous monoculture	Low	Low	Low	Low to moderate
Multiline wheat cultivar: continuous monoculture	Low	Low	Low	Low but targeted
Wheat landrace: continuous monoculture	Low	Low	Low	Low to moderate
Blend of wheat cultivars: continuous monoculture	Low	Low	Low to high	Low
Two-crop rotation	Low	Moderate	Low	Low
Two-crop intercrop: continuous monoculture	Moderate	Low	Low	Moderate
Complex intercrop with rotation	Moderate to high	Moderate to high	Low to high	Moderate
Perennial polyculture: 5- to 10-year cycle	High	Low to moderate	High	High

versions, or isolines, of a single cultivar, each carrying a different gene for resistance to a disease or other pest. A monoculture of a wheat multiline, for example, is intended to be uniform at all levels, with variation restricted, in theory, to a single important genetic locus. The third example – a field sown to a wheat landrace – is similar to the first two, except that diversity within cultivars is often (but not always) greater. Wheat farmers also may blend the seeds of two or more cultivars to sow a field, resulting in a varietal mixture with greater among-cultivar diversity. The magnitude of overall diversity depends on the degree of diversity among and within the cultivars blended.

Next, consider four systems that incorporate among-species diversity (Table 3.3). A two-crop rotation has low diversity at all levels in any one growing season, but has some interspecific diversity between seasons. Ecological benefits of the interspecific diversity are expected to be greater genetically for distant species (e.g. wheat rotated with a legume) than with more closely related ones (e.g. wheat with another cereal). A continuously intercropped field of two species would have interspecific diversity in space but not between seasons. More complex intercropping involving more than two species occurs primarily in more traditional, non-mechanized agroecosystems and can include considerable genetic diversity at all levels (Francis, 1986).

One experimental kind of complex intercropping system is a self-supporting, perennial polyculture, which is being investigated by the Land Institute in Kansas, USA (Soule and Piper, 1992). It represents perhaps the only current attempt to create a crop ecosystem that incorporates genetic diversity approaching that of a natural ecosystem (in this case, a North American prairie). In a perennial polyculture, the species diversity over time will depend on the number of years between interventions (ploughing or interseeding) and natural fluctuations of species composition within and between seasons.

Deploying Crop Diversity in Agroecosystems

Measuring diversity between and within crops is needed to allow an initial understanding of the function of diversity, and then a deployment of optimum diversity to meet objectives. However, the objectives of growing diversity may differ sharply between different farmers and different agroecologies; there may be conflicts of interests between local, national and global needs for diversity; and there may be a present cost to maintaining diversity whose future role in crop evolution is uncertain. There are three broad and overlapping groups of objectives for managing crop diversity in agroecosystems and these will be treated in different chapters. Firstly, diversity as a raw material for evolutionary change through environmental, farmer and breeder selection is considered briefly here and in more detail in Chapter 10. Secondly, crop diversity for agroecosystem services –

resource-use efficiency, nutrient cycling, and pest and disease relations, is treated by Edwards *et al.* in Chapter 8. Thirdly, crop diversity to meet the socio-economic needs of the farmer is further considered in Chapter 15.

Genetic diversity for environmental selection

Genetic diversity originates through mutation, and is exposed to, or protected from, environmental selection by modifications to the breeding system of the plant population. Environmental selection will include attacks by a diversity of pests and diseases, competition with a diversity of weeds, and interaction with a diversity of biotic and abiotic soil stresses. The degree of natural selection will depend on the strength of these effects, and also on the variation present in the plant population (a result of previous selection).

For example, an outcrossing population will continually expose genetic variation to selection (although phenotypic plasticity – itself under the influence of selection – may reduce the intensity of this selection). Environmental conditions may at any one time favour the progeny of some individuals over others. In contrast, with populations which propagate through strict self-fertilization or clonally, selection will favour some populations over others.

The potential for natural selection to result in change tends to be greater in traditional systems which generally contain more genetic variability. Also, in the often agronomically marginal conditions of traditional farming, strength of selection may be greater. However, there may be compensating effects: crop and varietal mixtures may buffer genetic variation from selection – as with crop mixtures protecting against pests (see Polaszek *et al.*, Chapter 11, this volume), and environmental conditions may change unpredictably – as when selection is pushed one way in an unusually dry season, and the other way in an unusually wet season.

Human influence on selection pressure: changing the environment

It can be said that farmers try to protect their crop from selection – which always implies partial loss of the crop. Since crop domestication, humans have profoundly changed the environment in which selection operates, with the aim of reducing selection pressures. Delimited fields remove crops physically from contact with wild relatives and exposure to alien pollen and sources of pests and disease: diversity and selection pressure are thereby both reduced. Agronomic management practices, such as irrigation, manuring and weeding, change physical and biological selection pressures, favouring some crops and varieties over others (see also Edwards *et al.*, Chapter 8, and Thurston *et al.*, Chapter 9, this volume). The higher the predictability of growing conditions, the less the need for variation to adapt to changing or diverse conditions. Planting will be timed to avoid exposing the crop to seasonal pest and disease pressures. Where micro-habitats exist on farms, these will be matched with appropriate varieties, lessening the environmental selection pressure (see Lenné and Wood, Chapter 18, this volume). Inter-cropping and the use of varietal mixtures buffers the crop from selection.

Direct human influence on diversity

Intensity of natural selection may be low: a 'better' variant only survives if the seed containing the variation finds suitable conditions to germinate and grow. In contrast, human selection may be intense. Modern plant breeding applies intense selection pressure on research stations to remove unwanted characters, in the hope of protecting farmers from the effects of environmental selection (and therefore crop loss) in fields (see Witcombe, Chapter 10, this volume).

A breeder or a farmer can select one morphologically obvious variant from an entire field. This variant can then be almost guaranteed survival by being nurtured in a uniform and optimum environment, protected from competition by spacing and weeding. In two generations a single seed of a desired variant can fill an entire field with offspring. However, farmers will only be able to apply this level of selection pressure on variation which they can clearly recognize. For resistance to pathogens, a farmer may recognize – through its effect on the crop – race-specific resistance, and select resistant plants. However, the inability of farmers to recognize a useful variant, such as race-non-specific resistance to a pest, combined with low natural selection pressure on an initially rare variant, could result in random loss of a potentially valuable character from the crop population.

Plant breeders affect interspecific diversity indirectly through the adaptive characteristics of the cultivars they develop (see Witcombe, Chapter 10, this volume). They affect diversity among available cultivars within species but have little influence on land area occupied by different cultivars or their pattern of deployment. A major purpose of formal seed production is to ensure that the seed used by farmers is as like as possible to the variety produced by breeders (see Wright and Turner, Chapter 13, this volume). Breeders also have great control over intracultivar diversity and may limit the genetic difference between generations. This can be ensured by hybrid seed production, and now by genetically engineering varieties so that the commercial crop produces only sterile seed (USA patent 5,723,765), or is apomictic (USA patent 5,710,367). Increased biodiversity within or among cultivars has rarely been an explicit goal of crop breeding programmes, but rather a by-product.

Intense selection pressure by farmers (or breeders) – at the level of favouring one plant in 10,000 – will reduce the overall genetic diversity of the population (Wood and Lenné, 1997). Breeders, as indicated above, now use genetic markers as an aid in recognizing useful variants, so that strong selection can be applied (see Witcombe, Chapter 10, this volume). However, selection pressure and evolutionary change in varieties may be countered by farmer varietal management. Clonal propagation (see Wood and Lenné, Chapter 2, this volume) is common in perennial crops such as fruit and nut trees, and tuber crops. With clonal propagation there is no genetic diversity of offspring and therefore no possibility for evolutionary change. This may continue for centuries, as with the classical varieties of wine grape, which have been distributed unchanged as clones throughout the world. Farmers may go to great trouble to maintain favoured varieties through clonal propagation – both for vegetative propagation itself, as with grafting, and often with the subsequent need to ensure cross-pollination for fruit set – as with date palm, where inflorescences have to be taken from male trees to manually pollinate the female inflorescences.

Diversity in reserve

Both farmers and breeders can maintain diversity in reserve (latent, apparent, reserve and resource diversity) to prevent selection and to provide future diversity to cope with changing conditions. Breeders maintain diversity in their advanced lines and in extensive genebanks (Duvick, 1984; see Witcombe, Chapter 10, and Engels and Wood, Chapter 14, this volume), whence varieties can be deployed as needed. Modern wheat varieties may encompass 'latent diversity' (Souza *et al.*, 1994). Traditional farmers may maintain diversity against selection by maintaining a range of varieties in intercrops and varietal mixtures, where selection pressure is lowered. Farmers may also routinely source their seed widely, with distant areas acting as a reserve of diversity beyond the influence of local selection pressures (e.g. for Andean potato; Zimmerer, 1991) (see Wright and Turner, Chapter 13, this volume). As indicated above, clonal propagation is common in traditional farming. This enables genetically uniform varieties to be maintained against environmental selection, but allows the deployment of diversity for selection through occasional cycles of sexual reproduction (Quiros *et al.*, 1992, where potato farmers report the use of true seed when tubers become 'tired'). The value of fixed varietal characteristics over diversity appears to be so great that many crops have partly or totally lost the ability to reproduce sexually: they are locked into uniform clonal propagation (e.g. many sweet potato varieties, triploid bananas).

References

Aldrich, P.R. and Doebly, J. (1992) Restriction fragment variation in the nuclear and chloroplast genomes of cultivated and wild *Sorghum bicolor*. *Theoretical and Applied Genetics* 85, 293–302.

Aldrich, P.R., Doebly, J., Schertz, K.F. and Stec, A. (1992) Patterns of allozyme variation in cultivated and wild *Sorghum bicolor*. *Theoretical and Applied Genetics* 85, 451–460.

Alexander, H.M. and Bramel-Cox, P.J. (1991) Sustainability of genetic resistance. In: Sleper, D.A., Barker, T.C. and Bramel-Cox, P.J. (eds) *Plant Breeding for Sustainable Agriculture: Considerations for Objectives and Methods*. Special Publication 18, Crop Science Society of America, Madison, Wisconsin, pp. 11–27.

Bennett, E. (ed.) (1968) *FAO/IBP Technical Conference on the Exploitation, Utilization and Conservation of Plant Genetic Resources, Rome, September, 1967*. FAO, Rome.

Bisby, F.A. (1995) Characterization of biodiversity. In: Heywood, V.H. (executive ed.) *Global Biodiversity Assessment*. Cambridge University Press, Cambridge, pp. 21–106.

Boster, J.S. (1985) Selection for perceptual distinctness: evidence from Aguaruna cultivars of *Manihot esculenta*. *Economic Botany* 39, 310–325.

Boyle, T. and Lenné, J.M. (1997) Defining and meeting needs for information: agriculture and forestry perspective. In: Hawksworth, D.L., Kirk, P.M. and Clarke, S.D. (eds) *Biodiversity Information: Needs and Options*. CAB International, Wallingford, pp. 31–54.

Briand, L., Brown, A.E., Lenné, J.M. and Teverson D.M. (1998) Random amplified polymorphic DNA variation within and among bean landrace mixtures (*Phaseolus vulgaris* L.) from Tanzania. *Euphytica* 102, 371–377.

Brockway, L.H. (1979) *Science and Colonial Expansion: the Role of British Royal Botanic Gardens*. Academic Press, London.

Brown, A.H.D. (1978) Isozymes, plant population genetic structure, and genetic conservation. *Theoretical and Applied Genetics* 52, 145–157.

Brown, A.H.D. (1995) The core collection at the crossroads. In: Hodgkin, T., Brown, A.H.D., van Hintum, T.J.L. and Morales, E.A.V. (eds) *Core Collections of Plant Genetic Resources*. John Wiley & Sons, Chichester, pp. 3–19.

Burdon, J.J. (1987) *Diseases and Plant Population Biology*. Cambridge University Press, Cambridge.

Chakraborty, R. and Rao, C.R. (1991) Measurement of genetic variation for evolutionary studies. In: Rao, C.R. and Chakraborty, R. (eds) *Handbook of Statistics*, Vol. 8. Elsevier Science, Amsterdam.

Clayton, W.D. (1983) Tropical grasses. In: McIvor, J.G. and Bray, R.A. (eds) *Genetic Resources of Forage Plants*. CSIRO, Melbourne, pp. 39–46.

Cox, T.S. (1990) The contribution of introduced germplasm to the development of U.S. wheat cultivars. In: Shands, H.L. and Wiesner, L. (eds) *The Use of Plant Introductions in Cultivar Development*. Crop Science Society of America, Madison, Wisconsin, pp. 114–144.

Cox, T.S. (1998) Deepening the wheat gene pool. *Journal of Crop Production* 1, 1–25.

Cox, T.S., Murphy, J.P. and Rodgers, D.M. (1986) Changes in genetic diversity in the red winter wheat regions of the United States. *Proceedings of the National Academy of Sciences, USA* 83, 5583–5586.

Crossa, J., DeLacy, I.H. and Taba, S. (1995) The use of multivariate methods in developing a core collection. In: Hodgkin, T., Brown, A.H.D., van Hintum, T.J.L. and Morales, E.A.V. (eds) *Core Collections of Plant Genetic Resources*. John Wiley & Sons, Chichester, pp. 77–92.

Cui, Y.X., Xu, G.W., Magill, C.W., Schertz, K.F. and Hart, G.E. (1995) RFLP-based assay of *Sorghum bicolor* (L.) Moench genetic diversity. *Theoretical and Applied Genetics* 90, 787–796.

Doggett, H. (1988) *Sorghum*. John Wiley & Sons, New York.

Duvick, D.N. (1984) Genetic diversity in major farm crops on the farm and in reserve. *Economic Botany* 38, 161–178.

Evans, A.M. (1976) Beans: *Phaseolus* spp. (Leguminosae – Papilionatae). In: Simmonds, N.W. (ed.) *Evolution of Crop Plants*. Longman, Harlow, pp. 168–172.

Francis, C.A. (ed.) (1986) *Multiple Cropping Systems*. Macmillan, New York.

Frankel, O.H., Brown, A.H.D. and Burdon, J.J. (1995) *The Conservation of Plant Biodiversity*. Cambridge University Press, Cambridge.

Groombridge, B. (ed.) (1992) *Global Biodiversity: Status of the Earth's Living Resources*. Chapman & Hall, London.

Hammond, R. (ed.) (1998) *1997 in Review*. CAB International, Wallingford.

Hargrove, T.R., Cabanilla, V.L. and Coffman, W.R. (1988) Twenty years of rice breeding. *BioScience* 38, 675–681.

Harlan, J.R. (1965) The possible role of weed races in the evolution of cultivated plants. *Euphytica* 14, 173–176.

Harlan, J.R. (1981) Ecological settings for the emergence of agriculture. In: Thresh, J.M. (ed.) *Pests, Pathogens and Vegetation*. Pitman, London, pp. 3–22.

Harlan, J.R. and de Wet, J.M.J. (1971) Toward a rational classification of cultivated plants. *Taxon* 20, 509–517.

Harlan, J.R. and Stemler, A.B.L. (1976) The races of sorghum in Africa. In: Harlan, J.R., de Wet, J.M.J. and Stemler, A.B.L. (eds) *The Origins of African Plant Domestication*. Mouton Press, The Hague, pp. 465–478.

Hawkes, J.G. (1986) Problems of taxonomy and nomenclature in cultivated plants. *Acta Horticulturae* 182, 41–52.

Hawksworth, D.L. and Kalin-Arroyo, M.T. (1995) Magnitude and diversity of biodiversity. In: Heywood, V.H. (executive ed.) *Global Biodiversity Assessment*. Cambridge University Press, Cambridge, pp. 107–191.

Heywood, V.H. and Baste, I. (1995) Introduction. In: Heywood, V.H. (executive ed.) *Global Biodiversity Assessment*. Cambridge University Press, Cambridge, pp. 1–19.

Hughes, J.B., Daily, G.C. and Ehrlich, P.R. (1997) Population diversity, its extent and extinction. *Science* 278, 689–692.

Kew (1887) *Bulletin of Miscellaneous Information of the Royal Botanic Gardens*, Kew.

Kihara, H. (1944) Discovery of the DD-analyser, one of the ancestors of *vulgare* wheats. *Agriculture and Horticulture (Tokyo)* 19, 889–890.

Ladizinsky, G. (1984) Founder effect in crop plant evolution. *Economic Botany* 39, 191–199.

Lagudah, E.S. and Halloran, G.M. (1988) Phylogenetic relationships of *Triticum tauschii*, the D genome donor to hexaploid wheat. *Theoretical and Applied Genetics* 75, 592–598.

Li, H.-L. (1974) Plant taxonomy and the origin of cultivated plants. *Taxon* 23, 715–724.

Louette, D., Charrier, A. and Berthaud, J. (1997) *In situ* conservation of maize in Mexico: genetic diversity and maize seed management in a traditional community. *Economic Botany* 51, 20–38.

Lubbers, E.L., Gill, K.S., Cox, T.S. and Gill, B.S. (1991) Variation of molecular markers among geographically diverse accessions of *Triticum tauschii*. *Genome* 34, 354–361.

McFadden, E.S. and Sears, E.R. (1946) The origin of *Triticum speltoides* and its free-threshing hexaploid relatives. *Journal of Heredity* 37, 81–89, 107–116.

McIntosh, R.A. (1991) Alien sources of disease resistance in bread wheats. In: Saskuma, T. and Kinoshita, T. (eds) *Proceedings of Dr. Kihara Memorial International Symposium on Cytoplasmic Engineering in Wheat: Nuclear and Organellar Genomes of Wheat Species.* Kyoto University Press, Yokohama, pp. 320–332.

McIntosh, R.A., Wellings, C.R. and Park, R.F. (1995) *Wheat Rusts: an Atlas of Resistance Genes.* CSIRO, Melbourne.

Morden, C.W., Doebly, J. and Schertz, K.F. (1989) Allozyme variation in Old World races of *Sorghum bicolor* (Poaceae). *American Journal of Botany* 76, 247–255.

Morden, C.W., Doebly, J. and Schertz, K.F. (1990) Allozyme variation among spontaneous species of *Sorghum* section *Sorghum* (Poaceae). *Theoretical and Applied Genetics* 80, 296–304.

Morris, M.L. and Heisey, P.W. (1997) Achieving desirable levels of crop genetic diversity in farmers' fields: factors affecting the production and use of improved seed. *Proceedings of an International Conference on Building the Basis for the Economic Analysis of Genetic Resources of Crop Plants.* CIMMYT and Stanford University, Palo Alto, California.

Murphy, J.P., Cox, T.S. and Rodgers, D.M. (1986) Cluster analysis of red winter wheat cultivars based on coefficients of parentage. *Crop Science* 26, 672–676.

Nishikawa, K., Furuta, Y. and Wada, T. (1980) Genetic studies on α-amylase isozymes in wheat. III. Intraspecific variation in *Aegilops squarrosa* and birthplace of hexaploid wheat. *Japanese Journal of Genetics* 55, 325–336.

de Oliveira, A.C., Richter, T. and Bennetzen, J.L. (1996) Regional and racial specificities in sorghum germplasm assessed with DNA markers. *Genome* 39, 579–587.

Quiros, C.F., Ortega, R., van Rammsdonk, L., Herrera-Montoya, M., Cisneros, P., Schimidt, E. and Brush, S.B. (1992) Increase of potato genetic resources in their center of diversity: the role of natural outcrossing and selection by the Andean farmer. *Genetic Resources and Crop Evolution* 39, 107–113.

Raeburn, P. (1995) *The Last Harvest.* Simon & Schuster, New York.

Reaka-Kudla, M.L., Wilson, D.E. and Wilson, E.O. (1996) *Biodiversity II.* Joseph Henry Press, Washington, DC.

Rejesus, R.M., Smale, M. and van Ginkel, M. (1996) Wheat breeders' perspectives on genetic diversity and germplasm use: findings from an international survey. *Plant Varieties and Seeds* 9, 129–147.

Simmonds, N.W. (1962) Variability in crop plants, its use and conservation. *Biological Reviews* 37, 442–465.

Skovmand, B., Varughese, G. and Hettel, G.P. (1992) *Wheat Genetic Resources at CIMMYT: Their Preservation, Enrichment and Distribution.* CIMMYT, Mexico.

Smartt, J. and Simmonds, N.W. (1995) *Evolution of Crop Plants.* Longman, London.

Soule, J.D. and Piper, J.K. (1992) *Farming in Nature's Image.* Island Press, Washington, DC.

Souza, E., Fox, P.N., Byerlee, D. and Skovmand, B. (1994) Spring wheat diversity in irrigated areas of two developing countries. *Crop Science* 34, 774–783.

Sprecher, S.L. (1988) Isozyme genotype differences between large seeded and small seeded gene pools in *Phaseolus vulgaris* L. *Bean Improvement Cooperative* 31, 36–37.

Voss, J. (1992) Conserving and increasing on-farm genetic diversity: farmer management of varietal bean mixtures in Central Africa. In: Moock, J. and Rhoades, R.E. (eds) *Diversity, Farmer Knowledge, and Sustainability.* Cornell University Press, Ithaca, New York, pp. 34–51.

de Wet, J.M.J., Harlan, J.R. and Brink, D.E. (1986) Reality of infraspecific taxonomic units in domesticated cereals. In: Styles, B.T. (ed.) *Infraspecific Classification of Wild and Cultivated Plants.* Clarendon Press, Oxford, pp. 211–222.

Wood, D. and Lenné, J.M. (1997) The conservation of agrobiodiversity on-farm: questioning the emerging paradigm. *Biodiversity and Conservation* 6, 109–129.

Xu, S.J. and Joppa, L.R. (1995) Mechanisms and inheritance of first division restitution in hybrids of wheat, rye, and *Aegilops squarrosa. Genome* 38, 607–615.

Yang, Ning and Smale, M. (1996) *Indicators of Wheat Genetic Diversity and Germplasm Use in the People's Republic of China.* CIMMYT Natural Resources Group, Paper 96–04, CIMMYT, Mexico.

Zimmerer, K.S. (1991) The regional biogeography of native potato cultivars in highland Peru. *Journal of Biogeography* 18, 165–178.

Zohary, D. and Hopf, M. (1988) *Domestication of Plants in the Old World.* Clarendon Press, Oxford.

Biodiversity in Domesticated Animals

4

D. Steane

Conservation and Use of Animal Genetic Resources in Asia and the Pacific, FAO Regional Office, Maliwan Mansion, 39 Phra Atit Road, Bangkok 10200, Thailand

This chapter will be restricted to the biodiversity in those species which are regarded as important livestock used mainly for food and other products such as skins, wool, feathers, fuel, fertilizer, draught power and a bank/cash reserve. It will not address other domesticated species. The term 'farm animal' is frequently used and usually refers to all livestock although some (e.g. yak) will not normally be kept on farms in the usual sense.

The Nature of Domesticated Animal Diversity

Domestication is a process which, dependent upon the animal species, has been continuing for some 12,000 years. While there may be debate as to further species domestication, certainly domesticated animal species, like plant species, have been developed to suit a wide range of conditions and requirements. This process is expected to continue as human requirements and management environments change. Consideration of farm animal biodiversity has only arisen in the last 50 years. The relative time scales of the development of diversity and the recognition of its importance are therefore disconcerting. We now have techniques by which a breed (or breeds) can be replaced completely by another within less time than one human generation.

Documented breed records have only been developed in the last 200 years or so although breeds have existed and records have been maintained for centuries by memory and orally. However, the term 'breed' needs defining. The following definition by Turton (1974) is preferred by the Food

and Agriculture Organization of the United Nations (FAO) as indicated
in the World Watch List for Domestic Animal Diversity (WWLDAD)
(FAO/UNEP, 1995).

> A breed is either a homogenous, subspecific group of domestic livestock with
> definable and identifiable external characteristics that enable it to be separated
> by visual appraisal from other similarly defined groups within the same
> species; or, it is a homogenous group for which geographical separation from
> phenotypically similar groups has led to general acceptance of its separate
> identity.

The WWLDAD adds:

> Thus, breeds have been developed according to geographic and cultural
> differences, and to meet human food and agricultural requirements. In this
> sense, breed is not a technical term, but certainly the differences, both visual
> and otherwise, between breeds account for much of the diversity associated
> with each domestic animal species. Breed is accepted as a cultural rather than a
> technical term, i.e. to emphasize ownership.

It is also noted that the term 'population' is a generic term but, when used in
a genetic sense, defines an interbreeding group, and may refer to all animals
within a breed.

During the long period of domestication and development, species of
animals, like plants, have undergone considerable change as a result of
selection pressures – see Peel and Tribe (1983) for a useful outline (see
Chapter 2, this volume, for plants). Some of these pressures were natural
– for example, the climate changes occurring in Europe around 3000 BC
when an essentially hot and dry climate became cold and wet. Genetic
variation also arose as humans migrated with their domestic animals to
different environmental/climatic conditions. These influences continue
today.

This chapter will not describe the development of diversity in farm
animals in detail – suffice it to say that diversity developed over centuries.
The gene combinations developed provide a vast array of resources available
for exploitation for the benefit and survival of humans. Clearly there are
breeds of one or more animal species in most human environments which
perform according to human requirements (see Plates 5–8). The vast array of
environments in which animals are used is itself impressive – from yaks in
the Hindu Kush–Himalayan region, to camels in arid deserts (both being
multipurpose species), to cattle from Siberia to tropical zones, to chickens in
the tropics and the high hills of the Himalayan region and so on (see
Thurston *et al.*, Chapter 9, this volume).

While the most important criterion for evaluating breeds in
developed countries is performance (usually output), breeds in
developing countries are usually recognized by other criteria. Such
breeds have low output because they often have little or no inputs and
grow and survive in conditions which are far from ideal. Output is not,

therefore, a reasonable criterion for valuing breeds but continues to be quoted by the livestock production industry. It is well recognized that many breeds in difficult environments exhibit highly desirable traits. These include reproduction under severe nutritional stress, disease tolerance or resistance, and resistance to heat or cold stress. Many examples are given in the *Proceedings of the World Congresses on Genetics Applied to Livestock Production* (WCGALP) (various citations in the reference list), the *World Watch List for Domestic Animal Diversity* (WWLDAD; FAO/UNEP 1993, 1995) and, more particularly, the breed database of the Domestic Animal Diversity Information System (DAD-IS), which is available through the Internet (http://www.fao.org/dad-is) or on a CD-ROM from FAO.

Domesticated animal diversity and world food resources

Domestic animal diversity is an important part of world food resources (see Thurston *et al.*, Chapter 9, this volume). Animal products are produced in many different systems and areas due to this diversity. Livestock contributes 41% of agricultural gross domestic product (GDP) (Steinfeld, Bangkok, 1998, personal communication), with about half of this being food (Hammond and Leitch, 1995). The contributions of the non-food outputs from livestock are in some cases more valuable than the food itself. The social and cultural contributions of animals are also extremely important even though they cannot easily be valued in economic terms.

During the years between 1989 and 1996, global animal production increased by 27% and total global meat production by an average of 2.9% per annum. A dramatic 7.7% annual increase in meat production has occurred in Asia (FAO, 1996a, 1997). However, over 800 million people still suffer from chronic malnutrition and over 200 million children lack sufficient food for full development (FAO, 1996b). Food production will have to double over the next 30 years if minimum nutritional standards are to be met. Available natural resources will be threatened unless drastic action is taken to reverse present trends. It is estimated that the demand for meat and milk will rise by 300% and 155%, respectively, by 2020 (Rosengrant *et al.*, 1995).

The pressures on animal production systems are increasing rapidly. In 1996, the FAO World Food Summit published the *Rome Declaration on World Food Security and World Food Summit Plan of Action* (FAO, 1996c), which emphasized the need to promote the conservation and utilization of animal genetic resources to combat threats to food security. It is essential that the genetic diversity available is fully exploited in a sustainable manner to meet this increasing demand for meat and milk from livestock.

The need for balance

Animal production is a crucial component of agriculture and the sustainability of many farming systems relies as much on animals as on plants. Neither can be ignored or eroded without serious consequences. Similarly, agriculture is itself only one component of the global ecosystem and, to some, it is an intrusion on the original biodiversity. Clearly balance is needed between the human needs and the maintenance of biodiversity for future survival. The maintenance of biodiversity is in the long-term interests of all in so far as it contributes to sustainability, food security and survival.

There are cases, however, where livestock production may create problems. For example, intensive pig and/or poultry units can cause major pollution problems, while the global movement of livestock feed causes massive shifts in resources in the long term: feedstuff producers fail to replenish their land with nutrients while the recipients suffer nutrient pollution which may seriously affect crop production, water quality, etc. There is already good evidence to support this (de Haan et al., 1997). Nevertheless, it is likely that future animal production will involve further intensification (Steinfeld et al., 1997) and solutions have to be found. Livestock has been blamed for desertification and deforestation (especially in the Amazon) and yet, 'Livestock do not move, produce or reproduce without our wanting it. They are completely dependent upon us and inseparable. Livestock do not degrade the environment, humans do' (Steinfeld et al., 1997). Steinfeld et al. (1997) and de Haan et al. (1997) fully assessed the balance between livestock and the environment, reaching a consensus supported by donors globally. Crucial issues for sustainable livestock production and agriculture over the next few years are also briefly discussed by Blackburn et al. (1998) and de Haan and Setschwaelo (1998).

Extent of farm animal diversity

As the overall extent of genetic diversity within an animal species is not yet known, the scale of the problem of maintenance and use of farm animal resources for future productivity is difficult to assess. Until recently, there was limited information on the number of breeds globally and even less on performance coupled to production environment. Until recently also, the number of breeds was considered to be a reasonable indication of diversity although, as for plant varieties, this has been questioned (see Cox and Wood, Chapter 3, and Witcombe, Chapter 10, this volume). Through domestication, humans divided animal genetic diversity into two basic components – that which is common across all breeds and that which is specific to certain breeds. Present estimates of the variation in these two components will vary according to the traits being studied but, overall, the variation is shared about equally (Hammond and Leitch, 1995). Therefore, any loss of breeds could result in a serious depletion of the available genetic diversity. Some (e.g. Land, 1986) believe that such losses are generally less serious than has been suggested by others (e.g. Gregory, 1986).

One of the first breed surveys (Maijala *et al.*, 1984) showed that the farm animal genetic resources in Europe consisted of 737 breeds across five species of which 240 were at serious risk. In 1991/92, FAO carried out a global survey which formed the basis of the first publication of the WWLDAD (FAO/UNEP, 1993). This showed 2719 breeds for seven species globally of which 27% were estimated to be at risk. As part of a continuous surveying, monitoring and updating process, the second edition of the WWLDAD (FAO/UNEP, 1995) covered 3882 breeds of 28 animal species, with 501 of 3019 mammalian and 372 of 863 avian breeds at risk, giving an overall level of about 30% of breeds. This figure may rise further as countries assess their resources more effectively.

Why maintain domesticated animal biodiversity?

Domesticated animal diversity is needed to ensure that variation is accessible for future use (see Plates 5–8). The demands and desires of humans are constantly changing, presenting new challenges to producers. In addition, changes are occurring in feed availability, disease, climate, and scope of human interventions. These changes, coupled with the increasing demands for livestock products, exert pressures to achieve the most efficient production system possible.

Many indigenous breeds have been discriminated against due to low outputs under inadequate feed levels and/or other stresses. This is unfortunate and, indeed, contrary to basic genetic evidence which suggests that selection for improved performance should be done under the same conditions as those of the performance environment (Falconer, 1961). After decades of neglect, local breeds are now considered to be sources of useful genetic variation (Frankham, 1994). Often, the value of a local breed has only been realized after it has either been reduced to low levels or disappeared. Identification and evaluation of useful attributes is extremely important so that source breeds are not lost (see sections on Characterization and Conservation). Even where there is adequate genetic variation, selection within a breed is usually slow, with rates of change being around 1–2% per annum. It is often advantageous to cross with a desired breed and to select for the desired trait. With modern biotechnological techniques, the opportunities for introgression are far greater than before.

The prevention of genetic erosion is paramount and it is essential to develop a strategy for the maintenance of farm animal genetic diversity (FAO, 1992b). This is not the same as maintaining all breeds! The need for action has been confirmed by the governing body of the FAO (Conference) and supported by the Conference of the Parties (COP) of the Convention on Biological Diversity (UNEP, 1992).

Causes and mechanisms of genetic erosion

Genetic erosion is continually taking place in animals, even under natural conditions. Selection from one generation to the next means that gene frequencies may alter and some genes may disappear, potentially reducing additive genetic variation over time. Hill (1998), however, did not fully support this – genetic drift is a well-recognized phenomenon in the context of genetic change. In addition, in all normal populations, there is constant change due to selection, mutation and migration. Selection for specific traits may also affect other traits if they are correlated or linked. Such correlated responses may be taken into account, but unexpected deterioration in desirable attributes may occur. The development of breeds is not the main issue here although some aspects may need consideration.

In almost all cases of genetic erosion, economic pressure at farm level (usually politically based) persuades the producer to change to genotypes or breeds which yield higher immediate returns. This is of significance in poorer societies where both the pressure for survival and the need to make short-term gains are great. Such decisions are, however, almost always independent of overall efficiency and maximization of long-term sustainable use of resources. They are often related to subsidies, whether direct or indirect (externalizing costs is an indirect subsidy in this context), or directly to policies. Farmers are often aware of the unsustainability of the system adopted but must follow the policy due to economic reasons. At the same time, some decisions (e.g. decreased milk prices) may result in increased levels of human nutrition even if these cause problems in the longer term (e.g. the EU Common Agricultural Policy produced lakes of milk and mountains of beef and butter). The problem is that the time-frame for achieving political gains is not in tune with that for maximizing the efficiency of the whole production system.

Rather less obvious but no less problematical is the general failure to take into account all aspects of the system when calculating economic values. There is often a failure to internalize all costs. An example is the considerable increase in mature size of cattle when exotics are crossed with local breeds in the tropics. The resultant F_1 adult requires considerably more feed for maintenance. Another example is costing of intensive units of exotic stock without any regard to the processing of effluent. The subsidies provided to encourage artificial insemination (AI) or mechanization also have effects on genetic erosion and the environment. For example, mechanization will lead to the reduction of draught animals which may affect soil structure and accumulation of crop residues.

One of the obvious means of erosion is breed replacement. This has happened several times throughout history. In the UK, for example, the Shorthorn took over from the Longhorn only to be replaced by the Dutch Friesian which has, more recently, essentially been substituted by the Holstein–Friesian from North America. Breed substitution is facilitated by technologies such as AI and, more recently, embryo transfer (ET), which allows immediate breed replacement, potentially increasing the rate of genetic erosion. However, the same biotechnologies can be used to improve the maintenance of genetic diversity (see later sections). The technology is not at fault, only its misapplication.

Bringing desirable genes into a population by crossbreeding has become a major practice in some parts of the world. In developing countries, the driving force has been the potential to substantially increase food output. Steane (1996) pointed out that while such programmes may have initial benefits, the main danger to animal diversity is the continuing trend to cross-breed *only* with exotic semen and/or animals. This practice of backcrossing only to the exotic breed is often (incorrectly) referred to as 'upgrading' and appears to be based on the premise that developing country breeds are un-improved and thus not useful. There is no doubt that valuable genetic improvements in farm animals in the developed world are of real value in crossing programmes in developing countries. However, the question has to be asked: under what conditions did the improvement take place? In many cases, there appears to be little or no consideration of the contribution of the selection environment (e.g. high feed availability, low stress levels, different market demands, etc.) and no consideration of the longer-term genetic implications.

Crossbreeding is an essential tool for improving food production efficiency but programmes must be properly designed to maximize gains and minimize problems. In dairy cattle, there appears to be almost universal support for the Holstein–Friesian breed as, in the right environment, it has the highest milk yield of all breeds. But, its butterfat levels are among the lowest. In a continuous backcrossing programme, use of the Holstein–Friesian can spell disaster in developing countries. Indeed, problems have already developed due to its poor reproductive performance. Why does this situation occur? From a peasant farmer's point of view, the first-cross is almost a miracle. The local indigenous cow which produces little milk (on no additional feed) suddenly, as a crossbred, produces several times more milk (but with additional feed). The exotic breed is given full credit for the transformation and the farmer requests more of the same, leading to erosion of the local breed.

Further genetic erosion can take place during the choice of traits for selection. The phenomenon of correlated response can be used to advantage but can equally cause problems. There is always a tendency to select those traits that are highly heritable and easy to measure since these will give the greatest response. An example has been the almost universal selection for milk yield based on the first lactation (standardized to 305 days). While this allows early assessment of sires, due to an associated gradual increase in calving interval, in some circumstances the overall effect has been detrimental usually under both stressful and good conditions (Simm and Pryce, 1998). In many circumstances, average lactation per day per year or, better still, per lifetime will provide a more correct evaluation (see Veerkamp *et al.*, 1995 for more information). Similarly, selection for growth rate in beef cattle has been in vogue in many countries for many years with some impressive results. However, the 'value' of growth rate has not taken into account the associated response in mature adult body size and the consequent increased feed requirement for maintenance (and greater susceptibility to feed deprivation with its concomitant reduction in fertility). In many environments where feed is a limiting factor, the additional maintenance requirement may have to be met with imported subsidized concentrate feed.

Disease can also play an important role in genetic erosion, particularly when the population is small or relatively restricted. Many breeds developed through geographical isolation. If the geographic area suffers a major epidemic, the breed can be quickly lost.

Public awareness of the consequences of genetic erosion can help, but success will differ between countries. In developing countries, the priorities of the farm family may only extend over 1 year and concerns with genetic erosion will not become important until they directly affect livelihoods. At the other end of the spectrum, households in the developed world are made aware of the dangers of genetic erosion to livestock production by non-government organizations (NGOs) which have sprung up since biodiversity attained a higher profile.

The basic causes and the various expressions of genetic erosion together with the mechanisms for erosion are only beginning to be addressed in many countries as the difficulties for some livestock production systems are identified. Without full consideration of each aspect of erosion, the development of livestock production systems which fully and sustainably exploit the resources available will not be achieved. Some of these aspects will be addressed further under Conservation.

Efficient use of resources

Efficiency concerns the use of combinations of resources and yet, all too often, animal breeding is regarded as a solution in itself, which is far from the reality. Animal breeds have been developed under specific conditions for specific products and any changes may mean that that particular breed is no longer the most effective. The common factor linking all aspects of future livestock production must be efficient use of resources for sustainable food production. The challenge is to provide adequate food from resources which also must be maintained for future generations. In this context, efficiency is a measure of input/output relationships. It is interesting that the recent financial meltdown in Asia (1998) has stimulated a reappraisal of the value and role of agriculture and of the use of local resources.

Franklin (1986) commented that quantitative genetics as applied to farm animals is essentially a statistical tool with little concern for biology. Certainly the scientific literature on animal breeding has paid more attention to statistical challenges than the biological implications of selection schemes.

One suspects that many regard the animal as a component of molecular genetics, an estimated breeding value (EBV) or whatever, almost ignoring the whole animal as a complex biological organism. Much interest has also been given to economic weights for different traits and much less to the effect of environmental factors, although Brascamp *et al.* (1998) attempt to bring both into consideration.

Tess *et al.* (1983) targeted biological efficiency in pig breeding, but limited attention has been directed at farm animals. Korver *et al.* (1991) and Luiting (1991) further developed the concept of feed efficiency and, in an attempt to approach the biological objective, used residual feed intake (RFI) or net feed intake (NFI) in poultry and dairy heifers, respectively. Studies with beef cattle (Archer *et al.*, 1998) confirmed that useful genetic variation existed and that genetic change for efficiency was possible. The value of feed efficiency has also been appreciated in pig breeding since the early 1970s. It is interesting to note that studies by Kempster *et al.* (1982), Southgate *et al.* (1982 a,b) and Thiessen *et al.* (1985) indicated that in cattle genetic variation for growth was mainly between breeds while variation for feed efficiency was mainly within breeds; this has considerable implications for breeding programmes. However, most studies of production efficiency have been limited in scope and have not addressed overall system efficiency. It is incumbent upon advisers, extensionists, aid agencies and recipients to consider the whole system, including production efficiency, rather than to continue to support 'output per single production cycle'. In the future, the equating of output with efficiency must cease and a better balance between system and breed is needed.

Given the difficulty of identifying the real economic situation for production, it is not surprising that the economics of biodiversity loss are even more difficult to specify. This is addressed by Perrings *et al.* (1995) and emphasizes several crucial issues regarding the value of ecosystem resilience. They point out that 'since all the general models (of climate change) predict an increase in the range of environmental conditions within which economic and ecological systems will have to function in the future, the loss of resilience in key ecosystems must be a matter of concern' (p. 307). There is no reason to believe that this does not apply to agroecosystems. It may be necessary to deal with an even greater range of production environments than experienced to date. The approach taken by the Perrings *et al.* study needs to be extended to provide the required practical methodologies to conserve domesticated animal biodiversity.

Characterization of Domesticated Animal Biodiversity

The characterization of breeds is currently of great interest and importance as indigenous breeds have genes of value (Frankham, 1994). Characterization can be done at several different levels, including biological (phenotypic and genetic) as well as social, religious and cultural/heritage levels. Each level needs to be addressed. Phenotypic measurements are only limited by cost and imagination. The more common production traits of economic value are usually recorded. However, the environment also needs to be described in order for the production data to be of any value. The development and use of microsatellite markers has greatly facilitated genetic analysis of diversity. By using genetic distance estimation, a strategy for the maintenance of diversity could be developed (FAO, 1993). But, it is important to appreciate that other characterization criteria are also needed to aid decision-making on the maintenance of genetic diversity. Social, religious and cultural measures need careful consideration and the implications of any loss must be fully acknowledged. Often social costs can outweigh financial benefits.

Because of the different reasons for maintenance of animal breeds, different characterization strategies may be needed at national, subregional and regional levels. Under the CBD, countries are clearly responsible for the characterization and maintenance of diversity at the national level. However, national decisions may well not take global issues into consideration as suggested in Perrings *et al.* (1995) in the chapter 'Unanswered Questions' – 'How can the structure of payoffs to the individual users be varied to assure a sustainable outcome?' This will be discussed further under Conservation.

Since there are at least 4000 recognized animal breeds, maintaining all breeds for all time is neither practical nor realistic in spite of the study by Smith (1984) showing the high benefit to cost ratio of breed maintenance. However, there is little more information than name and some estimate of population size for many breeds. In most countries, there is no active monitoring system and any census normally deals only with livestock numbers and ignores basic breed and crossbred status. In surveys of genetic resources, countries need to establish procedures which enable proper monitoring and reporting of numerical and performance data. This should be accompanied by data on the production environment since, without such information, performance data can be seriously misleading. FAO is initiating a system to include production environment parameters alongside all performance data, but even this information will only serve to ameliorate the poor situation.

Phenotypic data, production environment parameters for each breed in all major production systems, and physical data on breeds can form the basis for deciding which breeds may be able to contribute specific performance traits within stated environments. The same data will also provide clues as to the present selection pressures on the breed. However, such data will always be limited since many traits, even important ones such as disease tolerance and stress susceptibility, are not easily quantified and there may be differences in either the method of measurement or the definition of trait which need to be resolved (see section on the FAO Global Strategy).

Over the last 50–60 years, a great deal of phenotypic evaluation has been carried out, initially in pigs and poultry and, more recently, in cattle and sheep by national research organizations such as the US Department of

Agriculture and the Animal Breeding Research Organization, UK. Trials were usually carried out in one environment only. Such data assist in assessing some breed characters but fail to provide information on either the genetic content or the genetic diversity of a species. The data should be maintained in an easily accessible form. The FAO database, one component of the Domestic Animal Diversity Information System (DAD-IS), is a good example and continues to be developed for use by all countries both as a national as well as a global database. Other databases exist for breed information; for example, the Nordic countries have developed their own system (Danell *et al.*, 1998), but also contribute data to DAD-IS.

Until recently, there were no real measures of farm animal genetic diversity. Blood typing and protein polymorphisms became available but provided variable information. These were followed by restriction fragment length polymorphisms (RFLPs) and then DNA fingerprinting. Recent major advances allow mapping of whole-animal genomes, as for humans (Sutherland, 1996). The possibility of comprehensively characterizing farm animal genetic diversity is now a reality.

FAO has taken the lead on the need to characterize farm animal diversity initially through recommendations from an Expert Consultation (FAO, 1992b) and later by the publication of the Report of a Working Party (FAO, 1993). This report sets out an integrated global programme to establish the genetic relationships among the breeds of each domestic animal species. It outlines the basic requirements which will provide acceptable levels of accuracy for genetic distance estimates, and provides the scientific community with a system which, if followed, will allow the development of national, regional and global strategies for the maintenance of farm animal genetic diversity. Another working group prepared a proposal for sampling breeds, animals within breeds and the logistics of measuring animal samples for the assessment of global farm animal genetic diversity (see FAO DAD-IS). The common set of microsatellite markers to be used is recommended to FAO by the International Society of Animal Genetics (ISAG), and is subject to review (FAO, 1993). It is imperative that the recommended markers be used so that the vast amount of work now being undertaken can contribute to global knowledge rather than only to local information. FAO will provide ongoing coordination of these studies so that, over time, a larger database will be available into which new results can then contribute to overall conservation strategies.

Conservation of Farm Animal Diversity

The term 'conservation' probably creates more difficulties than any other within the vocabulary used for biodiversity issues. It clearly means very different things to different interests. For the purpose of this chapter, the definition used is that recommended by the FAO Informal Panel of Experts

Developing the Global Strategy for the Management of Animal Genetic Resources (FAO DAD-IS). This reads :

> 'Conservation of farm animal genetic resources' refers to all human activities (for example, strategies, plans, policies, actions and other measures) that are undertaken to ensure that the diversity of farm animal genetic resources is being maintained to contribute to food and agricultural production and productivity, now and in the future.

Conservation through utilization

The primary purpose of conservation is utilitarian. Domesticated livestock breeds have been developed for continued use although some of the traits selected may not have direct economic value (e.g. colour unless it is a specific colour marking attribute). Comprehensive characterization is crucial and care must be taken to ensure that the value and implications of useful traits are considered. This has been the basis of developed country breeding programmes but some programmes have discounted traits based either on externalized economic values or on political decisions which are changeable at the proverbial 'drop of a hat'.

The crucial step is to develop a breed improvement plan for the breed(s) involved. Breeding goals, traits to be measured, economic values (internalized) and the optimum strategy for implementation of the scheme together with optimization of dissemination methods for the improved material must then be identified before any breeding takes place. For breeds that will be utilized under severe stress conditions, selection for useful traits should take place under those conditions. A useful practical guide is to use a production environment one to two generations ahead of the present average environment – not an easy matter to predict. Planning must also include studies to determine whether a crossing programme or a pure breeding production system is the correct one for overall efficiency and sustainability. Once this is clear, then the selection strategy for the breed (or breeds) can be decided. The plan must accurately attribute production improvements so that the correct level of exotic versus indigenous genes for a particular situation can be introgressed and the relevant improvements made.

It is usually assumed that where heritability is low, the rate of change is likely to be low. This is, however, a false assumption since overall progress depends on total additive genetic variation. The use of non-additive variance can create change by heterosis (hybrid vigour) but this is neither cumulative nor permanent with further generations. It is often forgotten that, in the long term, crossing programmes depend upon pure breed improvement to improve all required traits.

Rege (1998) reviewed the dairy crossing information available for the tropics (Africa, Asia and Latin America). He noted that 'Friesian crosses were no better than crosses involving Jersey, Brown Swiss, Ayrshire or Red

Dane. F_2 performance was lower than that of F_1, but subsequent generations (F_3 and F_4) were not any worse than the F_2'. In many Asian countries, exotic genes (whether buffalo, cattle, goats, horses, pigs or sheep) have been widely distributed without any follow-up (Steane, 1996). While this situation is not good for the individual farmer, the genetic consequences for the country are even more dubious. A move to reduced diversity (without understanding the merits of the indigenous breed) means that potential for genetic changes is reduced and, thus, the ability to meet changed conditions may be compromised. Food security could also be compromised with alarming consequences.

Trail (1986) also commented on the paucity and/or absence of data in Africa and there is a similar lack of records in Latin America. In one exceptional study of crossbreeding for dairy production in Latin America, Madalena (1990) showed that the first cross was better than all others, particularly in low-input/management systems. Madalena (1998) also described a system where the first-cross dairy females were produced by beef breeders and then sold to dairy farmers. This practice seems to achieve the balance required for an efficient system. The general conclusion for dairy production in tropical, stressful conditions must be that, overall, the first cross is the most desirable animal. However, in breeds with low reproductive rates, it is difficult to maintain a first-cross programme, especially where severe stress factors are likely to reduce reproductive capacity even further. Similar evidence can be produced for almost all exotic breeds of farm animals: while contributing to a valuable crossbred, they are not themselves suited to the stressful conditions found in the tropics and subtropics of most developing countries (see Khan and Taneja, 1996, for a review of sheep and goats in Asia).

The fact that a crossing programme may provide the most efficient and sustainable production system can often be the means by which both conservation and use of a local resource are ensured. The local breed is adapted and can often provide some benefits – for example, the use of a local pig breed as the dam component of a hybrid female would use the resource, provide hybrid vigour, produce an animal reasonably adapted to local conditions and familiar to the local farmer, ensuring conservation of the local breed as well as optimum use of the genes available from other breeds with desired traits.

Where population size is 'large' in a genetic context, as much selection pressure as possible can be applied with little consequence on the rate of inbreeding. Effective population size (or Effective Number, Ne) is the criterion by which population size can best be determined (Falconer, 1961). This is an important parameter since the rate of increase of inbreeding is the most important factor in maintaining genetic diversity (Woolliams *et al.*, 1998). An Ne of 50 or more will allow selection to proceed effectively for more than ten generations without any concern. Where the Ne is small, inbreeding becomes a major concern and criteria other than selection of the best animals have to take priority (see later).

Actual population numbers are not necessarily a good guide to Ne. In the dominant breed, Holstein–Friesian, effective population size is giving rise to some concern at present (Wickham and Banos, 1998) as ongoing selection under near-optimal conditions has resulted in a great change for first lactation yield with correlated response in subsequent lactations and other traits. Wickham and Banos (1998) declare 'These trends attest to a reduction of the effective size of the breeding Holstein–Friesian population around the world. This clearly needs to be monitored so that breeders can be advised of trends which will affect both long and short term selection responses'.

The results of various dairy crossing programmes clearly show that, while the first cross with either Holstein or other popular dairy breeds provides better performance (Rege, 1998), subsequent crosses do not. The major problem, in stressful conditions, is obtaining reproductive rates which would allow a scheme using only first crosses. It remains to be seen whether the use of modern technologies such as *in vitro* maturation (IVM) and *in vitro* fertilization (IVF), together with sexing of sperm and improved conception with ET, will provide a sustainable system to allow the continuous production, to order, of replacement F_1 heifers at an acceptable cost. This is a good example of the benefits of maintenance of diversity and how biotechnologies may contribute to efficient livestock production.

The main difference between developed and developing country breeds is the level of selection which has been undertaken. Selection intensity, accuracy of breeding value estimation and generation interval are the effective components of genetic change. If breeds are to be used to full advantage then selection needs to take place (whether natural or artificial). Selection is the main cause of the genetic diversity now available. One of the major difficulties in the developing world is that, for many breeds, the parentage is not known or not recorded. Where animals are left to find their own food, casual matings take place. In such situations, it is important to try to limit the males capable of producing progeny. Even within a village, matings are rarely controlled. Indeed there is usually selection bias in large ruminants as the best animals are castrated for draught leaving the production of progeny to the lesser males. Over generations, the size of males decreases, which then limits the usefulness of subsequent animals for work. A similar picture exists for sheep and goats where the largest are selected for slaughter for religious festivals.

Whenever selection in developing countries and/or difficult environments arises, the Open Nucleus Breeding System (ONBS) is mentioned (Fig. 4.1; Jackson and Turner, 1972; James, 1977). The system uses a large base population and, at high selection intensity (but usually lower accuracy of measurement), identifies animals to be transferred to a 'Nucleus'. Selection within the nucleus is accurately carried out (often taking measurements not possible in a less controlled system), maximizing genetic change and then disseminating the improved material to the base population. The ONBS can lead to increases in rates of progress of 10–15% and there are examples of

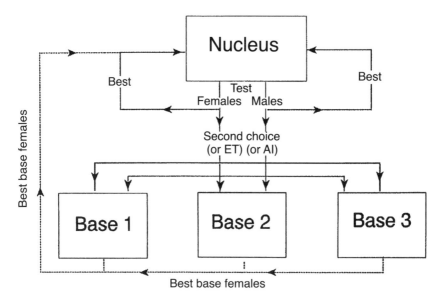

Fig. 4.1. Open Nucleus Breeding Scheme (ONBS).

highly successful 'Group Breeding Schemes', for example, the Welsh Mountain sheep in the UK (Pollott *et al.*, 1994) and Djallonke sheep both in Côte d'Ivoire (Oya, 1992) and in Togo (Pessinaba, 1992). Progress by selection is slow but additive (up to 2% per annum) and gains, although relatively small, are important. The lack of progress in many schemes (Steane *et al.*, 1982) is due to the lack of attention to the basic principles. In ONBS, sophisticated biotechnologies (e.g. ET) are unlikely to be of benefit.

The general argument for use of ONBS in developing countries is that it does not require extensive infrastructure or full recording and it is easier to control and measure a central nucleus. The major requirement for a successful ONBS in developing countries is full cooperation and coordination of all interested parties. ONBS schemes rely on trust and full cooperative efforts, particularly regarding the provision of animals to the nucleus and the equitable and effective dissemination of improved material. In many developing countries, the nucleus is likely to be located on an institutional farm where production levels are often lower than local farms and management is poorer. In many cases, the institutional farm tries to emulate developed country conditions which are not relevant to local needs and often not sustainable.

A normal closed nucleus system is usually adequate as long as it is large enough and/or has access to other genetic material. There is considerable commercial evidence to show that a closed nucleus system will make good genetic progress (e.g. pig or poultry breeding companies in Europe and the USA). Such nucleus units are much less complex than ONBS and usually under single management which determines policy and practice, increasing the likelihood of success. The 'dispersed nucleus' of the type generally used in dairy improvement is another successful means of achieving genetic change. It uses records to compare animals across the population and then nominates the best for further breeding (e.g. Holstein breeding in North America).

In large populations, where standard methods of measurement and evaluation are used and breeding values are estimated using best linear unbiased prediction (BLUP), considerable attention is being paid to the use of marker assisted selection (MAS) and quantitative trait loci (QTL). Studies have indicated that rates of genetic change can be increased by up to 10% using MAS (Van der Beek and van Arendonk, 1994). However, Gibson (1994) has shown in simulation studies that the short-term additional gain was at the cost of greater overall genetic change in the longer term. This result was confirmed by others (Dekkers and van Arendonk, 1998). It is also recognized that there are difficulties and errors in estimation methods (Franklin and Mayo, 1998; Mayo and Franklin, 1998). Nevertheless, the technique has potential. Numerous QTL have been identified and some are already in use in some breeding programmes (e.g. Rothschild, 1998). The long-term affects of reducing overall diversity are of concern to conserving diversity. They reflect the commercial importance of short-term gains in a competitive world. It remains to be seen how such reduction of genetic diversity within private programmes will be addressed by nations in the context of their responsibilities under the CBD to maintain diversity.

The difficulties for improvement in desired traits become more serious when the effective population size is small. This is even worse in species with low reproductive rates and long generation intervals. Fortunately, work has recently been initiated to consider optimal strategies for genetic change and reduction in the rate of inbreeding. Studies by Meuwissen (1997), Grundy *et al.* (1998) and Woolliams (1998) should contribute to the development of improved selection methods and reducing inbreeding in the future. Where problems of extremely small population size (low Ne) exist, it is important to take action as soon as possible. The simplest option is to try to select within a family, that is a sire should be replaced by one of his sons and a dam by one of her daughters. Where this is possible, a structured mating pattern can then ensure the lowest rate of inbreeding. However, selection intensity is usually severely limited by this structure.

Cryoconservation

Cryoconservation (or cryopreservation) is *ex situ* conservation as semen, ova, embryos, cells or DNA. However, cryoconservation has constraints for certain species (see Table 4.1) as a sole means of conservation. Obviously, only the storage of embryos allows the breed to be directly reproduced. Cells can now be used to produce a live animal (Wilmut *et al.*, 1997), but the

Table 4.1. State of the art in the freezing of gametes and embryos.

Species	Semen	Oocytes	Embryos
Cattle	+	0	+
Buffalo	+	?	0/+
Goat	+	0	+
Sheep	+	0	+
Pigs	+	–	0
Horse	+	?	0
Lamoids/camelids	0	?	?
Rabbit	+	0	+
Poultry	0	*	–

Reproduced from Table 5.1 in FAO (1998).
+, routine technique available; 0, positive research results; –, not feasible in the present state of the art; ?, unknown; *, some research hypotheses.

technique is not yet reliable. Ova and semen represent only half the animal, while DNA is a useful store of genes but cannot be used to reproduce an animal. While cryoconservation is regarded as a means for preserving a breed in perpetuity, the same technique can also be used simply to reduce rates of inbreeding within a live population. By storing semen, for example, for some time, the generation interval can be extended which can assist in reducing the rate of inbreeding.

It is crucial that when cryoconservation is considered for any species that an adequate sample of animals is taken to avoid inbreeding problems and that sufficient material is stored to allow good safety margins for regeneration. The technique should never be regarded as the sole technique for the maintenance of the breed unless there is no other possible option. Normally it should be used alongside *in vivo* maintenance whether *in situ* or *ex situ*. Even for breeds with large population size and under good selection systems, it is sensible to develop a scheme for regular sampling of material for cryogenic storage. This ensures that if there are genetic problems it is possible to move back and utilize material unaffected by those problems.

It is also essential to duplicate cryoconservation locations and one of these should preferably be outside the country so that if a disaster occurs, the breeds are still preserved. This aspect needs considerable planning and is an area requiring international attention. In particular, the recognition of national ownership (and, therefore, control) will need careful documentation to avoid future problems. There is an urgent need to establish programmes to properly cover the risks involved, to enable stores to be used as necessary, and to ensure proper replacement procedures for stores. FAO has developed comprehensive Guidelines for the Management of Small Populations of Farm Animal Genetic Resources at Risk (FAO, 1998).

Other methods

Many people are involved in conservation of farm animals and the role of
farmers is crucial, especially in the short term (see Thurston *et al.*, Chapter 9,
this volume). In some cases, it is relatively easy to persuade farmers to maintain breeds by assisting with a marketing scheme which builds on demand
for a special product, such as the Meat Marketing Scheme and Accredited
Butchers by the Rare Breeds Survival Trust in the UK (Alderson, Stoneleigh,
UK, 1998, personal communication) or one recently adopted in Vietnam for
the Ac breed of chicken (Nguyen van Thien, Hanoi, 1998, personal communication). In other cases, farmers are encouraged to maintain rare breeds through
subsidies to cover the difference in profit between the popular breed and the
rare breed, based on the local economic structure, for example in Italy.

The formation of clubs can also be of real value either in the context of
an association for a specific breed or NGOs such as the Rare Breeds Survival
Trust in the UK and other similar organizations. Encouragement of ownership of a breed by those using it is likely to assist in the long-term protection
of the resource. The development of farm parks as leisure attractions is
playing an important role both in conserving populations and in education.
In some cases, public institutions are used to maintain rare breeds. Zoos have
played this role for wildlife and some now include farm animal species. In
some species, commercial breeding companies are taking action to maintain
populations of breeds or of gene pools which contain material which may be
of future value.

The role of education in conservation is one which has not been recognized adequately until quite recently although there have been efforts in
some countries for many years – as exemplified by the Cotswold Farm Park
in the UK with its arrangements for school visits and training (FAO, 1992a).
The need for education has recently received some further attention. Notably,
at the 5th WCGALP FAO Symposium on Animal Genetic Resources,
Malmfors *et al.* (1994) and Vangen and Mukherjee (1994) presented excellent
considerations of the conceptual and applied aspects, respectively. Several
courses have been designed to consider the role of education in the management of genetic resources alongside the more traditional teaching of animal
breeding. This will, in time, have an effect, but whether this is sufficient to
safeguard present resources is questionable.

Communication

Increasing awareness of threats to farm animal diversity and stimulating
debate on the actions necessary to combat the problem have not been tackled
in a fully coordinated manner at global, regional or national levels. However,
there are now several groups attempting to develop some communication
strategies for the different levels of need and a global strategy is being drafted

(Hammond, Rome, 1998, personal communication). In this respect, the role of DAD-IS is crucial.

The CBD has raised the profile of biodiversity and, by emphasizing national ownership and responsibility, has created greater awareness – at least regarding the need for training, if not, as yet, for major activities. There is a priority case for training of senior administrators who are responsible for policy since few will be fully aware of the issues and options particularly for farm animals. The CBD has raised all the major issues concerning farm animal genetic diversity within the convention itself and Agenda 21 (the action programme). However, given the vastness of the CBD mandate, it is unlikely that urgent action will occur on all fronts. The CBD uses a Subsidiary Body for Scientific, Technical and Technology Advice (SBSTTA) and various specialist working groups. At the political level, control is through the Conference of the Parties (COP). The CBD has therefore raised the stakes in the context of the maintenance of biodiversity and nations have responded with, to date, over 170 countries ratifying the CBD (the USA being one notable exception).

Laing *et al.* (1998) have addressed the implications, issues and opportunities for the FAO Global Strategy and its linkages with the CBD. The authors point out the need to link the animal genetic resource focal points to those for biodiversity. For example, the regional programme for Asia has ensured that those drafting national Biodiversity Action Plans are well briefed on the farm animal genetic resource management plans.

The role of the new World Trade Organization (WTO), given its basic remit of free trade between all countries, clearly needs raising into greater open debate (Korten, 1995). Given that precise laws based on the CBD have not yet been developed, conflict between the recognition of national sovereignty (coupled to protection of resources) and totally unrestricted trade may arise unless positive steps are taken to ensure that the CBD requirements are paramount.

The FAO Global Strategy for animal genetic resources

FAO has, since its inception, been involved in the consideration of genetic variation and adaptation. From 1980 onwards, the activities of FAO in animal genetic resources (AnGR) have slowly increased. Based on the recommendations of the 1992 Expert Consultation on the Global Management of AnGR (FAO, 1992b), FAO has developed a Global Strategy. This Strategy continues to develop and can be described as a series of activities to assist the world to best use and maintain its farm animal genetic resources.

The Strategy incorporates four fundamental components:

- an intergovernmental technical support mechanism for enabling direct government involvement and ensuring continuity of policy advice;
- a technical programme of interdependent activities;

- a geographically distributed and country-based global structure, supported by regional and national focal points (Fig. 4.2), to assist and coordinate national actions; and
- a cadre of experts to guide the programme and maximize its cost-effectiveness.

*i*DAD: a global management entity

To ensure realization of the successful management of AnGR, FAO is establishing a management entity, the initiative for Domestic Animal Diversity (*i*DAD), based in Rome. *i*DAD's mission is to establish and maintain the global programme for the management of unique animal genetic resources and the conservation of diversity in each domestic animal species of interest to food and agriculture. To accomplish this mission, *i*DAD has been mandated to:

- assist countries to develop and implement effective management strategies;
- develop uniform basic procedures, guidelines and protocols;
- maintain the Global Early Warning System for Farm AnGR;
- bolster capacity building and upgrade training;
- increase awareness and communicate the issues globally;
- serve as the secretariat for the global intergovernmental mechanism for AnGR;
- coordinate activities regionally and globally.

The global structure which is being put into place aims to reflect that and will consist of three basic elements:

- a *national focus* for each country, comprising a coordinating institution strongly linked to the regional focus (Fig. 4.2) and a country technical contact nominated and supported by the Member Government. The National Focal Point (NFP) and the National Coordinator (NC) are the point of contact for the country's involvement in the FAO AnGR programme and are responsible for establishing, developing and maintaining the essential within-country network. The NC is the person responsible for all coordination both within and outside the country and for the maintenance of updated breed information for the country in the DAD-IS system. The NC is responsible for the development of the National Animal Genetic Resource Management Action Plan (NAGMAP), while the NFP essentially provides the support to the whole functioning of the AnGR work.
- a *regional focus* in each major genetic region of the world, aims to help to develop effective national coordinators, design and implement effective regional networks as integral components of the global structure, help to achieve early and wide introduction of national strategies, and trigger a

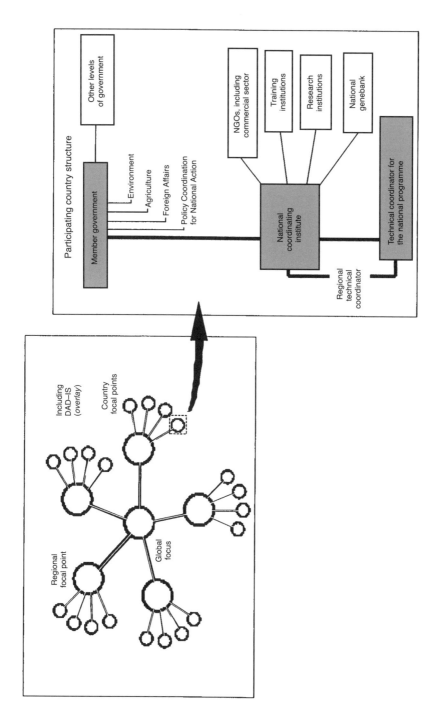

Fig. 4.2. Structure for FAO's Global Programme for the Management of Farm Animal Genetic Resources.

range of most effective projects covering the conservation complex for domestic animals. The Regional Focal Points (RFPs) will need to be established with the assistance of extra-budgetary funds. RFPs are planned for Asia and the Pacific, Europe, the Americas and Caribbean, Africa, the Near East, and the Mediterranean.

- a *global focus* for the programme is established at FAO headquarters to facilitate, communicate and coordinate the global effort, involve the range of governmental and non-governmental parties essential for the programme's success, maintain the global databank for AnGR (centred on global surveys) and the early warning system (the WWLDAD), to develop DAD-IS, provide the secretariat for the intergovernmental mechanism for AnGR, coordinate development of the necessary guidelines, procedures and assistance to countries in establishing their management strategies for AnGR, and to seek the essential extra-budgetary funding for the programme.

The RFP for Asia and the Pacific is operational; three other focal points are being established in Africa. National Focal Points (or National Co-ordinators) have been nominated for 12 countries in Asia, 11 in the Americas, 11 in the Near East, 37 in Africa and for 37 countries of Europe. Each country is further developing its own programme of work. The coordination is facilitated by FAO's development of DAD-IS. This has a series of modules which include the global database containing information on all known breeds.

DAD-IS already provides a valuable link for all involved in AnGR work through its list of participants, bulletin board, list of publications, modelling of improvement schemes and crossbreeding systems, etc., and is developing some other modules – training, access to software for experimental design and analysis, etc. Several expert cadres have provided FAO with invaluable assistance and guidance in the development of the strategy to date.

The intergovernmental mechanism is now formally in place. The link with the CBD already exists as the COP3 supports the Global Strategy and welcomes the meeting of the intergovernmental technical group and the Commission on Genetic Resources for Food and Agriculture (CGRFA). The guidelines for various crucial elements of the strategy are likely to be published in 1999.

The Future

The FAO Global Strategy is endeavouring to assist countries, organizations and all interested parties to better manage domestic animal genetic resources. There are many different groups involved and it is essential that they work together to best use the limited resources for overall effect. Countries are now starting to address their responsibilities for the resources and to develop National Animal Genetic Resource Management Action Plans as well as

Biodiversity Action Plans. The difficulty for implementation is the time-frame of return on investment. The achievement of sustainable exploitation of resources is not going to be easy or quick – these are 'uncharted waters' and there are few data and/or research results available.

It is important that activities are started in countries at the same time as the Global Strategy is being developed so that the whole process can grow together. The coordination must ensure that investments are used in an optimal manner – at global, regional and national levels. These levels should be regarded as integral components of an essential activity and not as competitors for funds. The relationship between a country's responsibilities under the CBD and those as a member of the WTO needs early and clear definition, and the Commission on Genetic Resources for Food and Agriculture (CGRFA) may need to assist the FAO in this respect.

We need to ensure that countries have sufficient human resources to carry out the responsibilities placed on them by the CBD. Many countries need assistance in formulating their national management plans since the basic essentials are often not known. In addition, because it is a relatively new area, there is little expertise available although the increase in awareness at all levels is encouraging and suggests that, with the right investment, this could be corrected.

The crucial question is not whether countries are aware of the problem or, indeed, conscious of the needs but whether all countries will be prepared to adjust their priorities to enable the important activities to take place and to ensure that there is sufficient diversity to enable the world to feed itself – even in 2020, let alone for a longer time than that. *The question is more one of long-term human food security than of anything else.*

Acknowledgements

The author wishes to thank colleagues for their assistance in the preparation of this chapter.

The presentation of material in this chapter expresses the opinion of the author and does not imply the expression of any opinion whatsoever on behalf of the Food and Agriculture Organization of the United Nations.

References

Archer, J.A., Arthur, P.F., Herd, R.M. and Richardson, E.C. (1998) Genetic variation in feed efficiency and its component traits. *Proceedings of the 6th World Congress of Genetics Applied to Livestock Production* 25, 81–84.

Blackburn, H., Lebbie, S.H.B. and van der Zjipp, A.J. (1998) Animal genetic resources and sustainable development. *Proceedings of the 6th World Congress on Genetics Applied to Livestock Production* 28, 3–10.

Brascamp, E.W., Groen, A.F., de Boer, I.J.M. and Udo, H. (1998) The effect of environmental factors on breeding goals. *Proceedings of the 6th World Congress on Genetics Applied to Livestock Production* 27, 129–136.

Danell, B., Vigh-Larsen, F., Maki-Tanila, A., Eythorsdottir, E. and Vangen, O. (1998) A strategic plan for Nordic co-operation in management of animal genetic resources. *Proceedings of the 6th World Congress on Genetics Applied to Livestock Production* 28, 111–114.

Dekkers, S.C.M. and van Arendonk, J.A.M. (1998) Optimum selection on identified quantitative trait loci. *Proceedings of the 6th World Congress on Genetics Applied to Livestock Production* 26, 361–364.

Falconer, D.S. (1961) *Introduction to Quantitative Genetics.* Agricultural Research Council's Unit of Animal Genetics, University of Edinburgh.

FAO (1992a) *In situ* conservation of livestock and poultry. *FAO Animal Production and Health Paper No. 99*, Henson, E.L. (ed.). FAO, Rome.

FAO (1992b) The management of global animal genetic resources. *FAO Animal Production and Health Paper No. 104*, Hodges, J. (ed.). FAO, Rome.

FAO (1993) *An Integrated Global Programme to Establish the Genetic Relationships Among the Breeds of Each Domestic Animal Species.* Report of a Working Group, June 1993. Animal Production and Health Division, FAO, Rome.

FAO (1996a) *Selected Indicators of Food and Agriculture Development in Asia–Pacific Region,* 1985–95. RAP publication: 1996/32.

FAO (1996b) *World Food Summit Information Sheet, Agriculture and Food Security.* FAO, Rome.

FAO (1996c) *World Food Summit – Rome Declaration on World Food Security and World Food Summit Plan of Action.* FAO, Rome.

FAO (1997) *Selected Indicators of Food and Agriculture Development in Asia–Pacific Region,* 1986–96. RAP publication: 1997/23.

FAO (1998) *Secondary Guidelines for Development of National Farm Animal Genetic Resources Management Plans: Management of Small Populations at Risk.* Domestic Animal Diversity Information System, Stage 2.0, 7 September (http:/ dad. fao. org/dad-is/refer/library/guidel/in/aml. popn. pdf). FAO, Rome.

FAO/UNEP (1993) *World Watch List for Domestic Animal Diversity,* 1st edn, Loftus, R. and Scherf, B.D. (eds). FAO/UNEP, Rome.

FAO/UNEP (1995) *World Watch List for Domestic Animal Diversity,* 2nd edn, Scherf, B.D. (ed.). FAO/UNEP Rome.

Frankham, R. (1994) Conservation of genetic diversity for animal improvement. *Proceedings of the 5th World Congress on Genetics Applied to Livestock Production* 21, 385–392.

Franklin, I.R. (1986) Breeding ruminants for the tropics. *Proceedings of the 3rd World Congress on Genetics Applied to Livestock Production* XI, 451–461.

Franklin, I.R. and Mayo, O. (1998) The place of QTL in the basis of quantitative genetics. II. Mapping QTL from a wide cross. *Proceedings of the 6th World Congress on Genetics Applied to Livestock Production* 26, 273–276.

Gibson, J.P. (1994) Short-term gain at the expense of long-term response with selection of identified loci. *Proceedings of the 5th World Congress on Genetics Applied to Livestock Production* 21, 201–204.

Gregory, K.E. (1986) Preservation and management of genetic resources: maintaining adaptation and diversity. *Proceedings of the 3rd World Congress on Genetics Applied to Livestock Production* XII, 492–499.

Grundy, B., Villanueva, B. and Woolliams, J.A. (1998) Dynamic selection procedures for constrained inbreeding and their consequences for pedigree development. *Proceedings of the 6th World Congress on Genetics Applied to Livestock Production* 25, 355–358.

de Haan, C. and Setshwaelo, L.L. (1998) Sustainable intensification of livestock production. *Proceedings of the 6th World Congress on Genetics Applied to Livestock Production* 28, 59–67.

de Haan, C., Steinfeld, H. and Blackburn, H. (1997) '*Livestock and the Environment*': *Finding a Balance*. EU Development Policy Sustainable Development and Natural Resources Report, WRENmedia, Wye, UK.

Hammond, K. and Leitch, H.W. (1995) The FAO Global Program for the Management of Farm Animal Genetic Resources. *Beltsville Symposia in Agricultural Research* XX, 24–42.

Hill, W.G. (1998) Inferences from evolutionary biology to livestock breeding. *Proceedings of the 6th World Congress on Genetics Applied to Livestock Production* 23, 32–39.

Jackson, N. and Turner, H. (1972) Optimal structure for a co-operative nucleus breeding system. *Proceedings of Australian Society of Animal Production* 9, 55–64.

James, J.W. (1977) Open nucleus breeding systems. *Animal Production* 24, 287–306.

Kempster, A.J., Cook, G.L. and Southgate, J.R. (1982) A comparison of the progeny of British Friesian dams and different sire breeds in 16- and 24-month beef production systems. 2. Carcass characteristics, and rate and efficiency of meat gain. *Animal Production (Journal of the British Society of Animal Production)* 34, 167–178.

Khan, B.U. and Taneja, V.K. (1996) Current development in management of small ruminant genetic resources: South and South-East Asia. *Proceedings of IGA/ FAO Round Table on the Global Management of Small Ruminant Genetic Resources*, Beijing, China. FAO, Bangkok.

Korten, D.C. (1995) *When Corporations Rule the World*. Earthscan Publications, London.

Korver, S., van Eekelen, E.A.M., Vos, H., Nieuwhof, G.J. and van Arendonk, J.A.M. (1991) Genetic parameters for feed intake and feed efficiency in growing dairy heifers. *Livestock Production Science* 29 (1), 49–60.

Laing, R., Fall, A., Han, G.J. and Martyniuk, E. (1998) The convention on biological diversity and farm animal genetic resources: implications, issues and opportunities. *Proceedings of the 6th World Congress on Genetics Applied to Livestock Production* 28, 51–58.

Land, R.B. (1986) Genetic resources requirements under favourable production marketing systems: priorities and organization. *Proceedings of the 3rd World Congress on Genetics Applied to Livestock Production* XII, 486–491.

Luiting, P. (1991) The value of feed consumption data for breeding in laying hens. PhD thesis, Wageningen Agricultural University, the Netherlands.

Madalena, F.E. (1990) Crossbreeding effects in tropical dairy cattle. *Proceedings of the 4th World Congress on Genetics Applied to Livestock Production* XIV, 310–319.

Madalena, F.E. (1998) Hybrid F_1 *Bos taurus* × *Bos indicus* dairy cattle production in the state of Minas Gerais, Brazil. *Proceedings of the 6th World Congress on Genetics Applied to Livestock Production* 27, 199–202.

Maijala, K., Cherekaev, A.V., Devillard, J.-M., Reklewski, Z., Rognoni, G., Simon,

D.L. and Steane, D.E. (1984) Conservation of animal genetic resources in Europe. Final report of an EAAP Working Party. *Livestock Production Science* 11, 3–22.

Malmfors, B., Philipsson, J. and Haile-Mariam, M. (1994) Education and training in the conservation of domestic animal diversity – student needs and field experience. *Proceedings of the 5th World Congress on Genetics Applied to Livestock Production* 21, 485–492.

Mayo, O. and Franklin, I.R. (1998) The place of QTL in the basis of quantitive genetics. I. General considerations. *Proceedings of the 6th World Congress on Genetics Applied to Livestock Production* 26, 77–80.

Meuwissen, T.H.E. (1997) Maximising the response of selection with a predefined rate of inbreeding. *Journal of Animal Science* 75, 934–940.

Oya, A. (1992) Le programme national de sélection ovin (PNSO) de Côte d'Ivoire: mise en place du contrôle de performances en ferme (abstract). In: Ray, B., Lebbie, S.H.B. and Reynolds, L. (eds) *Small Ruminant Research and Development in Africa. Proceedings of the 1st Biennial Conference of the African Small Ruminant Research Network,* ILRAD, Nairobi, Kenya, 10–14 December 1990. ILRI (International Livestock Centre for Africa), Nairobi, Kenya, pp. 561–562.

Peel, L. and Tribe, D.E. (1983) *World Animal Science, Domestication, Conservation and Use of Animal Resources.* Elsevier Science, Amsterdam.

Perrings, C., Maler, K.G., Folke, C., Holling, C.S. and Jansson, B.O. (1995) *Biodiversity Loss, Economic and Ecological Issues.* Cambridge University Press, New York.

Pessinaba, I.Y. (1992) Programme de développement de l'élevage des petits ruminants au Togo: essai de mise au point des possibilités d'association de la production vivriere et de l'élevage ovin. In: Ray, B., Lebbie, S.H.B. and Reynolds, L. (eds) *Small Ruminant Research and Development in Africa. Proceedings of the 1st Biennial Conference of the African Small Ruminant Research Network,* ILRAD, Nairobi, Kenya, 10–14 December 1990. ILRI (International Livestock Centre for Africa), Nairobi, Kenya, pp. 545–554.

Pollott, G.E., Croston, D. and Guy, D.R. (1994) Genetic progress in the CAMDA group breeding scheme nucleus. *Animal Production* 58, 431.

Rege, J.E.O. (1998) Utilization of exotic germplasm for milk production in the tropics. *Proceedings of the 6th World Congress on Genetics Applied to Livestock Production* 25, 193–200.

Rosengrant, M.W., Agcaoili-Sombilla and Perez, N.D. (1995) *Global Food Projections to 2020. Implications for Investment.* Food, Agriculture and Environment Discussion Paper 5, International Food Policy Research Institute, Washingon, DC.

Rothschild, M.F. (1998) Identification of quantitative trait loci and interesting candidate genes in the pig: progress and prospects. *Proceedings of the 6th World Congress on Genetics Applied to Livestock Production,* 26, 403–409.

Simm, G. and Pryce, J.E. (1998) Responding to future market requirements: implications for breeding strategies. *CAS Conference: 'Future Global, EU and UK Markets for Milk and Milk Products',* Reading, UK.

Smith, C. (1984) Estimated costs of genetic conservation in farm animals. Animal genetic resources conservation by management, data banks and training. *FAO Animal Production and Health Paper* 44(1), 21–30.

Southgate, J.R., Cook, G.L. and Kempster, A.J. (1982a) A comparison of the progeny of British Friesian dams and different sire breeds in 16- and 24-month beef production systems. 1. Live-weight gain and efficiency of food utilization. *Animal Production (Journal of the British Society of Animal Production)* 34, 155–166.

Southgate, J.R., Kempster, A.J. and Cook, G.L. (1982b) A comparison of the progeny of British Friesian dams and different sire breeds in 16- and 24-month beef production systems. 2. Carcass characteristics, and rate and efficiency of meat gain. *Animal Production (Journal of the British Society of Animal Production)* 34, 167–178.

Steane, D.E. (1996) Minimizing genetic erosion : farm animal genetic resources. In: McNeely, J.A. and Somchevita, S. (eds) *Biodiversity in Asia: Challenges and Opportunities for the Scientific Community.* OEPP, MoSTE, Bangkok, pp. 96–105.

Steane, D.E., Guy, D.R. and Smith, C. (1982) *Group Breeding Schemes for Beef. Group Breeding Schemes for Sheep.* MLC (Meat and Livestock Commission), UK.

Steinfeld, H., Haan, C. de and Blackburn, H. (1997) *Livestock–Environment Interactions – Issues and Options.* EU Development Policy Sustainable Development and Natural Resources, WRENmedia, Wye, UK.

Sutherland, G.R. (1996) The Human Genome Project. *Proceedings of the Second National Congress on Genetics – Genetics into the Next Millennium.* Genetics Society of Malaysia, UKM Bangi, pp. 1–3.

Tess, M.W., Bennet, G.L. and Dickerson, G.E. (1993a–c) Simulation of genetic changes in life cycle efficiency of pork production. I, II and III Effects of components on efficiency. *Journal of Animal Science* 56, 336–379.

Thiessen, R.B., Taylor, St C.S. and Murray, J. (1985) Multibreed comparisons of British cattle. Variation in relative growth rate, relative food intake and food conversion efficiency. *Animal Production (Journal of the British Society of Animal Production)* 41, 193–199.

Trail, J.C.M. (1986) Animal breeding in developing country situations in Africa. *Proceedings of the 3rd World Congress on Genetics Applied to Livestock Production* XI, 474–485.

Turton, J.D. (1974) The collection, storage and dissemination of information on breeds of livestock. *Proceedings of the 1st World Congress on Genetics Applied to Livestock Production* II, 61–74.

UNEP (1992) *The Convention on Biological Diversity.* UNEP, Nairobi.

Van der Beek, S. and van Arendonk, J.A.M. (1994) Marker assisted selection in a poultry breeding program. *Proceedings of the 5th World Congress on Genetics Applied to Livestock Production* 21, 237–240.

Vangen, O. and Mukherjee, T. (1994) Conceptual approach to integrating education in animal breeding and in conservation genetics. *Proceedings of the 5th World Congress on Genetics Applied to Livestock Production* 21, 477–484.

Veerkamp, R.F., Hill, W.G., Stott, A.W., Brotherstone, S. and Simm, G. (1995) Selection for longevity and yield in dairy cows using transmitting abilities for type and yield. *Animal Science* 61, 189–197.

Wickham, B.W. and Banos, G. (1998) Impact of international evaluations on dairy cattle breeding programmes. *Proceedings of the 6th World Congress on Genetics Applied to Livestock Production* 23, 315–322.

Wilmut, I., Schnieke, A.E., McWhir, J., Kind, A.J. and Campbell, K.H.S. (1997) Viable offspring derived from fetal and adult mammalian cells. *Nature* 385, 810–813.

Woolliams, J.A. (1998) A recipe for the design of breeding schemes. *Proceedings of the 6th World Congress on Genetics Applied to Livestock Production* 25, 427–430.

The Regulation and Functional Significance of Soil Biodiversity in Agroecosystems

5

D.A. Wardle[1], K.E. Giller[2]* and G.M. Barker[3]

[1] Landcare Research, PO Box 69, Lincoln 8152, New Zealand; [2] Department of Biological Sciences, Wye College, University of London, Wye, Ashford, Kent TN25 5AH, UK; [3]Landcare Research, Private Bag 3127, Hamilton, New Zealand

Introduction

In most terrestrial ecosystems, the majority of the biodiversity present occurs below the soil surface, not above it (Beard, 1991; Giller *et al.*, 1997). This is especially the case for agricultural systems in which emphasis is usually placed on maintaining a low diversity of plant species, as well as small populations of those above-ground consumer organisms associated with them. Below-ground organisms are critical in determining the functioning of agroecosystems; subsets of the soil biota are responsible for providing a range of ecosystem services through regulating such processes as decomposition, nutrient mineralization, energy flow and various transformations of the main nutrient cycles (Fig. 5.1). These processes (and the organisms that control them) determine plant growth and thus sustain the productivity of the agroecosystem in the long term (Ingham *et al.*, 1985; Lavelle, 1994). The soil also hosts a range of herbivorous invertebrates, microbial pathogens and rhizosphere organisms which can exert more direct effects on plant growth. While manipulations of plants above ground and of the soil itself have long been known to affect the soil biota, there is increasing evidence to confirm the linkage hypothesized between the composition of the soil biota and plant growth and organic matter return (Setälä and Huhta, 1991; Wardle and Lavelle, 1997).

While agricultural practices have the potential to alter components of the soil biota, agricultural sustainability depends upon the ability of soil organisms to provide adequate resources for plant growth. For this reason, two important questions emerge when soil biodiversity in agricultural

---

*See Contributors list for current address.

© 1999 CAB *International. Agrobiodiversity*
(D. Wood and J.M. Lenné)

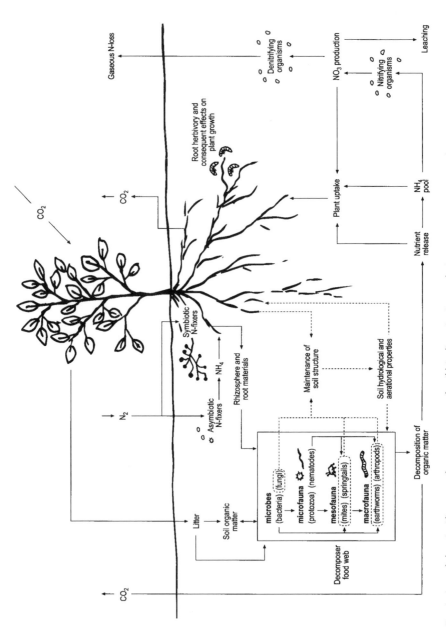

Fig. 5.1. Depiction of the significance of components of soil biodiversity in regulating ecosystem properties.

systems is considered. Firstly, how does agricultural practice, and in particular intensification, impact upon components of the soil biota, and the biodiversity of the below-ground system? Secondly, do shifts in soil biodiversity have important implications for the processes governed by soil organisms, the overall functioning of the below-ground subsystem, and ultimately plant (particularly crop) productivity? The aim of this chapter is to address both of these issues. Effects of tillage practices on soil biodiversity are treated in detail in Chapter 12.

Regulation of Soil Biodiversity

Stress, disturbance and plant influences

In agroecosystems, the principal aim of management is to produce harvestable animal or plant biomass, primarily through enhanced production of specific crop species. This is inevitably accompanied by a range of disturbance factors (i.e. those involving rapid changes in environmental conditions) and stress factors (i.e. those involving non-varying harsh conditions), which emerge from such practices as preparation of seedbeds, controlling organisms which compete with or consume crop plants, ensuring adequate plant nutrition and moisture, management of plant residues and harvesting of crop plants. These all have the potential to influence soil organisms. Both disturbance and stress are widely recognized in ecology as important determinants of biodiversity (Grime, 1979; Huston, 1994). In the absence of stress and disturbance in a given habitat, biodiversity is characteristically poor because a small number of very successful species monopolize most of the resources and effectively exclude others. As stress or disturbance intensifies, the dominant organisms are suppressed to an extent that other organisms can coexist with them, enhancing diversity. Diversity is then reduced further if stress or disturbance cause conditions to become adverse to the extent that more species are lost than can colonize. This can result in the classic 'hump-back' relationship hypothesized to exist between diversity and both stress (Grime, 1979; Austin, 1987) and disturbance (Connell, 1978), with maximal diversity occurring at intermediate intensities of these factors. It is likely, however, that other factors are also important in regulating diversity, since patterns of diversity along disturbance gradients can be taxon-specific (Lawton et al., 1998). In light of the above, we hypothesize that soil biodiversity need not necessarily be adversely affected by agricultural intensification.

In addition to the (usually) more immediate effects of stress and disturbance, agricultural practice is also likely to exert longer-term influence on the soil biota (and soil biodiversity) through determining the composition of the plant community. These influences, which are manifested through determining the quantity and quality of organic matter entering the soil, and possibly its diversity, can serve as powerful determinants of the soil biota, since adequate resource input is required to maintain soil organic matter (the basic substrate for soil organisms) in the long term (Parton et al., 1987).

In the remainder of this section we evaluate aspects of biodiversity in soils which are relevant to agricultural practice (e.g. stress, disturbance and plant influences). This will be done by considering three, functionally different, components of the soil biota: soil-associated food webs which regulate decomposition processes, microorganisms involved in nutrient transformations, and soil-associated herbivores. The research conducted on each of these three components with regard to biodiversity issues has been relatively independent of each other, resulting in different conceptual approaches and perceived issues of importance across the three components. We will now consider each component in turn.

Decomposer food webs

Decomposition of organic matter and related processes such as mineralization of nutrients are regulated by the decomposer food web, which is constituted of the resource base, the primary saprophytes (bacteria and fungi; comprising the 'microbial biomass') and the soil animals which consume them and each other. The fauna operate at three levels of resolution (Lavelle et al., 1993), namely: (i) the 'microfood-webs' involving nematodes, protozoa and their predators; (ii) 'litter-transformers' involving mesofauna (principally mites and springtails) and some macrofauna, in which organic matter is consumed and transformed into organic structures (faecal pellets); and (iii) 'ecosystem engineers' (e.g. earthworms) which build organo-mineral structures that create habitats for smaller organisms. The interactions which occur at each of these three spatial scales are critical in determining microbial activity, which is in turn important for regulating the functions of the below-ground subsystem.

Agricultural practices affect organisms at each of these three levels. As such, comparison of field plots subjected to differing levels of agricultural intensification often reveal large compositional differences in the soil biota (e.g. Paustian et al., 1990; De Ruiter et al., 1994; Yeates et al., 1997). One of the principal disturbances associated with intensive agriculture (and arguably the most studied with regard to soils), tillage, has been widely documented as exerting important effects on soil food webs (see Hendrix et al., 1986; Wardle, 1995), and comparisons of conventional tillage (CT) versus non-tillage (NT) systems have considerable utility for evaluating how disturbance affects food webs (Beare et al., 1992; see also Edwards et al., Chapter 12, this volume). There is evidence that larger organisms, many of which are higher-level consumers and occupy higher trophic levels, are more adversely affected by tillage than are smaller ones (Wardle, 1995), and that they are generally less resistant (and probably also less resilient) to disturbance than smaller ones. Since larger organisms are more likely to occupy higher trophic levels, this means that the significance of 'top-down' forces whereby organisms in lower trophic levels are controlled by predation and other activities from larger organisms assume a lesser significance in CT systems (Wardle, 1995). However, even within trophic levels there can be important effects of tillage on composition, and in a study involving long-term comparisons of CT and NT practices, Hendrix *et al.* (1986) showed that tillage favoured organisms in the bacterial-based, rather than the fungal-based compartment of soil food webs. These effects in turn had important consequences for decomposition rates, nutrient mineralization and soil organic matter loss, all of which were greater in CT than NT systems. These differences were supported by subsequent litter-bag studies conducted at the same site (Beare *et al.*, 1992), which showed that fungal-based food webs contributed to decomposition in NT systems, while bacterial-based food webs contributed in CT systems.

The addition of pesticides (especially herbicides) and mineral fertilizers are also important components of agricultural intensification, but effects of stress and disturbance resulting from these practices on soil food webs are less well understood. Part of the difficulty here is separating the direct effect of the added material on the organisms from indirect effects resulting from modification of plant productivity and community structure, and consequent modification of resource input to the soil (Mahn and Kastner, 1985; Wardle, 1995). Many *in vitro* studies claiming non-target effects are clouded by problems of methodological artefact, for example through using unnaturally high concentrations of the chemical under investigation (Domsch *et al.*, 1983). However, there are instances in which applications of herbicides have been shown to induce compositional shifts in components of the soil fauna, particularly those associated with plant litter (Hendrix and Parmelee, 1985; see also Edwards *et al.*, Chapter 12, this volume).

Given that the soil microflora (and especially the fungal component) is governed primarily by resource inputs (i.e. 'bottom-up' effects) (Wardle and Yeates, 1993), it is reasonable to expect that the amount and quality of resource inputs resulting from selection of plant species, addition of mulches and removal of weeds, all have important effects on microbial biomass and activity, and to expect this effect to exert some influence at higher trophic levels of soil food webs. Indeed, studies indicate that biomass of microbes (Groffman *et al.*, 1996), populations and composition of nematodes (Wasilewska, 1995), and composition of the soil food web (Griffiths *et al.*, 1992) are all highly responsive to the species composition of herbaceous plants. However, responses of soil food web components to increasing net primary productivity (NPP) appear to be inconsistent, with some studies showing positive relationships between NPP and biomass of microbes (e.g. Zak *et al.*, 1994) and soil fauna (Yeates, 1979), while others fail to find such effects (Yeates and Coleman, 1982; Wardle *et al.*, 1995). Part of the reason is that, in addition to adding resources which stimulate microbes, plants also compete with microbes for nutrients; it is the balance between these opposing effects which governs how plants may affect soil organisms (Van Veen *et al.*, 1989). Furthermore, in soil food webs (like most food webs), organisms can be affected by either bottom-up or top-down forces, with the result that some organisms may respond to alterations in resource input while others (through being regulated by predation) might not (Wardle and Yeates, 1993; De Ruiter *et al.*, 1995). This is especially apparent where, for example, organisms in higher trophic levels have been shown to suppress those in lower trophic levels along primary productivity or resource addition gradients (Mikola and Setälä, 1998a), or where trophic cascades have been identified in which a predator may induce the regulation of several trophic levels (Santos *et al.*, 1981; Kajak *et al.*, 1993, but see Mikola and Setälä, 1998b).

The effects of plants on soil food webs are also apparent when the effects of above-ground herbivory are considered (Table 5.1). In agroecosystems based on grazing (e.g. pastoral systems), intensification involves management practices aimed at increasing the livestock carrying capacity, as well as reducing consumption by herbivorous organisms such as invertebrate pests. Grazing may affect below-ground food webs by determining the quantity and quality of organic matter returned to the soil. In the short term, grazing can significantly alter the flow of carbon to the below-ground subsystem, and it has been demonstrated that heavy grazing by sheep favours 'fast' cycles dominated by labile substrates and bacteria, while light grazing favours 'slow cycles' dominated by resistant substrates and fungi (Bardgett, 1996). In the longer term, grazing also affects root mass, but both positive and negative effects are possible (Milchunas and Lauenroth, 1993) with consequences for both the soil microflora (Seagle *et al.*, 1992) and populations of key groups of soil animals (Bardgett *et al.*, 1998). Further, selective foliar herbivory may induce shifts in plant community structure and thus organic matter quality (Wardle and Barker, 1997), although these effects are likely to be idiosyncratic and ultimately dependent upon the traits of the organisms that are favoured. Although it has been shown that selective herbivory may favour unpalatable species with poor quality litter (and which apparently adversely affect soil organisms) in forests (Pastor *et al.*, 1988), this has yet to be shown for grazing in agroecosystems.

The vast majority of studies investigating the impacts of agricultural intensification on decomposer organisms have focused on broad functional or taxonomic groups (e.g. total microbial biomass; total numbers of springtails). However, such studies yield information only on broad compositional shifts in the decomposer biota, and the biomass of such groups may remain unchanged by agricultural practice despite important shifts in both composition and diversity within the groups. Those studies which have considered one or more groups of organisms at a finer level of resolution have provided evidence of important shifts in community structure in response to agricultural intensification (e.g. Rovira *et al.*, 1987; Freckman and Ettema, 1993; Yeates *et al.*, 1997). These shifts are, however, not closely linked to

Table 5.1. Mechanisms by which herbivory (e.g. grazing) may alter components of soil food webs through altering plants in grazed agroecosystems.

Means of effect	Mechanism involved	Probable effects on soil biota	Representative references
Root carbon flow	Grazing alters short-term patterns of root C flow and chemistry	Usually positive effects on microbial communities	Bardgett and Lehmans (1995); Holland (1996)
	Grazing results in shifts in root morphology, architecture and biomass	Idiosyncratic, depends on intensity of herbivory	Ingham and Detling (1984); Seagle et al. (1992)
Litter quality	Grazing enhances foliar nutrient concentration	Likely to benefit soil organisms	Yet to be tested
	Grazing enhances root litter quality	Probably benefits soil organisms	See Seastedt et al. (1988); Merrill et al. (1994)
	Cellular damage by grazing induces polyphenols	Probably harmful to soil organisms	See Findlay et al. (1996)
Plant community structure	Alteration of structure of plant community	Idiosyncratic, depends on plant species involved	Wardle and Barker (1997)
	Stimulation of unpalatable weed species	Idiosyncratic, depends on plant species involved	Wardle et al. (1995)

shifts in taxonomic diversity. For example, those studies presenting community structure data for specific groups of decomposer organisms in CT
and NT plots provide little consistent evidence of harmful effects of cultivation. Calculation of diversity indices from each of 20 such studies (Fig. 5.2)
demonstrates that for microfaunal organisms cultivation generally has little
effect, while for larger organisms effects on diversity can be either positive or
negative, depending on the study. Similarly, there is little evidence of applications of agrochemicals consistently reducing soil diversity (Wardle, 1995; see
also Edwards *et al.*, Chapter 12, this volume). This evidence is consistent with
the 'intermediate disturbance' hypothesis (Connell, 1978), in that although
agricultural disturbance may affect diversity, the intensity of disturbance is
likely to determine the direction of the effect. Further, even in disturbed
ecosystems, soil organisms are often subjected to significant and repeated
disturbances, for example through wetting and drying cycles and variations in
input of substrates. These are sometimes much more significant than factors
associated with agricultural intensification (Domsch *et al.*, 1983; Wardle and
Parkinson, 1991), and therefore the response of soil diversity to agricultural
disturbance may be overshadowed by shifts which occur naturally. In total, it
is apparent that agricultural intensification can induce shifts in the composition of the soil food web and the community structure of specific groups, as
well as select for organisms possessing specific traits; however, there is little
evidence to support the view that factors associated with the intensification of
agriculture have predictable, harmful consequences for the diversity of organisms in the decomposer subsystem.

Microorganisms associated with nutrient transformation

Although biodiversity is often considered in terms of the number of species
present in a given habitat, discussion of soil microbial biodiversity is clouded
by definition of groups on the basis of function and the rather bewildering
problems of its assessment. Many of the major functional groups in soil are
those which govern essential steps in the cycling of major nutrients such as
nitrogen and sulphur. Cycling of major cations (Ca^{2+}, Mg^{2+}, K^+) and most
micronutrients can, however, be described without resort to biological
transformations (although oxidation–reduction equilibria in anaerobic
environments are effectively balanced by redox couplets where the degree of
reduction depends on microbial activity). Phosphorus is a distinct case in
that the concentrations are strongly determined by chemical equilibria, and
particularly by the insolubility of many phosphate complexes. Cycling of P
through organic matter provides a mechanism which prevents chemical
precipitation of P and this is of great importance in forest ecosystems
(Gressel and McColl, 1997), but this mechanism generally plays a minor role
in agriculture. Organisms which mineralize P may well be the same as those
which mineralize N and S during decomposition of organic residues. Plant

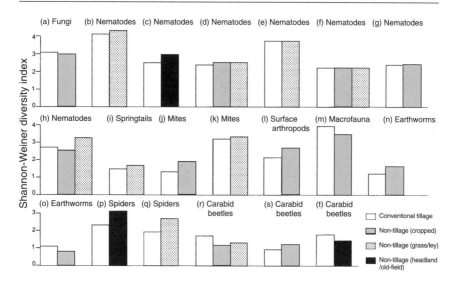

Fig. 5.2. Summary of studies from which diversity indices for specific groups of decomposer organisms have been extracted from the data presented, under conventionally tilled and non-tilled systems. References: (a) Wacha and Tiffany (1979); (b) Bostrom and Sohlenius (1986); (c) Domurat and Kozlowska (1974); (d) Freckman and Ettema (1993); (e) Sohlenius *et al.* (1987); (f) Yeates and Hughes (1990); (g) Yeates *et al.* (1993a); (h) Yeates and Bird (1994); (i) Lagerlöf and Andrén (1991); (j) Emmanuel *et al.* (1985); (k) Lagerlöf and Andrén (1988); (l) Blumberg and Crossley (1983); (m) Robertson *et al.* (1994); (n) Andersen (1987); (o) Haukka (1988); (p) Doane and Dondale (1979); (q) Huhta and Raatikainen (1974); (r) House and All (1981); (s) Stassart and Gregorie-Wibo (1983); (t) Von Klinger (1987). Note: many of these calculations are presented in Table 2 of Wardle (1995).

uptake is, of course, strongly dependent on mycorrhizal fungi to increase the surface area of contact for absorption of P (Smith and Read, 1997).

Microbial communities represent a 'last frontier' for description of biodiversity. Description of decomposition has been successfully achieved by assigning roles to the microbial biomass as a 'black box' (Powlson, 1994), and discussion of particular transformations of the major nutrient cycles are well explained in terms of functional groups of microorganisms. The term soil microbial biomass 'conjures up ... a shapeless mass of translucent protoplasm oozing indefinitely through the gaps between minerals' (Young, 1994), and description of the structure and components of the biomass is hampered by both its inherent complexity and methodological problems (Fig. 5.3). Classical methods for the study of soil microorganisms were centred on direct microscopical observation and the use of specific substrates to cultivate particular groups (a largely functional approach), which have been used together with physiological and biochemical methods to describe

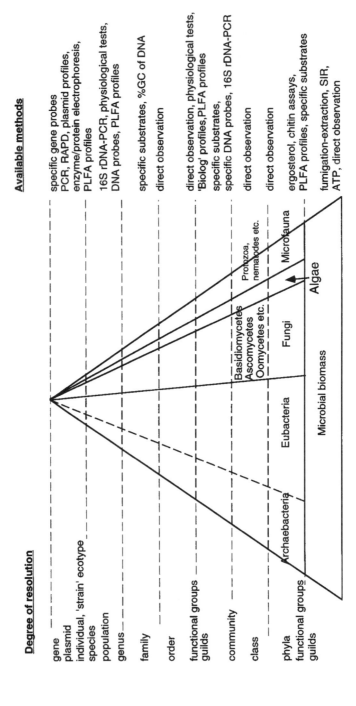

Fig. 5.3. Scales of resolution at which microbial biodiversity can be considered in soil, and the methods which can be used to determine biodiversity at these different scales. Note that several methods can be applied in different ways to examine various scales of resolution.

taxonomic groupings. The application of tools of molecular biology has allowed development of a phylogenetic classification of bacteria (Woese, 1987), and study of DNA isolated directly from soils has totally changed our understanding of the diversity of bacteria present in soils (Torsvik *et al.*, 1990, 1994). Methods now available for description of components of the soil microbial biomass, which can be applied without the bias of cultivation using specific substrates, include direct DNA extraction and sequencing, use of specific gene probes and phospholipid–fatty acid profiles (PLFAs) all of which can be used to describe distinct scales of resolution within the biomass. Specific DNA probes have been targeted for distinguishing major groups within the Archaebacteria and Eubacteria (Stahl, 1997; Wintzingerode *et al.*, 1997) but DNA probes could equally be designed to recognize and count individual species, or to target key genes required for a specific function. A wide range of methods is thus available for distinguishing groups within the microbial biomass at scales of resolution from the single gene to the kingdom.

Having peered inside the black box, our modified view of the microbial biomass is both unsettling and unsatisfactory. Estimates put the number of bacterial species in a gram of soil at several thousand (Torsvik *et al.*, 1994), of which a small percentage can be cultivated (Stahl, 1997). There are no reliable estimates of the comparable number of fungal species in soil. Even the measurement of the relative contributions of bacteria and fungi to the microbial biomass is problematic. Bacterial phylogenies based on molecular clocks confirm some earlier groupings but are highly divergent from classifications based on function. For example, among the autotrophic nitrifying bacteria, the ammonia oxidizers are found in both the beta and gamma proteobacteria (Head *et al.*, 1993). The enzyme nitrogenase, which confers capability to fix atmospheric N_2, is present in widely divergent groups throughout the Archaebacteria and the Eubacteria (Young, 1992). However, the occurrence of particular metabolic functions in organisms which are widely separated in the new molecular phylogenies can be explained by horizontal gene transfer between bacteria, now a widely recognized natural phenomenon. Young (1994) argues that this does not mean that the species concept should be discarded for microorganisms as a suite of mechanisms ensure that genetic information is more likely to be exchanged between closely related species.

One group of bacteria which has been studied in great detail because of their importance in agriculture, collectively called rhizobia, are the bacteria which fix N_2 in symbiosis with legumes. Phylogenetic analysis reveals that the rhizobia all fall within the alpha Proteobacteria, which is reassuring until the wider membership of this cluster is examined (Fig. 5.4). The slow-growing rhizobia (of the genus *Bradyrhizobium*) are shown as being closely related to the free-living N_2-fixing phototroph *Rhodopseudomonas palustris*, an observation considered surprising until a phototrophic, symbiotic strain belonging to this cluster was described from stem nodules of a tropical forage legume, *Aeschynomene* (Young, 1992). The close relationships between

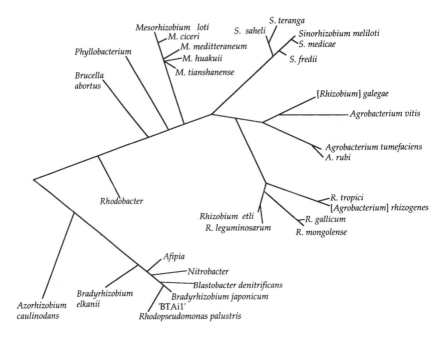

Fig. 5.4. Phylogenetic relationships between the rhizobia (bacteria which nodulate and fix nitrogen in association with legumes) and other soil bacteria (updated from Van Berkum *et al.*, 1988, and Martínez-Romero, 1994). Distances not to scale.

the soilborne bacterium *Afipia* (named after the US Armed Forces Institute of Pathology), which causes human infection following cat-scratches, within the *Bradyrhizobium* cluster, and the relatedness of *Brucella abortus* to the fast-growing rhizobia was certainly unexpected. The similarity between the plant pathogens of the genus *Agrobacterium* and the fast-growing rhizobia has long been recognized by their joint classification into the Rhizobiaceae, but now it is clear that the species which cause the hairy root phenotype on plants (*Agrobacterium rhizogenes*) are more closely related to *Rhizobium* than to the bacteria which cause 'crown galls' (*Agrobacterium tumefaciens*) (Young and Haukka, 1996). The number of rhizobial species described which can nodulate a given legume host is no doubt partly related to the plant's 'promiscuity' for nodulation, but is also a function of the focus of interest and activity of researchers on particular groups of legumes rather than a reflection of the diversity of the rhizobia *per se*. The vast diversity of soil microorganisms means that hundreds of years of work will be required before the degree of saturation is sufficient to have confidence of reasonable coverage of the prokaryotes or the fungi.

This discussion can do little more than highlight the problems and pitfalls of studying microbial diversity in soil, as little is known about the factors which regulate microbial biodiversity in soil. It is tempting to assume that biodiversity of the bacteria and fungi which transform nutrients is related to the diversity of substrates, which in turn would be determined by the diversity of the above-ground plant community providing substrates to the soil through senescence and litter-fall. One rare study lends support to this argument. McArthur *et al.* (1988) found that the diversity of the soil bacterium *Burkholderia cepacia*, which is often associated with rhizospheres of plants, increased in relation to the diversity of substrates. Rapid fluctuations in population size of particular groups are likely to mask underlying trends in species as, although a species or group may be vanishingly rare in relation to the total number of microbial propagules, they could still be represented by hundreds or thousands of individuals. It is difficult to detect marked bacteria when they are present in numbers less than 100 g^{-1} soil (Thompson *et al.*, 1992). Yet groups such as the nitrifying bacteria can be highly responsive to addition of substrates. When NH_4-N is added to soil, a strong lag phase of several days is often seen before substantial nitrification activity occurs (Macdonald, 1979, 1986). Nitrifiers can be isolated by cultivation on selective media, but a community analysis of extracted DNA or an immediate physiological assay on the soil (such as the Biolog method) would not detect such organisms. Equally, given the small sample sizes in relation to the total number of species present, measured responses in diversity of bacteria in the face of stress may reflect the change in dominance of particular species rather than changes in the total species diversity.

In one example for which repeated additions of sewage sludge led to accumulation of heavy metals in soil, diversity of bacteria determined by use of recalcitrant substrates was strongly reduced with increasing heavy metal concentration (Reber, 1992). Across the same gradient of heavy metal concentration, diversity within a species of *Rhizobium* exhibited a 'humpbacked' relationship with increasing diversity associated with slight increases in metal concentrations followed by strong reductions in diversity only when metal concentrations caused populations of rhizobia to crash (A. Lakzian, A. Turner, P. Murphy, J.L. Beynon and K.E. Giller, unpublished observations). At the two poles of this gradient, diversity of bacterial species (estimated using reannealling curves on extracted DNA) indicated a fall in species numbers from 5000 to less than 500 (R.A. Sandaa and V. Torsvik, 1997, unpublished results). In contrast, a study in which the hypothesis tested was that the diversity of rhizobia nodulating *Phaseolus vulgaris* would be decreased in an acid soil compared with a soil of near neutral pH, revealed that the biodiversity in the acid soil was actually greater, but the dominant species differed between the two soils (Anyango *et al.*, 1995). We now have an abundance of new methods emerging which can be used for study of biodiversity of nutrient-transforming bacteria and fungi, but a very limited number of instances where such methods have been applied to test specific hypotheses on the role of biodiversity.

Soil-associated herbivores

In the context of this chapter, a herbivore may be defined as a species with one or more life stages that reside in the soil for the primary purpose of nutrition by feeding on (vascular) plant roots or other underground plant-associated structures. As such, a herbivorous species may be soil dwelling for its entire life cycle, but commonly in groups such as arthropods one or more life stages occur above ground. Considering the richness, abundance and diversity of life-styles in above-ground systems, relatively few insect species have exploited the underground parts of plants as food resources. Only 10 of the 26 orders of insects are well represented as below-ground herbivores. Even in these large orders, below-ground herbivory is only well developed in restricted families or subfamilies, as Ritcher (1958) has shown in the Scarabaeidae. This paucity of soil-dwelling herbivorous insects is remarkable given that in many ecosystems NPP is greater below ground than above ground (Caldwell, 1987; Eissenstat and Yanai, 1997). Other arthropod groups represented among the herbivore guild in soil include Acarina and Symphyla, which, like their insect counterparts, are generally more common above ground. Similarly, Gastropoda occur primarily above ground, but members from some families, such as Agriolimacidae, Milacidae and Vaginulidae, are represented in the soil herbivore guild.

Insect species in the soil-dwelling herbivore guild have narrower geographical ranges than microfaunal elements, which tend to have greater passive dispersal tendencies and are thus widely distributed. As a consequence, there is usually high turnover in species and functional composition of arthropod herbivores over relatively short geographic distances and/or short segments of environmental gradients. The net result is that community-level response documented for one particular place and management situation may not readily be extrapolated to other places and situations. Furthermore, the lower species diversity in below-ground mesofaunal and macrofaunal herbivore communities, relative to those above ground or in the soil microfauna, may indicate low functional redundancy. Disturbance effects on one or a small number of species are therefore often manifested at the community level.

In contrast with the above groups, richness and abundance in nematodes is higher below ground than above ground in terrestrial ecosystems, primarily because the free-living stages of these organisms depend on water films for survival. Root herbivory is, however, confined to only two of the 17 recognized nematode orders (Hooper, 1978), or three according to a recent classification (Blaxter *et al.*, 1998). As with soil insect herbivores, diversity is usually low in the nematode herbivore guild, relative to detritivores, unless special conditions favour proliferation of one or more herbivore species. However, herbivorous nematodes numerically dominate the soil herbivore fauna and are major consumers of NNP in many ecosystems (Scott *et al.*, 1979). Furthermore, nematodes tend to be widely dispersed, at least at the generic level, which greatly facilitates comparative studies across regions or ecosystems and management systems.

Most community or food web studies on the disturbance effects of agricultural practices have not adequately addressed the below-ground herbivore component. Even when this guild is recognized, data are often only gathered for selected species, usually in the macrofaunal insect or microfaunal nematode component. Commonly, changes in the abundance in one or more life stages, relative to untreated reference sites, is interpreted and extrapolated to suggest that the management practice is adverse or beneficial to the organism in question. Little consideration is given to the distribution of mortality through the life cycle and little appreciation that marked changes in mortality in a particular part of the life cycle may have no real impact on the population trend of the animal if variation in that stage is not critical to population regulation (i.e. an upper limit imposed on their growth via density-dependent feedback). Disturbance effects on population dynamics, and thus intergenerational trends in abundance, are dependent on the level of irreplaceable mortality (Thompson, 1955; Morris, 1965) imposed by that disturbance. Disturbance that alters the abundance of one or more natural enemies may, for example, have profound effects on the temporal stability of abundance in a herbivore whose population is naturally regulated by top-down trophic interactions, either singly or in concert with other regulatory processes. Conversely, such changes in natural enemy abundance will have little impact on a herbivore whose populations are primarily regulated by food resource. A key concept, often overlooked in discussions on soil bio-diversity, is that the intergenerational trend in the abundance of the soil-dwelling stage may, in animal groups such as insects, be entirely regulated by population processes acting on above-ground life stages. Interestingly, the most robust data on impacts of agricultural practice on insect herbivores come from the plant protection literature. The influence of agricultural practices on individual invertebrate pest species, or selected groups of pest species, has often been quantified through life-table analyses and manipulation experiments during the process of development of biological control and integrated pest management programmes. The strength of these quantitative approaches is in their explicit recognition that population processes in insect herbivores are strongly influenced by multitrophic interactions and that changes in abundance of an individual species, in response to agricultural practice, will invariably be a manifestation of changes in species interactions.

Multitrophic interactions are also central to the ecology of herbivorous nematodes. However, because of technical difficulties with the identification of distinct generational cohorts and the mortalities acting on various stages within these cohorts, most studies have inferred effects on population processes simply by comparison of life-stage abundances at differentially treated sites. Such approaches are less informative about the processes pivotal to population dynamics and consequently there is only skeletal understanding of the trophic interactions regulating intergenerational change in species abundance (Freckman and Caswell, 1985; Yeates, 1996; Yeates and Wardle, 1996). However, the concept of community symmetry/maturity, not strongly featured in soil insect ecology, has long been applied to soil nematology. Indices developed from perceived position of taxa in functional groups (Yeates *et al.*, 1993b) and on an *r–K* strategy gradient (Bongers, 1990; Yeates, 1994) have been applied in order to document shifts in community structure under various agricultural management regimes (e.g. Freckman and Ettema, 1993; Wasilewska, 1995; Todd, 1996; McSorley, 1997).

The multitrophic interactions of interest to our discussion are those that affect rate of feeding, growth and survival of the herbivores. It has long been contested that the (apparent) low herbivory loads in most natural systems are partly a consequence of the array of chemical defence strategies adopted by plants: antifeedants, digestibility-reducing compounds and toxic secondary compounds. Thus, herbivores, particularly phytophagous insects, were regarded as being dominated by vertical intertrophic controls, rather than horizontal intratrophic interactions such as competition (Hairston *et al.*, 1960; Fowler and Lawton, 1985). Insect herbivores were vividly described as caught 'between the Devil and the deep blue sea' (Lawton and McNeill, 1979), attacked by predators and parasites from the trophic level above, but also constrained by nutritionally poor or robustly defended food supply. There is increasing evidence that invertebrate herbivores are resource limited (Faeth, 1987; Ohgushi, 1992). Experiments and quantitative field ecology over the last decade have shown that it is often wrong to measure the herbivores' food simply in terms of biomass; we now know that much of this 'super-abundant resource' is highly variable in space and time and that much of it is effectively unavailable to the herbivore (e.g. Barker *et al.*, 1989). In these resource-limited systems, intraspecific competition can be a key component of population regulation, with 'contest' being strongly stabilizing and 'scramble' competition being relatively destabilizing (May, 1973; Hassell, 1978).

Theory on plant–herbivore interactions has been based almost entirely on the above-ground component, simply because observation and robust data have not been available for the below-ground herbivores. Root-feeding insects display a range of host specificity similar to their foliar-inhabiting counterparts, and a complex guild of herbivores that include both generalists and specialists is often associated with the underground structures of an individual plant species in a region (e.g. Müller, 1989; Müller *et al.*, 1989). The available data suggest that, in agricultural settings, generalists dominate herbivore guilds in soil, but many exhibit preferences, as is well known in Scarabaeidae (Ueckert, 1979) and tychenchid nematodes (Yeates, 1987). Fine roots, whose primary function is the acquisition of water and nutrients with little role in storage and support, are in many ways similar to leaves, whose role is primarily acquisition of carbon. Like leaves, most fine roots typically exhibit determinant growth, extending only a few centimetres after emerging from woody laterals, and never undergo secondary thickening. It is these fine roots that constitute the greater part of underground NPP and the food source for the majority of soil-dwelling herbivores. Fine roots generally have short life spans (Eissenstat and Yanai, 1997) and their ephemeral character is illustrated by the fact that fine roots often have lower longevity than leaves. The ephemeral nature of the resource and the dominance of generalist herbivores are thus in accord with theory on plant–herbivore interactions.

Plants respond chemically and physiologically to herbivore attack and there is now ample evidence that feeding-induced changes in the host plant reduce herbivore 'fitness' (Green and Ryan, 1972; Wratten *et al.*, 1984; Edwards *et al.*, 1992). While spatial segregation is generally believed to moderate competitive effects, herbivores occupying different parts of the plant may still compete by exploiting a common resource such as phloem sap. Aphids feeding on roots, for example, compete with other aphid species galling the leaves without ever coming directly into contact with them (Mcran and Whitham, 1990). Likewise, herbivores that are separated in time are potential competitors too, if they exploit a common resource or induce a persistent response in the plant (Strauss, 1991). Most interactions are asymmetrical, with one species gaining substantial benefit and the other being adversely affected. Root herbivores mostly increase the host plant's stress response and may thereby affect the abundance of some foliar feeders (Masters *et al.*, 1990; Müller-Schärer, 1991). Conversely, defoliation by herbivores above ground can affect below-ground herbivores by inducing shifts in C allocation that may be reflected in new equilibria in root:shoot ratio, root tissue turnover, and root quality (C:N ratio, allelochemicals). Even more complex interactions may occur, such as when foliar herbivory suppresses or stimulates infection of roots by mycorrhizal fungi (Masters *et al.*, 1990; Gehring and Whitham, 1991) that can confer resistance to nematodes or root-feeding insects (Dehne, 1982). Clavicipitaceous fungi that commonly occur as endophytic mutualists in grasses (Clay, 1988, 1996), may produce alkaloids and induce phytoalexins that provide, depending on the particular grass–plant association, defence against above-ground and/or below-ground invertebrate and vertebrate herbivores. Interspecific competition is an important force structuring herbivore communities (Denno *et al.*, 1995) and these interspecific interactions are sensitive to environmental conditions imposed by different agricultural management regimes. A major difficulty in determining the importance of interspecific competition in structuring communities in agroecosystems is that patterns of resource partitioning may reflect more the outcome of historical competitive episodes (the 'ghost of competition past'; Connell, 1980) than current interactions in these new, synthetic communities.

There is increasing interest in diversification of within-field and whole landscapes in an attempt to farm in 'nature's image'. Agricultural diversification is considered to reduce the irruptive population dynamics of herbivores through effects on the movement and reproductive behaviour of the herbivores (resource concentration hypothesis) and, alternatively or additionally, through an enhancement of populations of their natural enemies by providing them with alternative food and shelter (enemies hypothesis) (see also Polaszek et al., Chapter 11, this volume). However, the effects on pest outbreaks of multiple cropping, as a common approach to diversification, have proved to be unpredictable. The enemies hypothesis, as formulated by Root (1973), is too simplistic and does not take into account the diversity of behavioural ecologies that occur among natural enemies and which profoundly influence their effectiveness in different types of habitat (Sheehan, 1986; Vet and Dicke, 1992). Also, in human-modified landscapes the herbivore guilds are often comprised of species 'released' from their natural enemies, either because the habitat requirements for the natural enemy species are no longer met, or the herbivores have been introduced from other biogeographic regions without their enemies. The resource concentration hypothesis predicts that specialist herbivores are more likely to find, remain in, and build up populations in simple habitat patches such as monocultures where their host plants are concentrated, while polyphagous generalists, on the other hand, are more likely to drift away from such patches into surrounding vegetation. While this often holds true, many arthropod herbivores that are pests in agricultural crops evolved in unstable systems and are well adapted to locating host plants in systems more complex than may be achieved in agricultural cropping situations. Many facets of these resource concentration and enemies hypotheses have not been adequately investigated for the below-ground herbivore guild, which is comprised predominantly of generalist species and whose natural enemies are also predominantly generalists.

The population dynamics of individual species are influenced by a complexity of species interactions, within and between trophic levels. It is to be expected that agricultural practices will impact on soil herbivore species and communities in an idiosyncratic manner because the outcome is highly sensitive to the faunal composition and the nature and strength of interactions among species. Often the indirect effects of agricultural practice are larger than direct effects. The actual abundance and diversity of herbivore species on plants, in this case the root system, is not related to breadth of diet. Rather, differences in the properties of plant species or individuals cause variations in their herbivorous faunas (Futuyma and Gould, 1979). These properties are highly sensitive to agricultural practice.

Functional Significance of Soil Biodiversity

In relating the significance of soil biodiversity to production-based agricultural systems, a key question which emerges is whether maintenance and enhancement of the biodiversity of soil organisms is important for maintaining the key ecosystem-level functions (e.g. decomposition, nutrient mineralization) required for crop (and livestock) productivity. This question is allied to the currently topical area in ecology regarding whether enriched biodiversity is essential in maintaining ecosystem properties and services. Although there has been much speculation (and much written) as to whether biodiversity is 'beneficial' for ecosystems, comparatively few experimental studies have directly tackled this issue and the available data are rather ambiguous. A handful of studies have investigated the effects of diversity of above-ground organisms, principally plants, on various ecosystem properties (mostly those directly or indirectly linked to plant productivity); some claim generally beneficial effects (e.g. Naeem *et al.*, 1994; Tilman *et al.*, 1996) while others indicate that ecosystem properties are driven mainly by plant traits and compositional effects (Grime, 1997; Hooper and Vitousek, 1997; Wardle *et al.*, 1997b,c); further the claim has been made that those studies claiming beneficial ecosystem-level effects of diversity have experimental design flaws which weaken their conclusions (Grime, 1997; Huston, 1997; see Wood and Lenné, Chapter 17, this volume).

Relatively little is known about how biodiversity of the components of decomposer subsystem and soil food webs affects soil processes, and this is largely reflective of difficulties in methodology; experiments manipulating diversity of soil organisms are inherently more difficult to set up than those manipulating plant diversity. However, experiments in which diversity of plant residues or litter is varied may help to address this issue. It is widely recognized in above-ground communities that habitat or resource heterogeneity usually contributes to a greater diversity of organisms per unit area (Pimm, 1991; Huston, 1994), and this is supported by two studies which have explicitly investigated this issue in relation to taxonomic diversity of decomposer groups, namely Cryptostigmatid mites (Anderson, 1978) and litter-feeding land snails (Barker and Mayhill, 1999). It is probable, therefore, that agroecosystems which promote a greater heterogeneity of plant residues or organic matter inputs contribute to maintaining a greater diversity of soil organisms.

The impact of resource heterogeneity (and associated soil biodiversity) on key decomposer-related ecosystem-level processes can be investigated by what are becoming known as 'litter-mix' experiments, usually involving litter-bags, in which two or more litter types are placed together in the same litter-bag, as well as corresponding litter-bags containing litter of the component species in monoculture. Such approaches provide some indication about the possible effects of overall soil biodiversity (including diversity of the resource base) on soil processes. Only a handful of studies using such an approach have been performed (and only three prior to 1994) (Table 5.2), and while the majority have focused on forested ecosystems, they provide clear evidence that litter biodiversity effects on ecosystem processes are idiosyncratic, with increasing litter diversity either increasing or decreasing both decomposition and nutrient mineralization depending on the plant species present. This provides support for the view that soil ecosystem function is

Table 5.2. Effects of plant litter diversity on functioning of the decomposer subsystem.

Vegetation types	Effect tested	Response to enriched diversity	Reference
Tree leaf litter	Mixed vs. monoculture litter on decomposition and soil biota	Either positive or negative effects depending on plant species	Chapman *et al.* (1988)
Tree leaf litter	Mixed (2 or 3) species vs. monocultures on processes and soil organisms	No response of decomposition rates but enhanced initial N release. Mixed responses of decomposer organisms	Blair *et al.* (1990)
Components of wheat residue	Mixtures of components of residue vs. residue components on monoculture	CO_2-C release from residues enhanced by mixing components	Collins *et al.* (1990)
Tree and shrub litter	Mixed vs. monoculture litter on decomposition rate	Either positive or negative effects (depending on plant species)	Fyles and Fyles (1993)
Tree litter	Mixed vs. monoculture litter on processes	No response of decomposition rate or N loss	Rustad (1994)
Residues of crop species	Mixtures of stems and leaves vs. components in monoculture	Higher microbial activity and consequent N immobilization	Quemada and Cabrera (1995)
Tree leaf litter	Mixed vs. monoculture litter on processes	Enhanced decomposition rate and concentration of nutrients	Salamanca *et al.* (1995)
Tree leaf litter	Mixed vs. monoculture litter on decomposition and nutrients	Enhanced CO_2-C release from litter; idiosyncratic effects on nutrient status	Briones and Ineson (1996)
Tree prunings	Mixed species vs. monoculture prunings on N release	Idiosyncratic patterns of N mineralization	Handanyato *et al.* (1997)
Tree leaf litter	Mixed vs. monoculture litter on C and N release	Idiosyncratic patterns of C and N release but more positive than negative effects	McTiernan *et al.* (1997)
Tree leaf litter	Mixed vs. monospecific litter on decomposition rates	No effect of litter mixing	Snowcroft (1997)
Tree, grass and weed litter	Mixed (2–8 species) vs. monospecific litter on processes	Idiosyncratic response of decomposition rate and N loss	Wardle *et al.* (1997a)

governed largely by the traits of the dominant plant species present (and thus the dominant contributor of litter input) (Grime, 1997; Wardle *et al.*, 1997a,b) rather than by the species richness of the vegetation. This is probably because litter mixing can have dual effects, resulting in either stimulation or retardation of one litter type by the other, depending on the relative resource qualities of the component species (Seastedt, 1984; Wardle *et al.*, 1997a). It is also relevant here that residue diversity does not necessarily equate to plant taxonomic diversity; Collins *et al.* (1990) and Quemada and Cabrera (1995) both demonstrated that mixing different residues (of vastly differing qualities) from the same crop species could have important non-additive effects on decomposer-related processes. The issue here is that any diversity effects which occur within the decomposer sub-system are more likely to emerge from heterogeneity of resources (i.e. diversity of substrates or forms of organic matter input) than from diversity of plant species *per se.*

To determine the role of soil organisms themselves in decomposition-related processes requires the use of manipulative experiments. This has usually been achieved most reliably by the use of microcosms or mesocosms to which components of the soil fauna have been added. An alternative approach, adding supposedly selective biocides to reduce the contribution of subsets of the decomposer fauna, has limitations because of artefacts created by non-target effects. Several microcosm or mesocosm studies have shown that decomposition, nutrient mineralization and ultimately plant growth are very responsive to all the principal soil animal groups, namely protozoa (e.g. Clarholm, 1985; Alphei *et al.*, 1996), nematodes (e.g. Ingham *et al.*, 1985), microarthropods (e.g. Setälä, 1995) and earthworms (e.g. Lavelle, 1994). This may create important feedbacks in which faunal activity also causes increases in the nutrient concentrations of plant tissues (Setälä and Huhta, 1991), resulting in the production of litter of improved quality, which may in turn promote soil organisms (Wardle and Lavelle, 1997). Such studies demonstrate the potential importance of soil organisms, and the functional significance of structurally diverse soil food webs in maintaining the properties of ecosystems (including agroecosystems) such as net ecosystem productivity. Field manipulation studies of larger organisms, for example through the addition of earthworms (Lavelle, 1994) or exclosure of macroinvertebrate predators in decomposer food webs (Kajak *et al.*, 1993), have also been shown to indirectly influence decomposition-related processes.

While diversity of soil organisms at the functional group level probably has a critical role in maintaining the productivity of most agroecosystems, the situation is less clear when the diversity of organisms within functional groups is considered. However, Andrén *et al.* (1995) presented evidence from a litter-bag experiment that barley straw decomposition could be explained in terms of simple models which did not explicitly incorporate the dynamics of decomposer organisms; during the course of the experiment there were important shifts in the structure and diversity of the decomposer community. Andrén *et al.* interpreted this to mean that the diversity and composition of the decomposer community was therefore unimportant in determining decomposer-related processes. Support for the view of considerable functional redundancy of organisms within trophic levels also emerges from experimental work in microcosms. Mikola and Setälä (1998c) found that manipulation of diversity of bacterial-feeding nematodes was unimportant in determining rates of CO_2 release and mineral nitrogen dynamics in soil, and the patterns which were observed were instead most likely due to the differences in ecophysiological traits across the nematode species. It has also been shown in soil microcosms that reduction of microbial functional diversity, achieved through the imposition of perturbations, does not have predictable consequences for decomposition rates of added litter (Degens, 1998).

Diversity–ecosystem function relationships also remain unclear when microorganisms associated with nutrient transformations are considered. Although we have argued that soil ecology could benefit by embracing and testing ecological theories developed for plants and animal communities (Wardle and Giller, 1996), it is clear that the roles of biodiversity of microorganisms in maintaining ecosystem properties are vastly different from those of above-ground organisms. In plant communities, the main effects on the ecosystem are exerted by dominants rather than by subordinates (Grime, 1998), whereas among the microbial communities species which are extremely rare can confer important functions. For example, only 50 cells g^{-1} soil of compatible rhizobia are required to ensure effective nodulation of a legume (i.e. less than $1/10^8$ of the total bacteria based on plate counts). Other species may be present in very small numbers, but are able to multiply rapidly in reponse to added substrates and hence are able to exert a functional role beyond all proportion to their representation in the total microbial biomass. A major contrast is that plant productivity is dominated by autotrophy whereas soil communities are driven by heterotrophy – activity of photoautotrophs is limited to surface films on the soil, and activity of other autotrophic organisms, such as the nitrifiers, is often limited by lack of substrates.

Increased resilience in the face of perturbations or stress is often assumed to be a major benefit resulting from the 'redundancy' of organisms able to conduct specific tasks – yet there is little empirical evidence to support this. A testable and highly plausible hypothesis is 'the more general the function, the greater the degree of redundancy associated with it'. Thus the likelihood of a functional change resulting from decreased biodiversity would be much greater when a function is the sole responsibility of a specific group of organisms. Extreme cases of this would be the absence of autotrophic nitrification in acid organic soils in temperate regions (Killham, 1986) (but note that active autotrophic nitrification occurs rapidly in many ancient acid soils in the tropics). Poor growth of white clover sown in soils treated with sewage sludge 20 years earlier was shown to be due to lack of N_2 fixation, although the clover plants were nodulated (McGrath *et al.*, 1988).

In this case the diversity of the rhizobial population was reduced to the extent that only a single 'strain' survived, which was ineffective in N_2 fixation with white clover (Giller *et al.*, 1989; Hirsch *et al.*, 1993). In further studies, N_2 fixation in legumes was absent only when toxicity of heavy metals resulted in complete loss of compatible rhizobia from the soils (Chaudri *et al.*, 1993; A. Lakzian, A. Turner, P. Murphy, J.L. Beynon and K.E. Giller, unpublished observations).

Root herbivory also makes a major potential contribution to ecosystem processes. Root production exceeds above-ground productivity in a range of ecosystems, and in many terrestrial ecosystems below-ground herbivores consume a greater proportion of total NPP than do above-ground herbivores (Coleman *et al.*, 1976; Scott *et al.*, 1979). In temperate grasslands, the below-ground standing crop consumed by insects is two to ten times greater than the above-ground mass, although the effects of below-ground herbivory remain largely unseen unless productivity is severely reduced over large areas (French *et al.*, 1979; Masters *et al.*, 1990). Despite this, little is understood as to how the diversity and composition of soil herbivores affects ecosystem functioning and few studies have addressed the ecological impact of root herbivory beyond the immediate impact on the plant being attacked. There is some evidence that root herbivores may influence foliar nutrient contents (e.g. Yeates *et al.*, 1977; Godfrey *et al.*, 1987), which may potentially affect the quality of subsequently produced litter as has often been demonstrated with foliar herbivory (Bardgett *et al.*, 1998). Further, herbivorous invertebrates can substantially reduce the efficacy of nitrogen fixation in the legume–rhizobium association, resulting in reduced plant productivity and reduced ecosystem N input (Watson *et al.*, 1985). Damage to root structures may facilitate transfer of nutrients between companion plant species. Murray and Hatch (1994) demonstrated experimentally that *Sitona* larvae feeding on the nodules and root tips of *Trifolium repens* resulted in marked increases in the nitrogen content of companion *Lolium perenne*, presumably through leakage of N from damaged legume tissues. A major role of soil herbivores is the indirect regulation of microbially dominated mineralization processes through floral–microfloral interactions. An important way of influencing microbial communities is by producing faeces, which are 'resource hot spots' in the soil. High microbial biomass and activity in the faecal material contribute to high carbon and nutrient turnover (Theenhaus and Scheu, 1996a,b). Starting from these 'resource hot spots' microorganisms may spread and accelerate decomposition processes of the surrounding substrate, for example litter material (Herlitzius and Herlitzius, 1977). These hot spots also increase the heterogeneity of the soil system which might be of considerable importance for maintenance of a diverse soil microflora and fauna (Schaefer, 1991).

An important aspect of how biodiversity affects ecosystems is whether enhanced diversity is critical in maintaining the stability of ecosystem-level properties, in terms of ensuring low temporal variability across years, as well as resistance and resilience to external perturbations (see Tilman, 1996). The effects of soil biodiversity on ecosystem stability remain entirely unexplored, although two recent studies based on microcosm experiments (McGrady-Steed *et al.*, 1997; Naeem and Li, 1997) claim to have found evidence that taxonomic diversity of aquatic microorganisms (probably functionally analagous to the bacterial-based components of soil food webs which occupy the aqueous component of soils) enhance stability and reduce spatial and temporal variability of key processes such as CO_2 evolution. However, the difficulties associated with the experimental designs of such studies (see Huston, 1997; Wardle, 1998) still makes these types of results ambiguous. Furthermore, key questions such as the extent to which the composition and diversity of the soil biota may affect such things as the stability of aspects of nutrient cycles, consistency of crop production across years, and crop resistance and resilience to disturbances (e.g. drought, invertebrate pest attack) remain to be investigated.

Conclusions

It is apparent that, for each of the three components of the soil biota that we have considered, aspects of agricultural intensification can alter both the composition and the diversity of the organisms which constitute these components. This is especially apparent when agricultural practices alter the inputs of substrates, or when disturbance or stress regimes are significantly altered. However, these effects tend to be idiosyncratic, and there are numerous instances of both stimulation and reduction of soil taxonomic entities and soil biodiversity resulting from components of agricultural intensification. There is little consistent support from the literature for the view that agricultural practice and intensification has consistently detrimental consequences for soil biodiversity (see also Edwards *et al.*, Chapter 12, and Lenné and Wood, Chapter 18, this volume).

Effects of agricultural intensification on soil organisms are likely to impact upon the processes that these organisms regulate, including decomposition, nutrient turnover and transfer, and consumption of below-ground NPP. However, while there is clear evidence for the role of key functional components (and composition) of the soil biota in affecting ecosystem properties, there is little evidence that diversity of the soil biota *per se* is a determinant of below-ground ecosystem function or the provision of ecosystem services. Indeed, there is emerging evidence that at finer taxonomic resolutions there is considerable redundancy within most conventional functional groupings of decomposer organisms. The role of diversity might, however, be more important for some groups of nutrient-transforming organisms in which only a small number of species are capable of performing a specific function. We recognize the power that comparative ecological studies are likely to have in unravelling the relative degrees of resolution within soil biota at which biodiversity plays a major role. This could be achieved by detailed analysis of different components across the same experimental treatments or environmental gradients. The extent to which soils can be abused, and yet still continue to produce yields, indicates the robust nature of below-ground biodiversity. How much of this below-ground biodiversity is needed to guarantee the provision of nutrients and a favourable physical environment for root growth is yet to be determined.

Acknowledgements

Many thanks to Drs D.E. Bignall and G.W. Yeates for helpful comments on this manuscript. Preparation of this paper was supported by funding from the New Zealand Marsden Fund.

References

Alphei, J., Bonkowski, M. and Scheu, S. (1996) Protozoa, Nematoda and Lumbricidae in the rhizosphere of *Hordelymus europaeus* (Poaceae): faunal interactions, response of microorganisms and effects on plant growth. *Oecologia* 106, 111–126.

Andersen, A. (1987) Regnorme uddrevet med strøm i forsøg med direkte såning og pløjning. *Tidsskrift Planteavlsforsøg* 91, 3–14.

Anderson, J.M. (1978) Inter- and intra-habitat relationships between woodland Cryptostigmata species diversity and the diversity of soil and litter micro-habitats. *Oecologia* 32, 341–348.

Andrén, O., Clarholm, M. and Bengtsson, J. (1995) Biodiversity and species redundancy among litter decomposers. In: Collins, H.P., Robertson, G.P. and Klug, M.J. (eds) *The Significance and Regulation of Soil Biodiversity*. Kluwer, Dordrecht, pp. 141–151.

Anyango, B., Wilson, K., Beynon, J.L. and Giller, K.E. (1995) Diversity of rhizobia nodulating *Phaseolus vulgaris* in two Kenyan soils of contrasting pH. *Applied and Environmental Microbiology* 61, 4016–4021.

Austin, M.P. (1987) Models for the analyses of species' response to environmental gradients. *Vegetatio* 69, 35–45.

Bardgett, R.D. (1996) Potential effects on the soil mycoflora of changes in the UK agricultural policy for upland grasslands. In: Frankland, J.C., Magan, N. and Gadd, G.M. (eds) *Fungi and Environmental Change*. Cambridge University Press, Cambridge, pp. 163–183.

Bardgett, R.D. and Lehmans, D.L. (1995) The effects of cessation of fertiliser application, liming and grazing on microbial biomass and activity in a reseeded upland pasture. *Biology and Fertility of Soils* 19, 148–154.

Bardgett, R.D., Wardle, D.A. and Yeates, G.W. (1998) Linking above-ground and below-ground interactions: how plant responses to foliar herbivory influence soil organisms. *Soil Biology and Biochemistry* 30, 1867–1878.

Barker, G.M. and Mayhill, P.C. (1999) Patterns of diversity and habitat relationships in terrestrial mollusc communities of the Pukeamaru Ecological District, northeastern New Zealand. *Journal of Biogeography* (in press).

Barker, G.M., Pottinger, R.P. and Addison, P.J. (1989) Population dynamics of the Argentine stem weevil (*Listronotus bonariensis*) in pastures of Waikato, New Zealand. *Agriculture, Ecosystems and Environment* 26, 79–115.

Beard, J. (1991) Woodland soil yields a multitude of insects. *New Scientist* 131(1784), 14.

Beare, M.H., Parmelee, R.W., Hendrix, P.F., Cheng, W., Coleman, D.C. and Crossley, D.A. (1992) Microbial and faunal interactions and effects on litter nitrogen and decomposition in agro-ecosystems. *Ecological Monographs* 62, 569–591.

Blair, J.M., Parmelee, R.W. and Beare, M.H. (1990) Decay rates, nutrient fluxes and decomposer communities in single and mixed-species foliar litter. *Ecology* 71, 1976–1985.

Blaxter, M.L., De Ley, P., Garey, J.R., Liu, L.X., Scheldeman, P., Vierstraete, A., Vanfleteren, J.R., Mackey, L.Y., Dorris, M., Frisse, L.M., Vida, J.T. and Thomas, W.K. (1998) A molecular evolutionary framework for the phylum Nematoda. *Nature* 392, 71–75.

Blumberg, A.Y. and Crossley, D.A. (1983) Comparison of soil surface arthropod populations in conventional tillage, no-tillage and oldfield systems. *Agro-Ecosystems* 8, 247–253.

Bongers, T. (1990) The maturity index: an ecological measure of environmental disturbance based on nematode species composition. *Oecologia* 83, 14–19.

Bostrom, S. and Sohlenius, B. (1986) Short-term dynamics of nematode communities in arable soil: influence of a perennial and an annual cropping system. *Pedobiologia* 29, 345–357.

Briones, M.J.I. and Ineson, P. (1996) Decomposition of *Eucalyptus* leaves in litter mixtures. *Soil Biology and Biochemistry* 28, 1381–1388.

Caldwell, M.M. (1987) Competition between roots in natural communities. In: Gregory, P.J., Lake, J.V. and Rose, D.A. (eds) *Root Development and Function*. Cambridge University Press, New York, pp. 167–185.

Chapman, K., Whittaker, J.B. and Heal, O.W. (1988) Metabolic and faunal activity in litters of tree mixtures compared with pure stands. *Agriculture, Ecosystems and Environment* 34, 65–73.

Chaudri, A.M., McGrath, S.P., Giller, K.E., Rietz, E. and Sauerbeck, D. (1993) Enumeration of indigenous *Rhizobium leguminosarum* biovar *trifolii* in soils previously treated with metal-contaminated sewage sludge. *Soil Biology and Biochemistry* 25, 301–309.

Clarholm, M. (1985) Interactions of bacteria, protozoa and plants leading to mineralisation of soil nitrogen. *Soil Biology and Biochemistry* 17, 181–187.

Clay, K. (1988) Clavicipitaceous fungal endophytes of grasses: coevolution and the change from parasitism to mutualism. In: Hawksworth, D.L. and Pirozynski, K. (eds) *Co-evolution of Fungi with Plants and Animals*. Academic Press, London, pp. 79–105.

Clay, K. (1996) Interactions among fungal endophytes, grasses and herbivores. *Researches on Population Ecology* 38, 191–201.

Coleman, D.C., Andrews, R., Ellis, J.E. and Singh, J.S. (1976) Energy flow and partitioning in selected man-managed and natural ecosystems. *Agro-Ecosystems* 3, 45–54.

Collins, H.P., Elliott, L.F., Rickman, R.W., Bezdicek, D.F. and Papendick, R.I. (1990) Decomposition and interactions among wheat residue components. *Soil Science Society of America Journal* 54, 780–785.

Connell, J.H. (1978) Diversity in tropical rain forests and coral reefs. *Science* 199, 1302–1310.

Connell, J.H. (1980) Diversity and the coevolution of competitors, or the ghost of competition past. *Oikos* 35, 131–138.

Degens, B. (1998) Decreases in microbial functional diversity do not result in corresponding changes in decomposition function under different moisture conditions. *Soil Biology and Biochemistry* 30, 1989–2000.

Dehne, H.W. (1982) Interaction between vesicular–arbuscular mycorrhizal fungi and plant pathogens. *Phytopathology* 72, 1115–1119.

Denno, R.F., McClure, M.S. and Ott, J.R. (1995) Interspecific interactions in phytophagous insects: competition re-examined and resurrected. *Annual Review of Entomology* 40, 297–331.

De Ruiter, P.C., Neutel, A.-M. and Moore, J.C. (1994) Modelling food webs and nutrient cycling in agro-ecosystems. *Trends in Ecology and Evolution* 9, 378–383.

De Ruiter, P.C., Neutel, A.-M. and Moore, J.C. (1995) Energetics, patterns of interaction strengths and stability in real ecosystems. *Science* 269, 1257–1260.

Doane, J.F. and Dondale, C.D. (1979) Seasonal captures of spiders (Araneae) in a wheat field and its grassy borders in central Saskatchewan. *Canadian Entomologist* 111, 439–445.

Domsch, K.H., Jagnow, G. and Anderson, T.-H. (1983) An ecological concept for the assessment of side-effects of agrochemicals on microorganisms. *Residue Reviews* 86, 65–105.

Domurat, K. and Kozlowska, J. (1974) Comparative investigations on the communities of nematodes of cultivated fields and neighbouring barrens. *Zeszyty Problemowe Postepow Nauk Rolniczych* 154, 191–199.

Edwards, P.J., Wratten, S.D. and Parker, L. (1992) The ecological significance of rapid wound-induced changes in plants: insect grazing and plant competition. *Oecologia* 91, 266–272.

Eissenstat, D.M. and Yanai, R.D. (1997) The ecology of root lifespan. *Advances in Ecological Research* 27, 1–60.

Emmanuel, N., Curry, J.P. and Evans, G.O. (1985) The soil *Acari* of barley plots with different cultural treatments. *Experimental and Applied Acarology* 1, 101–113.

Faeth, S.H. (1987) Community structure and folivorous insect outbreaks: the roles of vertical and hortizontal interactions. In: Barbosa, P. and Schultz, J.C. (eds) *Insect Outbreaks*. Academic Press, London, pp. 135–171.

Findlay, S., Carreiro, M., Krishik, V. and Jones, C.G. (1996) Effects of damage to living plants on leaf litter quality. *Ecological Applications* 6, 269–275.

Fowler, S.V. and Lawton, J.H. (1985) Rapidly induced defences and talking trees: the Devil's Advocate position. *American Naturalist* 126, 181–195.

Freckman, D.W. and Caswell, E.P. (1985) The ecology of nematodes in agroecosystems. *Annual Review of Phytopathology* 23, 275–296.

Freckman, D.W. and Ettema, C.H. (1993). Assessing nematode communities in agroecosystems of varying human disturbance. *Agriculture, Ecosystems and Environment* 45, 239–265.

French, N.R., Steinhorst, R.K. and Swift, D.M. (1979) Perspectives in grassland ecology. In: French, N.R. (ed.) *Perspectives in Grassland Ecology*. Springer-Verlag, New York, pp. 59–87.

Futuyma, D.J. and Gould, F. (1979) Associations of plants and insects in a deciduous forest. *Ecological Monographs* 49, 33–50.

Fyles, J.W. and Fyles, I.H. (1993) Interaction of Douglas fir with red alder and salal foliage litter during decomposition. *Canadian Journal of Forest Research* 23, 358–361.

Gehring, C.A. and Whitham, T.G. (1991) Herbivore-driven mycorrhizal mutualism in insect-susceptible pinyon pine. *Nature* 353, 556–557.

Giller, K.E., McGrath, S.P. and Hirsch, P.R. (1989) Absence of nitrogen fixation in clover grown on soil subject to long term contamination with heavy metals is due to survival of only ineffective *Rhizobium*. *Soil Biology and Biochemistry* 21, 841–848.

Giller, K.E., Beare, M.H., Lavelle, P., Izac, A.M.-N. and Swift, M.J. (1997) Agricultural intensification, soil biodiversity and agroecosystem function. *Applied Soil Ecology* 6, 3–16.

Godfrey, L.D., Yeargan, K.V. and Muntifering, R.B. (1987) Digestibility, protein content and nutrient yields of alfalfa stressed by selected early season insect pests and diseases. *Journal of Economic Entomology* 80, 257–262.

Green, T.R. and Ryan, C.A. (1972) Wound-induced proteinase inhibitor in plant leaves: a possible defence mechanism against insects. *Science* 175, 776–777.

Gressel, N. and McColl, J.G. (1997) Phosphorus mineralisation and organic matter decomposition: a critical review. In: Cadisch, G. and Giller, K.E. (eds) *Driven by Nature. Plant Litter Quality and Decomposition*. CAB International, Wallingford, pp. 297–309.

Griffiths, B.S., Welschen, R., Van Arendonk, J.J.C.M. and Lambers, H. (1992) The effect of nitrate-nitrogen on bacteria and the bacterial-feeding fauna in the rhizosphere of different grass species. *Oecologia* 91, 253–259.

Grime, J.P. (1979) *Plant Strategies and Vegetation Processes*. John Wiley & Sons, Chichester.

Grime, J.P. (1997) Biodiversity and ecosystem function: the debate deepens. *Science* 277, 1260–1261.

Grime, J.P. (1998) Benefits of plant diversity to ecosystems: immediate, filter and founder effects. *Journal of Ecology* 86, 902–910.

Groffman, P.M., Eagan, P., Sullivan, W.M. and Lemunyon, J. (1996) Grass species and soil type effects on microbial biomass and activity. *Plant and Soil* 183, 61–67.

Hairston, N.G., Smith, F.E. and Slobodkin, L.B. (1960) Community structure, population control and competition. *American Naturalist* 44, 421–425.

Handanyanto, E., Cadisch, G. and Giller, K.E. (1997) Regulating N mineralisation from plant residues by manipulation of quality. In: Cadisch, G. and Giller, K.E. (eds) *Driven by Nature. Plant Litter Quality and Decomposition*. CAB International, Wallingford, pp. 175–185.

Hassell, M.P. (1978) *The Dynamics of Arthropod Predator–Prey Systems*. Princeton University Press, Princeton, New Jersey.

Haukka, J. (1988) Effects of various cultivation methods on earthworm biomass. *Annales Agriculturae Fenniae* 27, 263–269.

Head, I.M., Horns, W.D., Embley, T.M., McCarthy, A.J. and Saunders, J.R. (1993) The phylogeny of autotrophic ammonia-oxidising bacteria as determined by 16S ribosomal DNA sequence analysis. *Journal of General Microbiology* 139, 1147–1153.

Hendrix, P.F. and Parmelee, R.W. (1985) Decomposition, nutrient loss and microarthropod densities in herbicide-treated grass litter in a Georgia piedmont agroecosystem. *Soil Biology and Biochemistry* 17, 421–428.

Hendrix, P.F., Parmelee, R.W., Crossley, D.A., Coleman, D.C., Odum, E.P. and Groffman, P.M. (1986) Detritus food webs in conventional and no-tillage agroecosystems. *BioScience* 36, 374–380.

Herlitzius, R. and Herlitzius, H. (1977) Streuabbau in Laubwäldern. *Oecologia* 30, 147–171.

Hirsch, P.R., Jones, M.J., McGrath, S.P. and Giller, K.E. (1993) Heavy-metals from past applications of sewage sludge decrease the genetic diversity of *Rhizobium leguminosarum* biovar *trifolii* populations. *Soil Biology and Biochemistry* 25, 1485–1490.

Holland, J.N. (1996) Effects of above-ground herbivory on soil microbial biomass in conventional and no-tillage agroecosystems. *Applied Soil Ecology* 2, 275–279.

Hooper, D.J. (1978) Structure and classification of nematodes. In: Southey, J.F. (ed.) *Plant Nematology.* HMSO, London, for Ministry of Agriculture, Fisheries and Food, pp. 3–45.

Hooper, D.U. and Vitousek, P.M. (1997) The effects of plant composition and diversity on ecosystem processes. *Science* 277, 1302–1305.

House, G.J. and All, J.N. (1981) Carabid beetles in soybean ecosystems. *Environmental Entomology* 10, 194–196.

Huhta, V. and Raatikainen, M. (1974) Spider communities of leys in winter cereal fields in Finland. *Annales Zoologisches Fennici* 11, 97–104.

Huston, M.A. (1994) *Biological Diversity. The Coexistence of Species on Changing Landscapes.* Cambridge University Press, Cambridge.

Huston, M.A. (1997) Hidden treatments in ecological experiments: evaluating the ecosystem function of biodiversity. *Oecologia* 110, 449–460.

Ingham, R.E. and Detling, J.K. (1984) Plant–herbivore interactions in a North American mixed grass prairie. III. Soil nematode populations and root biomass on *Cynomys ludovicianus* colonies and adjacent uncolonised areas. *Oecologia* 63, 307–313.

Ingham, R.E., Trofymow, J.A., Ingham, E.R. and Coleman, D.C. (1985) Interactions of bacteria, fungi and their nematode grazers on nutrient cycling and plant growth. *Ecological Monographs* 55, 119–140.

Kajak, A., Chmielewski, K., Kaczmarek, M. and Rembialkowska, E. (1993) Experimental studies on the effects of epigeic predators on organic matter decomposition processes in managed peat grasslands. *Polish Ecological Studies* 17, 289–310.

Killham, K. (1986) Heterotrophic nitrification. In: Prosser, J.I. (ed.) *Nitrification.* IRL Press, Oxford, pp. 117–126.

Lagerlöf, J. and Andrén, O. (1988) Abundance and activity of soil mites (*Acari*) in four cropping systems. *Pedobiologia* 32, 129–145.

Lagerlöf, J. and Andrén, O. (1991) Abundance and activity of *Collembola, Protura* and *Diplura* (Insecta, Apterygota) in four cropping systems. *Pedobiologia* 35, 337–350.

Lavelle, P. (1994) Faunal activities and soil processes: adaptive strategies that determine ecosystem function. In: *XV ISSS Congress Proceedings,* Vol. 1, *Introductory Conferences.* Acapulco, Mexico, pp. 189–220.

Lavelle, P., Blanchart, E., Martin, A., Martin, S., Spain, A.V., Toutain, F., Barois, I.

and Schaefer, R. (1993) A hierarchical model for decomposition in terrestrial ecosystems: applications to soils of the humid tropics. *Biotropica* 25, 130–150.

Lawton, J.H. and McNeill, S. (1979) Between the Devil and the deep blue sea: on the problem of being a herbivore. In: Anderson, R.M., Turner, B.D. and Taylor, L.R. (eds) *Population Dynamics*. Blackwell, London, pp. 223–244.

Lawton, J.H., Bignall, D.E., Bolton, B., Bloemers, G.F., Eggleton, P., Hammond, P.M., Hodda, M., Holt, R.D., Larsen, T.B., Mawdsley, N.A., Stork, N.E., Srivastava, D.S. and Watt, A.D. (1998) Biodiversity inventories, indicator taxa and effects of habitat modification in tropical forest. *Nature* 391, 72–76.

Macdonald, R.M. (1979) Population dynamics of the nitrifying bacterium *Nitrosolobus* in soil. *Journal of Applied Ecology* 16, 529–535.

Macdonald, R.M. (1986) Nitrification in soil: an introductory history. In: Prosser, J.I. (ed.) *Nitrification*. IRL Press, Oxford, pp. 1–16.

Mahn, E.-G. and Kastner, A. (1985) Effects of herbicide stress on weed communities and soil nematodes in agro-ecosystems. *Oikos* 44, 185–190.

Martínez-Romero, E. (1994) Recent developments in *Rhizobium* taxonomy. *Plant and Soil* 161, 11–20.

Masters, G.J., Brown, V.K. and Gange, A.C. (1990) Plant mediated interactions between above- and belowground insect herbivores. *Oikos* 66, 148–151.

May, R.M. (1973) *Stability and Complexity in Model Ecosystems*. Princeton University Press, Princeton, New Jersey.

McArthur, J.V., Kovacic, D.A. and Smith, M.H. (1988) Genetic diversity in natural populations of a soil bacterium across a landscape gradient. *Proceedings of the National Academy of Sciences, USA* 85, 9621–9624.

McGrady-Steed, J., Harris, P.M. and Morin, P. (1997) Biodiversity regulates ecosystem predictability. *Nature* 390, 162–165.

McGrath, S.P., Brookes, P.C. and Giller, K.E. (1988) Effects of potentially toxic metals in soil derived from past applications of sewage sludge on nitrogen fixation by *Trifolium repens* L. *Soil Biology and Biochemistry* 20, 415–424.

Mcran, N.A. and Whitham, T.G. (1990) Interspecific competition between root-feeding and leaf-galling aphids mediated by host-plant resistance. *Ecology* 71, 1050–1058.

McSorley, R. (1997) Relationship of crop and rainfall to soil nematode community structure in perennial agroecosystems. *Applied Soil Ecology* 6, 147–159.

McTiernan, K.B., Ineson, P. and Coward, P.A. (1997) Respiration and nutrient release from tree leaf litter mixtures. *Oikos* 78, 527–538.

Merrill, E.H., Stanton, N.L. and Hak, J.C. (1994) Response of bluebunch wheatgrass, Idaho fescue and nematodes to ungulate grazing in Yellowstone National Park. *Oikos* 69, 231–240.

Mikola, J. and Setälä, H. (1998a) Productivity and trophic-level biomasses in a microbial-based soil food web. *Oikos* 82, 158–168.

Mikola, J. and Setälä, H. (1998b) No evidence of trophic cascades in an experimental microbial-based soil food web. *Ecology* 79, 153–164.

Mikola, J. and Setälä, H. (1998c) Relating species diversity to ecosystem functioning – mechanistic backgrounds and an experimental approach with a decomposer food web. *Oikos* 83, 180–194.

Milchunas, D.G. and Lauenroth, W.K. (1993) Quantitative effects of grazing on vegetation and soils over a global range of environments. *Ecological Monographs* 63, 327–366.

Morris, R.F. (1965) Contemporaneous mortality factors in population dynamics. *Canadian Entomologist* 97, 1173–1184.

Müller, H. (1989) Structural analysis of the phytophagous insect guilds associated with the roots of *Centaurea maculosa* Lam., *C. diffusa* Lam. and *C. vallesiaca* Jordan in Europe. 1. Field observations. *Oecologia* 78, 41–52.

Müller, H., Stinson, C.S.A., Marquardt, K. and Schroeder, D. (1989) The entomofaunas of roots of *Centaurea maculosa* Lam., *C. diffusa* Lam. and *C. vallesiaca* Jordan in Europe. *Journal of Applied Entomology* 107, 83–95.

Müller-Schärer, H. (1991) The impact of root herbivores as a function of plant density and competition: survival, growth, and fecundity of *Centaurea maculosa* in field plots. *Journal of Applied Ecology* 28, 759–776.

Murray, P.J. and Hatch, D.J. (1994) *Sitona* weevils (Coleoptera: Curculionidae) as agents for rapid transfer of nitrogen from white clover (*Trifolium repens* L.) to perennial ryegrass (*Lolium perenne* L.). *Annals of Applied Biology* 125, 29–33.

Naeem, S. and Li, S. (1997) Biodiversity enhances ecosystem reliability. *Nature* 390, 507–509.

Naeem, S., Thompson, L.J., Lawler, S.P., Lawton, J.H. and Woodfin, R.M. (1994) Declining biodiversity can alter the performance of ecosystems. *Nature* 368, 734–737.

Ohgushi, T. (1992) Resource limitation on insect herbivore populations. In: Hunter, M.D., Ohgushi, T. and Price, P.W. (eds) *Effects of Resource Distribution on Animal–Plant Interactions*. Academic Press, San Diego, pp. 199–241.

Parton, W.J., Schimel, D.S., Cole, C.V. and Ojima, D.S. (1987) Analysis of factors controlling soil organic matter level in Great Plains grasslands. *Soil Science Society of America Journal* 51, 1173–1179.

Pastor, J., Dewey, R.J., Naiman, R.J., McInnes, R.J. and Cohen, Y. (1988) Moose, microbes and the Boreal forest. *BioScience* 38, 770–777.

Paustian, K., Andrén, O., Clarholm, M., Hansson, A.-C., Johansson, G., Lagerlöf, J., Lindberg, T., Pettersson, R. and Sohlenius, B. (1990) Carbon and nitrogen budgets in four agro-ecosystems with annual and perennial crops, with and without N-fertilisation. *Journal of Applied Ecology* 27, 60–84.

Pimm, S. (1991) *The Balance of Nature*. University of Chicago Press, Chicago.

Powlson, D.S. (1994) The soil microbial biomass: before, beyond and back. In: Ritz, K., Dighton, J. and Giller, K.E. (eds) *Beyond the Biomass. Compositional and Functional Analysis of Soil Microbial Communities*. John Wiley & Sons, Chichester, pp. 3–20.

Quemada, M. and Cabrera, M.L. (1995) Carbon and nitrogen from leaves and stems of four cover crops. *Soil Science Society of America Journal* 59, 471–477.

Reber, H.H. (1992) Simultaneous estimates of the diversity and the degradative capability of heavy-metal-affected soil bacterial communities. *Biology and Fertility of Soils* 13, 181–186.

Ritcher, P.O. (1958) Biology of the Scarabaeidae. *Annual Review of Entomology* 3, 311–334.

Robertson, L.N., Kettle, B.A. and Simpson, G.B. (1994) The influence of tillage practices on soil macrofauna in a semi-arid agroecosystem in northeastern Australia. *Agriculture, Ecosystems and Environment* 48, 149–156.

Root, R.B. (1973) Organization of a plant–arthropod association in simple and diverse habitats: the fauna of collards (*Brassica oleracea*). *Ecological Monographs* 43, 95–124.

Rovira, A.D., Smetten, K.R.J. and Lee, K.E. (1987) Effect of rotation and conservation tillage on earthworms in a red-brown earth under peat. *Australian Journal of Agricultural Research* 38, 829–834.

Rustad, L. (1994) Element dynamics along a decay continuum in a red spruce ecosystem in Maine, U.S.A. *Ecology* 75, 867–879.

Salamanca, E., Kaneko, N. and Katagiri, S. (1995) Mass loss and release of mixed-oak, pine and dwarf bamboo leaf litters. In: *Driven by Nature. Plant Litter Quality and Decomposition* (International Conference, Wye College, University of London), Abstracts, p. 32.

Santos, P.F., Phillips, J. and Whitford, W.G. (1981) The role of mites and nematodes in early stages of buried litter decomposition in a desert. *Ecology* 62, 664–669.

Schaefer, M. (1991) The animal community: diversity and resources. In: Rödrig, E. and Ulrich, M. (eds) *Temperate Deciduous Forests (Ecosystems of the World, IV)*. Elsevier, Amsterdam, pp. 51–120.

Scott, J.A., French, N.R. and Leetham, J.W. (1979) Patterns of consumption in grasslands. In: French, R.N. (ed.) *Perspectives in Grassland Ecology*. Springer-Verlag, New York, pp. 89–105.

Seagle, S.W., McNaughton, S.J. and Ruess, R.W. (1992) Simulated effects of grazing on soil nitrogen and mineralisation in contrasting Serengeti grasslands. *Ecology* 73, 1105–1123.

Seastedt, T.R. (1984) The role of microarthropods in decomposition and mineralisation processes. *Annual Review of Entomology* 29, 25–46.

Seastedt, T.R., Ramundo, R.A. and Hayes, D.C. (1988) Maximisation of densities of soil animals by foliage herbivory: empirical evidence, graphical and conceptual models. *Oikos* 51, 243–248.

Setälä, H. (1995) Growth of birch and pine seedlings in relation to grazing by soil fauna on ectomycorrhizal fungi. *Ecology* 76, 1844–1851.

Setälä, H. and Huhta, V. (1991) Soil fauna increases *Betula pendula* growth: laboratory experiments with coniferous forest floor. *Ecology* 72, 665–671.

Sheehan, W. (1986) Response by specialists and generalist natural enemies to agroecosystem diversification: a selective review. *Environmental Entomology* 15, 456–461.

Smith, S.E. and Read, D.J. (1997) *Mycorrhizal Symbiosis*. Academic Press, London.

Snowcroft, P.G. (1997) Mass and nutrient dynamics of decaying litter from *Passiflora mollissima* and selected native species in a Hawaiian montane rain forest. *Journal of Tropical Ecology* 13, 407–426.

Sohlenius, B., Bostrom, S. and Sandor, A. (1987) Carbon and nitrogen budgets of nematodes in arable soils. *Biology and Fertility of Soils* 6, 1–8.

Stahl, D.A. (1997) Molecular approaches for the measurement of density, diversity and phylogeny. In: Hurst, C.J. (ed.) *Manual of Environmental Microbiology*. ASM Press, Washington, DC, pp. 102–114.

Stassart, P. and Gregorie-Wibo, C. (1983) Influence du travail du sol sur des populations de Carabides en grande culture, resultats preliminaires. *Mededelingen Rijsfaculteit Landbouwwetenschappen – Ghent* 48, 465–474.

Strauss, S.Y. (1991) Direct, indirect, and cumulative effects of three native herbivores on a shared host plant. *Ecology* 72, 543–558.

Theenhaus, A. and Scheu, S. (1996a) Successional changes in microbial biomass, activity and nutrient status in faecal material of the slug *Arion rufus* (Gastropoda) deposited after feeding on different plant materials. *Soil Biology and Biochemistry* 28, 569–577.

Theenhaus, A. and Scheu, S. (1996b) The influence of slug (*Arion rufus*) mucus and cast material addition on microbial biomass, respiration, and nutrient cycling in beech leaf litter. *Biology and Fertility of Soils* 23, 80–85.

Thompson, I.P., Young, C.S., Cook, K.A., Lethbridge, G. and Burns, R.G. (1992) Survival of two ecologically distinct bacteria (*Flavobacterium* and *Arthrobacter*) in unplanted and rhizophere soil: field studies. *Soil Biology and Biochemistry* 24, 1–14.

Thompson, W.R. (1955) Mortality factors acting in a sequence. *Canadian Entomologist* 87, 264–275.

Tilman, D. (1996) Biodiversity: population versus ecosystem stability. *Ecology* 77, 350–363.

Tilman, D., Wedin, D. and Knops, J. (1996) Productivity and sustainability influenced by biodiversity in grasslands. *Nature* 379, 718–720.

Todd, T.C. (1996) Effects of management practices on nematode community structure in tallgrass prairie. *Applied Soil Ecology* 3, 235–246.

Torsvik, V., Goksøyr, J. and Daae, F.L. (1990) High diversity in DNA of soil bacteria. *Applied and Environmental Microbiology* 56, 782–787.

Torsvik, V., Goksøyr, J., Daae, F.L., Sørheim, R., Michalsen, J. and Salte, K. (1994) Use of DNA analysis to determine the diversity of microbial communities. In: Ritz, K., Dighton, J. and Giller, K.E. (eds) *Beyond the Biomass. Compositional and Functional Analysis of Soil Microbial Communities*. John Wiley & Sons, Chichester, pp. 39–48.

Ueckert, D.N. (1979) Impact of a white grub (*Phyllophaga crinita*) on a short-grass community and evaluation of selected rehabilitation practices. *Journal of Range Management* 32, 445–448.

Van Berkum, P., Beyene, D., Bao, G., Cambell, T.A. and Eardly, B.D. (1998) *Rhizobium mongolense* sp. nov. is one of three rhizobial geneotypes identified which nodulate and form nitrogen-fixing symbioses with *Medicago ruthenica* (L.) Ledebour. *International Journal of Systematic Bacteriology* 48, 13–22.

Van Veen, J.A., Merckx, R. and Van de Geijn, S.C. (1989) Plant and soil related controls of the flow of carbon from the soil microbial biomass. *Plant and Soil* 115, 179–188.

Vet, L.E.M. and Dicke, M. (1992) Ecology of infochemical use by natural enemies in a tritrophic context. *Annual Review of Entomology* 37, 141–172.

Von Klinger, K. (1987) Auswirkungen eingesater randstreifen an einem winterweizen-field raubarthropodenfauna und getreideblattlaudefall. *Journal of Applied Entomology* 104, 47–58.

Wacha, A.G. and Tiffany, K.H. (1979) Soil fungi isolated from fields under different tillage and weed control regimes. *Mycologia* 71, 1215–1226.

Wardle, D.A. (1995) Impacts of disturbance on detritus food webs in agro-ecosystems of contrasting tillage and weed management practices. *Advances in Ecological Research* 26, 105–185.

Wardle, D.A. (1998) A more reliable design for biodiversity study? *Nature* 394, 30.

Wardle, D.A. and Barker, G.M. (1997) Competition and herbivory in establishing grasslands: implications for plant biomass, species diversity and soil microbial activity. *Oikos* 80, 570–580.

Wardle, D.A. and Giller, K.E. (1996) The quest for a contemporary ecological dimension to soil biology. *Soil Biology and Biochemistry* 28, 1549–1554.

Wardle, D.A. and Lavelle, P. (1997) Linkages between soil biota, plant litter quality and decomposition. In: Cadisch, G. and Giller, K.E. (eds) *Driven by Nature. Plant Litter Quality and Decomposition.* CAB International, Wallingford, pp. 107–124.

Wardle, D.A. and Parkinson, D. (1991) Relative importance of the effects of 2,4-D, glyphosate and environmental variables on the soil microbial biomass. *Plant and Soil* 122, 29–37.

Wardle, D.A. and Yeates, G.W. (1993) The dual importance of competition and predation as regulatory forces in terrestrial ecosystems: evidence from soil food-webs. *Oecologia* 93, 303–306.

Wardle, D.A., Nicholson, K.S. and Rahman, A. (1995) Ecological effects of the invasive weed species *Senecio jacobaea* L. (ragwort) in a New Zealand pasture. *Agriculture, Ecosystems and Environment* 56, 19–28.

Wardle, D.A., Bonner, K.I. and Nicholson, K.S. (1997a) Biodiversity and plant litter: experimental evidence which does not support the view that enhanced species richness improves ecosystem function. *Oikos* 79, 247–258.

Wardle, D.A., Zackrisson, O., Hörnberg, G. and Gallet, C. (1997b) The influence of island area on ecosystem properties. *Science* 277, 1296–1299.

Wardle, D.A., Zackrisson, O., Hörnberg, G. and Gallet, C. (1997c) Biodiversity and ecosystem properties. *Science* 278, 1867–1869.

Wasilewska, L. (1995) Differences in development of soil nematode communities in single- and multi-species grass experimental treatments. *Applied Soil Ecology* 2, 53–64.

Watson, R.N., Yeates, G.W., Littler, R.A. and Steele, K.W. (1985) Responses in nitrogen fixation and herbage production following pesticide applications on temperate pastures. *Proceedings of the 4th Australasian Conference on Grassland Invertebrate Ecology*, pp. 103–113.

Wintzingerode, F. van, Göbel, U.B. and Stackebrandt, E. (1997) Determination of microbial diversity in environmental samples: pitfalls of PCR-based rRNA analysis. *FEMS Microbiology Reviews* 21, 213–229.

Woese, C.R. (1987) Bacterial evolution. *Microbiological Reviews* 52, 221–271.

Wratten, S.D., Edwards, P.J. and Dunn, I. (1984) Wound-induced changes in the palatability of *Betula pubescens* and *B. pendula. Oecologia* 61, 372–375.

Yeates, G.W. (1979) Soil nematodes in terrestrial ecosystems. *Journal of Nematology* 11, 213–229.

Yeates, G.W. (1987) How plants affect nematodes. *Advances in Ecological Research* 17, 61–113.

Yeates, G.W. (1994) Modification and qualification of the nematode maturity index. *Pedobiologia* 38, 97–101.

Yeates, G.W. (1996) Nematode ecology. *Russian Journal of Nematology* 4, 71–75.

Yeates, G. and Bird, A.F. (1994) Some observations on the influence of agricultural practices on the nematode faunae of some South Australian soils. *Fundamental and Applied Nematology* 17, 133–145.

Yeates, G.W. and Coleman, D.C. (1982) Role of nematodes in decomposition. In: Freckman, D.W. (ed.) *Nematodes in Soil Ecosystems.* University of Texas Press, Austin, Texas, pp. 55–80.

Yeates, G.W. and Hughes, K.A. (1990) The effects of three tillage regimes on plant and soil nematodes in an oats/maize rotation. *Pedobiologia* 34, 379–387.

Yeates, G.W. and Wardle, D.A. (1996) Nematodes as predators and prey: relationships to biological control and soil processes. *Pedobiologia* 40, 43–50.

Yeates, G.W., Ross, D.J., Bridger, B.A. and Visser, T.A. (1977) Influence of the nematodes *Heterodera trifolii* and *Meloidogyne hapla* on nitrogen fixation by white clover under glasshouse conditions. *New Zealand Journal of Agricultural Research* 20, 401–413.

Yeates, G.W., Bongers, T., de Goede, R.G.M., Freckman, D.W. and Georgieva, S.S. (1993a) Feeding habits in soil nematode families and genera – an outline for soil ecologists. *Journal of Nematology* 25, 315–331.

Yeates, G.W., Wardle, D.A. and Watson, R.N. (1993b) Relationships between nematodes, soil microbial biomass and weed management strategies in maize and asparagus cropping systems. *Soil Biology and Biochemistry* 25, 869–876.

Yeates, G.W., Bardgett, R.D., Cook, R., Hobbs, P.J., Bowling, P.J. and Potter, J.F. (1997) Faunal and microbial diversity in three Welsh grassland soils under conventional and organic management regimes. *Journal of Applied Ecology* 34, 453–470.

Young, J.P.W. (1992) Phylogenetic classification of nitrogen fixing organisms. In: Stacey, G., Burris, R.H. and Evans, H.J. (eds) *Biological Nitrogen Fixation*. Chapman & Hall, New York, pp. 43–85.

Young, J.P.W. (1994) Sex and the single cell: the population ecology and genetics of microbes. In: Ritz, K., Dighton, J. and Giller, K.E. (eds) *Beyond the Biomass. Compositional and Functional Analysis of Soil Microbial Communities*. John Wiley & Sons, Chichester, pp. 101–107.

Young, J.P.W. and Haukka, K.E. (1996) Diversity and phylogeny of rhizobia. *New Phytologist* 133, 87–94.

Zak, D.R., Tilman, D., Parmenter, R.R., Rice, C.W., Fisher, F.M., Vose, J., Milchunas, D. and Martin, C.W. (1994) Plant production and soil microorganisms in late-successional ecosystems: a continental-scale study. *Ecology* 75, 2333–2347.

6

Pathogen Biodiversity: Its Nature, Characterization and Consequences

D.J. Allen[1], J.M. Lenné[2] and J.M. Waller[3]

[1]Higher Quantock, Stockland, Honiton, Devon EX14 9DX, UK; [2]Agrobiodiversity International, 13 Herons Quay, Sandside, Milnthorpe, Cumbria LA7 7HN, UK; [3]CABI Bioscience UK Centre, Bakeham Lane, Egham, Surrey TW20 9TY, UK

Introduction

A pathogen is an organism, often a microorganism, virus or viroid, which causes disease (Holliday, 1989). For the purposes of this chapter, we will concern ourselves only with plant pathogens among which fungi and viruses are the most important; other groups of plant pathogens include bacteria, nematodes, phytoplasmas, viroids and a few parasitic algae and angiosperms. Predominant among animal pathogens are bacteria, viruses, mycoplasmas and protozoa; the reader is directed to Hall (1977) for discussion of interactions between pathogens and domesticated animals in agroecosystems. Pathogens constitute only a small fraction of the overall diversity of microorganisms which, with invertebrates, are estimated to comprise about 88% of all species on Earth (Hawksworth and Ritchie, 1993; O'Donnell et al., 1994; see LaSalle, Chapter 7, this volume). The economic significance of plant pathogens lies in their destructive effect on the most important product of agroecosystems: human food. Because of this potential to decrease crop productivity, the importance of pathogen diversity is much better appreciated in agroecosystems than in wild ecosystems.

In this chapter we make no attempt to provide a comprehensive review of the literature. Our scope is necessarily highly selective, to bring out what we believe are the key points while remaining reasonably succinct. We begin by tracing the evolution of disease in plants, drawing attention to contrasts between agroecosystems and wild ecosystems, and between coevolved host–parasite relations and new encounters between pathogens and plants. We then discuss pertinent aspects of host range and host specificity before focusing on the levels of diversity expressed by pathogens within species. Emphasis is given to the concept of physiologic race as the principal unit of infraspecific diversity, before considering the mechanisms by which diversity arises. A brief review of the methods used in the characterization of diversity leads into consideration of functional diversity, the fitness of races and its cost, and the interactions in which races are involved in agriculture. We go on to discuss the dynamics of plant pathogen diversity in cropping systems as a prelude to our appraisal of the implications of diversity for effective management of disease in the world's crops (see Polaszek *et al.*, Chapter 11, this volume).

The Nature and Origin of Diversity Among Pathogens

The nature and extent of species diversity displayed by pathogens

Both the diversity displayed among plant pathogens themselves and in their differing strategies for effective parasitism is immense (Burdon, 1992, 1993), and species-dependent associations are common (Oberwinkler, 1992). Recognition of this extensive diversity is crucial to an understanding of how pathogens and plant hosts interact, especially in agroecosystems. Plant pathogenicity is found among most groups of microorganisms as well as in higher plants, and pathogen diversity can be described in terms of taxonomy, morphology, genetics and function. Life histories display variation in modes of parasitism, of dispersal, of reproduction, and of survival (Burdon, 1992, 1993).

Much has been written on the taxonomic and morphological diversity of microorganisms, including pathogens (Hawksworth, 1991; Hawksworth and Colwell, 1992). Morphology often reflects adaptive and functional strategy (Oberwinkler, 1992), whereas taxonomically and morphologically distinct pathogens often have similar biophysiological strategies, expressed for instance as either specialization or opportunism found in saprophytes, necrotrophs, hemi-biotrophs or biotrophs (Burdon, 1992, 1993; Meyer, 1993), displayed as adaptation to a particular habitat (Durbin, 1959) or temperature regime, or similar symptoms (Table 6.1).

The fungi are the largest group of plant pathogens, although only some 5% of the 7727 plant genera contain them (Table 6.2). Many are relatively obscure and their behaviour unknown, so that the 400 genera of recognized fungal pathogens is probably an underestimate of the parasitic habit among them. For example, of some 350 species of fungi recorded from *Musa* spp. in the herbarium of the International Mycological Institute, only about 20% can be considered to be pathogens in the broad sense, although for many of these the true parasitic nature of their association has not been investigated and most are of little current consequence in terms of damage to their host.

Table 6.1. Diversity in disease symptoms in plants infected by pathogens.

Types of plant disease (after Wheeler, 1969)	Typical pathogens
Postharvest disease: tuber and fruit rots; soft rots	*Penicillium, Rhizopus, Sclerotinia* spp.; *Erwinia carotovora*
Damping-off and seedling blights	*Pythium, Phytophthora, Aspergillus, Rhizoctonia, Fusarium, Helminthosporium* spp.
Root and foot rots, take-all	*Armillaria mellea, Valsa eugeniae, Gaeumannomyces graminis, Corticium rolfsii, Fusarium solani, Helicobasidium purpureum, Aphanomyces* spp.
Wilts	*Ophiostoma novo-ulmi, Fusarium oxysporum, Verticillium albo-atrum: Burkholderia solanacearum, Xanthomonas campestris* pv. *vasculorum, Erwinia stewartii: Striga hermonthica, Orobanche crenata*
Downy mildews	*Peronospora, Plasmopara, Bremia* spp.; *Sclerospora graminicola*
Powdery mildews	*Erysiphe, Phyllactinia* spp. *Leveillula taurica*
Rusts	*Puccinia, Uromyces, Melampsora* spp.; *Hemileia vastatrix*
Smuts	*Ustilago, Urocystis, Tilletia, Entyloma* spp.
Blights	*Phytophthora infestans, Ascochyta rabiei: Erwinia amylovora, Xanthomonas campestris*
Anthracnose	*Colletotrichum, Gloeosporium* spp.
Leaf spots	*Botrytis fabae: Alternaria, Septoria, Diplodia, Cercospora, Helminthosporium* spp; *Pseudomonas syringae*
Leaf curl, witches' broom, club root, blister blight	*Plasmodiophora brassicae; Taphrina deformans; Exobasidium vexans; Crinipellis perniciosa;* tomato leaf curl geminivirus
Galls	*Agrobacterium tumefaciens; Meloidogyne* spp., wound tumour phytoreovirus
Cankers and scab	*Nectria galligena; Venturia inaequalis; Elsinoe phaseoli; Spongospora subterranea; Clavibacter michiganensis*
Mosaic, green vein-banding and yellows	Tobacco mosaic tobamovirus, bean common mosaic potyvirus, cucumber mosaic cucumovirus, beet western yellows luteovirus, and numerous other viruses

However, changes in selection pressure, in host genotype and in agroecology can lead to the emergence of tomorrow's epidemics.

Table 6.2. The distribution of plant pathogenicity in the Fungi *sensu lato* (after Hawksworth *et al.,* 1995).

Classes	No. of genera	No. of plant pathogenic genera
Protozoa		
Acrasiomycota	4	0
Dictyosteliomycota	4	0
Myxomycota (slime moulds)	74	0
Plasmodiophoromycota	15	2 (*Plasmodiophora* and *Spongospora*)
Chromista		
Hyphochytridiomycota	7	0
Labyrinthulomycota	13	0
Oomycota	95	19 (especially in *Peronosporales* and *Pythiales*)
Fungi *sensu stricto*		
Ascomycota	3255	60 (in some 13 orders of which the most important are *Dothidiales*, *Diaporthales*, *Erysiphales* and *Hypocreales*)
Basidiomycota	1428	249 (of which 19 are *Basidiomycetes*, 167 are *Teliomycetes*, chiefly rusts, and 63 are *Ustomycetes*, mostly smut fungi)
Chytridiomycota	112	3 (including *Synchytrium*)
Zygomycota	173	6 (including *Rhizopus* and *Choanephora*)
Mitosporic fungi	2547	70
Total	7727	407

The evolution of disease

Pathogenicity is broadly and unevenly scattered across the taxonomic spectrum of fungi (Table 6.2). Some orders such as the *Erysiphales* (the powdery mildews) and the *Uredinales* (the rusts) consist entirely of specialized pathogens, but many others also contain pathogenic species. Often within a single genus there are species which exhibit a wide range of trophic behaviour; *Fusarium* is a good example (Waller and Brayford, 1990). The trophic variation in this and many other genera of mitosporic fungi (including *Cladosporium, Alternaria* and *Colletotrichum*) perhaps suggests evolution of the more specialized parasitic state from saprotrophism, as is conventionally considered to be the case. However, at least partial endophytism is also commonly exhibited within these genera, suggesting that the endophytic habit represents a transitional phase. Endophytes are harmless, often beneficial, but under favourable conditions some species of fungal endophyte like *Discula umbrinella* can become pathogenic (Petrini, 1998). Furthermore, there is now evidence that biotrophy is actually the ancestral trophic state, as we discuss on pages 129–130 and 138.

Studies on diseases in plant communities have found that interactions which occur between coevolved plants and pathogens in both natural ecosystems and agroecosystems range from minor to epidemic (Burdon, 1987, 1993; Dobson and Crawley, 1994). In some natural plant communities, epidemics rarely occur; in others, epidemics are common. In some agroecosystems, epidemics commonly ravage crops; in others, years may go by in which disease has little, if any, effect on crop production (Burdon, 1993). There is no doubt that the evolution of disease and pathogen variability has played an important role in fostering diversity in ecosystems (Gillett, 1962; see Polaszek *et al.*, Chapter 11, this volume). Pathogens may change the outcome of inter- and intraspecific competition, or change the genetic structure of plant populations and the diversity of the plant community (Burdon, 1987). The widespread occurrence of plant defence compounds and batteries of resistance genes in wild relatives of crops are evidence of powerful selective forces in the past (Harper, 1990).

In many ancient plant communities, long-term genetic interaction between host and pathogen populations results in the pathogen being held in check by both individual and population resistance (Browning, 1974; Harlan, 1976; Burdon and Shattock, 1980; Burdon, 1987). Gene centres of plants have long been recognized as rich sources of resistance (Leppik, 1970), especially to diseases against which the host and pathogen coevolved, perhaps in a gene-for-gene relationship (Frank, 1993). Advocacy of plant conservation *in situ* for disease resistance (Dinoor and Eshed, 1997) seems paradoxical: the effective long-term conservation of wild relatives of crops for their potential as sources of disease resistance actually requires effective conservation of the pathogen against which the resistance is being sought! (Wills, 1996; Burdon, 1998; Lenné, 1998). Wild and weedy relatives can be important sources of inoculum for crops (Lenné and Wood, 1991) and there are numerous examples of pathogens from wild ecosystems moving to introduced crops (see Thresh, 1981), as we discuss below.

In most cases of extreme susceptibility in the host and extreme virulence in the pathogen epidemics are the consequence of human activities. Buddenhagen (1977) has examined the origin of epidemic diseases in tropical crops, concluding that all have one thing in common: that epidemics arise from an intensified union of crop and pathogen following some type of separation. Three broad categories can be distinguished:

1. Epidemics that result from a long-term coevolution between host and pathogen, either:
(a) with gradual intensification of relationships, or
(b) being rejoined spatially in a re-encounter after recent interruption.
2. Epidemics that arise from the pathogen's ability to overcome a genetic separation created through plant breeding.
3. Epidemics of genetically new-encounter diseases that follow reunion of two long-separated components of isolated evolutionary systems, being either:

(a) a result of intercontinental or regional movement of a crop plant into a new environment, or

(b) the movement of a pathogen into a new region where it infects a related host species.

Type 1a epidemics are nicely illustrated by recent work (Guzman *et al.*, 1995) on the common bean (*Phaseolus vulgaris*) and its angular leaf spot pathogen *(Phaeoisariopsis griseola)*. Type 1b is well shown by the recent spread of the groundnut rust fungus (*Puccinia arachidis*) after coevolution with its host in Latin America (Allen, 1983; Subrahmanyam *et al.*, 1985). Type 2 epidemics have often caused the greatest concern over the classic cycles of 'boom and bust' in cultivar production (Robinson, 1976). Examples of new-encounter diseases as a result of crop movement (Type 3a epidemics) include maize streak, rice yellow mottle, groundnut rosette, cocoa swollen shoot, African cassava mosaic and black root of common bean, among virus diseases (Thresh, 1980; Allen and Lenné, 1998; Allen *et al.*, 1998). Moko disease of banana, caused by the bacterium *Burkholderia solanacearum*, and red leaf blotch of soybean, caused by the fungus *Dactuliochaeta glycines*, are also new-encounter diseases, and there are many other examples that we have discussed elsewhere (Allen *et al.*, 1998). Plant quarantine has played an important role in restricting the international movement of pathogens (Sheffield, 1968; Neergaard, 1977), many of which are effectively seedborne and widely distributed despite such procedures. Examples of new-encounter diseases as a result of pathogen movement (Type 3b) include Dutch elm wilt, caused by *Ophiostoma novo-ulmi,* chestnut blight in the USA caused by *Cryphonectria parasitica* (Brasier, 1990), jarrah dieback caused by *Phytophthora cinnamomi* in southwestern Australia (Newhook and Podger, 1972; Weste and Marks, 1987), and late blight of potato, which is notorious for the disastrous social and economic consequences of its epidemics in Ireland in the 1840s.

There has been much speculation on the origin of the first inoculum of the European population of the late blight fungus (*Phytophthora infestans*). Most of the available evidence points to a Mexican origin of the pathogen: Mexico seems the centre of maximum diversity of *P. infestans* as well as the country in which the greatest number of R genes are found, especially in *Solanum demissum* (Fry *et al.*, 1993). Although it is possible that *P. infestans* moved directly from Mexico to Europe, evidence more strongly supports an indirect route, via either the USA (Fry *et al.*, 1993), or South America followed by migration to Europe with or without passage through the USA (Tooley *et al.*, 1989; Andrivon, 1996). Whichever theory is correct, it is clear that *P. infestans* newly encountered the potato in either South America or the USA (see also p. 134).

It is sometimes found that secondary centres of genetic diversity of a crop are useful hunting grounds for resistance against new-encounter diseases. Surprisingly though, there is an increasing number of cases wherein resistance is found in germplasm from areas free of the disease, a type of resistance best considered allopatric (Harris, 1975), a topic that we have expanded upon elsewhere (Allen and Lenné, 1998; Allen *et al.*, 1998).

Host range and specificity

There is great diversity in the degree of host specificity shown by plant pathogens. Those with narrow host ranges are well illustrated by biotrophic fungi like the rust, smut and powdery mildew pathogens of cereals that are obligate parasites (Scott *et al.*, 1980). Some leaf spot fungi, like the groundnut pathogens *Cercosporidium arachidicola* and *Phaeoisariopsis personata*, are essentially host genus-specific and the bacterium *Pseudomonas syringae* pv. *phaseolicola* can be regarded as host tribe-specific (Taylor *et al.*, 1996; Allen and Lenné, 1998). Necrotrophs, that kill tissue as they grow through it such that they are always colonizing dead substrate, are opportunistic pathogens with very wide host ranges; the facultative parasite *Botrytis cinerea* is one such example. The diversity in modes of parasitism exemplified by fungi is illustrated in Table 6.3. It has been suggested that there is perhaps an evolutionary progression from obligate saprophytism and facultative parasitism to obligate parasitism and, possibly, mutualistic symbiosis (Tarr, 1972). However, a recent review (Parbery, 1996) presents strong evidence to

Table 6.3. Diversity in trophic specialization among crop-associated fungi.

Trophic specializaton	Disease/association	Pathogen examples
Symbionts	Mycorrhizae	*Glomus* spp.
Epiphytes	Inflorescence disease	*Balansia* spp.
	Sooty moulds	*Capnodium* spp.
Endophytes	Anthracnose of beech	*Discula umbrinella*
Biotrophs	Powdery mildews	*Erysiphe graminis*
	Downy mildews	*Peronospora parasitica*
		Bremia lactucae
	Rusts	*Puccinia graminis*
		Uromyces appendiculatus
	Smuts	*Ustilago scitaminea*
		Tilletia caries
Hemibiotrophs	Bean anthracnose	*Colletrichum lindemuthianum*
	Late blight	*Phytophthora infestans*
	Rice blast	*Pyricularia oryzae*
	Early leaf spot	*Mycosphaerella arachidis*
	Vascular wilt	*Fusarium oxysporum*
Necrotrophs	Southern leaf blight	*Cochliobolus heterostrophus*
	White mould	*Sclerotinia sclerotiorum*
	Crown rot	*Aspergillus niger*
	Southern blight	*Corticium rolfsii*

the contrary: that the biotrophic character is ancient and ancestral to necro-
trophism and saprophytism (see p. 138). Host recognition by pathogens
(Callow, 1977) presumably in part governs host range and, in cases apparently
illustrated by the legume xanthomonads (Allen and Lenné, 1998), taxonomic
confusion among host genera seems likely to have led to confusion in the
identity of host-specific bacterial pathogens. It has been shown (Savile,
1954) that fungi are themselves useful aids to host plant taxonomy.

A final comment concerning pathogen host range brings us back to the
rust fungi, which are noteworthy for their diversity of spore stages in the life
cycle in which, in heteroecious species, there is a host alternation. In *Puccinia
graminis*, for instance, the diploid stages are specific to *Gramineae* (*Poaceae*)
whereas the haploid stages are specific to woody members of *Berberidaceae*.
The haploid stage and the hypha-like structure of cereal rust fungi in tissue
of alternate hosts are simpler, and phylogenetically older, and therefore more
adaptable than the diploid stage and the haustoria that form on the primary
host (Wahl and Anikster, 1982). The course of evolution of heteroecism
(Olive, 1911; Wilson and Henderson, 1966) presumably has important implications both for pathogen survival and for generation of novel diversity
within species, a topic we shall return to later.

The nature of infraspecific diversity in pathogens

Within species of plant pathogens, variants that differ morphologically in
addition to variation in other characters that may include pathogenicity are
recognized as varieties or, perhaps even subspecies. For example, eight
varieties are recognized within *Puccinia coronata* within which at least one
(var. *gibberosa*) differs from var. *coronata* in uredospore and teliospore
dimensions (Wilson and Henderson, 1966), and *Uromyces appendiculatus* var.
crassitunicatus is similarly regarded as distinct from the type of the species
(Irwin, 1988). Caten's concept of biologic race exemplified by *Ophiostoma
novo-ulmi* (Caten, 1987) perhaps may be regarded as equivalent; biologic races
differ by many characters beyond pathogenicity and are frequently geographically or reproductively isolated. Vegetative compatibility groups are
recognized within species of *Fusarium* (Katan *et al.*, 1994; Gordon *et al.*,
1996) and anastomosis groups, which appear to be distinct evolutionary
units that warrant taxonomic recognition, have long been used to partition
diversity within *Rhizoctonia solani* (Vilgalys, 1988). Ecotypes, between
which there is some relation to grouping of hyphal anastomosis within *R.
solani* (Durbin, 1959), have also been separated in *Phytophthora drechsleri*
(Shepherd and Pratt, 1973) and *Puccinia graminis* (Green, 1973). Host-
specific morphotypes have been distinguished within species of parasitic
witchweeds (Musselman *et al.*, 1991; see Polaszek *et al.*, Chapter 11, this
volume).

Within these groups, variants that differ in either host-specific and host non-specific pathogenicity have been defined, so that diversity is found at different levels within species of pathogens. Thus, formae speciales are characterized only in physiological or biochemical terms, particularly in pathogenicity or host adaptation, and are infraspecific taxa of fungi. The term pathovar is synonymous, being used for bacteria (Holliday, 1989). *P. graminis* comprises at least six formae speciales of which *tritici* is the most important. More than 75 formae speciales are recognized within *Fusarium oxysporum*, and 136 pathovars of *Xanthomonas campestris* have been described (Bradbury, 1986; Holliday, 1989). Within these and at a lower systems level, physiologic races are distinguishable. A physiologic race may be defined as a taxon of parasites characterized by pathogenic specialization to different cultivars of one host species. Races have been recognized particularly in fungi, but are reported also among bacteria, viruses (as 'strains'; Gibbs and Harrison, 1976) and parasitic species of angiosperm (Allen and Lenné, 1998; Gagne *et al.*, 1998; see Polaszek *et al.*, Chapter 11, this volume). Formae speciales of fungi, pathovars of bacteria, as well as anastomosis groups and varieties of pathogens, frequently comprise several to many races (Caten, 1987). A race can also be seen as a population in which all individuals carry the same combination of virulence genes (Parlevliet, cited by Holliday, 1989).

Whereas biotrophic fungal parasites tend to vary most conspicuously in a qualitative manner in their virulence, or host-specific pathogenicity, necrotrophic fungal parasites more often vary quantitatively (Scott *et al.*, 1980) in what is best regarded as degrees of aggressiveness. Aggressive races which do not interact differentially with host cultivars (van der Plank, 1968) have been identified in *F. oxysporum* f. sp. *lycopersici* (Caten, 1987), in rust fungi both in wild populations (Prakash and Thielges, 1987) and in agriculture (Katsuya and Green, 1967; Ogle and Brown, 1970), and in numerous other cases revealed by quantitative assay. The underlying distribution of variation in aggressiveness is continuous, although apparent discontinuities arise through sampling and suggest the existence of discrete groups interpreted spuriously as physiologic races (Caten, 1987; Robinson, 1987). If more isolates are examined, the continuous nature of the variation becomes clear, and attempts to recognize separate aggressive races become futile (Caten, 1987). The term pathotype, which is essentially synonymous with race, is preferred by some authors (e.g. Robinson, 1976), in part because the term distinguishes between horizontal pathotypes (that differ in aggressiveness) and vertical pathotypes (or physiologic races, that vary in virulence). Since the discovery of physiologic specialization, first in *Puccinia* (Stakman, 1914) and *Colletotrichum* (Barrus, 1918), the race concept has become one of the main tools for understanding the structure and dynamics of pathogen populations, and further progress was made with the gene-for-gene theory which has enabled simultaneous examination of both host and pathogen (Flor, 1956; Crute *et al.*, 1997).

Pathogens also vary at the infraspecific level in their sensitivity to chemicals, including sensitivity to host defensive chemicals in wild populations. Ennos and Swales (1991) have investigated the evolutionary response of the pine canker fungus (*Crumenulopsis sororia*) to variation in the proportion of monoterpenes which are released into resin of natural *Pinus sylvestris* populations in response to infection. Considerable differences were found between isolates of the pathogen, suggesting the potential for the evolution of chemically specialized pathogen sub-populations. In agriculture, the development of resistance to agrochemicals in pathogen populations is well known, with examples from the fungi *Pyrenophora avenae* (Greenaway and Cowan, 1970), *Mycosphaerella arachidis* (Littrell, 1974) and *Colletotrichum kahawae* (Ramos and Kamidi, 1982; Masaba and Waller, 1992); variants of the bacterium *P. syringae* pv. *phaseolicola* have also been shown to develop antibiotic resistance (Russell, 1975), and many other examples are discussed by Georgopoulos (1987).

The origin of diversity

Population shifts in plant pathogens result either from a readjustment of frequency of the components of an original population, or from the infusion into the latter of a new character that gives selective advantage to those races possessing it (Watson, 1970). Changes in virulence gene frequencies sometimes appear to be independent of pathogen exposure to host resistance (Alexander *et al.*, 1985), but more often changes in virulence are reported in the wake of the commercial release of new cultivars in which resistance then proves transient, because of the emergence of a matching variant of the pathogen (van der Plank, 1968; Marshall, 1977; Burdon, 1987). Large-scale application of a new fungicide can provide a similar selective advantage for resistant mutants which may then increase in the pathogen population. Thus, microbial populations have a remarkable ability to adapt to changed circumstances, whether to changes in the genetic structure of populations of their hosts or to changes in usage of crop protection chemical; but how does this come about?

Large changes in virulence can take place in asexual populations of a rust fungus like *Uromyces appendiculatus* over five generations (Alexander *et al.*, 1985), and the emergence of new races has been attributed to mutation in other species of rust reproducing asexually in the absence of alternate hosts (Luig and Watson, 1970; Watson, 1970), as well as in the barley scald pathogen, *Rhynchosporium secalis* (Allard, 1990). The anthracnose pathogen of stylo, *Colletotrichum gloeosporioides*, provides a nice example of clonal evolution through mutation and chromosome exchange (Masel *et al.*, 1996), and clonal evolution through mutation is thought also to account in part for the origin of new races in *Phytophthora sojae* (Forster *et al.*, 1994; Drenth *et al.*, 1996). The origin of fungicide tolerance in the asexual *Colletotrichum kahawae* is also due to mutation (Ramos and Kamidi, 1982). However, in certain instances the degree of variation found in an asexually reproducing pathogen is difficult to explain on the basis of gene mutation alone. Heterokaryosis, the condition in which a fungal hypha or cell has two or more genetically distinct haploid nuclei (Parmeter *et al.*, 1963), and parasexuality (Pontecorvo, 1956), that permits the segregation and recombination of genetic factors outside the sexual stage and plays a part in virulence changes in various plant pathogens, have been demonstrated in *F. oxysporum* (Buxton, 1956), *Rhizoctonia solani* (McKenzie *et al.*, 1969), *Phoma medicaginis* (Sanderson and Srb, 1965) and various other species (Pontecorvo, 1956; Tinline and MacNeill, 1969), possibly including certain asexual populations of rust fungi (Watson, 1970).

Among viruses, new variants may arise as a result of interspecific recombination, as appears to be the case with severe cassava mosaic in Uganda (Zhou *et al.*, 1997). There is now powerful evidence that banana streak badnavirus has evolved through conjugation between virus and host genomes (La Fleur *et al.*, 1996; Frison and Sharrock, 1998).

In populations of parasitic angiosperms, host-specific races have evolved in strongly inbreeding species like *Striga gesnerioides* (Lane *et al.*, 1996; see Polaszek *et al.*, Chapter 11, this volume) and *Orobanche cumana* (Gagne *et al.*, 1998), but apparently not in an obligate outcrosser such as *Striga hermonthica*. *S. hermonthica* exhibits considerable diversity within populations whereas *S. gesnerioides* exhibits great variability among populations but is more uniform within a given population, as would be expected from the distinct breeding systems of these two species (Musselman *et al.*, 1991).

Populations of plant pathogens in which sexual reproduction occurs in general tend to display greater diversity than asexually propagating ones (Burdon and Roelfs, 1985; Groth and Roelfs, 1987; Drenth *et al.*, 1994). In heteroecious rust fungi, the presence of the alternate host (on which the sexual stage develops) can have important implications for the joint evolution of the pathogen and its host by presenting opportunities for recombination (Simons *et al.*, 1979; Wahl and Anikster, 1982; Burdon and Roelfs, 1985). Similarly, the presence of both mating types in heterothallic fungi tends to accelerate the evolution of genetic diversity (Gustafsson *et al.*, 1983; Drenth *et al.*, 1994). The course of events in late blight of potato (*P. infestans*) in Europe nicely illustrates this point. Before 1842, late blight was known only in the Toluca valley of Mexico. Later that year, the disease appeared in the USA in New England then, a year or two later, it appeared in Belgium from where the disease spread throughout Europe. Despite the relative genetic uniformity of the *P. infestans* population then in Europe, its consequences were devastating. Probably only a single clone had reached Europe, and the fungus's very limited ability to evolve without sexual propagation gave potato breeders a great advantage. Whereas the single clone (called US-1) of the A1 mating type had become widespread globally, the A2 mating type remained restricted to Central and South America. Furthermore, asexually produced conidia are not well adapted to survival. This remained the case until 1976, when drought led to crop failure and so also to increased pressure to import, and novel, varied genotypes of *P. infestans*, including both the A2 type and novel A1 types, were introduced into Europe from the Americas (Fry and Goodwin, 1995; Shattock and Day, 1996). The A2 mating type was also introduced into the USA, Canada and East Asia in the late 1980s to early 1990s (Fry and Goodwin, 1995). In Europe, the new types largely displaced the old, uniform A1 type (Shattock and Day, 1996). Sexual reproduction has been found between the new A1 and A2 types and may have also occurred between the A2 and the old A1 type before the latter was displaced. This has not only dramatically increased the potential for rapid evolution of the late blight fungus but also improved the pathogen's ability to survive, as sexually produced oospores (Drenth *et al.*, 1994; Andrivon, 1996; Shattock and Day, 1996). Although the new strains of *P. infestans* have caused more serious outbreaks of late blight than have been seen for some time in the USA, this has not occurred in Europe where blight continues to be successfully controlled at least 15 years after the new strains became widespread (Fry and Goodwin, 1995; Pearce, 1997; L.R. Cooke, DANI, Belfast, 1998, personal communication).

Selection, migration and recombination may be the most important factors shaping race structure and evolution in populations of sexually reproducing powdery mildews, in which mutation is perhaps of limited significance (Andrivon and de Vallavieille-Pope, 1993). The success of the barley powdery mildew pathogen (*Erysiphe graminis* f. sp. *hordei*), in the UK at least, lies in its life strategy. During the late summer, the one period in the year when there is little or no living host tissue available, the pathogen passes through its resting stage as cleistothecia. These thick-walled fruiting bodies germinate in the autumn, nicely timed to take advantage of the newly emerging autumn-sown crop, and with a high chance of success since the ascospores inside them carry the maximum release of genetic variability of the pathogen (Wolfe, 1987).

In a range of other cases, the role of the teleomorph in the maintenance of genetic diversity in a plant pathogen remains ill-understood; examples include *Didymella* spp. (Jellis and Punithalingam, 1991; Trapero-Casas and Kaiser, 1992) and *Elsinoe phaseoli* (Phillips, 1996). Even though a teleomorph may occur rarely in nature, rare outcrossing can be a major contributor to the origin of new races (Forster *et al.*, 1994).

Infraspecific variants, by whatever means they arise, under some circumstances continue to evolve and speciate. If we trace the history of coffee berry disease, which arose in western Kenya in 1922, we see that the pathogen (*C. kahawae*) has unique biochemical and ecological properties, that populations are still nearly clonal, and that the species is closely related evolutionarily to *C. gloeosporioides*. The rather rapid evolution and speciation of this pathogen appears to have been driven by its encounter with an expanding area of susceptible *Coffee arabica* with which it had not co-evolved (Waller and Bridge, 1999). The relatively sudden emergence of the virulent Dutch elm wilt pathogen seems comparable: in this case, an initial, clonal population evolved into a genetically diverse species (*Ophiostoma novo-ulmi*) as it spread, causing a widespread pandemic (Brasier, 1988).

The Characterization of Diversity

A range of techniques is available. Much information at the species level is reviewed in Hawksworth (1994), and Burdon (1987) has addressed diversity at the infraspecific level.

Differential hosts

The most widely used means of distinguishing physiologic races, as 'units' of biodiversity within plant pathogens, is by comparison of reactions on a set of host cultivars carrying different sources of qualitative resistance. The races first described in *P. graminis* f. sp. *tritici* were identified in this manner, and differential sets of host cultivars have since been used widely, with other rusts, with powdery and downy mildews, and with numerous other fungi (Scott *et al.*, 1980; Caten, 1987) as well as with bacteria (Taylor *et al.*, 1996) and viruses (Spence and Walkey, 1995). The potential number of races that can be recognized is determined by the number of different sources of resistance used (Caten, 1987). This is well illustrated by the halo blight pathogen, *Pseudomonas syringae* pv. *phaseolicola*, in which only two races had long been recognized on the basis of the differential reactions of *Phaseolus vulgaris* cv. Red Mexican UI3 (Walker and Patel, 1964). When intensive studies that involved alternate probing for variation in host and pathogen led to the development of a set of eight differential cultivars, a more complex race structure was revealed (Taylor *et al.*, 1996). As the number of differential hosts is increased, so more and more races can be identified, and the limits to racial classification appear to depend as much upon the diligence of the workers as upon the biology of the system (Caten, 1987).

With widespread and important pathogens, there is often a need for standardization of differential sets for international use so that patterns of race distribution can be determined accurately (Drijfhout and Davis, 1989). Ideally, each differential host should possess only one resistance gene (Robinson, 1976). Application of a method of naming races using binary notation (Habgood, 1970) can help in international comparison, but implicit in the system is a degree of inflexibility which is a disadvantage: valid results are obtained only if all the differentials are used. Scott *et al.* (1980) stress the importance of maintaining flexibility in systems of racial classification. Race determinations often have been carried out on standard sets of differentials giving virtually no information relevant to commercial crops; it is often preferable to select important commercial cultivars themselves as differentials, especially if evidence suggests that a virulence shift has occurred in the pathogen population (Scott *et al.*, 1980). Robinson (1976) points out that there are several taxonomic categories of host differentials, and it is only if they are all members of one host plant species that they differentiate physiologic races (or vertical pathotypes). If the various differentials are different species, they differentiate formae speciales within the parasite species. Occasionally, it is found that cultivars within the principal host species cannot differentiate races and the interaction is essentially quantitative. However, by use of a set of cultivars all of a closely related host species, physiologic races can be differentiated among the same set of isolates, as found the case in a xanthomonad pathogen (Opio *et al.*, 1996). The point to be made here is that, whatever the method used to detect virulence genes in a pathogen population, if *the host itself* cannot detect those genes then the durability of its resistance seems unlikely to be threatened.

Other biological methods

Assessment of differences in pathogenicity is most often documented in race surveys, and variation in virulence is more commonly measured than a pathogen's ability to multiply rapidly, a character embracing aggressiveness (Burdon, 1993). Morphological markers like growth rate, colony colour, texture and sporulation are useful for species that can be cultured on artificial media, and vegetative compatibility grouping is a powerful tool for assessment of variation in fungal genera including *Fusarium* and *Rhizoctonia*.

Diversity indices

There is a range of ecological diversity indices that have been applied to the assessment of plant pathogen diversity at the infraspecific level, and the reader is referred to reviews including Lebeda (1982) and Groth and Roelfs (1987) for details (see also Cox and Wood, Chapter 3, Wardle *et al.*, Chapter 5, and Witcombe, Chapter 10, this volume).

Biochemical and molecular methods

A range of biochemical, molecular and physiological tests have been used to detect pathogenicity and its variability within populations of microorganisms. Many of these have been linked, often indirectly, to pathogenic characteristics and can be used in diagnostic tests (Schots *et al.*, 1994), but there is no consistency of association across pathogens. Certain constant characteristics can be used as molecular or biochemical markers to enable closer analysis of pathogen variability and dynamics, as used, for example, in tracing the source of inoculum in epidemics or the geographic spread of pathogens. Isozymes have proved to be relatively poor tools for the assessment of variation in pathogen populations (Burdon, 1993). Specialized biotrophs like smut and rust fungi, as well as some hemibiotrophs, appear to possess limited isozymic variation, even in populations wherein considerable variation in virulence has been demonstrated (Lenné and Burdon, 1990). Molecular markers, such as restriction fragment length polymorphisms (RFLPs), random amplified polymorphic DNA (RAPDs), amplified fragment length polymorphisms (AFLPs), and highly repeated sequences (HRS; fingerprints), offer opportunities for the study of pathogen diversity (Burdon, 1993; Schots *et al.*, 1994), and all are being used widely in plant pathology. Their appropriateness to studies of pathogen diversity depends on the nature and scope of the questions addressed. For instance, there have been cases in which the genetic structure of a pathogen population does not correlate with diversity for pathogenicity, as shown in *C. kahawae* (Waller and Bridge, 1999) and *Colletotrichum gloeosporioides* (Chakraborty *et al.*, 1997). In each case the very close genetic relatedness between isolates did not reflect their considerable variation in pathogenicity.

Most modern molecular methods detect major genes that, in host–parasite relationships, are most often associated with discontinuous variation as revealed by the clear-cut reactions of host cultivars with race-specific resistance when challenged by matching physiologic races. The further development and use of molecular tools that can be applied to polygenic systems (e.g. Edwards, 1992) will help to redress this imbalance, with possible implications for the improved durability of host plant resistance, despite current optimism (e.g. Staskawicz, 1998) over prospects for the development of synthetic resistance by molecular cloning of resistance genes.

Functional Diversity

In this section, we discuss the population dynamics of pathogens, their survival, fitness, and both temporal and spatial changes in their structure. A measure of natural biocontrol of plant pathogens occurs in many agricultural systems, perhaps especially those including perennial crops, and is associated with diversity among microorganisms on plant surfaces. These natural systems can be disturbed, for instance by injudicious use of fungicides that can exacerbate disease. In order to manipulate such systems to our advantage, there is a need to improve our understanding of the functional diversity of organisms involved in pathogen suppression (Waller, 1991).

Plant pathogens can be described according to taxonomic group, mode of parasitism, and various ecological criteria (Burdon, 1992). Although modes of parasitism have been described for most pathogens, detailed studies on the strategies which synchronize their life cycles with those of their hosts are limited (Parbery, 1996). Although it is widely believed that the first fungal pathogens were saprophytes from which necrotrophs and biotrophs evolved (Leppik, 1965; Luttrell, 1974), much evidence supports the contrary view, that the first fungi were biotrophs from which necrotrophs and saprophytes developed (Savile, 1990; White, 1992; Parbery, 1996). Biotrophy is widespread among fungi, and fungal pathogens often switch from biotrophy to necrotrophy during pathogenesis, but the reverse has not been found (Parbery, 1996). Biotrophy is also the main mode of parasitism of viruses, nematodes, phytoplasmas and some bacteria.

A valuable means of comparing functional characteristics, like reproduction, survival and dispersal, of plant pathogens has been demonstrated by Burdon (1992). Rust fungi are described as *r*-strategists (see Wardle *et al.*, Chapter 5, Edwards *et al.*, Chapter 8, and Wood and Lenné, Chapter 17, this volume), typically obligate biotrophs, with poor off-season survival, high fecundity, short generation time, efficient dispersal of propagules, a narrow host range and pathogenic on annuals and ephemeral plant parts. Conversely, soil pathogens like *Rhizoctonia* and *Pythium* species are considered *K*-strategists typified by necrotrophic hemibiotrophy or facultative saprophytism, with good off-season survival, low fecundity, long generation time, inefficient propagule dispersal, a wide host range and pathogenic on perennial hosts.

The relation between survival and diversity

Many serious epidemics occurred well before the advent of modern agriculture: the famines mentioned in biblical times have been attributed to epidemics of rust on the wheat crop (Large, 1940; Carefoot and Sprott, 1969), before the time of modern plant breeding but when considerable variability in the crop existed. *Phytophthora cinnamomi* is among the world's most destructive pathogens of diverse native vegetation (Newhook and Podger, 1972; Weste and Marks, 1987; Tommerup, 1998). Much of the diversity for disease resistance present in crop plants today represents the ghosts of diseases past (Harper, 1990). But how *long* past? Riley (1973) suggested that a 'ghost resistance' may continue to be conferred by vertical resistance genes after they have been overcome, and Dinoor and Eshed (1997) draw attention to the 'trail of defeated genes' that may be a useful reservoir of host resistance against pathogens in other regions. The existence of rust resistance in *Glycine* in desert areas of Australia where the pathogen does not occur (Burdon, 1998) may perhaps be a case in point; other examples of such allopatric resistance are discussed elsewhere (Allen *et al.*, 1998). Wild-type races of plant pathogens can be considered as those that existed before people began to control them by use of resistant cultivars. Survival of such races, on native susceptible hosts and local landraces, presumably would depend on the fitness of those races themselves, as well as on the environment and the suitability of the host tissue (Watson, 1970). In rust fungi with an alternation between hosts, heteroecism is perhaps a strategy that encourages spatial diversity through improved survival. In many plant pathogenic fungi, survival between seasons depends on the production of hard-walled fruiting structures, like oospores, that are products of sexual reproduction so that, in heterothallic species of *Phytophthora* for example, the local presence of both mating types governs successful carry-over. Under certain circumstances, oospores may be induced to form by other species of *Phytophthora* (Shepherd, 1978; Boccas, 1981) or by antagonists (Brasier, 1971).

Environment clearly has profound effects not only on pathogen survival but also on the evolutionary dynamics of host–pathogen interactions. We consider several examples on poplar and birch. Populations of rust (*Melampsora occidentalis*) on poplar under the dry continental climate east of the Cascade mountains in the Pacific Northwest of the USA have an annual cycle. Here, the rust fungus exhibits physiologic specialization, indicating pathogen adaptation to the less favourable environment. Conversely, the moist maritime climate west of the Cascade mountains provides a very favourable environment for rust development throughout the season, conditions under which the evolution of physiological specialization in the pathogen is not favoured (Hsiang and Chastagner, 1993). No simply inherited resistance against canker (*Hypoxylon mammatum*) has evolved in poplar (*Populus tremuloides*) on which the pathogen remains throughout the host's life (Burdon *et al.*, 1996). On birch in Sweden, populations of rust (*Melampsoridium betulinum*) on leaves, which are deciduous and annual, risk extinction at the end of the season; this possibly favours the evolution of major genes for resistance. Conversely, the stem canker pathogen of birch (*Inonotus obliquus*), which is perennial, apparently favours the development of minor gene resistance (Burdon and Thrall, 1998). Indeed, Burdon *et al.* (1996) suggest that the evolution of gene-for-gene systems may be most likely to develop where susceptible host tissues are ephemeral: in annual species, on deciduous leaves, and in above-ground plant parts that die back at the end of the season. On the one hand, the randomness of encounters between a pathogen and its host seems likely to enhance the advantage of retaining 'unnecessary' genes for resistance whereas, on the other hand, where pathogen and host are in close contact for prolonged periods, the host might be expected to possess few major genes for resistance. Such findings break new ground. It seems vitally important now that the understanding gained of host–pathogen relations in essentially wild ecosytems is examined for its relevance to disease management in agriculture.

Infected seed plays an important role both in long-distance dispersal of pathogens to new areas and in survival between seasons; the seed transmissibility of viruses in legumes is notorious (Allen and Lenné, 1998). The extent to which seed transmissibility has a bearing upon pathogen biodiversity is unclear, but it seems probable that the absence of seed transmission would constitute a barrier to long-distance spread of virus and so to the evolution of distinct local populations of the pathogen. Evidence to this effect comes from work on geminiviruses, a family of whitefly and leafhopper-transmitted viruses that are not seedborne. The emerging pattern of relationship among some whitefly-transmitted geminivirus species is that those from the same area are closely related whereas those from different continents are more distantly related, irrespective of their host species (Hong and Harrison, 1995; Natesham *et al.*, 1996).

Virulence, fitness and fitness cost

The relationship between the relative survival ability and virulence expressed by two races of *Puccinia striiformis* in mixtures on a susceptible wheat cultivar was investigated by Brown and Sharp (1970). They found that the more virulent race predominated regardless of the original proportion of uredospores in the initial mixtures, although the poorer competitor persisted at a frequency of about 2% after seven successive generations. In a parallel study on *P. graminis*, Ogle and Brown (1970) drew similar conclusions. Conversely, Katsuya and Green (1967) concluded that the effects of temperature coupled with variation in aggressiveness, or reproductive potential, accounted for the predominance of one race of *P. graminis* f. sp. *tritici* over another, and fitness of races of powdery mildew (*E. graminis* f. sp. *hordei*) has also been attributed to a temperature dependence (Welz *et al.*, 1990). Nelson (1972) has suggested that a race which increases in distribution and frequency in a population must exhibit two fundamental characters: it must possess the necessary genes for virulence, and it must possess certain fitness. Van der Plank (1968) has proposed that there is a negative relationship between virulence and the fitness of a race to survive: the possession of unnecessary genes for virulence decreases fitness, and complex races with many virulence genes are at a selective disadvantage over simple races with few virulence genes. A stabilizing selection is exerted against a particular variant in a population when selection pressure for that variant is removed. There are, however, many exceptions to this hypothesis, and clear evidence for the existence of stabilizing selection was not found by Parlevliet (1981) who dismissed it as an empty concept. In some cases (Luig and Watson, 1970), there is strong evidence that a negative relationship does exist between the number of genes for virulence and aggressiveness, but unnecessary genes for virulence are not always lost from the population. Unexpectedly high frequencies of 'unnecessary' genes for virulence may be due to genetic linkage with 'necessary' virulence genes (Gustafsson *et al.*, 1983). More recent work, on racial diversity within populations of the barley mildew pathogen (Andrivon and de Vallavieille-Pope, 1993), failed to provide convincing evidence for the existence of stabilizing selection, whereas results from work on the racial structure of flax rust populations in wild flax indicate that virulence does carry a fitness cost (Burdon and Jarosz, 1992).

Temporal and spatial changes in population structure

Wild populations of plant pathogens are genetically diverse. A collection of 45 Andean wild isolates of *Colletotrichum lindemuthianum* from five populations of wild common bean was shown to consist of at least 15 races (Sicard *et al.*, 1997), eight races were identified among 19 isolates of *Melampsora medusae* from natural stands of poplar (*Populus deltoides*) along the lower Mississippi river (Prakash and Thielges, 1987), and 13 races were detected among 96 isolates of *Melampsora lini* from natural populations of wild flax (*Linum marginale*) in New South Wales (Burdon and Jarosz, 1991). In the wild, shifts in population structure can occur with time, even during the course of a single season. Burdon and Jarosz (1991) found that the race of the flax rust fungus that predominated in the population early in the season increased over the course of the growing season from 67% of early season samples to 78% of late season samples. Short-term changes in population structure occur too in agriculture, as found in a single epidemic of powdery mildew, *E. graminis* f. sp. *hordei* (Welz *et al.*, 1990), but the shifts that occur across seasons are the better documented, and have been associated sometimes with the catastrophic consequences of a breakdown of resistance in a crop cultivar. Stakman *et al.* (1943) were among the first to monitor population trends of pathogen races, in the USA in wheat stem rust (*P. graminis* f. sp. *tritici*). The effects of changes in commercial use of wheat cultivars on the evolution of virulence in the wheat stem rust pathogen have been traced by Luig and Watson (1970) over a period of about 50 years in Australia and New Zealand. They recognized three eras: the first when no rust-resistant cultivars were grown, the second when those with a single gene for resistance were common, and the third in which wheat cultivars with resistance controlled by three or more genes became available. As more genes were accumulated into resistant cultivars, the pathogen adapted by gaining corresponding genes for virulence.

In Europe, the influence of host resistance on the population of the barley powdery mildew pathogen has received substantial attention in the classic studies of Wolfe (Wolfe and Schwarzbach, 1978; Wolfe and Barrett, 1979; Wolfe, 1987), and recent work (Andrivon and de Vallavieille-Pope, 1993) further reveals the rapidity with which evolving patterns of races of *E. graminis* f. sp. *hordei* become adapted to newly introduced host resistance genes. The selective effect of a host on a pathogen population is also well illustrated by the notorious epidemic of southern leaf blight of maize in the USA in 1970. At that time, about 85% of the US maize crop contained Texas male sterile cytoplasm so that the host population would have been highly selective for the increase of the race of *Cochliobolus heterostrophus* with matching virulence of this type. Its advance and consequences are well known (Ullstrup, 1972).

Assessment of the genetic diversity and population structure of the bacterial blight pathogen of rice (*Xanthomonas campestris* pv. *oryzae*) suggests that pathogen variability had been consistently high over a period of 15 years. The genetic diversity of the total population was 0.93, of which 42% was due to genetic differentiation of races (Leach *et al.*, 1992). Broad diversity was also indicated by investigations on the anthracnose pathogen (*Colletotrichum gloeosporioides*) in a stylo pasture in northern Queensland. Seven races were identified in 1987 and 16 races were found in 1989; about 40% of isolates were shown to be virulent on only two of the differentials and no evidence was found that more complex races predominated either in the greenhouse or in the field, and all isolates contributed evenly to the number of lesions produced. Davis *et al.* (1994) suggest that the continual natural diluting of field inoculum by the large diversity of other races may prevent the rapid development of severe epidemics in this pasture.

Virulence surveys may not only reveal temporal changes in population structure but also indicate that regional variations exist, with marked spatial changes in the degree of diversity between pathogen populations, as have been demonstrated with crown rust in oats between different zones in Australia (Brouwer and Oates, 1986). Relative uniformity within a population in a given area can suggest a recent introduction of the pathogen, as indicated by investigation of the population structure of the pitch canker fungus (*Fusarium subglutinans* f. sp. *pini*) in California (Gordon *et al.*, 1996), of *Phytophthora sojae* in Australia (Drenth *et al.*, 1996) and *C. gloeosporioides* on stylo in Australia (Chakraborty *et al.*, 1997). Differences in the structure of pathogen populations from genetically very similar host populations in natural ecosystems may imply that, in addition to resistance genes in the host, other evolutionary forces are also important in determining the genetic structure of individual pathogen populations (Jarosz and Burdon, 1991). Individual populations of *Melampsora lini* were found generally to be composed of one or two predominant races in addition to which there was a variable number of other races at very low incidences. The diversity of populations showed little variation either within populations across seasons or between populations within seasons, but at the level of individual populations, significant variation in the racial structure of four populations between years was detected. When compared in terms of virulence alone, significant year-to-year variation occurred in only two populations. Burdon and Jarosz (1992) suggested that the pathogen exists as an overall 'metapopulation', a set of local populations, with high degrees of migration occurring between individual populations. Sicard *et al.* (1997) also found indications of migration of the pathogen (*Colletotrichum lindemuthianum*) between wild populations. Recent developments in understanding of coevolved host and pathogen interactions suggest that many such systems survive as a consequence of natural fragmentation in both time and space. In such metapopulations, the incidence and severity of disease caused by a particular pathogen may vary substantially between populations spatially as well as within populations in time (Thrall and Burdon, 1997). There is now evidence, especially from the study of rusts and smuts, that this epidemiological asynchrony occurs over a range of distances from as little as only a few hundred metres (Burdon, 1998).

Much of the foregoing refers to pathogen population shifts governed by host resistance. Changes in population structure also occur in response to fungicide usage that exerts selection pressure favouring fungicide-resistant mutants. Such mutants exist as sub-populations at some low but finite frequency before exposure of the entire population to the new chemical. Their existence at a low frequency suggests that, in the absence of fungicide, they are less fit than fungicide-sensitive individuals which predominate in the population (Skylakakis, 1987). With increasing application of fungicide, insensitive mutants increase in the population because of their selective advantage compared with the wild-type, allowing them to reproduce to a greater extent and so to compensate to some extent for their relatively poor reproduction on the untreated crop (Wolfe, 1987).

The Consequences of Pathogen Diversity

The existence of diversity in plant pathogens and their ability to evolve and generate new diversity in agroecosystems provides a continuing capacity for disease development and, as a consequence, crop loss. There is little wonder that in agroecosystems, pathogen diversity is seen by many as harmful, to be managed or even eliminated (Lenné, 1998) rather than to be conserved to generate useful disease resistance (Ingram, 1998; see Polaszek *et al.*, Chapter 11, this volume). In reality, trade-offs are needed between the two views: we need to develop cost-effective and efficient ways of managing diseases which reduce food production but at the same time develop strategies to allow secure generation of diversity for disease resistance for future use in crop improvement. As management of pests and pathogens in agroecosystems is the subject of Chapter 11, our treatment here is brief.

Farmers have practised both direct and indirect methods of disease management since the dawn of agriculture (Thurston, 1992; see Polaszek *et al.*, Chapter 11, this volume). These practices include seed selection, rotation, roguing, intercropping, manipulation of plant density and irrigation. All have directly influenced pathogen diversity. The development of modern plant breeding about 100 years ago initiated a process of genetic manipulation for improvement, including substantial effort on disease resistance. Modern plant breeding for disease resistance, in conjunction with widespread use of chemicals, has had profound effects on pathogen diversity in agroecosystems. Modern plant breeding is often criticized for reducing diversity and increasing vulnerability to pathogens (Brown, 1983; see Witcombe, Chapter 10, and Polaszek *et al.*, Chapter 11, this volume). There is no doubt that inappropriate use of single gene resistances to variable pathogens has often resulted in 'boom and bust' cycles when those genes are overcome by matching virulence genes in the pathogen population (Robinson,

1976). However, many serious epidemics have occurred in diverse crops and cropping systems: the ravages of wheat rusts, the potato famine in Ireland, and the Bengal rice famine (see Large, 1940; Carefoot and Sprott, 1969; Padmanabhan, 1973). In these cases, uniformity for one or several genes governing susceptibility to a particular pathogen variant led to the crops' vulnerability (Buddenhagen, 1977). But there are many examples too of successful, long-term control of disease through use of single gene resistance in uniform crop cultivars (Marshall, 1977; Johnson, 1984), so the problem now is how to determine when crop uniformity poses a risk. There may be cogent lessons to be learnt from the mechanisms governing stability in pure stands in natural ecosystems, and it is seldom appreciated that diversity is *not* mandatory for ecosystem stability (Wood, 1998).

In natural ecosystems, the diversity of the host population may influence the diversity of the pathogen population, and presumably vice versa. Evidence to this effect has been presented by Prakash and Thielges (1987) from work on wild populations of poplar rust in which the apparent stability of variable populations of *Melampsori lini* was attributed to the genetic diversity of the host. Similarly, the effects of genetic diversity in experimental stands of golden rod (*Solidago altissima*) on the population structure of its powdery mildew pathogen (*Erysiphe cichoracearum*) revealed the potential role of pathogens as selective agents in host populations (Schmid, 1994). In coevolved combinations of host and pathogen, the greatest diversity in both pathogen and host populations probably occurs in centres of origin of the host, among its wild relatives, where resistance genes are most concentrated (Leppik, 1970). For this reason, 'pathogen parks' are sometimes advocated to maintain coevolved relationships (Browning, 1974; see Thurston *et al.*, Chapter 9, this volume). But attention is drawn pertinently to the paradox that long-term *in situ* conservation of wild relatives requires parallel, and potentially dangerous, conservation of the pathogen against which resistance is being sought (Burdon, 1998).

Some 34,000 plant species amounting to about 12.5% of the world's flora face extinction. Because plants are the primary producers in natural ecosystems, each species has dependent upon it about 30 other species, and it follows that for every plant species that becomes extinct, all others that are exclusively dependent upon it will be lost as well (Ingram, 1998). A plea for the conservation of plant pathogens may seem at best paradoxical, yet pathogens are key components of the biodiversity of all natural ecosystems. In temperate woodland, for example, *Heterobasidion annosum* kills one or more trees which in falling leave gaps in the canopy admitting sufficient light for the maintenance of flowering herbs in the field layer beneath (Ramsbottom, 1953), and in tropical forests everything from nutrient cycling to maintenance of biodiversity in the ecosystem depends on the action of tree pathogens (Gilbert, 1998). In agroecosystems, cases for the conservation of diversity among pathogens include their value in revealing sources of disease resistance (see page 127) and as the raw material for basic research on pathogen variation, evolution and population dynamics (Ingram, 1998).

Acknowledgements

We should like to thank Dr Louise Cook, of the Department of Agriculture in Northern Ireland, for information and recent references on potato late blight.

References

Alexander, H.M., Groth, J.V. and Roelfs, A.P. (1985) Virulence changes in *Uromyces appendiculatus* after five asexual generations on a partially resistant cultivar of *Phaseolus vulgaris*. *Phytopathology* 75, 449–453.

Allard, R.W. (1990) The genetics of host–pathogen coevolution: implications for genetic resource conservation. *Journal of Heredity* 81, 1–6.

Allen, D.J. (1983) *The Pathology of Tropical Food Legumes: Disease Resistance in Crop Improvement*. John Wiley & Sons, Chichester.

Allen, D.J. and Lenné, J.M. (1998) *The Pathology of Food and Pasture Legumes*. CAB International, Wallingford.

Allen, D.J., Lenné, J.M. and Wood, D. (1998) New encounter diseases and allopatric resistance. *Seventh International Congress of Plant Pathology*, Edinburgh, abstract 4.1.5.

Andrivon, D. (1996) The origin of *Phytophthora infestans* populations present in Europe in the 1840s: a critical review of historical and scientific evidence. *Plant Pathology* 45, 1027–1035.

Andrivon, D. and de Vallavieille-Pope, C. (1993) Racial diversity and complexity in regional populations of *Erysiphe graminis* f. sp. *hordei* in France over a 5-year period. *Plant Pathology* 42, 443–464.

Barrus, M.F. (1918) Varietal susceptibility of beans to strains of *Colletotrichum lindemuthianum* (Sacc. & Magn.) B. & C. *Phytopathology* 8, 589–614.

Boccas, B.R. (1981) Interspecific crosses between closely related heterothallic *Phytophthora* species. *Phytopathology* 71, 60–65.

Bradbury, J.F. (1986) *Guide to Plant Pathogenic Bacteria*. CAB International, Kew.

Brasier, C.M. (1971) Induction of sexual reproduction in single A2 isolates of *Phytophthora* by *Trichoderma viride*. *Nature New Biology* 231, 283.

Brasier, C.M. (1988) Rapid changes in the genetic structure of epidemic populations of *Ophiostoma ulmi*. *Nature* 332, 538–541.

Brasier, C.M. (1990) The unexpected element: mycovirus involvement in two recent pandemics, Dutch elm disease and chestnut blight. In: Burdon, J.J. and Leather, S.R. (eds) *Pest, Pathogens and Plant Communities*. Blackwell Scientific Publications, Oxford, pp. 289–307.

Brouwer, J.B. and Oates, J.D. (1986) Regional variation of *Puccinia coronata avenae* in Australia and its implications for oat breeding. *Annals of Applied Biology* 109, 269–277.

Brown, J.F. and Sharp, E.L. (1970) The relative survival ability of pathogenic types of *Puccinia striiformis* in mixtures. *Phytopathology* 60, 529–533.

Brown, W.L. (1983) Genetic diversity and genetic vulnerability: an appraisal. *Economic Botany* 37, 4–12.

Browning, J.A. (1974) Relevance of knowledge about natural ecosystems to development of pest management programs for agro-ecosystems. *Proceedings of the American Phytopathological Society* 1, 191–199.

Buddenhagen, I.W. (1977) Resistance and vulnerability of tropical crops in relation to their evolution and breeding. *Annals of the New York Academy of Sciences* 287, 309–326.

Burdon, J.J. (1987) *Diseases and Plant Population Biology.* Cambridge University Press, Cambridge.

Burdon, J.J. (1992) The growth and regulation of pathogenic fungal populations. In: Carroll, G.C. and Wicklow, D.T. (eds) *The Fungal Community: its Organisation and Role in the Ecosystem.* Marcel Dekker, New York, pp. 173–181.

Burdon, J.J. (1993) The structure of pathogen populations in natural plant communities. *Annual Review of Phytopathology* 31, 305–323.

Burdon, J.J. (1998) Biodiversity and the conservation of wild relatives of crops for disease resistance. *Seventh International Congress of Plant Pathology,* Edinburgh, abstract 4.1.2S.

Burdon, J.J. and Jarosz, A.M. (1991) Host–pathogen interactions in natural populations of *Linum marginale* and *Melampsora lini*: I. Patterns of resistance and racial variation in a large host population. *Evolution* 45, 205–217.

Burdon, J.J. and Jarosz, A.M. (1992) Temporal variation in the racial structure of flax rust (*Melampsora lini*) populations growing on natural stands of wild flax (*Linum marginale*): local versus metapopulation dynamics. *Plant Pathology* 41, 165–179.

Burdon, J.J. and Roelfs, A.P. (1985) The effect of sexual and asexual reproduction on the isozyme structure of populations of *Puccinia graminis. Phytopathology* 75, 1068–1073.

Burdon, J.J. and Shattock, R.C. (1980) Disease in plant communities. *Applied Biology* 5, 145–219.

Burdon, J.J. and Thrall, P.H. (1998) Patterns in the evolutionary dynamics of host–pathogen interactions. *Seventh International Congress of Plant Pathology,* Edinburgh, abstract 2.3.1S.

Burdon, J.J., Wennstrom, A., Elmquist, T. and Kirkby, G.C. (1996) The role of race specific resistance in natural plant populations. *Oikos* 76, 411–417.

Buxton, E.W. (1956) Heterokaryosis and parasexual recombination in pathogenic strains of *Fusarium oxysporum. Journal of General Microbiology* 15, 133–139.

Callow, J.A. (1977) Recognition, resistance and the role of plant lectins in host-parasite interactions. *Advances in Botanical Research* 4, 1–49.

Carefoot, G.L. and Sprott, E.R. (1969) *Famine on the Wind: Plant Diseases and Human History.* Angus & Robertson, London.

Caten, C.E. (1987) The concept of race in plant pathology. In: Wolfe, M.S. and Caten, C.E. (eds) *Populations of Plant Pathogens: their Dynamics and Genetics.* Blackwell, Oxford, pp. 21–37.

Chakraborty, S., Perrott, R., Charchar, M.J.D'A., Fernandes, C.D. and Kelemu, S. (1997) Biodiversity, epidemiology and virulence of *Colletotrichum gloeosporioides.* II. Genetic and pathogenic diversity in isolates of *Colletotrichum gloeo-*

sporioides from eight species of *Stylosanthes. Tropical Grasslands* 31, 393–401.

Crute, I.R., Holub, E.B. and Burdon, J.J. (1997) *The Gene-for-Gene Relationship in Plant–Parasite Interactions.* CAB International, Wallingford.

Davis, R.D., Boland, R.M. and Howitt, C.J. (1994) The developing relationship between *Stylosanthes* and anthracnose after 14 years in a North Queensland pasture. 2. Diversity in the pathogen population. *Australian Journal of Experimental Agriculture* 34, 621–626.

Dinoor, A. and Eshed, N. (1997) Plant conservation *in situ* for disease resistance. In: Maxted, N., Ford-Lloyd, B.V. and Hawkes, J.G. (eds) *Plant Genetic Conservation.* Chapman & Hall, London, pp. 323–336.

Dobson, A. and Crawley, M. (1994) Pathogens and the structure of plant communities. *Trends in Ecology and Evolution* 9, 393–398.

Drenth, A., Tas, I.C.Q. and Govers, F. (1994) DNA fingerprinting uncovers a new sexually reproducing population of *Phytophthora infestans* in the Netherlands. *European Journal of Plant Pathology* 100, 97–107.

Drenth, A., Whisson, S.C., Maclean, D.J., Irwin, J.A.G., Obst, N.R. and Ryley, M.J. (1996) The evolution of races of *Phytophthora sojae* in Australia. *Phytopathology* 86, 163–169.

Drijfhout, E. and Davis, J.H.C. (1989) Selection of a new set of homogeneously reacting bean (*Phaseolus vulgaris*) differentials to differentiate races of *Colletotrichum lindemuthianum. Plant Pathology* 38, 391–396.

Durbin, R.D. (1959) Factors affecting the vertical distribution of *Rhizoctonia solani* with reference to CO_2 concentration. *American Journal of Botany* 46, 22–25.

Edwards, M. (1992) Use of molecular markers in the evaluation and introgression of genetic diversity for quantitative traits. *Field Crops Research* 29, 241–260.

Ennos, R.A. and Swales, K.W. (1991) Genetic variation in a fungal pathogen: response to host defensive chemicals. *Evolution* 45, 190–204.

Flor, H.H. (1956) The complementary genic systems in flax and flax rust. *Advances in Genetics* 8, 29–54.

Forster, H., Tyler, B.M. and Coffey, M.D. (1994) *Phytophthora sojae* races have arisen by clonal evolution and by rare outcrosses. *Molecular Plant–Microbe Interactions* 7, 780–791.

Frank, S.A. (1993) Coevolutionary genetics of plants and pathogens. *Evolutionary Ecology* 7, 45–75.

Frison, E.A. and Sharrock, S.L. (eds) (1998) Banana streak virus: a unique virus – *Musa* interaction? *Proceedings of a Workshop of the PROMUSA Virology Working Group*, Montpellier, France, 19–21 January 1998, IPGRI, Rome, and INIBAP, Montpellier.

Fry, W.E. and Goodwin, S.B. (1995) Recent migrations of *Phytophthora infestans.* In: Dowley, L.J., Bannon, E., Cooke, L.R., Keane, T. and O'Sullivan, E. (eds) *Phytophthora infestans.* European Association for Potato Research and Boole Press, Dublin, pp. 89–95.

Fry, W.E., Goodwin, S.B., Dyer A.T., Matussak, J.M., Drenth, A., Tooley, P.W., Sujkowski, L.S., Koh, Y.J., Cohen, B.A., Spielman, L.J., Deahl, K.L., Inglis, D.A. and Sandlan, K.P. (1993) Historical and recent migrations of *Phytophthora infestans*: chronology, pathways and implications. *Plant Disease* 77, 653–661.

Gagne, G., Roeckel-Drevet, P., Grezes-Besset, B., Shindrova, P., Ivanov, P., Crand-Ravel, C., Vear, F., Tourvielle de Labrouhe, D., Charmet, G. and Nicolas, P. (1998) Genetic diversity of *Orobanche cumana* populations in different

European countries. *Seventh International Congress of Plant Pathology*, Edinburgh, abstract 2.2.94.

Georgopoulos, S.G. (1987) The development of fungicide resistance. In: Wolfe, M.S. and Caten, C.E. (eds) *Populations of Plant Pathogens: their Dynamics and Genetics*. Blackwell, Oxford, pp. 239–251.

Gibbs, A. and Harrison, B.D. (1976) *Plant Virology: the Principles*. Edward Arnold, London.

Gilbert, G. (1998) Importance of tree diseases to the biodiversity and conservation of forest ecosystems. *Seventh International Congress of Plant Pathology*, Edinburgh, abstract 4.1.4S.

Gillett, J.B. (1962) Pest pressure, an underestimated factor in evolution. *Systematics Association Publications* 4, 37–46.

Gordon, T.R., Storer, A.J. and Okamoto, D. (1996) Population structure of the pitch canker pathogen, *Fusarium subglutinans* f. sp. *pini*, in California. *Mycological Research* 100, 850–854.

Green, G.J. (1973) Environmental races of *Puccinia graminis tritici* in North America and their epidemiological potential. *Second International Congress of Plant Pathology*, Minneapolis, abstract 0720.

Greenaway, W. and Cowan, J.W. (1970) The stability of mercury resistance in *Pyrenophora avenae*. *Transactions of the British Mycological Society* 54, 127–138.

Groth, J.V. and Roelfs, A.P. (1987) Analysis of virulence diversity in populations of plant pathogens. In: Wolfe, M.S. and Caten, C.E. (eds) *Populations of Plant Pathogens: their Dynamics and Genetics*. Blackwell, Oxford, pp. 63–74.

Gustafsson, M., Arhammer, M. and Gustafsson, I. (1983) Linkage between virulence genes, compatibility types and sexual recombination in the Swedish population of *Bremia lactucae*. *Phytopathologische Zeitschrift* 108, 341–354.

Guzman, P., Gilbertson, R.L., Nodari, R., Johnson, W.C., Temple, S.R., Mandala, D., Mkandawire, A.B.C. and Gepts, P. (1995) Characterization of variability in the fungus *Phaeoisariopsis griseola* suggests coevolution with the common bean (*Phaseolus vulgaris*). *Phytopathology* 85, 600–607.

Habgood, R.M. (1970) Designation of physiological races of plant pathogens. *Nature* 227, 1268–1269.

Hall, H.T.B. (1977) *Diseases and Parasites of Livestock in the Tropics*. Longman Scientific and Technical, London.

Harlan, J.R. (1976) Disease as a factor in plant evolution. *Annual Review of Phytopathology* 14, 31–51.

Harper, J.L. (1990) Pest, pathogens and plant communities: an introduction. In: Burdon J.J. and Leather, S.R. (eds) *Pest, Pathogens and Plant Communities*. Blackwell Scientific Publications, Oxford, pp. 3–14.

Harris, M.K. (1975) Allopatric resistance: searching for sources of insect resistance for use in agriculture. *Environmental Entomology* 4, 661–669.

Hawksworth, D.L. (1991) *The Biodiversity of Micro-organisms and Invertebrates: its Role in Sustainable Agriculture*. CAB International, Wallingford.

Hawksworth, D.L. (1994) *Biodiversity: Measurement and Estimation*. The Royal Society in association with Chapman & Hall, London.

Hawksworth, D.L. and Colwell, R.R. (1992) Biodiversity amongst microorganisms and its relevance. *Biodiversity and Conservation* 1, 221–226.

Hawksworth, D.L. and Ritchie, J.M. (1993) *Biodiversity and Biosystematic Priorities: Micro-organisms and Invertebrates*. CAB International, Wallingford.

Hawksworth, D.L., Kirk, P.M., Sutton, B.C. and Pegler, D.N. (1995) *Dictionary of the Fungi*, 8th edn. CAB International, Wallingford.

Holliday, P. (1989) *A Dictionary of Plant Pathology*. Cambridge University Press, Cambridge.

Hong, Y.G. and Harrison, B.D. (1995) Nucleotide sequences from tomato leaf curl viruses from different countries: evidence for three geographically separate branches in evolution of the coat protein of whitefly-transmitted geminiviruses. *Journal of General Virology* 76, 2043–2049.

Hsiang, T. and Chastagner, G.A. (1993) Variation in *Melampsora occidentalis* rust on poplars in the Pacific Northwest. *Canadian Journal of Plant Pathology* 15, 175–181.

Ingram, D.S. (1998) Everything in the garden's lovely. *New Scientist* 2154, 44.

Irwin, J.A.G. (1988) *Uromyces appendiculatus* var. *crassitunicatus* var. nov. from *Macroptilium atropurpureum*. *Australian Systematic Botany* 1, 363–367.

Jarosz, A.M. and Burdon, J.J. (1991) Host–pathogen interactions in natural populations of *Linum marginale* and *Melampsora lini*. II. Local and regional variation in patterns of resistance and racial structure. *Evolution* 45, 1618–1627.

Jellis, G.J. and Punithalingam, E. (1991) Discovery of *Didymella fabae* sp. nov., the teleomorph of *Ascochyta fabae*, on faba bean straw. *Plant Pathology* 40, 150–157.

Johnson, R. (1984) A critical analysis of durable resistance. *Annual Review of Phytopathology* 22, 309–322.

Katan, T., Katan, J., Gordon, T.R. and Pozniak, D. (1994) Physiologic races and vegetative compatibility groups of *Fusarium oxysporum* f. sp. *melonis* in Israel. *Phytopathology* 84, 153–157.

Katsuya, K. and Green, G.J. (1967) Reproductive potentials of races 15B and 56 of wheat stem rust. *Canadian Journal of Botany* 45, 1077–1091.

La Fleur, D.A., Lockhart, B.E.L. and Olszewski, N.E. (1996) Portions of the banana streak badnavirus genome are integrated in the genome of its host. *Phytopathology* 86, 100 (abstract).

Lane, J.A., Moore, T.H.M., Child, D.V. and Cardwell, K.F. (1996) Characterization of virulence and geographic distribution of *Striga gesnerioides* on cowpea in West Africa. *Plant Disease* 80, 299–301.

Large, E.C. (1940) *The Advance of the Fungi*. Jonathan Cape, London.

Leach, J.E., Rhoads, M.L., Cruz, C.M.V., White, F.F., Mew, T.W., Leung, H. and Vera-Cruz, C.M. (1992) Assessment of genetic diversity and population structure of *Xanthomonas oryzae* pv. *oryzae* with a repetitive DNA element. *Applied and Environmental Microbiology* 58, 2188–2195.

Lebeda, A. (1982) Measurement of genotypic diversity of virulence in populations of phytopathogenic fungi. *Zeitschrift für Pflanzenkrankheiten und Pflanzenschutz* 89, 88–95.

Lenné, J.M. (1998) The biodiversity and conservation of crops for disease resistance. *Seventh International Congress of Plant Pathology*, Edinburgh, abstract 4.1.1S.

Lenné, J.M. and Burdon, J.J. (1990) Virulence and isozymic variation within and between natural populations of *Colletotrichum gloeosporioides* from *Stylosanthes guianensis*. *Phytopathology* 80, 728–731.

Lenné, J.M. and Wood, D. (1991) Plant diseases and the use of wild germplasm. *Annual Review of Phytopathology* 29, 35–63.

Leppik, E.E. (1965) Some viewpoints on the phylogeny of fungi. V. Evolution of biological specialisation. *Mycologia* 57, 6–22.

Leppik, E.E. (1970) Gene centers of plants as sources of disease resistance. *Annual*

Review of Phytopathology 8, 323–344.

Littrell, R.H. (1974) Tolerance in *Cercospora arachidicola* to benomyl and related fungicides. *Phytopathology* 64, 1377–1378.

Luig, N.H. and Watson, I.A. (1970) The effect of complex genetic resistance in wheat on the variability of *Puccinia graminis* f. sp. *tritici*. *Proceedings of the Linnean Society of New South Wales* 95, 22–45.

Luttrell, E.S. (1974) Parasitism of fungi on vascular plants. *Mycologia* 66, 1–15.

Marshall, D.R. (1977) The advantages and hazards of genetic homogeneity. *Annals of the New York Academy of Sciences* 287, 1–20.

Masaba, D. and Waller, J.M. (1992) Coffee berry disease: the current status. In: Bailey, J.A. and Jeger, M.J. (eds) *Colletotrichum: Biology, Pathology and Control*. CAB International, Wallingford, pp. 237–249.

Masel, A.M., He, C.Z., Poplawski, A.M., Irwin, J.A.G. and Manners, J.M. (1996) Molecular evidence for chromosome transfer between biotypes of *Colletotrichum gloeosporioides*. *Molecular Plant–Microbe Interactions* 9, 339–348.

McKenzie, A.R., Flentje, N.T., Stretton, H.M. and Mayo, M.J. (1969) Heterokaryon formation and genetic recombination within one isolate of *Thanatephorus cucumeris*. *Australian Journal of Biological Science* 22, 895–904.

Meyer, O. (1993) Functional groups of micro-organisms. In: Schulze, E.D. and Mooney, H.A. (eds) *Biodiversity and Ecosystem Function*. Springer-Verlag, Berlin, pp. 67–96.

Musselman, L.J., Bhrathalakshmi, Safa, S.B., Knepper, D.A., Mohamed, K.I. and White, C.L. (1991) Recent research on the biology of *Striga asiatica, S. gesnerioides* and *S. hermonthica*. In: Kim, S.K. (ed.) *Combating Striga in Africa*. International Institute of Tropical Agriculture, Ibadan, Nigeria, pp. 31–41.

Natesham, H.M., Muniyappa, V., Swanson, M.M. and Harrison, B.D. (1996) Host range, vector relations and serological relationships of cotton leaf curl virus from southern India. *Annals of Applied Biology* 128, 233–244.

Neergaard, P. (1977) Quarantine policy for seed in transfer of genetic resources. In: Hewitt, W.B. and Chiarappa, L. (eds) *Plant Health and Quarantine in International Transfer of Genetic Resources*. CRC Press, Cleveland, Ohio, pp. 309–314.

Nelson, R.R. (1972). Stabilizing racial populations of plant pathogens by use of resistance genes. *Journal of Environmental Quality* 1, 220–227.

Newhook, F.J. and Podger, F.D. (1972) The role of *Phytophthora cinnamomi* in Australian and New Zealand forests. *Annual Review of Phytopathology* 10, 299–326.

Oberwinkler, F. (1992) Biodiversity amongst filamentous fungi. *Biodiversity and Conservation* 1, 293–311.

O'Donnell, A.G., Goodfellow, M. and Hawksworth, D.L. (1994) Theoretical and practical aspects of the quantification of biodiversity amongst microorganisms. In: Hawksworth, D.L. (ed.) *Biodiversity: Measurement and Estimation*. The Royal Society and Chapman & Hall, London, pp. 65–73.

Ogle, H.J. and Brown, J.F. (1970) Relative ability of two strains of *Puccinia graminis tritici* to survive when mixed. *Annals of Applied Biology* 66, 273–279.

Olive, E.W. (1911) Origin of heteroecism in the rusts. *Phytopathology* 1, 139–149.

Opio, A.F., Allen, D.J. and Teri, J.M. (1996) Pathogenic variation in *Xanthomonas campestris* pv. *phaseoli*, the causal agent of common bacterial blight in *Phaseolus* beans. *Plant Pathology* 45, 1126–1133.

Padmanabhan, S.Y. (1973) The great Bengal rice famine. *Annual Review of Phytopathology* 11, 11–26.

Parbery, D.G. (1996) Trophism and the ecology of fungi associated with plants. *Biological Reviews* 71, 473–527.

Parlevliet, J.E. (1981) Stabilizing selection in crop pathosystems: an empty concept or a reality? *Euphytica* 30, 259–269.

Parmeter, J.R., Snyder, W.C. and Reichle, R.E. (1963) Heterokaryosis and variability in plant pathogenic fungi. *Annual Review of Phytopathology* 1, 51–76.

Pearce, F. (1997) The famine fungus. *New Scientist* 2079, 32–36.

Petrini, O. (1998) What are endophytes anyway? *Seventh International Congress of Plant Pathology*, Edinburgh, abstract 2.9.1S.

Phillips, A.J.L. (1996) Variation in pathogenicity among isolates of *Elsinoe phaseoli* from *Phaseolus* species. *Annals of Applied Biology* 128, 209–218.

van der Plank, J.E. (1968) *Disease Resistance in Plants*. Academic Press, New York.

Pontecorvo, G. (1956) The parasexual cycle in fungi. *Annual Review of Microbiology* 10, 393–400.

Prakash, C.S. and Thielges, B.A. (1987) Pathogenic variation in *Melampsora medusae* leaf rust of poplars. *Euphytica* 36, 563–570.

Ramos, A.H. and Kamidi, R.E. (1982) Determination and significance of the mutation rate of *Colletotrichum coffeanum* from benomyl sensitivity to benomyl tolerance. *Phytopathology* 72, 181–185.

Ramsbottom, J. (1953) *Mushrooms and Toadstools*. Collins New Naturalist, London.

Riley, R. (1973) Genetic changes in hosts and the significance of disease. *Annals of Applied Biology* 75, 128–132.

Robinson, R.A. (1976) *Plant Pathosystems*. Springer-Verlag, Berlin.

Robinson, R.A. (1987) *Host Management in Crop Pathosystems*. Macmillan, New York.

Russell, P.E. (1975) Variation in the virulence of some streptomycin resistant mutants of *Pseudomonas phaseolicola*. *Journal of Applied Bacteriology* 39, 175–180.

Sanderson, K.E. and Srb, A.M. (1965) Heterokaryosis and parasexuality in the fungus *Ascochyta imperfecta*. *American Journal of Botany* 52, 72–81.

Savile, D.B.O. (1954) The fungi as aids in the taxonomy of flowering plants. *Science* 120, 583–585.

Savile, D.B.O. (1990) Coevolution of Uredinales and Ustilaginales with vascular plants. *Report of the Tottori Mycological Society* 28, 15–24.

Schmid, B. (1994) Effects of genetic diversity in experimental stands of *Solidago altissima* – evidence for the potential role of pathogens as selective agents in plant populations. *Journal of Ecology* 82, 165–175.

Schots, A., Dewey, F.M. and Oliver, R.P. (1994) *Modern Assays for Plant Pathogenic Fungi: Identification, Detection and Quantification*. CAB International, Wallingford.

Scott, P.R., Johnson, R., Wolfe, M.S., Lowe, H.J.B. and Bennett, F.G.A. (1980) Host-specificity in cereal parasites in relation to their control. *Applied Biology* 5, 349–393.

Shattock, R.C. and Day, J.P. (1996) Migration and displacement; recombinants and relicts: twenty years in the life of potato late-blight (*Phytophthora infestans*). In: *Proceedings of the Brighton Crop Protection Conference, Pests and Diseases*, Vol. 3, 1129–1136.

Sheffield, F.M.L. (1968) Closed quarantine procedures. *Review of Applied Mycology* 47, 1–8.

Shepherd, C.J. (1978) Mating behaviour of Australian isolates of *Phytophthora* species. I. Inter- and intraspecific mating. *Australian Journal of Botany* 26, 123–138.

Shepherd, C.J. and Pratt, B.H. (1973) Separation of two ecotypes of *Phytophthora drechsleri* Tucker occurring in Australian native forests. *Australian Journal of Biological Sciences* 26, 1095–1107.

Sicard, D., Buchet, S., Michalakis, Y. and Neema, C. (1997) Genetic variability of *Colletotrichum lindemuthianum* in wild populations of common bean. *Plant Pathology* 46, 355–365.

Simons, M.D., Rothman, P.G. and Michel, L.J. (1979) Pathogenicity of *Puccinia coronata* from buckthorn and from oats adjacent to and distant from buckthorn. *Phytopathology* 69, 156–158.

Skylakakis, G. (1987) Changes in the composition of pathogen populations caused by resistance to fungicides. In: Wolfe, M.S. and Caten, C.E. (eds) *Populations of Plant Pathogens: their Dynamics and Genetics*. Blackwell, Oxford, pp. 227–237.

Spence, N.J. and Walkey, D.G.A. (1995) Variation for pathogenicity among isolates of bean common mosaic virus in Africa and a reinterpretation of the genetic relationship between cultivars of *Phaseolus vulgaris* and pathotypes of BCMV. *Plant Pathology* 44, 527–546.

Stakman, E.C. (1914) A study in cereal rusts: physiological races. *Minnesota Agricultural Experimental Station Bulletin* 138.

Stakman, E.C., Loegering, W.Q., Cassell, E.C. and Hines, L. (1943) Population trends of physiologic races of *Puccinia graminis tritici* in the United States for the period 1930–1941. *Phytopathology* 33, 884–898.

Staskawicz, B.J. (1998) Molecular biology of disease resistance: from genetic understanding to control. *Seventh International Congress of Plant Pathology*, Edinburgh, abstract 3.2.PL.

Subrahmanyam, P., Reddy, L.J., Gibbons, R.W. and McDonald, D. (1985) Peanut rust: a major threat to peanut production in the semiarid tropics. *Plant Disease* 69, 813–819.

Tarr, S.A.J. (1972) *The Principles of Plant Pathology*. Macmillan, London.

Taylor, J.D., Teverson, D.M., Allen, D.J. and Pastor-Corrales, M.A. (1996) Identification and origin of races of *Pseudomonas syringae* pv. *phaseolicola* from Africa and other bean growing areas. *Plant Pathology* 45, 469–478.

Thrall, P.H. and Burdon, J.J. (1997) Host–pathogen dynamics in a metapopulation context: the ecological and evolutionary consequences of being spatial. *Journal of Ecology* 85, 743–753.

Thresh, J.M. (1980) The origin and epidemiology of some important plant virus diseases. *Applied Biology* 5, 1–65.

Thresh, J.M. (1981) *Pests, Pathogens and Vegetation*. The Pitman Press, Bath.

Thurston, H.D. (1992) *Sustainable Practices for Plant Disease Management in Traditional Systems*. Westview Press, Boulder, Colorado.

Tinline, R.D. and MacNeill, B.H. (1969) Parasexuality in plant pathogenic fungi. *Annual Review of Phytopathology* 7, 147–170.

Tommerup, I.C. (1998) *Phytophthora*: an introduced hazard in forests, woodlands and woody heaths. *Seventh International Congress of Plant Pathology*, Edinburgh, abstract 4.4.1S.

Tooley, P.W., Therren, C.D. and Ritch, D.L. (1989) Mating type, race composition, nuclear DNA content and isozyme analysis of Peruvian isolates of *Phytophthora infestans*. *Phytopathology* 79, 478–481.

Trapero-Casas, A. and Kaiser, W.J. (1992) Development of *Didymella rabiei*, the teleomorph of *Ascochyta rabiei*, on chickpea straw. *Phytopathology* 82, 1261–1266.

Ullstrup, A.J. (1972) The impacts of the southern corn leaf blight epidemics of 1970–1971. *Annual Review of Phytopathology* 10, 37–50.

Vilgalys, R. (1988) Genetic relatedness among anastomosis groups in *Rhizoctonia* as measured by DNA/DNA hybridization. *Phytopathology* 78, 698–702.

Wahl, I. and Anikster, Y. (1982) Role of the alternate host in evolution of some cereal rusts. *Garcia de Orta, Estudios Agronomicos* 9, 21–27.

Walker, J.C. and Patel, P.N. (1964) Inheritance of resistance to halo blight of bean. *Phytopathology* 54, 952–954.

Waller, J.M. (1991) General discussion. In: Hawksworth, D.L. (ed.) *The Biodiversity of Micro-organisms and Invertebrates: its Role in Sustainable Agriculture.* CAB International, Wallingford, p. 197.

Waller, J.M. and Brayford, D. (1990) *Fusarium* diseases in the tropics. *Tropical Pest Management* 36, 181–194.

Waller, J.M. and Bridge, P.D. (1999) Recent advances in understanding *Colletotrichum* diseases of some tropical perennial crops. In: Dickman, M., Prusky, D. and Freeman, S. (eds) *Host Specificity, Pathology and Host–Pathogen Interactions of Colletotrichum.* American Phytopathological Society (in press).

Watson, I.A. (1970) Changes in virulence and population shifts in plant pathogens. *Annual Review of Phytopathology* 8, 209–230.

Welz, H.G., Nagarajan, S. and Kranz, J. (1990) Short-term virulence dynamics of *Erysiphe graminis* f. sp. *hordei* in a single epidemic on two susceptible barley cultivars. *Zeitschrift für Pflanzenkrankheiten und Pflanzenschutz* 97, 250–262.

Weste, G. and Marks, G.C. (1987) The biology of *Phytophthora cinnamomi* in Australasian forests. *Annual Review of Phytopathology* 25, 207–273.

Wheeler, B.J. (1969) *An Introduction to Plant Diseases.* John Wiley & Sons, London.

White, N.H. (1992) A case for the antiquity of fungal parasitism in plants. *Advances in Plant Pathology* 8, 31–37.

Wills, C. (1996) Safety in diversity. *New Scientist* 2002, 38–42.

Wilson, M. and Henderson, D.M. (1966) *British Rust Fungi.* Cambridge University Press, Cambridge.

Wolfe, M.S. (1987) Trying to understand and control powdery mildew. In: Wolfe, M.S. and Caten, C.E. (eds) *Populations of Plant Pathogens: their Dynamics and Genetics.* Blackwell, Oxford, pp. 253–273.

Wolfe, M.S. and Barrett, J.A. (1979) Disease in crops: controlling the evolution of plant pathogens. *Journal of the Royal Society of Arts* 127, 321–333.

Wolfe, M.S. and Schwarzbach, E. (1978) Patterns of race changes in powdery mildew. *Annual Review of Phytopathology* 16, 159–180.

Wood, D. (1998) Ecological principles in agricultural policy: but which principles? *Food Policy* 23, 371–381.

Zhou, X., Liu, Y., Calvert, L., Munoz, C., Otim-Nape, G.W., Robinson, D.J. and Harrison, B.D. (1997) Evidence that DNA-A of a geminivirus associated with severe cassava mosaic disease in Uganda has arisen by interspecific recombination. *Journal of General Virology* 78, 2101–2111.

7

Insect Biodiversity in Agroecosystems: Function, Value and Optimization

J. LaSalle

Unit of Parasitoid Systematics, CABI Bioscience UK Centre, Department of Biology, Imperial College at Silwood Park, Ascot, Berkshire SL5 7PY, UK

This chapter can only begin to cover everything that insects do in agro-ecosystems, and will concentrate on the function of beneficial, rather than detrimental, insects. It is taken as given that pests will invade agriculture systems and sometimes cause serious damage, but that these represent a relatively minor portion of the total insect biodiversity present. It is the purpose of a functional agrobiodiversity programme to manage an agro-ecosystem in order to get the optimum diversity within the system. The dual objectives of a functional agrobiodiversity programme are quite simple: increase the number of beneficial organisms, decrease the number of detri-mental organisms. However, it is often by performing the first objective that the second objective is achieved. Thus, the focus of this chapter is on the role and management of beneficial organisms. Similarly, this chapter will con-centrate on the elements of management strategies which employ biological control and/or integrated pest management (IPM), as these are seen as strategies which try to enhance and optimize biodiversity rather than limit it.

Insects are dominant components of most ecosystems, and agro-ecosystems are no exception. Insects represent some of the best and the worst things that can happen to plants: the best because they play a key role in pollination and natural pest control; the worst because most key agri-cultural pests are in fact insects. Insects in agroecosystems are as essential as they are destructive. In addition to the more visible pest and beneficial insects in agroecosystems, there are a myriad of often unnoticed 'ecosystem engineers'. Insects form a vital component of these mediators of a variety of ecosystem functions.

This chapter will cover the following main topics: the function of insects in agroecosystems; optimization of insect diversity in agroecosystems – theoretical and practical considerations; the importance of taxonomy to agrobiodiversity projects; insect agrobiodiversity in application; and examples of some critical issues in biological control and conservation biology. It will concentrate on insects, but will bear in mind that insects are only a part of the arthropod fauna present in agroecosystems. Many other important arthropod groups, and in particular spiders and mites, are present in agroecosystems. Examples from these groups will be used where appropriate.

This chapter will concentrate on species-level diversity, but will include discussions and examples of other levels of diversity. There are three main levels of diversity: genes, species and ecosystems (Reid and Miller, 1989; Solbrig, 1991), and diversity in each level is valuable in agroecosystems. Way and Heong (1994) discussed several components of diversity relevant to insect pest management in rice. They included within-species diversity, within-field (community) diversity, within groups of fields (compound community) diversity, and within groupings of rice and non-rice habitats (ecosystem diversity). Therefore, when discussing agrobiodiversity, it is essential that we not only consider the number of species in a system, but the genetic variation within each species and the interactions among the species (see Polaszek *et al.*, Chapter 11, this volume).

Natural control, and by extension biological control, will be an important focus of this chapter. Parasitoids and predators which regulate the population levels of herbivores are essential components in any ecosystem. The insect parasitoids and predators are of particular importance in agroecosystems, which are typified by having large populations of few plant species. This provides ideal conditions for producing large populations of herbivores, which can easily destroy a crop if not checked.

Function of Insect Biodiversity in Agroecosystems

Beneficial insects perform many functions in agricultural ecosystems. In addition to the type of functions which are generally easily visible, such as pollination and serving as natural enemies, there are a huge number of more hidden functions which are discussed briefly under the subject of ecosystem engineers.

Pollinators

Pollination is essential to agriculture. Although many of the world's most important crops (such as cereals) are wind pollinated, about 30% of human food is derived from bee-pollinated plants (McGregor, 1976). Bees are also essential to forage crops such as lucerne (alfalfa) and clover (O'Toole, 1993). Bees (honeybees, bumblebees, solitary bees) are the most important group of insect pollinators, although other insects, mainly Diptera, are important as supplementary pollinators (Free, 1993). O'Toole (1993) discussed the importance of native bees in agroecosystems. Honeybees (*Apis mellifera*) are important pollinators in agroecosystems, and have been credited with being responsible for as much as 80% of all insect pollination of crops (Robinson *et al.*, 1989); however, other authors (Kevan *et al.*, 1990; O'Toole, 1993) felt that this estimate was too high, and O'Toole (1993) offered several examples of crops which depend on native bees rather than honeybees for optimal pollination.

Examples of pollination by both honeybees and native bees have been well recorded (see reviews in Free, 1993; O'Toole, 1993). A good example of non-bee pollination is the oil palm, *Elaeis guineensis* (see Free, 1993). This palm is native to West Africa, but has since been moved to other tropical areas of the world. It had been assumed that the inflorescences were wind pollinated, supported by the plant's production of large quantities of small, smooth pollen grains. However, oil production in other areas never reached the levels of West Africa. It took detailed studies by Syed (1979, 1982) to demonstrate that oil palm was indeed insect pollinated, and that the pollinators were small weevils in the genus *Elaeidobus* (Curculionidae), with *E. kamerunicus* being one of the most important species. This species was imported into oil palm growing regions in Malaysia, where it resulted in a marked increase in kernel and oil yield. In under a year, oil yield in Peninsular Malaysia had risen by 20%, and in East Malaysia by 53% (Syed *et al.*, 1982). This species has since been moved to several other oil palm growing areas in Asia (Free, 1993). Other examples of non-bee pollinators include fig wasps (Agaonidae) for figs, and ceratopogonid midges for cocoa (Free, 1993).

Increasing crop yield is not the only benefit of insect pollinators (Free, 1993). Many plants, particularly those capable of self-pollination, can achieve some measure of pollination without the presence of insects. However, pollination by insect pollinators has many advantages. The presence of pollinators at the proper time can result in an earlier and more uniform crop. Self-pollination can result in inadequate pollination, and small, inferior fruits; pollinators can alleviate this and ensure good quality fruit. Insect pollination can increase the quality of fruits and seeds, including an increase in: oil content in lavender, sunflower and safflower, rubber content in *Parthenium argentatum*, and pyrethrin content in *Chrysanthemum cinerariifolium* (Free, 1993).

Bumblebees are an example of an introduced pollinator. New Zealand has an ideal climate for growing forage crops such as clovers (*Trifolium* spp.), but a very small indigenous fauna of bees which are short-tongued. Although clover grew well, it was inadequately pollinated (particularly the long-tubed cultivars), and seed had to be imported. In the 1880s three species of bumblebee were introduced into New Zealand from England. These bees became established, and were able to properly pollinate clover. In addition to providing the basis for cheap dairy produce and meat, this meant that within a few years New Zealand switched from being an importer to an exporter of clover seed (Gurr, 1962, 1972; O'Toole, 1993).

Parasitoids

Parasitoids are extremely important elements in agroecosystems, and have been successfully used many times in biological control programmes. Greathead (1986) recorded 393 species of parasitoids which had been established in biological control programmes. Parasitoids have been established more than twice as often as predators in biological control importations, and have been effective in the same ratio (Greathead, 1986). The most important group of parasitoids by far is the Hymenoptera, although several species of Diptera have also been used successfully.

An outstanding recent example of the use of parasitoids in classical biological control is the cassava mealybug in Africa (Neuenschwander and Herren, 1988; Herren and Neuenschwander, 1991). Cassava has been grown in Africa since the 16th century, and is the staple food crop for 200 million people. The cassava mealybug was first discovered in Congo and Zaire in 1973, and by 1986 had invaded 25 countries and covered about 70% of the 'cassava belt'. This mealybug resulted in defoliation, stunted growth and severe tuber yield losses. A parasitic wasp, *Epidinocarsis lopezi*, was imported from South America to Nigeria in 1981, and proved to be not only an effective natural enemy, but also a rapid disperser. It has since moved both by dispersal and by introduction to several other countries in the region, and in many areas is providing complete biological control.

There are numerous other cases of complete biological control using parasitoids, and DeBach and Rosen (1991) reviewed several of the most prominent successes. These include: winter moth in Canada; gypsy moth and brown-tail moth in New England; alfalfa blotch leafminer in the eastern USA; cereal leaf beetle in the midwestern USA; purple scale, olive scale, California red scale, walnut aphid, woolly whitefly and citrophilus mealybug in California; rhodesgrass mealybug in Texas; citrus whitefly in Florida; citrus blackfly in Cuba and Mexico; sugar-cane borer in the Caribbean; Comstock mealybug in the Soviet Union; dictyospermum scale in Greece; Florida red scale and citriculus mealybug in Israel; coffee mealybug in Kenya; eucalyptus snout beetle in South Africa; arrowhead scale, red wax scale and spiny blackfly in Japan; sugar-cane leafhopper, sugar-cane beetle borer and Oriental fruit fly in Hawaii; coconut moth, coconut scale and coconut leafmining beetle in Fiji; rhinoceros beetle in the South Pacific; green bug in Australia. Truly the power of parasitoids in agroecosystems cannot be underestimated (see also Polaszek *et al.*, Chapter 11, this volume).

Economic savings due to parasitoids are enormous. Norrgard (1988a,b) calculated a 149 to 1 benefit–cost ratio for the control in Africa of cassava mealybug (*Phenacoccus manihoti*) by the encyrtid parasitoid *Epidinocarsis lopezi*, with annual savings as high as US$250 million. Estimated savings due to seven major biological control successes during a 50-year period in California were about US$250 million (van den Bosch *et al.*, 1982). These were due to six cases of insect biological control, and one case of using beetles for weed biological control. Figures were based only on savings in crop losses and insecticide applications, and did not take into account potential increases in pest distribution which would have occurred without control.

Predators

Predators differ from parasitoids in several ways. In general, they tend to be more polyphagous than parasitoids, and although specific predators have been used in biological control (see below), their main contribution to agroecosystems seems to be through the action of indigenous species. Due to their polyphagy, it should be easier for predators to persist in a given region under extreme conditions and invade, and rapidly colonize, new regions. Predators are important as natural indigenous components in agro-ecosystems, with several groups being of particular importance: carabids, staphylinids, coccinellids, chrysopids, syrphids, mites, and spiders (Hoy *et al.*, 1983; Riechert and Lockley, 1984; Booij and Noorlander, 1992; Dennis and Fry, 1992; Kromp and Steinberger, 1992).

Predators have also been used in classical biological control programmes. Although they have not been used as frequently in successful programmes as parasitoids, their effect has in some cases certainly been as spectacular. One of the most famous of all biological control programmes is that of the cottony cushion scale (*Icerya purchasi*) in California (DeBach and Rosen, 1991). Indeed, it is this project which really established biological control as a viable scientific method of pest control. The cottony cushion scale invaded California in the 1860s, and by the late 1880s had nearly devastated the California citrus industry. Two natural enemies were imported from Australia into California in 1888: a predatory beetle, *Rodolia cardinalis*, and a parasitic fly, *Cryptochaetum iceryae*. It was the beetle which was to win fame for the control of the cottony cushion scale. The beetles quickly became established, and reproduced in the millions. By the end of 1890, the scale had become almost completely decimated throughout the state and the citrus industry had been saved. The total cost of this entire project was less than US$5000.

Mites are another extremely important group of arthropod predators, and they have figured extensively in biological control programmes. Although mites have been predominantly used to control other mites, particularly spider mites, they have also been used in biological control programmes against nematodes, grasshoppers and locusts, scale insects, medically important Diptera, stored product pests, and dung breeding pests (Hoy *et al.*, 1983; Gerson and Smiley, 1990).

Spiders also play an important role as predators. They are some of the most ubiquitous of predaceous organisms; they feed almost exclusively on insects, and their importance to natural and biological control has been discussed by Riechert and Lockley (1984). A well-documented example of the importance of spiders in agroecosystems is as predators in rice in Southeast Asia. There are currently over 340 species of riceland spider known from South and Southeast Asia (Barrion and Litsinger, 1995), and they are important elements in control of various riceland pests, and particularly leafhoppers (Barrion and Litsinger, 1981, 1984, 1995; see also Polaszek *et al.*, Chapter 11, this volume).

Weed biological control agents

Up to the year 1985, 214 arthropod natural enemies had been introduced into 53 countries for the control of 89 weeds (Julien, 1987; DeBach and Rosen, 1991). Although 60–75% of these introductions were ineffective (Julien, 1987), weed biological control agents have been responsible for several striking successes with massive benefits to both agricultural and natural ecosystems. Again, mites as well as insects can act as weed biological control agents, and have been used in programmes against both terrestrial and aquatic weeds (Hoy *et al.*, 1983).

Probably the most famous weed biological control project is prickly pear cactus (*Opuntia* spp.) in Australia (DeBach and Rosen, 1991). Two species of *Opuntia* were introduced as ornamentals into Australia in the 1800s. These escaped cultivation and spread into wild areas, with ultimately 24 million hectares (Mha) of land heavily infested with the cactus, and 12 Mha so infested as to make the land useless for other activities (primarily grazing and farming). The larvae of a small moth, *Cactoblastis cactorum*, were collected in Argentina and first colonized in Australia in 1926. These rapidly increased in population, and destroyed large stands of cactus. By the early 1930s the general collapse and destruction of the cactus was virtually assured. A small regeneration of the cactus occurred in the mid 1930s, but the moth again responded and prickly pear has not been a problem since then.

Another notable case of weed biological control is the control of Klamath weed (*Hypericum perforatum*) in California (DeBach and Rosen, 1991). At its height, Klamath weed covered 2 Mha of rangeland in the northwestern USA. This plant displaced better forage plants, and produced weight loss in cattle and sheep because of its mild toxicity. Two species of beetles (*Chrysolina* spp.) were imported from Australia, which had a similar problem with the weed in 1945 and 1946. One of these, *Chrysolina quadrigemina*, quickly became the dominant natural enemy of the plant, and by 1955 the effect of the beetle was so spectacular that populations of the weed had been reduced by more than 99%. Estimated savings from this programme were at least US$20,960,000 for the period of 1953–1959, or about US$3,500,000 per year.

Ecosystem engineers

Functioning ecosystem services, and the organisms that produce them, are essential to all aspects of life on this planet (Schulze and Mooney, 1993; Chapin *et al.*, 1997, 1998; Costanza *et al.*, 1997). The previous examples of beneficial arthropods in agroecosystems are for the most part examples of insects which are manipulated by humans. However, a large group of insects exist, many in the soil, which are involved mainly in a variety of types of ecosystem engineering. These insects are involved in numerous ecosystem services, which include decomposition of organic matter, nutrient cycling, bioturbation, suppression of soil pests, and a variety of other activities necessary to ecosystem function. Reviews of these types of functions include: Schulze and Mooney (1993), Swift and Anderson (1993), Samways (1994) and Brussaard *et al.* (1997).

A curious example of an applied project using an 'ecosystem engineer' is a biological control programme against dung in Australia (Waterhouse, 1974). Cattle were imported into Australia in the late 18th century, and by the middle of the 20th century there were over 30 million animals. The problem was that the Australian fauna lacked any of the bovine dung beetles capable of breaking down cattle dung. It has been estimated that if the dung from 30 million cattle is not broken down, as much as 2,400,000 ha of productive pasture are effectively removed every year (Waterhouse, 1974; DeBach and Rosen, 1991). Additionally, the dung paddies provided a resource for the production of tall, rank herbage unsuitable to grazing, and harboured filth flies and other unwanted pests. To remedy this problem, a special biological control programme against dung was instigated in the 1960s. Several species of African bovine dung beetles were imported to break down the dung paddies. The programme was successful, and in some areas paddies which had previously remained on the ground for several months could disappear in as little as 48 hours. One excellent side-effect was the provision of soil nutrients from the decomposed dung positively influencing plant growth.

Optimization of Insect Biodiversity in Agroecosystems: Theoretical Aspects

The question of whether more species necessarily imply more stability is one of ongoing debate within the ecological literature. Lawton and Brown (1993) discussed two extreme possibilities. In one extreme, all species contribute in some way to the integrity of the ecosystem, and the removal of any species will have an effect on all others. They term this the 'rivet theory', after the analogy likening species to rivets in an airplane first used by Ehrlich and Ehrlich (1981). In this analogy, a certain amount of small species (rivets) can be removed from the system with only minor effect, but at some point removal of another small species will ultimately weaken the system to the point where it fails. They refer to the alternative view as the 'redundant species hypothesis', in which species richness is irrelevant, and all that is important is that sufficient biomass of organisms in the main functional classes (primary producers, consumers, decomposers) is maintained. In this view, many redundant species could be removed without impact on the system.

For the most part, this debate is out of the remit of this chapter (see Wardle *et al.*, Chapter 5, this volume). Here it is sufficient to say that in general, high and prolonged levels of diversity in an agroecosystem will permit the development of a higher level of internal links which will promote stability. However, it must be remembered that the nature and quality of the trophic interactions and responses (particularly density-dependent ones), rather than sheer numbers, may be the key factor for stability in agro-ecosystems. Further information on this debate, some particularly in relation to agricultural systems, has been provided by a variety of authors (Pimentel, 1961; Southwood and Way, 1970; Dempster and Coaker, 1974; van Emden and Williams, 1974; Goodman, 1975; Murdoch, 1975; Litsinger and Moody, 1976; Roger *et al.*, 1991; Swift and Anderson, 1993; Woodward, 1993; Altieri, 1994; Naeem *et al.*, 1994; Way and Heong, 1994; Chapin *et al.*, 1997, 1998; Grime, 1997; Lawton, 1997; Symstad *et al.*, 1998).

Even from an applied standpoint, there is controversy over this subject. For example, there has been debate within the biological control literature as to the preference for single or multiple species introductions (for a review of this subject see Ehler, 1990). The single species introduction school would favour extensive studies to ensure that the one (or few) species introduced best matched the requirements and would thus be the most effective natural enemies; the multiple introduction school would favour introducing all potentially effective parasitoids as the only effective way of finding out which species are best suited to the new environment. DeBach and Rosen (1991) provided the following support for multiple species introductions: multiple importations are the only practical manner of obtaining the best natural enemy, or combination of natural enemies, for a given habitat or host range; competition between various species of natural enemies is not normally detrimental to overall host population regulation, and although competing populations of natural enemies may affect each other's individual efficiency, their combined effect on the host population will be greater; although there is usually one best natural enemy for any particular habitat, climate or host stage, it might require additional species to provide control throughout the entire geographic and climatic range, or life cycle, of a pest; and the best natural enemy may not be found until all natural enemies are known. Although from a biological control point of view multiple introductions might make more sense (and even this is debatable), in light of biodiversity considerations it may become increasingly more difficult to implement indiscriminate multiple introductions in the future (see below).

One complicating factor in the diversity/stability debate is hyperparasitoids, which represent an addition to diversity which has traditionally been considered as adverse in agroecosystems. However, there is some indication that the addition of a hyperparasitoid might increase stability, and this might compensate for any adverse effect they may have in reducing the regulatory activity of primary parasitoids (see discussions in Bennett, 1981; Luck *et al.*, 1981; LaSalle, 1993).

One final theoretical point to consider is how to measure the effect of diversity in agroecosystems. An interesting attribute of the natural enemies used in biological control is that they react to the population size of their host in a density-dependent manner (Huffaker and Messenger, 1964; Huffaker *et al.*, 1976, 1984; LaSalle, 1993). Thus, as a host population increases, the intensity of the mortality producing actions of natural enemies increases until the host population is overpowered and begins to decrease. This results in a decrease in the natural enemy population due to limited resource. In this way, the two linked populations fluctuate between upper and lower limits which prevent both a massive increase in the host population size and a decrease in host population size to the level where it would result in natural enemy extinction. The importance of this is that a species may be rare in a system because it is: (i) unimportant and/or (ii) a very important and effective keystone natural enemy which is controlling its host or prey population to a very low level, and as a result surviving at a low level.

Optimization of Insect Biodiversity in Agroecosystems: Techniques

There are several methods by which agroecosystems can be manipulated to prevent pest build-up and enhance the diversity or efficacy of beneficial insects (Box 7.1). These include introductions, crop management and habitat management. As usual, there are not always clear distinctions between these categories. For example, breeding plant resistance to insects can involve crop management through selective breeding programmes as well as introductions by searching for and introducing genetic variation not already present in the system. The positive effects of having more than a single plant species in the system might be realized whether the plant is another crop species, a weed, or part of an adjacent patch of non-agricultural land.

Box 7.1. Optimization of insect biodiversity in agroecosystems: a model project.

How do you go about an applied agrobiodiversity project? Obviously, limitations of time and money will restrict what can actually be accomplished. The following steps are offered in what could be considered a fairly complete project with a practical or applied focus, centred on the study of insects (although they could apply in principle to a wide range of organisms).

1. The first step in any project is to gather all pertinent published information for your system or organisms. This will prevent the wasting of time and resources through performing studies to gain information which is already available. Today, literature surveys are facilitated through the availability of abstracting and indexing services and journals with searching capabilities which allow the rapid retrieval of a wealth of information in a very short period of time. Consultation with a taxonomist before conducting searches is imperative. Information on a given species can be stored under outdated names, and thus be irretrievable unless the taxonomic history of the species is known.

2. Building a knowledge base of the natural enemy complex for a given pest should combine field surveys with information already available from the literature and specimen-based information from museums and insect collections. Unfortunately, much of this specimen-based information is not readily available, but current efforts by many major museums towards information management within their collections will make such information more readily available in the future. Pertinent information would be: species of natural enemies, location, timing of presence and estimated impact.

3. One focus of biodiversity projects could be the sustainable use of preserved biodiversity. Surveys of the same pests and/or related pests and their natural enemies in native ecosystems would increase the potential for identification of previously unknown natural enemies which could be imported into agroecosystems for possible utilization in non-chemical control programmes.

4. The identification and use of resistant cultivars of plants is of tremendous importance in insect agrobiodiversity projects. Although not the primary subject of this chapter, many of the activities concerning insect biodiversity discussed here will become unnecessary if cultivars with stable resistance can be identified.

5. The following section discusses a wide variety of crop management and environment management and modification strategies which can help to control pest insects and conserve and augment beneficial insects. A review of these types of methods should be conducted to determine whether any are of potential value.

6. Careful documentation of the differences between various treatments, conditions and strategies can be invaluable in providing the tools for a clear decision-making process in agrobiodiversity programmes. The information gained from documentation of diversity levels in disturbed versus natural habitats, monocultures versus polycultures, and various conditions of within-field habitat modification can provide a solid scientific basis for management decisions by other workers in similar systems.

7. Many recent advances in information management and transfer mean that traditional publication is a rather inefficient way to widely disseminate information. Databases, web-sites and electronic publications can ensure that pertinent information becomes available quickly, efficiently and cost-effectively to a wide user community.

> **8.** Involving taxonomists in agrobiodiversity projects, and consulting them early so that they can contribute to project planning, will improve chances of project success in a cost-effective manner.

Species introductions

The introduction of insect species has played a very important role in agroecosystems for over a century now, and will continue to do so. For the most part these species introductions fall into two categories: biological control agents and pollinators. Both of these categories have been discussed above with examples of introduced species, and it is sufficient here to reiterate that this is one of the most widely used methods in sustainable agriculture. Good reviews of species introductions have been given for natural enemies (Clausen, 1978; Greathead, 1986), weed biological control agents (Julien, 1987) and pollinators (Free, 1993).

Mass rearing and release

Mass rearing of natural enemies and augmentative releases have long been used as a technique in situations where the actions of an introduced natural enemy do not supply sufficient control (DeBach and Hagen, 1964; Smith, 1966; Rabb *et al.*, 1976; DeBach and Rosen, 1991). This may be the result of a lack of effective natural enemies, or a breakdown of the efficacy of a normally effective natural enemy under given temporal or regional situations. In these instances, releases of large numbers of a species of natural enemy which will not persist in the system can be used in the absence of permanently established natural enemies.

One of the most frequently used insects in mass release programmes are species of tiny egg parasitoids in the genus *Trichogramma*. These tiny parasitoids are used somewhat like a living insecticide, and over 32 Mha of agriculture and forestry were treated annually with *Trichogramma* releases (Li, 1994). These have been used against a wide variety of Lepidoptera on over 20 different crops, and they have proven to be quite effective in many instances (Li, 1994).

Biotypes in natural enemies

Genetic diversity within a species can be of considerable importance to agrobiodiversity projects. Several examples of the importance of this intraspecific diversity to biological control projects are given by Unruh and Messing (1993). In most of these cases, significant levels of variation within a species were only discovered through attempts to use these species in biological control projects.

The best known example of applied use of genetic diversity in a biological control programme is the aphidiine braconid, *Trioxys pallidus*, which is a parasitoid of the walnut aphid, *Chromaphis juglandicola*. The aphid invaded California, quickly spread throughout all the walnut-producing regions of the state, and developed resistance to insecticide – in short it was a problem that needed solving through biological control. The parasitoid *Trioxys pallidus* was imported from France, and released in California where it readily established in the cooler, southern coastal regions, but, despite repeated attempts, did not establish in dry interior regions, with hot summers and cold winters, where most of California's walnut production actually takes place. Additional exploration for a suitable natural enemy was conducted, and the same species, *T. pallidus*, was imported from Iran where it lived in areas climatically similar to California's Central Valley. The result was the rapid establishment and colonization in all the major walnut growing areas of the state. Populations of *T. pallidus* from the cooler Mediterranean region were preadapted to exist under similar conditions in the coastal regions of California; only the Iranian populations of the same species, which had evolved to more extreme climatic conditions, could survive in the harsher interior regions.

Another classic example is that of the larch sawfly, *Pristophora erichsonii*, which invaded North America from Europe, and became a major forest pest (Turnock *et al.*, 1976). An ichneumonid, *Mesoleius tenthredinis*, was introduced from Europe, and provided effective control under most conditions. This control was negated by the appearance and spread of a resistant strain of the larch sawfly, which was able to encapsulate and kill the parasitoid eggs. This defence mechanism was overcome by the importation and colonization of a Bavarian strain of *Mesoleius tenthredinis* which could overcome the encapsulation reaction. What is more, this ability to overcome encapsulation was inherited by progeny in crosses of the Bavarian strains and the strains already present in North America, which facilitated the re-establishment of control. Further examples can be found in Polaszek *et al.*, Chapter 11, this volume.

Crop management

Pest management can be improved through crop management in two ways: (i) through methods which are beneficial to natural enemies, or (ii) through methods that are detrimental to the pests. These include crop species rotation; mixed and intercropping systems; perennial rather than annual crops; short-maturing crops; synchronous and discontinuous planting; planting in seasons unfavourable to the pest; high planting density; and small, scattered fields (Stern *et al.*, 1976; Powell, 1986; Altieri, 1994; see Polaszek *et al.*, Chapter 11, this volume). Powell (1986) listed several examples of cropping practices actually resulting in increased parasitism rates. Those practices included intercropping, timing of first harvesting of a crop, strip harvesting, the use of resistant cultivars, and mixed cropping.

In a now classic study, van den Bosch *et al.* (1967) determined that strip harvesting rather than solid harvesting lucerne resulted in lower populations of the aphid *Acyrthosiphon pisum* due to its parasitoid *Aphidius smithi*. Solid harvesting in the summer disrupted both species, but affected the parasitoid more by removing the suitable microclimate at the same time as making its host rare. This could result in the eradication of the parasitoid with an ensuing increase in pest population levels. Strip harvesting had a stabilizing effect on the system by providing both suitable microclimate and hosts for the parasitoid to survive in good numbers.

The subject of vegetational diversity in agroecosystems has received a great deal of attention (Dempster and Coaker, 1974; Litsinger and Moody, 1976; Perrin, 1977; Altieri and Letourneau, 1982; van Emden, 1990; Andow, 1991; Altieri, 1994; Barbosa and Benrey, 1998; Barbosa and Wratten, 1998). There is considerable evidence that increasing the diversity of plants in an agroecosystem (e.g. changing from a monoculture to a polyculture) will reduce the amount of herbivore damage on any given plant. There seem to be two reasons for this: polycultures encourage stable natural enemy populations, and specialized herbivores might choose against polycultures in favour of nearby monocultures which provide concentrated resources (Altieri, 1994).

Altieri (1994) listed several ways in which crop diversification could affect beneficial insects: providing alternative hosts or prey at times of pest host scarcity; providing food in the form of pollen or nectar for adult natural enemies; providing nesting sites, overwintering sites and other forms of refuge; maintaining acceptable populations of the pest host over extended periods to ensure survival of beneficials. In one example, population densities of the predatory minute pirate bug, *Orius tristicolor*, and its prey the western flower thrips, *Frankliniella occidentalis*, were monitored and compared between squash monocultures and tricultures of squash inter-cropped with corn and cowpea by Letourneau and Altieri (1983). They found that in the triculture the predator displayed a more rapid colonization rate than in the monoculture, and that the thrips densities on squash were lower and declined earlier in the season. It was felt that the earlier *Orius* colonization of the triculture was probably due to a greater attraction to a mixed-crop habitat rather than a monoculture (see also Polaszek *et al.*, Chapter 11, this volume).

Plant breeding

Plant breeding programmes are extremely important in sustainable agriculture, and can be used to achieve many goals (Murray, 1991; Poelham and Sleper, 1995; Hoy et al., 1998). Most of these are out of the remit of a chapter on insect agrobiodiversity; however, it must be remembered that a resistant crop cultivar can very effectively eliminate or neutralize pest species, and thus alleviate the need for many of the methodologies given in this chapter which are meant to reduce the population levels of pests. At the same time, partial host plant resistance integrated with biological control in an integrated pest management (IPM) programme can also successfully neutralize pest species (Thomas and Waage, 1996; Polaszek et al., Chapter 11, this volume).

Gatehouse (1991) gave several examples of plant cultivars being able to deter insect attack. In one, the introduction of pubescent cotton varieties in Africa and India was able to prevent the attack of cotton jassids (*Empoasca* spp.) and effectively eliminate this leafhopper as a pest of cotton (Hill, 1987). In addition to physical deterrents, plants also produce a variety of chemicals which can act as deterrents. Stored common beans (*Phaseolus vulgaris*) can be susceptible to serious attack by the bruchid beetle *Zabrotes subfasciatus* in Central and South America. Many wild collections of this plant were screened, with several lines showing almost complete resistance to this pest. The cause of this resistance was the presence of the polypeptide arcelin, which is toxic to *Z. subfasciatus* larvae. Breeding programmes produced commercial lines of beans with high arcelin content which display resistance to attack by this pest (Gatehouse, 1991).

Advances in genetic engineering are transforming the field of plant breeding. In one example, a proteinase inhibitor in cowpea (*Vigna unguiculata*), known as the cowpea trypsin inhibitor (CpTI), has insecticidal properties against a variety of economically important pests. It was possible to produce a synthetic gene encoding for CpTI, which was introduced into tobacco plants. Subsequent tests of CpTI tobacco strains showed a reduction in leaf damage and insect survival for several major tobacco pests, including tobacco budworm (*Heliothis virescens*), corn earworm (*Heliothis zea*), armyworm (*Spodoptera littoralis*) and tobacco hornworm (*Manduca sexta*) (Gatehouse, 1991). Similarly, the production of insect-resistant plants which contain genes encoding insecticidal proteins taken from the bacteria *Bacillus thuringiensis* has been documented on several occasions (Barton et al., 1987; Fischhoff et al., 1987; Vaeck et al., 1987; Krattiger, 1997; Polaszek et al., Chapter 11, this volume).

Environmental modification

Environmental modification is used in a broad sense to include any modifications to the composition or structure of an ecosystem or the species within it, except for the actual composition of the crop itself, which is treated above. This is by no means a new subject, and reviews of environmental modification in relation to biological control date back over 30 years. Even then, van den Bosch and Telford (1964) had already noted the importance of artificial shelters, supplementary food, alternative hosts, the improvement of pest–natural enemy synchronization, and modification of adverse agricultural practices. More recent reviews (Rabb *et al.*, 1976; Powell, 1986; Altieri, 1994; Wratten and van Emden, 1995; Barbosa, 1998) reiterate these themes and provide additional examples; see also Polaszek *et al.*, Chapter 11, this volume.

'Adjacent' habitats is a broad term for field margins, hedgerows, buffer zones and other small-scale refuge habitats for a variety of organisms. These habitats are important components of agroecosystems, as they have been shown to harbour both natural enemies (Pollard, 1968; Booij and Noorlander, 1992; Dennis and Fry, 1992; Kromp and Steinberger, 1992; Altieri, 1994) and pollinators (Lagerlöf *et al.*, 1992; Banaszak, 1992; O'Toole, 1993). The management of these areas can be important to agroecosystems. The previous studies also showed that in many cases the effectiveness of the beneficial insects in a given cropping system was directly related to the distance between the adjacent habitat and crop.

Dennis and Fry (1992) showed that field margins increase diversity of predators on farmland through providing a stable complex habitat in the summer for species which would not survive in a crop habitat alone, and providing refuges during the winter for many of the species which are active in crops in the summer. Vineyards planted near, and particularly downwind of, prune trees displayed improved biological control of the grape leafhopper, *Erythroneura elegantula*, by the parasitoid *Anagrus epos* (Murphy *et al.*, 1996, 1998). The parasitoid successfully overwintered in prune trees, which resulted in enhanced early-season parasitism of the leafhopper, and this resulted in a net-season increase in mortality. An earlier study on this same parasitoid showed that biological control of *E. elegantula* by *A. epos* in vineyards was improved in the presence of wild blackberries (*Rubus* spp.) (Doutt and Nakata, 1973). The reason for this was the presence of an alternative host leafhopper, *Dikrella cruentata*, which bred throughout the year on blackberry leaves, and provided a seasonal bridge to allow the parasitoid to maintain a high population level even in the absence of sufficient levels of the pest species, *E. elegantula*. Banaszak (1992) showed that refuge habitats are essential for the conservation of native bees, which are themselves essential in complementing the pollinating activities of the honeybee.

Weed management

It has long been known that the presence of weeds in agroecosystems can influence associated organisms, and affect both insect pests and their natural enemies (van Emden, 1965; Altieri et al., 1977; Altieri and Whitcomb, 1979 a,b; Thresh, 1981; Way and Cammell, 1981; William, 1981; Altieri and Letourneau, 1982; Norris, 1982; Altieri, 1994; see also Polaszek et al., Chapter 11, this volume). Unfortunately, weeds can have both negative and positive effects on insect diversity in agroecosystems.

As pest organisms, weeds compete with crops for essential requirements (sunlight, moisture, nutrients), and can serve as major reservoirs of insect pests. Altieri (1994) listed several examples of pest outbreaks due to the abundance of weeds near cropping systems. These included carrot fly moving from *Urtica dioica* to carrots, rosy apple aphid moving from plantain (*Plantago* spp.) to apples, dock sawfly moving from docks (*Rumex* spp.) to apples, and several examples of cereal pests moving from grassy weeds. In all these cases, failure to control the weed species can result in pest outbreaks.

As potentially beneficial organisms, weeds can provide resources for beneficial arthropods in much the same way as increased plant diversity due to polycultures (see above). Altieri (1994) gave an extensive table of examples where the presence of weeds in cropping systems resulted in enhanced pest control. The reasons for increased biological control were varied, but included disruption of crop colonization by pests through repellency (deterring insects away from both weeds and the crops) or crop trapping (weeds attracting pests which would normally go to crops), and provision of food, nectar, shelter or alternative hosts to natural enemies.

Powell (1986) listed several examples of increased parasitism due to the presence of nectar from weeds as a food source for adult parasitoids. In one example Topham and Beardsley (1975) found that various euphorbiaceous weeds (*Euphorbia* spp.) were important nectar sources for *Lixophaga sphenophori*, a tachinid parasitoid of the New Guinea sugar-cane weevil, *Rhabdoscelus obscurus*, in Hawaii. Herbicidal elimination of these weeds from field margins led to a decrease in population levels of *L. sphenophori*, and thus a decrease in the efficiency of this parasitoid as a biological control agent.

Artificial nesting sites

The provision of artificial nesting sites has been a very effective method in some cases. Two native bee species, *Nomia melanderi* and *Megachile rotundata*, are important pollinators of lucerne, and are essential to its production in North America. Both, however, require specialized nesting sites, and these have to be provided in or near the lucerne fields as part of an effective management programme. *Nomia melanderi* is a gregarious, ground-nesting bee, and needs 'bee beds' in the form of areas of compacted soil maintained at optimum levels of salinity and moisture. *Megachile rotundata* nests opportunistically in ready made cavities, and farmers provide 'bee boards' which are mobile and can be placed evenly throughout fields. The management of these bees has greatly increased productivity in lucerne and is a classic example of habitat modification (Bohart, 1972; Bitner, 1979; O'Toole, 1993).

Habitat connectivity

Large-scale environmental modifications can also affect agricultural ecosystems. Habitat fragmentation in general is a major threat to biological diversity, and Reid and Miller (1989) pointed out that habitat loss, degradation and fragmentation are the most important factors influencing species extinction rates. Habitat fragmentation produces isolated patches. Kreuss and Tscharntke (1994) showed that isolation in agricultural systems resulted in a decrease in total species number and reduced effects by natural enemies. Isolation affected natural enemies more than their phytophagous hosts, largely because it was more difficult for natural enemies to successfully colonize small patches than for herbivores. In their studies, herbivores in isolated patches experienced only 19–60% of the parasitism levels of non-isolated populations. As rates of parasitism are directly related to the success of biological control projects, they concluded that habitat connectivity was essential to maintaining good biological control in agroecosystems (Kreuss and Tscharntke, 1994).

The Importance of Taxonomy to Agrobiodiversity Projects

It is essential to involve taxonomists in any comprehensive agrobiodiversity programme. Schauff and LaSalle (1998) listed several ways in which taxonomists could support biological control programmes, and these apply equally to agrobiodiversity programmes. They include: authoritative identifications; access to literature (including alternative names under which information might be stored); providing information on biology, distribution or the native home of organisms; authoritative evaluation of information in the literature; providing methods for recognizing and distinguishing organisms; and providing training and workshops.

Schauff and LaSalle (1998) pointed out that lack of taxonomic input could result in loss of time and money, and perhaps even programme failure. They divided problems encountered from not having taxonomic input into two main categories.

- Type 1 problems were those which prevented you from doing what you want to do. These include: not being able to find the native home of a pest; rear species in culture (biology not known); find pertinent information hidden under obsolete species names; properly distinguish pests and/or natural enemies in their native homes; properly document programme results; and obtain permission to release organisms in quarantine.

- Type 2 problems were those in which time and resources were wasted by doing what you do not want (or need) to do. These include: importation of species already present; repeating biological studies which have already been done and published; spending effort and money on curating; shipping and trying to culture hyperparasitoids and/or parasitoids that do not attack the target host; spending effort and money on natural enemies which may never be approved for release under current concerns for the effect of generalists on non-target organisms; and using costly new technologies on wrong organisms.

Schauff and LaSalle (1998) concluded that, given the problems which could be encountered through lack of authoritative taxonomic input, it is cost-effective to include taxonomists in projects, and that they should be consulted early so that they can contribute to project planning to maximize efficiency.

Critical Issues at the Biological Control and Conservation Biology Interface: Examples from an Ongoing Programme

Increased interest in and knowledge of biodiversity and conservation biology issues will affect our attitudes towards a variety of applied practices. The positive and negative aspects of the interaction between biological control and biodiversity are of particular interest. Although biological control has been considered to be using biodiversity to protect biodiversity, if not done properly it can be a potential threat to diversity. The following section briefly discusses some of these issues, and relies heavily upon examples taken from recent work on the citrus leafminer (CLM, *Phyllocnistis citrella*).

Biodiversity as a biological control resource

Although many people feel that biodiversity should be preserved simply for its own sake, regardless of its worth or value, it is often difficult to convince some people of this need without providing realistic indications of its worth. One of the values of conserved biodiversity is that it represents a pool of potential biological control agents which could be used in applied programmes (Waage, 1991; LaSalle, 1993; LaSalle and Gauld, 1993). Newly introduced pest species are a common occurrence, with an estimated 11 potential pests entering the USA alone every year, of which seven are liable to be injurious (Sailer, 1983). Preserving the ability to control new pests in an environmentally and economically sound manner is one of the strongest arguments for preserving biodiversity.

Although the previous arguments are philosophically agreeable, there has been little actual evidence that fortuitous biological control by indigenous parasitoids occurs. The CLM, which has recently spread from its native Asia into both the Mediterranean Basin and the Americas (Hoy and Nguyen, 1997) has provided us with some direct evidence of this phenomenon. Although classical biological control programmes are under way against this pest, there are also a large number of indigenous parasitoids which have moved over on to CLM in these regions that it has recently invaded (Schauff *et al.*, 1998). Heppner (1993) recorded about 30 species of Asian parasitoids of *P. citrella* however, over 80 species of parasitoids have now been reared from *P. citrella* throughout the world, many of them having been recently recruited as the pest spread. Several of these newly recruited species appear to be capable of exerting substantial levels of control on the *P. citrella* populations, and in some cases may even preclude the need to introduce Asian parasitoids for biological control (LaSalle and Peña, 1997; Schauff *et al.*, 1998).

Valuing biodiversity

In order to convince a wide range of people of the need to conserve biodiversity, it is important to be able to provide direct measures of the value of preserving it (several chapters in Wilson, 1988; Orians *et al.*, 1990; Swanson, 1995; Kunin and Lawton, 1996; Costanza *et al.*, 1997). The ability to recognize indigenous parasitoid species, and distinguish them from imported parasitoids, will permit the evaluation of the effect of fortuitous biological control. The cost-effectiveness of fortuitous biological control can be measured in the same way as has been done in classical biological control programmes (e.g. Dean *et al.*, 1979; van den Bosch *et al.*, 1982; Norrgard, 1988a,b; DeBach and Rosen, 1991). Thus, the effects of indigenous parasitoids, such as the CLM parasitoids mentioned above, could be used to document one of the benefits of biodiversity conservation, the retention of species of potential value in sustainable agriculture.

Ownership of biological control agents

Who owns biological control agents? Do they represent genetic resources which are part of a nation's wealth and should not be exploited without compensation, or should they be made freely available to any workers who need them? The Convention on Biological Diversity (UNEP, 1992) clearly confirms sovereignty over all biodiversity. However, the complex issues associated with the interface between biological control and biodiversity continue to stimulate debate. Readers wishing to explore this subject further should refer to Altieri (1991, 1992) and Farrar (1992) for opposing viewpoints (see also Gisselquist, Chapter 16, this volume).

Effect of introduced biological control agents on non-target organisms

There is a concern about the possible effects of introduced biological control agents on non-target organisms (Howarth, 1991). The main worry is that introduced natural enemies may not be completely host specific, and they may switch over on to native hosts. Species introductions are indeed a serious threat to survival of native species. Reid and Miller (1989, p. 51) stated that 'Introduced species threaten 19 percent of all endangered, vulnerable and rare species of vertebrates', making them the third worst threat behind habitat loss and fragmentation (67%) and overexploitation (37%). In a specific study, Duan *et al.* (1996) found that three species of intentionally introduced parasitoids and six species of inadvertently introduced parasitoids were attacking non-pest species of fruit flies (Tephritidae) in Hawaii. Two of these species of fruit flies were introduced biological control agents for weeds, and two were Hawaiian endemics. It is clear that, in light of the potentially serious effects, careful consideration should be taken whenever one intends to introduce a species. However, this is an equivocal situation, and I would like to make two comments on this issue.

Firstly, the introduction of non-native species in biological control programmes is not a problem, it is one possible solution. Such action is stimulated by the actual problem, generally in the form of an introduced pest species, a potential threat both to agriculture and biodiversity. Possible solutions include: chemical treatments, doing nothing, introducing exotic natural enemies, and control strategies which minimize chemicals and employ other methodologies such as cultural control. There are potential problems with all possible solutions, which could include environmental damage, loss of food to hungry populations, or simply that they do not work effectively. Indeed, the failure to introduce a biological control agent might have a more devastating effect on biodiversity if it means that a starving population is forced to overexploit natural resources in search of food. So, despite the potential for non-target effects, it may be that introducing a natural enemy is still the most environmentally sound method of responding to a pest problem.

Secondly, much of the focus of non-target organisms has been placed on the wrong trophic level. Although Howarth (1991) mentioned upper trophic levels, the predominance of the literature is directly concerned with the threat of introduced natural enemies of herbivores to non-target herbivores. This may not be the most serious threat. The upper trophic levels, and particularly parasitoids, are more extinction-prone than their herbivorous hosts (LaSalle and Gauld, 1992 [1991]; LaSalle, 1993; Unruh and Messing,

1993). When introduced parasitoids move over on to non-target hosts, there may be a greater likelihood that they endanger other native parasitoids through competition than that they endanger non-target herbivores through regulatory pressure. Having shown that these native parasitoids represent a pool of potential biological control agents, we should take care not to expose them to undue threats.

A good example of the decision-making dilemma in species associated with non-target organisms again comes from the CLM. An undescribed species of *Quadrastichus* (Hymenoptera: Eulophidae) is known from Asia, where it has been considered to be a specific parasitoid of CLM. It was released, and is now apparently established, as an introduced natural enemy of CLM in Israel (Argov and Rössler, 1996). It was recently discovered that this species also attacks *Liriomyza* (Diptera: Agromyzidae) in Asia (LaSalle, 1998, unpublished data), and it is being reared on *Liriomyza* sp. in quarantine in Spain (J. Jacas, Valencia, 1998, personal communication). This presents an interesting dilemma. The fact that it can use *Liriomyza* as an alternative host makes it a very attractive parasitoid from the point of view of biological control. If weeds are allowed to persist in the vicinity of citrus groves, they will surely harbour populations of *Liriomyza* spp. which can serve as an alternate host to *Quadrastichus* sp., and this will benefit any attempts to control CLM using this parasitoid. This very attribute that makes *Quadrastichus* sp. so good from a pest management point of view makes it very bad from a conservation biology point of view. It is known that this is a polyphagous species, it is known that its introduction might threaten native parasitoids, and it is known that these native parasitoids are potentially very valuable as biological control agents. Balancing this type of conflicting information when making choices about which species of parasitoids should be introduced should become an important part of the decision-making process in biological control.

References

Altieri, M.A. (1991) Guest Editorial. Classical biological control and social equity. *Bulletin of Entomological Research* 81, 365–369.

Altieri, M.A. (1992) Guest Editorial: Response. Reply from Altieri. *Bulletin of Entomological Research* 82, 298.

Altieri, M.A. (1994) *Biodiversity and Pest Management in Agroecosystems*. Food Products Press, New York.

Altieri, M.A. and Letourneau, D.K. (1982) Vegetation management and biological control in agroecosystems. *Crop Protection* 1, 405–430.

Altieri, M.A. and Whitcomb, W.H. (1979a) Manipulation of insect patterns through seasonal disturbance of weed communities. *Protection Ecology* 3, 339–343.

Altieri, M.A. and Whitcomb, W.H. (1979b) The potential use of weeds in the manipulation of beneficial insects. *HortScience* 14, 12–18.

Altieri, M.A., Schoonhoven, A.V. and Doll, J.D. (1977) The ecological role of weeds in insect pest management systems: a review illustrated with bean (*Phaseolus vulgaris* L.) cropping systems. *PANS* 23, 195–205.

Andow, D.A. (1991) Vegetational diversity and arthropod population response. *Annual Review of Entomology* 36, 561–586.

Argov, Y. and Rössler, Y. (1996) Introduction, release and recovery of several exotic natural enemies for biological control of the citrus leafminer, *Phyllocnistis citrella*, in Israel. *Phytoparasitica* 24, 33–38.

Banaszak, J. (1992) Strategy for conservation of wild bees in an agricultural landscape. In: Paoletti, M.G. and Pimentel, D. (eds) *Biotic Diversity in Agroecosystems*. Elsevier, Amsterdam, pp. 179–192.

Barbosa, P. (ed.) (1998) *Conservation Biological Control*. Academic Press, San Diego.

Barbosa, P. and Benrey, B. (1998) The influence of plants on insect parasitoids: implications for conservation biological control. In: Barbosa, P. (ed.) *Conservation Biological Control*. Academic Press, San Diego, pp. 55–82.

Barbosa, P. and Wratten, S.D. (1998) The influence of plants on invertebrate predators: implications for conservation biological control. In: Barbosa, P. (ed.) *Conservation Biological Control*. Academic Press, San Diego, pp. 83–100.

Barrion, A.T. and Litsinger, J.A. (1981) The spider fauna of Philippine rice agroecosystems. I. Dryland. *Philippine Entomologist* 5, 139–166.

Barrion, A.T. and Litsinger, J.A. (1984) The spider fauna of Philippine rice agroecosystems. II. Wetland. *Philippine Entomologist* 6, 11–37.

Barrion, A.T. and Litsinger, J.A. (1995) *Riceland Spiders of South and Southeast Asia*. CAB International, Wallingford.

Barton, K.A., Whitely, H.R. and Yang, N.-S. (1987) *Bacillus thuringiensis*-endotoxin expressed in *Nicotiana tabacum* provides resistance to lepidopteran insects. *Plant Physiology* 85, 1103–1109.

Bennett, F.D. (1981) Hyperparasitism in the practice of biological control. In: *The Role of Hyperparasitism in Biological Control: a Symposium*. Division of Agricultural Sciences, University of California, Publication 4103, pp. 43–49.

Bitner, R.M. (1979) The current status of pollinating alfalfa by the leafcutting bee *Megachile rotundata* F. and the alkali bee, *Nomia melanderi* Ckll., in the Pacific Northwest. In: *Proceedings of the Fourth International Symposium on Pollination*. Maryland Agricultural Experiment Station Special Miscellaneous Publication 1, pp. 253–256.

Bohart, G.E. (1972) Management of wild bees for the pollination of crops. *Annual Review of Entomology* 17, 287–312.

Booij, C.J.H. and Noorlander, J. (1992) Farming systems and insect predators. In: Paoletti, M.G. and Pimentel, D. (eds) *Biotic Diversity in Agroecosystems*. Elsevier, Amsterdam, pp. 125–135.

van den Bosch, R. and Telford, A.D. (1964) Environmental modification and biological control. In: DeBach, P. (ed.) *Biological Control of Insect Pests and Weeds*. Chapman & Hall, London, pp. 459–488.

van den Bosch, R., Lagace, C.F. and Stern, V.M. (1967) The interrelationship of the aphid, *Acyrthosiphon pisum*, and its parasite, *Aphidius smithi*, in a stable environment. *Ecology* 48, 993–1000.

van den Bosch, R., Messenger, P.S. and Gutierrez, A.P. (1982) *An Introduction to Biological Control*. Plenum Press, New York.

Brussaard, L., Behan-Pelletier, V.M., Bignell, D.E., Brown, V.K., Didden, W., Folgarait, P., Fragoso, C., Freckman, D.W., Gupta, V.V.S.R., Hattori, T.,

Hawksworth, D.L., Klopatek, C., Lavelle, P., Malloch, D.W., Rusek, J., Söderström, B., Tiedje, J.M. and Virginia, R.A. (1997) Biodiversity and ecosystem functioning in soil. *Ambio* 26, 563–570.

Chapin, F.S., III, Walker, B.H., Hobbs, R.J., Hooper, D.U., Lawton, J.H., Sala, O.E. and Tilman, D. (1997) Biotic control over the functioning of ecosystems. *Science* 277, 500–504.

Chapin, F.S., III, Sala, O.E., Burke, I.C., Grime, J.P., Hooper, D.U., Lauenroth, W.K., Lombard, A., Mooney, H.A., Mosier, A.R., Naeem, S., Pacala, S.W., Roy, J., Steffen, W.L. and Tilman, D. (1998) Ecosystem consequences of changing biodiversity: experimental evidence and a research agenda for the future. *BioScience* 48, 45–53.

Clausen, C.P. (ed.) (1978) *Introduced Parasites and Predators of Arthropod Pests and Weeds: a World Review.* Agriculture Handbook 480, United States Department of Agriculture, Agricultural Research Service, Washington, DC.

Costanza, R., d'Arge, R., de Groot, R., Farber, S., Grasso, M., Hannon, B., Kimburg, K., Naeem, S., O'Neill, R.V., Paruelo, J., Raskin, R.G., Sutton, P. and van den Belt, M. (1997) The value of the world's ecosystem services and the natural capital. *Science* 387, 253–260.

Dean, H.A., Schuster, M.F., Boling, J.C. and Riherd, P.T. (1979) Complete biological control of *Antonina graminis* in Texas with *Neodusmetia sangwani* (a classical example). *Bulletin of the Entomological Society of America* 25, 262–267.

DeBach, P. and Hagen, K.S. (1964) Manipulation of entomophagous species. In: DeBach, P. (ed.) *Biological Control of Insect Pests and Weeds.* Chapman & Hall, London, pp. 429–458.

DeBach, P. and Rosen, D. (1991) *Biological Control by Natural Enemies.* Cambridge University Press, Cambridge.

Dempster, J.P. and Coaker, T.H. (1974) Diversification of crop ecosystems as a means of controlling pests. In: Jones, D.P. and Solomon, M.E. (eds) *Biology in Pest and Disease Control.* John Wiley & Sons, New York, pp. 106–114.

Dennis, P. and Fry, G.L.A. (1992) Field margins: can they enhance natural enemy population densities and general arthropod diversity on farmland. In: Paoletti, M.G. and Pimentel, D. (eds) *Biotic Diversity in Agroecosystems.* Elsevier, Amsterdam, pp. 95–115.

Doutt, R.L. and Nakata, J. (1973) The *Rubus* leafhopper and its egg parasitoid: an endemic biotic system useful in grape-pest management. *Environmental Entomology* 2, 381–386.

Duan, J.J., Purcell, M.F. and Messing, R.H. (1996) Parasitoids of non-target tephritid flies in Hawaii: implications for biological control of fruit fly pests. *Entomophaga* 41, 245–256.

Ehler, L.E. (1990) Introduction strategies in biological control of insects. In: Mackauer, M., Ehler, L.E. and Roland, J. (eds) *Critical Issues in Biological Control.* Intercept, Andover, pp. 111–134.

Ehrlich, P.R. and Ehrlich, A.H. (1981) *Extinction. The Causes and Consequences of the Disappearance of Species.* Random House, New York.

van Emden, H.F. (1965) The role of uncultivated land in the biology of crop pests and beneficial insects. *Scientific Horticulture* 17, 121–136.

van Emden, H.F. (1990) Plant diversity and natural enemy efficiency in agroecosystems. In: Mackauer, M., Ehler, L.E. and Roland, J. (eds) *Critical Issues in Biological Control.* Intercept, Andover, pp. 63–80.

van Emden, H.F. and Williams, G.F. (1974) Insect stability and diversity in agro-ecosystems. *Annual Review of Entomology* 19, 455–475.

Farrar, P. (1992) Guest Editorial: Response. Classical biological control and social equity: a reply to Altieri. *Bulletin of Entomological Research* 82, 297–298.

Fischhoff, D.A., Bowdish, K.S., Perlak, F.J., Marrone, P.G., McCormick, S.M., Niedermeyer, J.G., Dean, D.A., Kusano-Kretzmer, K., Mayer, E.J., Rochester, D.E., Rogers, S.G. and Fraley, R.T. (1987) Insect tolerant transgenic tomato plants. *BioTechnology* 5, 807–813.

Free, J.B. (1993) *Insect Pollination of Crops,* 2nd edn. Academic Press, London.

Gatehouse, J.A. (1991) Breeding for resistance to insects. In: Murray, D.R. (ed.) *Advanced Methods in Plant Breeding and Biotechnology.* CAB International, Wallingford, pp. 250–276.

Gerson, U. and Smiley, R.L. (1990) *Acarine Biocontrol Agents: an Illustrated Key and Manual.* Chapman & Hall, London.

Goodman, D. (1975) The theory of diversity–stability relationships in ecology. *Quarterly Review of Biology* 50, 237–266.

Greathead, D.J. (1986) Parasitoids in classical biological control. In: Waage, J. and Greathead, D. (eds) *Insect Parasitoids.* Academic Press, London, pp. 289–318.

Grime, J.P. (1997) Biodiversity and ecosystem function: the debate deepens. *Science* 277, 1260–1261.

Gurr, L. (1962) The role of insects as pollinators of forage crops in New Zealand. *Sheepfarming Annual* 1962, 67–74.

Gurr, L. (1972) The introduction of bumblebees into North Island, New Zealand. *New Zealand Journal of Agricultural Research* 15, 635–638.

Heppner, J.B. (1993) Citrus Leafminer *Phyllocnistis citrella,* in Florida (Lepidoptera: Gracillariidae: Phyllocnistinae). *Tropical Lepidoptera* 4, 49–64.

Herren, H.R. and Neuenschwander, P. (1991) Biological control of cassava pests in Africa. *Annual Review of Entomology* 36, 257–283.

Hill, D.S. (1987) *Agricultural Insect Pests of Temperate Regions and their Control* 2nd edn. Cambridge University Press, Cambridge.

Howarth, F.G. (1991) Environmental impacts of classical biological control. *Annual Review of Entomology* 36, 485–509.

Hoy, M.A. and Nguyen, R. (1997) Classical biological control of the citrus leafminer *Phyllocnistis citrella* Stainton (Lepidoptera: Gracillariidae): theory, practice, art and science. *Tropical Lepidoptera* 8 (Suppl. 1), 1–19.

Hoy, M.A., Cunningham, G.L. and Knutson, L. (1983) *Biological Control of Pests by Mites.* Division of Agricultural Sciences, University of California, Berkeley, Publication 3304.

Hoy, C.W., Feldman, J., Gould, F., Kennedy, G.G., Reed, G. and Wyman, J.A. (1998) Naturally occurring biological controls in genetically engineered crops. In: Barbosa, P. (ed.) *Conservation Biological Control.* Academic Press, San Diego, pp. 185–205.

Huffaker, C.B. and Messenger, P.S. (1964) The concept and significance of natural control. In: DeBach, P. (ed.) *Biological Control of Insect Pests and Weeds.* Chapman & Hall, London, pp. 74–114.

Huffaker, C.B., Simmonds, F.J. and Laing, J.E. (1976) The theoretical and empirical basis of biological control. In: Huffaker, C.B. and Messenger, P.S. (eds), *Theory and Practice of Biological Control.* Academic Press, New York, pp. 42–78.

Huffaker, C.B., Berryman, A.A. and Laing, J.E. (1984) Natural control of insect populations. In: Huffaker, C.B. and Rabb, R.L. (eds) *Ecological Entomology.* Wiley, New York, pp. 359–398.

Julien, M.H. (1987) *Biological Control of Weeds: a World Catalogue of Agents and their Target Weeds,* 2nd edn. CAB International, Wallingford.

Kevan, P.G., Clark, E.A. and Thomas, V.G. (1990) Insect pollinators and sustainable agriculture. *American Journal of Alternative Agriculture* 5, 13–22.

Krattiger, A.F. (1997) Insect resistance in crops: a case study of *Bacillus thuringiensis* (Bt) and its transfer to developing countries. *International Service for the Acquisition of Agri-biotech Applications* No. 2, 44 pp.

Kreuss, A. and Tscharntke, T. (1994) Habitat fragmentation, species loss, and biological control. *Science* 264, 1581–1584.

Kromp, B. and Steinberger, K.-H. (1992) Grassy field margins and arthropod diversity: a case study on ground beetles and spiders in eastern Austria (Coleoptera: Carabidae; Arachnida: Aranei, Opiolines). In: Paoletti, M.G. and Pimentel, D. (eds) *Biotic Diversity in Agroecosystems.* Elsevier, Amsterdam, pp. 71–93.

Kunin, W.E. and Lawton, J.H. (1996) Does biodiversity matter? Evaluating the case for conserving species. In: Gaston, K.J. (ed.) *Biodiversity: a Biology of Numbers and Difference.* Blackwell Science, Oxford, pp. 283–308.

Lagerlöf, J., Stark, J. and Svensson, B. (1992) Margins of agricultural fields as habitats for pollinating insects. In: Paoletti, M.G. and Pimentel, D. (eds) *Biotic Diversity in Agroecosystems.* Elsevier, Amsterdam, pp. 117–124.

LaSalle, J. (1993) Parasitic Hymenoptera, biological control, and biodiversity. In: LaSalle, J. and Gauld, I.D. (eds) *Hymenoptera and Biodiversity.* CAB International, Wallingford, pp. 197–215.

LaSalle, J. and Gauld, I.D. (1992 [1991]) Parasitic Hymenoptera and the biodiversity crisis. *Redia* 74 (Appendix), 315–334.

LaSalle, J. and Gauld, I.D. (1993) Hymenoptera: their diversity and their impact on the diversity of other organisms. In: LaSalle, J. and Gauld, I.D. (eds) *Hymenoptera and Biodiversity.* CAB International, Wallingford, pp. 1–26.

LaSalle, J. and Peña, J.E. (1997) A new species of *Galeopsomyia* (Hymenoptera: Eulophidae: Tetrastichinae): a fortuitous parasitoid of the citrus leafminer, *Phyllocnistis citrella* (Lepidoptera: Gracillariidae). *Florida Entomologist* 80, 461–470.

Lawton, J.H. (1997) The role of species in ecosystems: aspects of ecological complexity and biological diversity. In: Abe, T., Levin, S.A. and Higashi, M. (eds) *Biodiversity: an Ecological Perspective.* Springer-Verlag, New York, pp. 215–228.

Lawton, J.H. and Brown, V.K. (1993) Redundancy in ecosystems. In: Schulze, E.-D. and Mooney, H.H. (eds) *Biodiversity and Ecosystem Function.* Springer-Verlag, Berlin, pp. 255–270.

Letourneau, D.K. and Altieri, M.A. (1983) Abundance patterns of a predator, *Orius tristicolor* (Hemiptera: Anthocoridae), and its prey, *Frankliniella occidentalis* (Thysanoptera: Thripidae): habitat attraction in polycultures versus monocultures. *Environmental Entomology* 12, 1464–1469.

Li, L.-Y. (1994) Worldwide use of *Trichogramma* for biological control on different crops. In: Wajnberg, E. and Hassan, S. (eds) *Biological Control with Egg Parasitoids.* CAB International, Wallingford, pp. 37–53.

Litsinger, J.A. and Moody, K. (1976) Integrated pest management in multiple cropping systems. In: Sanchez, P.A. (ed.) *Multiple Cropping*. ASA Publication No. 27, Madison, Wisconsin, pp. 293–316.

Luck, R.F., Messenger, P.S. and Barbieri, J.F. (1981) The influence of hyperparasitism on the performance of biological control agents. In: *The Role of Hyperparasitism in Biological Control: a Symposium*. Division of Agricultural Sciences, University of California. Publication 4103, pp. 34–42.

McGregor, S.E. (1976) *Insect Pollination of Cultivated Crop Plants*. Handbook No. 496, US Department of Agriculture, Washington, DC.

Murdoch, W.W. (1975) Diversity, stability, complexity and pest control. *Journal of Applied Ecology* 12, 745–807.

Murphy, B.C., Rosenheim, J.A. and Granett, J. (1996) Habitat diversification for improving biological control: abundance of *Anagrus epos* (Hymenoptera: Mymaridae) in grape vineyards. *Environmental Entomology* 25, 495–504.

Murphy, B.C., Rosenheim, J.A., Dowell, R.V. and Granett, J. (1998) Habitat diversification tactic for improving biological control: parasitism of the western grape leafhopper. *Entomologia Experimentalis et Applicata* 87, 225–235.

Murray, D.R. (ed.) (1991) *Advanced Methods in Plant Breeding and Biotechnology*. CAB International, Wallingford.

Naeem, S., Thompson, L.J., Lawler, S.P., Lawton, J.H. and Woodfin, R.M. (1994) Declining biodiversity can alter the performance of ecosystems. *Nature* 368, 734–737.

Neuenschwander, P. and Herren, H.R. (1988) Biological control of the cassava mealybug, *Phenacoccus manihoti*, by the exotic parasitoid *Epidinocarsis lopezi* in Africa. *Philosophical Transactions of the Royal Society, Series B* 318, 319–333.

Norrgard, R.B. (1988a) Economics of the cassava mealybug [*Phenacoccus manihoti*; Hom.: Pseudococcidae] biological control program in Africa. *Entomophaga* 33, 3–6.

Norrgard, R.B. (1988b) The biological control of cassava mealybug in Africa. *American Journal of Agricultural Economics* 70, 366–371.

Norris, R.F. (1982) Interaction between weeds and other pests in the agroecosystem. In: Hatfield, J.L. and Thomason, I.J. (eds) *Biometeorology in Integrated Pest Management*. Academic Press, New York, pp. 343–406.

Orians, G.H., Brown, G.M., Jr, Kunin, W.E. and Swierzbinski, J.E. (eds) (1990) *The Preservation and Valuation of Biological Resources*. University of Washington Press, Seattle.

O'Toole, C. (1993) Diversity of native bees and agroecosystems. In: LaSalle, J. and Gauld, I.D. (eds) *Hymenoptera and Biodiversity*. CAB International, Wallingford, pp. 169–196.

Perrin, R.M. (1977) Pest management in multiple cropping systems. *Agro-ecosystems* 3, 93–118.

Pimentel, D. (1961) Species diversity and insect population outbreaks. *Annals of the Entomological Society of America* 54, 76–86.

Poehlman, J.M. and Sleper, D.A. (1995) *Breeding Field Crops*, 4th edn. Iowa State Press, Ames, Iowa.

Pollard, E. (1968) The effect of the removal of the bottom flora of a hawthorn hedgerow on the Carabidae of the hedge bottom. *Journal of Applied Ecology* 5, 125–139.

Powell, W. (1986) Enhancing parasitoid activity in crops. In: Waage, J. and Greathead, D. (eds) *Insect Parasitoids*. Academic Press, London, pp. 319–335.

Rabb, R.L., Stinner, R.E. and van den Bosch, R. (1976) Conservation and augmenation of natural enemies. In: Huffaker, C.B. and Messenger, P.S. (eds) *Theory and Practice of Biological Control*. Academic Press, New York, pp. 233–254.

Reid, W.V. and Miller, K.R. (1989) *Keeping Options Alive: the Scientific Basis for Conserving Biodiversity*. World Research Institute, Washington, DC.

Riechert, S.E. and Lockley, T. (1984) Spiders as biological control agents. *Annual Review of Entomology* 29, 299–320.

Robinson, W.S., Nowogrodski, R. and Morse, R.A. (1989) The value of honeybees as pollinators of US crops. *American Bee Journal* 128(6), 411–423; 129(7), 477–487.

Roger, P.A., Heong, K.L. and Teng, P.S. (1991) Biodiversity and sustainability of wetland rice production: role and potential of microorganisms and invertebrates. In: Hawksworth, D.L. (ed.) *The Biodiversity of Microorganisms and Invertebrates: Its Role in Sustainable Agriculture*. CAB International, Wallingford, pp. 117–136.

Sailer, R. (1983) History of insect introductions. In: Wilson, C.L. and Graham, C.L. (eds) *Exotic Plant Pests and North American Agriculture*. Academic Press, New York, pp. 15–38.

Samways, M.J. (1994) *Insect Conservation Biology*. Chapman & Hall, London.

Schauff, M.E. and LaSalle, J. (1998) The relevance of systematics to biological control: protecting the investment in research. In: Zaluki, M.P., Drew, R.A.I. and White, G.G. (eds) *Pest Management: Future Challenges*. Proceedings of the 6th Australasian Applied Entomological Research Conference, Brisbane, 29 September–2 October 1998, Vol. 1, pp. 425–436.

Schauff, M.E., LaSalle, J. and Wijesekara, G.A. (1998) The genera of chalcid parasitoids (Hymenoptera: Chalcidoidea) of citrus leafminer *Phyllocnistis citrella* Stainton (Lepidoptera: Gracillariidae). *Journal of Natural History* 32, 1001–1056.

Schulze, E.-D. and Mooney, H.H. (eds) (1993) *Biodiversity and Ecosystem Function*. Springer-Verlag, Berlin.

Smith, C.N. (ed.) (1966) *Insect Colonization and Mass Production*. Academic Press, New York.

Solbrig, O.T. (1991) *From Genes to Ecosystems: a Research Agenda for Biodiversity*. International Union of Biological Sciences, Paris.

Southwood, T.R.E. and Way, M.J. (1970) Ecological background to pest management. In: Rabb, R.L. and Guthrie, F.E. (eds) *Concepts of Pest Management*. North Carolina State University, Raleigh, pp. 6–29.

Stern, V.M., Adkisson, P.L., Beingolea, O. and Viktorov, G.A. (1976) Cultural controls. In: Huffaker, C.B. and Messenger, P.S. (eds) *Theory and Practice of Biological Control*. Academic Press, New York, pp. 593–613.

Swanson, T.M. (ed.) (1995) *The Economics and Ecology of Biodiversity Decline: the Forces Driving Global Change*. Cambridge University Press, Cambridge.

Swift, M.J. and Anderson, J.M. (1993) Biodiversity and ecosystem function in agricultural systems. In: Schulze, E.-D. and Mooney, H.H. (eds) *Biodiversity and Ecosystem Function*. Springer-Verlag, Berlin, pp. 15–41.

Syed, R.A. (1979) Studies on oil palm pollination by insects. *Bulletin of Entomological Research* 69, 213–224.

Syed, R.A. (1982) Insect pollination of oil palm: feasibility of introducing *Elaeidobius* spp. into Malaysia. In: Pushparajah, E. and Chew, P.S. (eds) *The Oil Palm in the Eighties: a Report of the Proceedings of the International Conference on Oil Palm in Agriculture in the Eighties, Kuala Lumpur, 17–20 June 1981,* Vol 1. Incorporated Society of Planters, Kuala Lumpur, pp. 263–289.

Syed, R.A., Law, I.H. and Coorley, R.H.V. (1982) Insect pollination of oil palm: introduction, establishment and pollinating efficiency of *Elaeidobius kamerunicus* in Malaysia. *Planter, Kuala Lumpur* 58, 547–561.

Symstad, A.J., Tilman, D., Wilson, J. and Knops, J.M.H. (1998) Species loss and ecosystem functioning: effects of species identity and community composition. *Oikos* 81, 389–397.

Thomas, M. and Waage, J. (1996) *Integrating Biological Control and Host-plant Resistance Breeding*. CTA, Wageningen.

Thresh, J.M. (ed.) (1981) *Pests, Pathogens and Vegetation: the Role of Weeds and Wild Plants in the Ecology of Crop Pests and Diseases*. Pitman, London.

Topham, M. and Beardsley, J.W. (1975) Influence of nectar source plants on the New Guinea sugarcane weevil parasite, *Lixophaga sphenophori* (Villeneuve). *Proceedings of the Hawaiian Entomological Society* 22, 145–155.

Turnock, W.J., Taylor, K.L., Schröder, D. and Dahlsten, D.L. (1976) Biological control of pests of coniferous forests. In: Huffaker, C.B. and Messenger, P.S. (eds) *Theory and Practice of Biological Control*. Academic Press, New York, pp. 289–311.

UNEP (1992) *Convention on Biological Diversity*. UNEP, Nairobi.

Unruh, T.R. and Messing, R.H. (1993) Intraspecific biodiversity in Hymenoptera: implications for conservation and biological control. In: LaSalle, J. and Gauld, I.D. (eds) *Hymenoptera and Biodiversity*. CAB International, Wallingford, pp. 27–52.

Vaeck, M., Reynaerts, A., Hofte, H., Jansens, S., De Beuckeleer, M.D., Dean, C., Zabeau, M., van Montagu, M.V. and Leemans, J. (1987) Transgenic plants protected from insect attack. *Nature* 328, 33–37.

Waage, J.K. (1991) Biodiversity as a resource for biological control. In: Hawksworth, D.L. (ed.) *The Biodiversity of Microorganisms and Invertebrates: Its Role in Sustainable Agriculture*. CAB International, Wallingford, pp. 149–162.

Waterhouse, D.F. (1974) The biological control of dung. *Scientific American* 230(4), 100–109.

Way, M.J. and Cammell, M.E. (1981) Effects of weeds and weed control on invertebrate pest ecology. In: Thresh, J.M. (ed.) *Pests, Pathogens and Vegetation: The Role of Weeds and Wild Plants in the Ecology of Crop Pests and Diseases*. Pitman, London, pp. 443–458.

Way, M.J. and Heong, K.L. (1994) The role of biodiversity in the dynamics and management of insect pests of tropical irrigated rice – a review. *Bulletin of Entomological Research* 84, 567–587.

William, R.D. (1981) Complementary interactions between weeds, weed control practices, and pests in horticultural cropping systems. *HortScience* 16, 508–513.

Wilson, E.O. (ed.) (1988) *Biodiversity*. National Academy Press, Washington, DC.

Woodward, F.I. (1993) How many species are required for a functional ecosystem. In: Schulze, E.-D. and Mooney, H.H. (eds) *Biodiversity and Ecosystem Function*. Springer-Verlag, Berlin, pp. 271–291.

Wratten, S.D. and van Emden, H.F. (1995) Habitat management for enhanced activity of natural enemies of insect pests. In: Glen, D.M., Greaves, M.P. and Anderson, H.M. (eds) *Ecology and Integrated Farming Systems*. John Wiley & Sons, Chichester, pp. 117–145.

Determinants of Agrobiodiversity in the Agricultural Landscape

8

P.J. Edwards[1], J. Kollmann[1] and D. Wood[2]

[1]Geobotanisches Institut ETH, Zürichbergstrasse 38, 8044 Zürich, Switzerland; [2]Agrobiodiversity International, 13 Herons Quay, Sandside, Milnthorpe, Cumbria LA7 7HN, UK.

Introduction

Agricultural communities are new assemblages of organisms which originate from a wide diversity of natural ecosystems and which find in agroecosystems the habitat conditions they need to survive. Even where agriculture has been practised continuously for more than 4000 years (as in the case of rice cultivation in India, Thailand and southern China), agroecosystems are young on an evolutionary time scale. Thus, agriculture represents an enormous unplanned experiment on how biodiversity responds to environmental change. Indeed, agricultural landscapes are excellent models for studying the relationships between biodiversity and the habitat templet, and the importance of biodiversity for ecosystem function (Swift *et al.*, 1996).

In this chapter, we are concerned with four main questions. The first is: what are the ecological characteristics of the organisms which live in agroecosystems? As a general hypothesis we propose that the factors which determine the flora and fauna of agroecosystems are the quality or adversity of the habitat for growth and reproduction, the temporal pattern of disturbance, for example through harvesting or ploughing, and the spatial arrangement of habitat elements within the landscape. In all of these factors, agroecosystems differ in important respects from natural systems. The second question concerns how much biodiversity can be supported in the agricultural landscape? We examine two examples of agroecosystems which are rich in species, before considering how intensification affects the level of biodiversity. Thirdly, we consider briefly the extent to which biodiversity is important for the functioning and sustainability of agroecosystems and make linkages to other chapters where specific issues are covered in more detail. Finally, we examine some of the underlying causes for specialization and intensification of agriculture, and discuss some of the problems of understanding the consequences for biodiversity of large-scale economic trends.

How Agroecosystems are Different

Agroecosystems are closely dependent on human choice and, then, on human management for their location and contents (Table 8.1; see Plates 9–12). Firstly, agroecosystems have not been converted from a random

Table 8.1. Comparison of agroecosystems and natural ecosystems.

	Agroecosystems	Natural ecosystems
Site characteristics	Selected sites suitable for agriculture; disturbance frequent and predictable; high levels of resources	Total range of sites with mostly continuous, long history; disturbance less frequent, unpredictable; resources often limiting
Biota	Planned and unplanned diversity; recent assemblage of organisms from diverse habitats	Only defined by habitat and biogeographical characteristics
Community interactions	Much simpler	Complex trophic structure, multiple interactions
Evolutionary processes	Recent evolutionary history important; rapid evolution as a result of strong selection pressures and short generation time; increasing control of evolution of crops and farm animals	Often slower evolution
Ecosystem processes	Strongly influenced by management; nutrient cycle open, erosion, pesticides	Nutrient cycle mostly closed, sustainable system
Landscape structure and dynamics	Planned structure, autogenic processes and environmental variation minimized through management; traditional systems highly diverse (patch size, species number, dispersal processes); modern systems often simplified	Hierarchical structure due to autogenic processes (habitat patch scale) and environmental heterogeneity; high structural diversity

sample of unmanaged ecosystems: the initial choice of which parts of natural ecosystems to convert into agroecosystems depends on the perceived value as farmland. Land that is of too poor quality for farming – too high, too steep, too wet, too nutrient-poor – remains in more natural vegetation. Subsequently this poor land may find itself in reserves; for example, many of the conserved areas of the USA are beyond the agricultural frontier in Alaska, and the English National Parks are predominantly upland areas of no use for arable farming. Secondly, the relatively slow biotic and physical changes in natural ecosystems are in marked contrast to the often rapid and abrupt changes in agroecosystem management, which in turn can have a dramatic effect on the organisms of the agroecosystem. Changes include intensification (see below), with irrigation, terracing, drainage and nutrient enrichment, and extensification, with conversion to managed woodlands or pasture, or even abandonment as the farm economy changes.

Despite profound differences between agroecosystems and natural ecosystems based on the intention and intensity of human management, some useful comparisons can be made. One emphasis has been on species diversity (Conway, 1997). Because some agroecosystems have a low diversity of species they are often regarded as being like simplified natural ecosystems. For example, if we compare a cereal monoculture with the forest vegetation which formerly grew on the same site then this seems trivially obvious. However, if we consider the full range of agroecosystems, including some traditional home gardens in the tropics which have a very high diversity of cultivated species, and especially if we consider the larger scale and include the non-agricultural elements of the landscape, then it is by no means so obvious that agroecosystems are simplifications of natural systems. Furthermore, by emphasizing simplification, we may miss the most important differences between agroecosystems and natural ecosystems, which lie in the processes determining their structure and function.

Natural ecosystems can be thought of as having a hierarchical structure. On a small scale we can recognize an individual patch of more or less uniform habitat. A larger area of a particular community type is made up of many such habitat patches which are a result of slight differences in abiotic conditions or the product of autogenic processes in the vegetation. For example, natural forest vegetation is often composed of a mosaic of patches which represent different stages of regeneration of the dominant tree species (Remmert, 1991). The various phases of this regeneration pattern provide distinct habitats for many subordinate organisms, and their life histories are finely tuned to the spatial and temporal dynamics of the mosaic cycle. For example, the dispersal ability of butterflies which inhabit forest gaps is likely to be closely related to the spatial pattern of gaps in the natural forest landscape and therefore the distance that insects must disperse to find another gap (cf. Sutcliffe and Thomas, 1996). On a yet larger scale the landscape is made up of distinct types of communities which are the product of differences in environmental conditions or ecological history.

Agroecosystems can also be described as having a hierarchical structure. The individual habitat patch is usually the field and the most obvious difference from a natural community is that the farmer determines which are the dominant species, that is, there is 'planned diversity' (Matson *et al.*, 1997). Although most fields contain a large number of other organisms, including the soil flora and fauna, insects, pathogens and weeds, the 'unplanned diversity' is often strongly influenced either by interactions with the crop or by direct management such as weed and pest control (Wood and Lenné, Chapters 2 and 18, and Polaszek *et al.*, Chapter 11, this volume). This is in marked contrast to natural systems in which internal processes such as competition and predation play a central role in determining the relative abundance of species. Similarly, in many natural systems ecosystem functions such as nutrient cycling are largely regulated by the activities of the biota whereas in agricultural systems the nutrient cycle is typically open and determined by inputs from the farmer (Swift and Anderson, 1996).

On a larger spatial scale agroecosystems are also a system of habitat patches, though the dynamics are largely determined by the farmer's selection of crops (see Plates 9–12). The patches or fields are often much larger and more internally uniform than natural patches, and are set in a matrix of non-agricultural habitat such as hedgerows and woodland. In this chapter, we are concerned chiefly with the unplanned diversity, including those organisms which live in non-agricultural habitats. As we will see in the following sections, the spatial organization and temporal dynamics of the various habitat elements in an agricultural landscape play a crucial role in determining the kinds of organisms which can live there.

Natural ecosystems are composed of organisms interacting with each other and with the physical environment: in other words we can understand them in terms of their physical and biological components and the resulting interactions. Because agroecosystems are systems strongly influenced by humans, they can be thought of as having an additional socio-economic or cultural component which must be considered if we wish to understand how these systems function. The evolution of agroecosystems reflects not only environmental conditions but also local knowledge, tradition and economic circumstances, as well as the influence of new ideas and research findings (Altieri, 1995). Thus, at the highest level of the agroecosystem hierarchy we can recognize the farming system, which encompasses not only the physical and biotic components of the agroecosystem, but also the historical, social and economic context, which determines how the system has developed and functions. The continued evolution of farming systems means that rapid and, in an ecological sense, unpredictable change is a feature of agroecosystems, and one which has important implications for biodiversity.

In the past few decades and in most parts of the world the pace of change in many agroecosystems has greatly accelerated: natural habitats are exploited in various ways, plants and animals are intentionally or unintentionally introduced into other regions, and agricultural practices change.

Through these processes new ecological niches are created, existing ones are destroyed, and the opportunity is provided for some species to extend their range into the new habitats while others become extinct. Thus we cannot expect agricultural communities to be as well defined or stable as natural ones which have had much longer time to reach some kind of equilibrium with their environment. For example, in communities which have been well studied, such as those of arable weeds and invertebrates in Europe, we know that they have changed greatly in the past few decades in response to changing management (Aebischer, 1990; Marshall and Hopkins, 1990). Furthermore, it is clear that rapid evolutionary change has been very important in fitting organisms to the new niches created by changing agricultural practices (Baker, 1974; Hodgson, 1987; Gould, 1990; Radosevich *et al.*, 1997).

The Agroecosystem as a Habitat Templet

The concept of the habitat templet

An important idea in ecology is that the characteristics of a particular habitat determine the kinds of organisms that live there. Southwood (1977, 1988) describes the habitat as a templet on which 'evolution forges characteristic life history strategies'. Similar ideas have been presented by several other authors (Grime, 1977; Greenslade, 1983; Sibly and Calow, 1985; Hildrew and Townsend, 1987). Although the habitat of each organism is unique (Begon *et al.*, 1996), two main axes of variation in abiotic conditions are usually recognized as important for life history evolution: one is the pattern of disturbance within a habitat and the other is the quality or adversity of the environmental conditions. Disturbance in this context has been the subject of many definitions. Southwood thinks of the disturbance axis as representing the durational stability of the habitat, and measures it for any particular species in terms of the length of time that the habitat is suitable for the organism in relation to the generation time of the organism. For example, only one generation of sepsid flies will develop in a cowpat and emerging adults must seek a new habitat that may be some distance away (Blackith and Blackith, 1993). In contrast, many generations of the beech scale insect can develop on its host plant, *Fagus sylvatica*, before the latter dies and dispersal becomes essential (Wainhouse and Howell, 1983). The adversity axis is also not easy to define, and authors have variously described it in terms of severity of the environment, stress, levels of particular resources, and productivity. The concept of the habitat templet is primarily concerned with abiotic factors, but biotic interactions are also important in determining life history evolution. Southwood (1988) includes a 'biotic interactions' axis which runs as a diagonal across the abiotic templet. This axis describes the impact of the biotic interactions (e.g. predation, parasitism, competition), and its position on the templet implies that the highest level of such interactions occurs in habitats with a low level of disturbance and adversity.

In thinking about the agricultural landscape as a habitat templet a number of points need to be made. Firstly, a very wide range of habitat conditions is represented amongst agricultural systems: in terms of the two major axes of the habitat templet, disturbance and adversity, they differ widely in the extent to which they resemble natural systems. For example, the traditional extensive pastoral system of the Mediterranean may not differ greatly from the natural system either in terms of the frequency of disturbance or in terms of the availability of resources for growth. However, at the other extreme, an intensive arable system has regular cycles of harvesting and ploughing which bear little resemblance to any natural disturbance regime.

Secondly, the agricultural landscape has been created by humans. Most organisms inhabiting an agroecosystem cannot be thought of as being adapted, in the sense that the habitat templet has shaped their evolution. Rather, certain organisms have been able to colonize the agricultural landscape because they had life history characteristics which preadapted them to live in such an environment (Harper, 1982). Very often we do not know exactly what the natural habitat of plants and animals inhabiting agroecosystems was. Probably many arable weeds are natives of regularly disturbed habitats such as scree slopes, river terraces and coastal beaches (Ellenberg, 1988). However, although the plants and animals of the unplanned 'diversity' have been recruited from other habitats, evolutionary processes are certainly important in matching their life histories more precisely to the conditions in which they find themselves (Baker, 1974). There is plenty of evidence for this evolutionary 'fine tuning' among weed species. For example, *Camelina sativa* is a crop mimic which occurs in flax fields while *Echinochloa crus-galli* is a mimic of rice; in both cases centuries of selection have produced a plant which matches the phenology of the crop very precisely (Gould, 1990). An entomological example of adaptation to cultural practices is provided by the northern corn rootworm, *Diabrotica longicornis barberi,* a major pest of maize in the midwestern USA. Krysan *et al.* (1986) showed how populations of this species evolved an extended diapause when maize was grown in a 2-year rotation with another crop as a cultural method of alleviating the rootworm problem.

Finally, we should recognize that although the habitat templet constrains the range of life history strategies which are possible, it does not impose uniformity (Southwood, 1988). There may be several possible life history strategies which are compatible with a particular environmental regime. For example, many arable weeds are successful because they have a very short life span and a high relative growth rate (e.g. *Senecio vulgaris, Chenopodium album*). As a result, they can complete their life cycle in the period either before the crop dominates the field or after it is harvested. In contrast, crop mimics such as *Avena fatua* or *Agrostema githago* are species which persist in the cereal field because their life histories match those of the crop so closely. Yet other pernicious arable weeds are successful because they are very competitive perennials which reproduce vigorously by vegetative means and are hard to eradicate (e.g. *Agropyron repens*, *Convolvulus arvensis*). However, despite the range of life histories which are possible within a particular habitat, it is evident that agriculture severely restricts the range of organisms which can persist. In general, it is true that the more intensive the agriculture practised, the smaller is the potential species pool.

The disturbance axis

Disturbances, regular and irregular, large and small, are a feature of all environments (Pickett and White, 1985). Examples include landslides, flooding, windthrow of trees, molehills and hoofprints. In general, agricultural environments are characterized by very distinctive and well-defined regimes of disturbance in terms of harvesting, ploughing or grazing, and this is reflected in the life history and dispersal abilities of the species which can live in agricultural landscapes (Hodgson and Grime, 1990).

In agroecosystems in which disturbance is very regular, we often find that the life history and phenology of organisms is closely attuned to that rhythm. For example, some traditional hay meadows in Europe have had a remarkably regular management for centuries, with the hay crop being cut at the same time each year. Such grasslands typically contain tall, early flowering species which reproduce freely from seed, such as *Fritillaria meleagris*, *Tragopogon pratensis* and *Silene flos-cuculi* (Harper, 1977; Ellenberg, 1988). The species composition contrasts strongly with grasslands which have been managed by grazing or by more frequent mowing, and which tend to contain perennial grasses such as *Lolium perenne* and *Cynosurus cristatus*, and low-growing, clonally reproducing dicotyledonous herbs such as *Prunella vulgaris* and *Ranunculus repens*. Similarly, among ground-nesting birds of arable land such as the skylark (*Alauda arvensis*) it is important that the breeding period is completed before harvest. When the established rhythm of disturbance is interrupted or altered then such species may be rapidly lost, as has occurred in recent years both with some ground-nesting birds, for example stone curlew (*Burhinus oedicnemus*) and corncrake (*Crex crex*), and many hay meadow species (Pain and Pienkowski, 1997). Indeed, farmers sometimes use such changes in management as a way of coping with a pest species. For example, an alternation of autumn and spring crops is an effective means of reducing the abundance of the annual grass *Alopecurus myosuroides*, which is a serious pest of arable fields in parts of England.

In contrast, weed species which persist in agroecosystems where the pattern of disturbance is irregular often show a high level of variability in their germination requirements, patterns of dormancy and growth phenology. This provides a kind of 'risk spreading' in an unpredictable environment. In particular, such species often produce long-persisting seedbanks or seeds which are polymorphic in their germination behaviour (e.g. *Chenopodium album*; Williams and Harper, 1965).

The adversity axis

Unlike many natural ecosystems, agroecosystems usually provide conditions which are favourable for growth; human intervention ensures that there are adequate supplies of water and high levels of essential nutrients for plant growth. Since conditions for growth are often very favourable (i.e. there is a low level of adversity), biotic interactions such as competition are important in determining the kinds of organisms which can persist. For plants, competitive ability is closely related to relative growth rate (Grime, 1979). Comparative studies show that species associated with disturbed but highly productive habitats such as arable land tend to have higher relative growth rates than plants from more stable or less productive habitats (Grime and Hunt, 1975; Poorter and Remkes, 1990; Radosevich *et al.*, 1997). Similarly, plant-feeding insect species of arable ecosystems benefit from the high nutritional content of crop plants and have short development times; their populations exhibit higher intrinsic rates of increase than insects in most other ecosystems, which accounts in part for their populations reaching epidemic proportions very rapidly (see Polaszek *et al.*, Chapter 11, this volume).

Templet theory predicts that favourable conditions for growth combined with high disturbance favours such characters as high migration or dispersal abilities, large reproductive output and short life span (Southwood, 1988). Exactly these characteristics are evident for many organisms in agroecosystems. Furthermore, it is clear that agriculture has influenced the evolution of these characteristics. De Wet and Harlan (1975) propose that most weeds have evolved directly from wild species that invaded human-disturbed habitats. 'Weediness' seems to be a characteristic which has evolved recently in many plant species associated with agroecosystems. For example, according to Oberdorfer (1994) the ubiquitous short-lived annual of arable fields and intensive pastures, *Poa annua* ($2n=28$), appears to be a recent allopolyploid derived from *Poa infirma* and the perennial, subalpine species *Poa supina* (both $2n=14$). Moreover, even within this species the history characteristics of individuals reflect the kind of habitat in which they occur; for example, populations from highly disturbed habitats are more ruderal than those from permanent pasture (Law *et al.*, 1977).

Landscape structure

Most agricultural landscapes consist of a mosaic of distinct habitat elements. These include the various crop types and non-crop elements such as hedgerows, uncultivated marginal strips, trackways and remnants of natural vegetation. The different elements are usually disturbed by management (e.g. ploughing, harvesting, herbicides, burning, etc.) with different frequencies and at different times of the year. Although Southwood (1988) did not specifically include the spatial organization of habitat patches in his templet model, it is clear that it has important implications for the persistence of species in the agricultural environment for various reasons. Firstly, many mobile animals such as birds require a variety of habitat elements (Tucker, 1997). For example, predatory birds such as the barn owl may nest in woodland fragments, and hunt their prey by moving along hedgerows (de Bruijn, 1979). Similarly, the skylark may nest in the crop but search for food in marginal vegetation where there are more insects to be found. Secondly, for more mobile species such as birds and mammals, undisturbed habitat elements represent refuges to which they can flee when the habitat they occupy is damaged or destroyed. Particularly important in this respect are permanent or semi-permanent structures such as hedgerows and fragments of grassland (Wratten and Thomas, 1990). Finally, the undisturbed elements represent a source of individuals for recolonization of habitats which have been disturbed. There have been many studies which show that populations of insects such as carabid beetles in cereal fields are greatly enhanced if there are semi-permanent grassland strips or hedgerows adjacent to a field. For example, Dennis and Fry (1992) show that field margins increased general arthropod diversity on farmland in summer by providing a stable, complex habitat for species that would not survive in the farm landscape with the presence only of crop habitats, and in winter by providing refuges for species active in the arable field during the summer (Arnold, 1995; see Polaszek *et al.*, Chapter 11, this volume).

The spatial arrangement of habitat elements is important for the persistence of a metapopulation, especially for species with a low dispersal ability, or for those which only disperse within a particular habitat type. However, the problems of dispersal within the agricultural landscape are different for every organism, and it is difficult to generalize about the significance of spatial structure (Burel, 1992). For example, some butterfly species only move along woodland edges and hedgerows, and an open area of as little as 50 m may be a significant barrier to migration. Similarly, several forest carabid beetles use hedgerows as dispersal corridors (Charrier *et al.*, 1997). For such species the connectivity of these linear structures within the landscape is an important factor in their ability to migrate. In contrast, linyphyiid spiders, which are important predators of pests in agroecosystems, can disperse aeronautically by 'ballooning'. These spiders are among the most sensitive indicators of pesticide side-effects in temperate arable crops and dispersal is clearly important if they are to persist in intensively managed agroecosystems. Using a modelling approach, Halley *et al.* (1996) demonstrated that field sizes of up to 4 km^2 had little effect on population density because of the large dispersal distances achieved by the spiders. None the less, because of their high sensitivity to pesticides, habitat landscape heterogeneity is important for their survival and abundance. The inclusion of small amounts of grassland in large areas of intensive cereal production provides population pools which can recolonize the agricultural environment, and dramatically increase the population of spiders in the landscape.

For plants there is a similarly wide range in dispersal ability. The seeds of some ubiquitous weeds are readily dispersed (e.g. *Cirsium arvense*, *Taraxacum officinale*), for example by wind, farm animals or farm machinery, and intensification of agriculture may even promote their dispersal (Strykstra *et al.*, 1997). Many other species, however, have a very limited ability to disperse in the modern agricultural landscape and, while traditional agriculture may have promoted their dispersal (Bonn and Poschlod, 1998), they have been negatively affected by intensification. Fragmentation of plant populations may also greatly reduce pollen dispersal by insect vectors, and small isolated populations of formerly widespread species suffer from genetic erosion and local extinction (Ouborg and van Treuren, 1994).

Summarizing the results from several studies of dispersal, Halley and Lawton (1996) noted that many organisms of farmland seem to be characterized by a maximum dispersal distance of the order of 1 km. They argue that if this result is more than an artefact of the size of the study sites used, it means that we could expect to see major changes in the flora and fauna of the countryside occurring as soon as the average size of fields or contiguous monocultures has reached the kilometre scale.

How Much Biodiversity Can an Agroecosystem Support?

Agroecosystems with high biodiversity: two case studies

It is generally assumed that diversity decreases as habitats change from the native vegetation through traditional agriculture to modern agriculture (Pimentel *et al.*, 1992; Swift *et al.*, 1996). Before discussing how intensification affects biodiversity, it is instructive to examine some of the features of agroecosystems which have a moderate to high level of unplanned biodiversity. We have chosen two very different agroecosystems: the first is the traditional agropastoral system in Spain known as the 'dehesa', and the second is paddy rice production as practised for thousands of years in Thailand. In Table 8.2 we attempt to identify the features of these two systems which are important in explaining their high biodiversity.

Case study I: dehesas, an agropastoral system in Spain

Dehesas (more accurately wooded dehesas) are an ancient agropastoral system which covers about 3.1 Mha mainly in the western and south-western parts of the Iberian peninsula (Campos, 1995; Díaz *et al.*, 1996). They are wooded pastures with a savannah-like cover of trees, mostly *Quercus ilex*, less frequently *Q. cerris* or *Q. suber* (see Plate 9). The roots of this agricultural system probably lay in the late Stone Age when the natural forests of the respective region were opened by first settlers. In Roman times the system

Table 8.2. Factors promoting biodiversity in two agricultural systems in southwest Spain and northeast Thailand.

	Dehesas	Paddy fields
Age of the system	2000–6000 years, continuity	4000–6000 years, continuity
Seasonality	Winter rain, summer drought	Monsoon rain late summer–autumn, dry season, in spring
Spatial heterogeneity	Trees, scrub, grassland, fallows, fields	Fields, ditches, dykes, pools
Polyculture	Pasture sheep, goats, cattle; acorns; mast for pigs; firewood, charcoal; cork; wheat and barley	Rice, wild vegetables, fish, prawns; pig feed; grazing by water buffaloes and ducks
Dispersal of species	High	High
Biomass harvested	Rather low, sustainable	Rather low, sustainable
External fertilizer	Low	Low
Pesticides	Not used	Not used
Mechanization	Not used	Not used

was already well developed, and has continued since then (Cabo Alonso, 1978).

Dehesas are characterized by a tight integration of forestal, pastoral and arable management. The pastures and fallows are used for grazing of sheep, goats and cattle in autumn and spring; in summer the herds are driven up into the mountains. In autumn the oak trees produce a rich crop of acorns which are used to feed pigs. The trees have to be pruned regularly to reach optimal yield; they are also used for production of charcoal and cork. Additionally, the trees provided shelter for the grazing animals and prevented soil erosion. An essential element of the system is low scrub ('matorral') which is used for hunting, bee-farming, as a source of firewood, and for regeneration of oak saplings. On about a quarter of a dehesa farm extensive arable fields are maintained in a rotational system. Typically, dehesas sustain a high diversity of animals which help to balance the economic risk of an individual farm.

Dehesas are famous for the rich wildlife they support (e.g. Díaz *et al.*, 1996). A host of resident and migrant birds are found throughout the seasons. Thus, any changes in the agricultural management have serious potential effects not only for the local biodiversity, as has been shown by Carrascal *et al.* (1993) for the white stork, but also for migrants from distant European regions. Most of the western European population of common cranes, for example, spends the winter in Iberian wooded dehesas where they feed on the acorn crop (Díaz *et al.*, 1996).

The system is highly adapted to the poor soil conditions and to a dry summer climate (Pulido and Escribano, 1995). It supports only a strictly local and rather low population density and has been sustainable for 2000–6000 years. The system consists of large farms (300–5000 ha) which are run either in common ownership or by rural nobility on large estates. Low costs and low productivity (Jiménez-Díaz *et al.*, 1978) have helped to conserve this system without mechanization over large areas. However, the dehesas have experienced a continued crisis since the mid-1960s due to rising wages and decreasing prices for pork and wool (Knuth, 1989). These problems have been added to by the introduction of African swine fever in the early 1970s and the emigration of younger people. The importance of sheep grazing has dropped dramatically, whereas the number of cattle has increased. Grazing, often by introduced breeds, is less sustainable than that by sheep since it needs external feed inputs. In addition, the production of charcoal is no longer important.

Supported by governmental and European Union subsidies, large areas of the dehesas have been cleared in the last 20 years for road construction, large reservoirs, irrigated fields, pine and eucalyptus plantations (cf. Plate 9), or have been abandoned. These policies destroy most of the former bio-diversity and cause increased frequency of wild fires, massive soil erosion, and contamination of rivers and groundwater with pesticides and fertilizer. The first reaction to these serious problems was to ban afforestation with eucalyptus due to its negative effects on groundwater levels and allelopathic interactions with other plants.

The originally closed and self-supporting system is now exposed to a larger market which leads to a net influx of seed grain, fertilizer, pesticides and feed for animals. Thus, the functional integration of the system has been eroded, and its future is uncertain. However, following Altieri (1990), it can serve as a model for sustainable agriculture supporting remarkably high biodiversity under difficult climatic and soil conditions.

Case study II: traditional rice farming in northeastern Thailand

The second case study focuses on typical paddy field cultivation near Udorn Thani in northeastern Thailand (Heckman, 1979). Although not as diverse as some tropical agroecosystems, this agroecosystem illustrates how much diversity may develop even under relatively intensive land use, given a long continuity of management. In fact, this part of the Mekong valley was probably the site of the first rice cultivation in 3500 BC, and rice cultivation in Udorn Thani and its neighbouring provinces has been carried on for several millennia, favoured by stable climatic conditions. This suggests that the agricultural system and its biodiversity are in a highly mature state.

This traditional paddy cultivation is a multispecies culture with rice and fish as the main crops. Additionally, semi-cultivated vegetables (*Ipomea aquatica*), edible snails, prawns, water bugs, frogs and wild birds are harvested. The products are sold in the local markets. Although the disadvantage of this polyculture may be a reduced yield of most species, yield stability is higher (Liebman, 1995). Besides supplying human food, the paddy field is a major food source for domestic ducks and water buffaloes which are released into the paddy fields after the harvest to graze on the stubble. These animals help to decompose the plant matter and contribute significantly to the eutrophication of the ponds. Some water plants are also used as pig feed (*Pistia stratoides*, *Hydrilla verticillata*). The fields are also visited by winter migrants, for example, several species of snipe. Heckman (1979) reports extensive species lists which demonstrate the high biodiversity of this type of traditionally managed paddy field (199 algae, 93 ciliata, 38 cormophyta and 269 animals in a 88 m² study field).

The paddy field receives no net gain of organic material since a great proportion of the biomass produced during the year is removed by the harvest and decomposed in the dry season. However, the system seems to be in equilibrium due to rapid demineralization of dead plant material in the shallow water and the nutrient influx from the nearby settlements and floods replenishing the export of biomass and nutrients with the harvest. Adjacent to the paddy fields, scattered trees shed allochthonous detritus into the fields.

The main feature of the paddy field is its seasonal succession which is driven by the monsoon rains in late summer to autumn alternating with a dry period in spring (Fig. 8.1). The rice planting modifies the habitat each year. In the early rainy season the fields are invaded by aquatic weeds. When the water fills the fields in early June to early August, the farmers plough the soil with water buffaloes. The rice produces spikelets after the rains cease in mid October. During the dry period, the water level of the fields falls rapidly until the harvest. The main fishing activities take place in late November to early December. The paddy field ecosystem shows a strictly seasonal development of floating (*Azolla pinnata*, *Lemna paucicostata*), submerged (*Hydrilla verticillata*), littoral (*Ludwigia octovalvis*) and fully terrestrial plants (*Echinochloa crus-galli*). Similar seasonal rhythms are found for animal species (Odonata, Hemiptera) depending on the height of the water level. There is no mass mortality during the dry season. Water enters the system through both precipitation and surface water from adjacent channels. These channels and ponds serve as refuges for aquatic organisms during the dry season. The biodiversity of the paddy fields depends on horizontal migration of species driven by the advancing or receding water level. Other species survive in the muddy soil or become dormant.

Some characteristics of agroecosystems with a high biodiversity

Examples of agricultural systems such as those presented above help us to identify general features of local agroecosystems which support a high level of biodiversity (Table 8.2). There is a general tendency for these features to change and for biodiversity to decline with intensification, though the reasons for the decline of particular species are often not obvious. Many different processes are usually at work at many different spatial scales (Tucker, 1997).

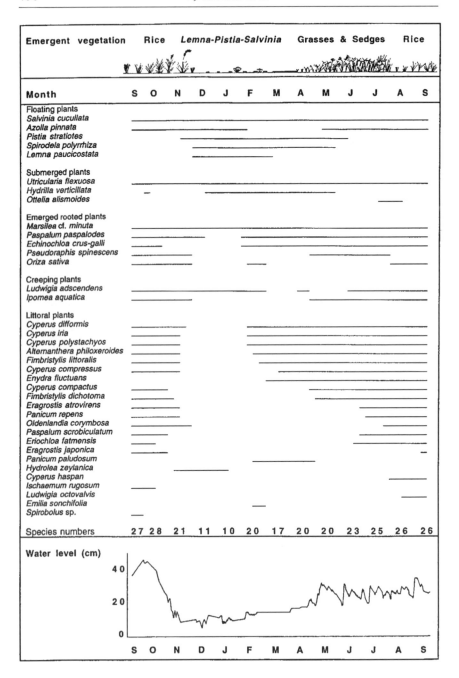

Fig. 8.1. Seasonal changes in the plant diversity of a paddy field in northwestern Thailand driven by fluctuations of the water level (modified after Heckman, 1979).

A long continuity of agricultural practice

It takes time for ecological communities to adjust to environmental change. Most successions involve a process of gradual species accumulation, and the rate of the process depends to some extent on the size of the available species-pool. The highly diverse agricultural landscapes of central Europe have developed over centuries (Ellenberg, 1988), which has been sufficient time for the appearance of new ecotypes and even new species (e.g. *Rubus* spp.; Matzke-Hajek, 1997). There has probably been a net gain in biodiversity, even though some species have certainly become extinct. Indeed, the need to protect domestic stock led to the deliberate extinction of predators such as wolves and bears which, given that top predators are usually keystone species, may have had a great impact on off-farm biodiversity.

The agricultural tradition in Europe is at least 1000 years older than in North America and as a consequence the number of invasive European weeds in North America is much higher than vice versa (see Wood and Lenné, Chapter 2, this volume). The rice paddy system in Thailand described in case study II developed over thousands of years. In comparison, paddy fields in Bangladesh, Singapore or southern Malaysia have less rich flora and fauna, partly because they have been cultivated for a little more than a century (Shajaat Ali, 1987).

Intensification is usually associated with rapid change in agricultural practice. For example, changes in agricultural practice in Europe during the last 50 years have been more rapid and more profound than any which have occurred in the past 2000 years. The great increases in the use of fertilizers have created conditions for which very few plant species are equipped. In this sense, new niches have been created which have been occupied by a few plant species capable of very rapid growth. As we have already seen, some of the most successful weeds of modern agriculture appear to be the product of recent evolutionary change, often associated with the formation of polyploids (e.g. *Galium aparine, Stellaria media*; Hodgson, 1987). The rapid evolution of plants and animals in response to changes in environmental conditions is also well demonstrated by the appearance of plant species resistant to herbicides (Radosevich *et al.*, 1997) and insects resistant to pesticides (Gould, 1990; Heong *et al.*, 1995; see Polaszek *et al.*, Chapter 11, this volume). However, the process of adjustment to new conditions has only recently begun and the numbers of species which have disappeared because they were unable to adapt to new conditions far exceeds those which have been able either to colonize the new agricultural landscape or to adapt to it by evolution.

A substantial proportion of non-crop land

If we consider only the areas which are cultivated, then a rapid loss of biodiversity seems probable whenever land is used for agriculture. However, if we consider the matrix of natural and semi-natural habitats which persist in many traditional agricultural landscapes, then the diversity may remain very high or become even higher than in the original landscape. For example, Janzen (1973) showed that casually managed agroecosystems in the tropics may actually promote more species diversity. He noted that forests showing traces of former subsistence agriculture seem to have a higher species diversity than those where such agriculture is missing.

Diversity of landscape elements

We have already seen how a mosaic of habitats is important for the persistence of many species. In different ways the dehesas and rice paddy fields provide a diversity of landscape elements. In the dehesa there is mix of different types of vegetation including woodland, pasture, arable land and scrub. The paddy field system is more intensive, in that most of the land is managed to permit the production of rice, but this involves a range of distinct habitats including ditches, dykes and paddy fields, each supporting its own distinctive flora and fauna.

Diversity of production elements

Most traditional agroecosystems involve some kind of polyculture. A single farm will often contain different production elements, with contrasting but complementary purposes. For example, in home gardens in the tropics 30–100 plant species may be cultivated including herbs, shrubs, vines and trees, as well as several different species of livestock (Gliessman, 1989; Nair, 1989; Landauer and Brazil, 1990; Ramakrishnan, 1992). Similarly a farm producing rice as a staple crop under varietal-poor monoculture will often have a home garden with very diverse fruit, vegetable, spice and medicinal plants: thus, total agrobiodiversity on the farm will be high. Even within rice production, a rice farmer may intensify production of one variety for market, but also grow a diversity of other rice varieties for social and cultural purposes (e.g. sticky rice for ritual cakes, rice varieties for brewing: see also Thurston *et al.*, Chapter 9, this volume for cassava, maize and rice examples). This distribution of diversity on the farm is mainly determined by economic rather than biological factors. Intensification may reduce the agrobiodiversity of one crop element of the farm, while having no impact on the diversity of other elements. Furthermore, such polycultures provide a wide range of niches for unplanned diversity because of their structural, phenological and taxonomic complexity.

Substantial periods when land is not used for production

Even systems with little remaining natural or semi-natural habitat can be diverse if there are adequate periods of time when the land is not dedicated to crop production. Shifting agriculture in the tropics and rotational systems with a fallow period provide, to different degrees, periods when the resources needed for growth are not used for production, and are therefore available for other organisms. Similarly, in the traditional paddy system with only one crop per year there are about 8 months of the year when the fields are not dominated by rice. Following the harvest in December, a succession of other organisms occupy the paddy fields.

Low use of biocides

There have been very few long-term studies to monitor changes in biodiversity in agroecosystems over a period of more than 25 years. One exception is a study of plant and invertebrate diversity on arable farmland in southern England (Aebischer, 1990) which reveals a general trend for loss of both plant and insect diversity over time. While it is very difficult to establish the exact causes for such declines, the use of biocides is certainly a factor. Pesticides can reduce insect diversity and abundance directly through their toxicity, but also indirectly by destroying their plant or fungal food species (see Polaszek *et al.*, Chapter 11, this volume). However, such effects are not inevitable, as herbivory by soil-dwelling insects can depress plant species richness and soil-applied insecticides could enhance plant diversity (Brown and Gange, 1989). There is also strong evidence that some pesticides are detrimental to birds. For example, it has been shown that pesticides reduce the supply of invertebrate food for the young of the grey partridge (Potts, 1986), and have probably been an important factor in the decline of several other farmland bird species in Europe (Tucker, 1997).

Moderate external inputs of fertilizer

According to Grime's (1979) 'hump-backed model', the number of plant species which occur together depends upon the productivity of the habitat, being greatest under conditions of moderate productivity. Only a few species are capable of surviving in the poorest habitats; in contrast, most species could potentially grow under very productive conditions but as a result of competition a small number of species come to dominate.

Trends in species diversity with intensification are consistent with ecological theory. For example, in traditional cereal systems, as they were practised until 50 years ago, there was often a very high diversity of weed species (Ellenberg, 1988). Crop breeding has resulted in the development of cereal varieties that are very responsive to applications of nitrogen greatly in excess of those that were formerly applied. Yields are much higher, and the competitive abilities of modern crop varieties are greater in relation to most weed species. Indeed, Grime (1979) has argued that the objective of many forms of arable farming, especially cereal cultivation, is to achieve weed control by creating conditions in which the crop plant attains the status of a dominant. These changes are certainly the reason for the decline of many arable weed species in Europe. It is interesting that those species which continue to be abundant, for example *Avena fatua*, *Galium aparine*, *Alopecurus myosuroides*, have high relative growth rates and can even respond better than the crop to an increased nitrogen supply (Firbank and Watkinson, 1985; Mahn, 1988). Similarly, it is well known that the increased use of fertilizer on pastures leads to a drastic reduction of plant diversity;

while there may be more than 40 plant species in a square metre of an unfertilized calcareous pasture, this can be reduced to less than ten by the application of high levels of nitrogenous fertilizer (Rodwell, 1992). These results suggest that competition may be increasingly important in structuring arable and pasture communities as intensification increases.

Does Biodiversity Matter?

Ecosystem processes such as primary production and nutrient cycling are obviously regulated by the organisms which make up their community. However, it is not clear how important the number and precise identity of the species present may be for these ecosystem functions. Can an ecosystem which is relatively poor in species have the same characteristics of productivity, nutrient turnover and stability as one with a much larger number of species? If not, what is the nature of the relationship between species richness and ecosystem level property? Is there a critical level above which additional species make no difference to ecosystem functioning, or does the performance of an ecosystem tend to improve as more species are added?

Recently, ecologists have begun to address these questions, mainly using simple model ecosystems such as microcosms or with field experiments in which grassland communities are sown with varying numbers of plant species of different functional groups (Lawton, 1994). The results do not all point in the same direction, but in general these experiments have shown that net primary production, and the ability of communities to recover following perturbation, tends to increase as the number of plant species increases. These questions are also of great relevance to agriculture, expecially in areas which have seen a massive reduction in species diversity associated with intensification. Furthermore, much can be learnt about the importance of biodiversity from agroecosystems.

Productivity

Although ecologists have recently become interested in the relationship between species diversity and productivity, this has been an important theme in agricultural research for many years. As a working hypothesis we might argue that a monoculture is unlikely to be able to fully use the resources of light, nutrients and water which are available for growth. We would therefore expect that a combination of two or more species, provided they were to some extent complementary in their ability to use resources, would be more efficient and thus have a higher yield.

In practice, this often appears to be the case. For example, in De Wit replacement experiments, in which a pair of species are grown alone and in varying proportions, a common, though not invariable, outcome is that the mixture has a higher yield than either of the monocultures. Similarly, growing two or more crops together (intercropping – a practice which is common in the tropics) often, though by no means always, produces a higher yield than would be obtained by the individual crops alone (Trenbath, 1976; Harris *et al.*, 1987; Swift *et al.*, 1996). In particular the benefits of intercropping in terms of yield occur when one of the crops is a leguminous species, presumably because the non-legumes benefit from the nitrogen which is fixed.

Soil properties and nutrient cycling

Cultivation leads to major changes in the soil biota which may result in a reduction in its diversity, as well as changes in the physical structure and organic content of the soil. Such changes may affect the biological regulation of decomposition and nutrient availability. The diversity of plant species can also significantly affect nutrient cycling. These issues are discussed in detail by Wardle *et al.*, Chapter 5, this volume. In general, the efficiency of nutrient use is likely to be lower in a monoculture than in a more diverse vegetation. Recent work has shown how leaching losses of nitrate ions from grassland are significantly lower in swards of high species richness (Tilman *et al.*, 1996). In arable systems there is interest in finding ways of reducing nitrate losses in maize and cereal crops by planting winter cover crops which can use nitrate during the fallow season.

Insect pests

A reduction in plant species diversity is generally associated with a corresponding reduction in the diversity and trophic complexity of the associated invertebrate communities (see LaSalle, Chapter 7, and Polaszek *et al.*, Chapter 11, this volume). There is also a tendency for densities of certain herbivorous insect species to be higher in monocultures than in mixed cultures and hence crop losses may also be higher. This is especially true for some specialist insects with restricted host ranges, which may become major pests. There are several reasons why losses may be lower in polycultures. These are related to the greater difficulty an insect may have in finding its host plant and also potentially higher parasitism and predation rates due to the more diverse structure of the invertebrate community (Altieri, 1994).

Non-crop areas also have an important influence on the control of pest populations (see Polaszek *et al.*, Chapter 11, this volume). In temperate arable systems, hedgerows and grassland strips can significantly enhance populations of natural enemies by providing overwintering sites, offering sources of food, etc. In home gardens and traditional plantation systems, trees and shrubs provide a suitable habitat for birds, which play a significant role in reducing insect pests (Matson *et al.*, 1997). However, the presence of non-crop habitats can be harmful; for example, slug damage to crops is often significantly higher near wild flower strips. There is increasing interest in introducing non-crop species, including trap crops, into agroecosystems to help control insect pests (Hokkanen, 1991; Swift *et al.*, 1996).

Pathogens

The influence of crop diversity on crop diseases is less clear than for insect pests. Particular pathogens are often less damaging in multiple cropping systems than in monocultures (Allen, 1990; Thurston, 1992; Smithson and Lenné, 1996), but other factors such as microclimatic conditions also play an important role (Matson *et al.*, 1997). There is abundant evidence that genetic uniformity in crops increases the likelihood that pathogens will adapt to the host genotype and cause serious losses (Buddenhagen, 1977). In natural systems, coevolutionary interactions between a pathogen and its host species commonly lead to the maintenance of a diversity of resistance factors within a plant population (Burdon, 1987). In recent years, there has been great interest in finding ways to introduce similar diversity into crops through the use of multiline cultures and varietal mixtures (Wolfe, 1985; Smithson and Lenné, 1996). Such mixtures are being increasingly used, and there is good evidence that they can help to reduce the rate at which pathogens spread through a crop and to stabilize yields. These issues are discussed in detail by Allen *et al.*, Chapter 6, and Polaszek *et al.*, Chapter 11, this volume.

Intensification: Its Causes and Its Impact on Biodiversity

Intensification implies that a greater share of environmental resources (light, water, nutrients, land) are appropriated for food production. The relation of agricultural intensification to biodiversity is both a major concern and a cause of confusion (Srivastava *et al.*, 1996). A recent World Bank policy paper (Pagiola and Kellenberg, 1997) identifies intensification as a major contributor to a reduction of biodiversity worldwide. However, the same paper also puts forward a basic principle of biodiversity-friendly policy reform: to 'encourage agricultural intensification'. This was also emphasized by Abdulai and Hazell (1995) who argue that 'Agricultural intensification will be essential for relieving the pressure on natural resources and containing further degradation'. Both views of the impact of intensification may be correct, depending on the circumstances.

Part of the reason for the confusion concerning the impact of intensification on biodiversity is the failure to define adequately the spatial scale at which the controlling processes operate. Ecologists tend to consider responses to management change at the scale of the individual field or farm. However, the social and economic processes which are responsible for the observed changes may be operating at a much larger spatial scale and at a higher level in the agroecosystem hierarchy (see p. 186). For example, the spatial scale of the impacts on biodiversity depends on the reasons driving farmers to intensify production. If intensification is driven by local land hunger, the response may include greater inputs of labour for irrigation, terracing, drainage, weeding, crop management and the application of high levels of organic fertilizer. This will have an impact on local biodiversity. However, there will be far wider impacts on biodiversity, both on-farm and in the wider agroecosystem, when intensification is driven by market opportunities and the resultant specialization of production systems.

In practice, the development of markets for crops is an increasingly common reason for intensification. However, market opportunities depend on improved transport, and both the opportunity itself and the transport can bring about a cascade of intensification, crop and varietal specialization, and great impacts on agrobiodiversity and biodiversity in the wild.

The move away from subsistence cropping to a trade economy means that it is no longer necessary to grow a full complement of crops on the farm. Markets allow crop specialization and cash cropping. But specialization depends on comparative advantages, which in turn depend on characteristics of the agroecosystem. Under competition from specialized farming elsewhere, a diversity of micro-environments on a farm may no longer be an advantage. Progressively, marginal areas of farms and marginal farms themselves may be abandoned. For example, over the past 150 years, land use in the lowlands of Vermont has changed from 85% arable farming to 85% forest as the agricultural frontier moved to the Mid-West with the building of railroads (cf. Foster, 1992). Distinct, but marginal, micro-environments probably contain a large part of agrobiodiversity – both planned and unplanned. Even if not abandoned, the diverse marginal habitats of subsistence farms may be greatly modified and homogenized, as markets permit intensification by investments in land-improvement: irrigation, terracing and drainage. These homogenized fields permit the use of fewer varieties within and between fields. An extreme example is glasshouse production, where the climate itself is homogenized towards an optimum.

'Green Revolution' technology is often exclusively blamed for causing genetic erosion. However, market-driven specialization dates back at least to the wheat production of provinces of the Roman Empire (Sicily, Egypt). There has been specialization in Southeast Asia for rice production for export from as early as the 1830s (Coclanis, 1995). Regional rice exports reached the very high level of 1 million tonnes (Mt) annually in the 1880s (Coclanis, 1995, p. 159). More than 100 years ago Voelcker (1894) recognized the move away from subsistence farming in India, where the farmer: 'formerly looked to his fields yielding him a crop which would provide grain for himself and his family, as well as straw for his cattle, the element of export has now entered into his calculations, and has marked changes in the kind and extent of crops grown'. In the Punjab, in the year 1888–89 alone, an increase of 11% was recorded in the area devoted to wheat growing (with the probable replacement of the genetic diversity of other crops). With market specialization, both the crop spectrum, and the varietal spectrum within a crop, can be reduced and there may be genetic erosion of local varieties (see Thurston *et al.*, Chapter 9, this volume). However, specialization of production may enhance the economic viability of the entire farm, and permit the continued management and persistence of the agrobiodiversity of all the other elements.

Increasing specialization of the local agroecosystem can also allow complementary specialization and division of labour regionally, with impacts on wild biodiversity. This complementarity can take many forms. For example, between pastoralists and arable farmers, as in West Africa, where the Fulani herders pasture their animals seasonally on crop stubble, this permits larger herds and a greater intensity of grazing on natural pasture. There may be complementarity between farmers and hunter–gatherers, with the trade in grain for bush-meat: this permits a more thorough and more damaging exploitation of wild animal biodiversity. When fire is used in hunting – as it often is in drier areas – the impact on plant biodiversity can be severe.

Perhaps the greatest impact of intensification and specialization is on 'land-saving'. For any level of production, doubling yields can 'save' 50% of the land for other uses. India provides a good example (Lele and Balu Bumb [n.d.]). Production of food grains in India increased from 48.1 Mt in 1951 to 157.4 Mt in 1993. Fifty per cent of the total increase in production in wheat and rice came from only 15% of the land (Surryanarayana, 1995). For wheat alone, to achieve the current production (59 Mt, 1994) at 1961 yield levels (851 kg ha^{-1}) India would need an additional 45 Mha of good land (FAO Production Yearbooks, see also Plucknett and Winklemann, 1995; Harrington, 1997). Similarly, global rice production has doubled during the past 30 years (from 260 to 520 Mt) although the area planted to rice has remained stable since 1980 (Evenson *et al.*, 1996). Achieving this production without intensification and specialization would have seriously eroded Asia's remaining forests and wild biodiversity. The highly contentious issue of land-saving illustrates once again how important it is to consider economic and ecological processes at compatible spatial scales: in this case losses of biodiversity associated with intensification must be viewed in the context of the protection given to biodiversity over larger areas.

On a global scale, intensification driven by markets can have great negative and positive impacts on both agrobiodiversity and wild biodiversity. For example, the global trade in wheat may reduce the economic viability of formerly highly genetically diverse wheat farming in countries such as Iran, which imported 3 Mt in 1995 (USDA, 1998: Canada and the USA together exported over 50 Mt of wheat in 1995). Across crops, low-price wheat can undermine sorghum and millet diversity in West Africa, and imports of sorghum to Mexico can damage marginal farmers who now maintain key agrobiodiversity of native maize in Mexico. Similarly, intensive plantation rubber production in Southeast Asia takes pressure off wild rubber in Brazil (formerly harvested by felling trees). The global trade in soybean meal and oil as commodities means that the efficiency of intensive soybean production in the USA determines the economic viability of soybean production in the Brazilian cerrados: the lower the efficiency and higher the price in the USA, the more cerrados are converted to soybean fields, with loss of distinctive endemic biodiversity.

Market specialization may be of great importance for agrobiodiversity, with an even higher impact in centres of origin of crops and livestock. Centres of origin and diversity of crops are usually 'marginal' in access and with mountainous or broken topography. A combination of increasing global trade in staple crops and increased local accessibility is a major threat to the economic viablity of the subsistence farms that have maintained a large part of global agrobiodiversity under non-mechanized cultivation.

Conclusions

The debate among ecologists about the function of biodiversity in natural ecosystems has only just begun (Grime, 1997). It is perhaps of greater global importance – and there may be greater opportunities through agricultural research – to understand the role of biodiversity within and between agro-ecosystems, and between agriculture and its surroundings. Grime (1997) argues that research on the significance of biodiversity should reassert a perspective 'in which high species-richness is viewed not as an attribute of certain ecosystems but instead as a function of population processes associated with special circumstances that hover precariously between two different forces for extinction (extreme habitat conditions and competitive dominance)'. However, in all agriculture these 'forces for extinction' are under almost continual human management: the habitat conditions by field improvement, irrigation and fertilization, and the control of dominance by tilling, planting distance and weeding. Our hope is that a greater under-standing of the functioning of the agroecosystem in the landscape can contribute to the understanding and management of wild biodiversity (see Wood and Lenné, Chapter 17, this volume).

References

Abdulai, A. and Hazell, P. (1995) The role of agriculture in sustainable economic development in Africa. *Journal of Sustainable Agriculture* 7, 101–119.

Aebischer, N.J. (1990) Assessing pesticide effects on non-target invertebrates using long-term monitoring and time-series modelling. *Functional Ecology* 4, 369–373.

Allen, D.J. (1990) The influence of intercropping with cereals on disease development in legumes. In: Waddington, S.R., Palmer, A.F.E. and Edje, O.T.

(eds) *Proceedings of a Workshop on Research Methods for Cereal/Legume Intercropping in Eastern and Southern Africa.* CIMMYT Report No. 17, pp. 62–67.

Altieri, M.A. (1990) Why study traditional agriculture? In: Carroll, C.R., Vandermeer, J.H. and Rosset, P.M. (eds) *Agroecology.* McGraw-Hill, New York, pp. 551–564.

Altieri, M.A. (1994) *Biodiversity and Pest Management in Agroecosystems.* Food Products Press, New York.

Altieri, M.A. (1995) *Agroecology: the Science of Sustainable Agriculture.* Westview Press, Boulder, Colorado.

Arnold, G.W. (1995) Incorporating landscape pattern into conservation programs. In: Hannson, L., Fahrig, L. and Merriam, G. (eds) *Mosaic Landscapes and Ecological Processes.* Chapman & Hall, London, pp. 309–337.

Baker, H.G. (1974) The evolution of weeds. *Annual Review of Ecology and Systematics* 5, 1–24.

Begon, M., Harper, J.L. and Townsend, C.R. (1996) *Ecology.* Blackwell Science, Oxford.

Blackith, R.E. and Blackith, R.M. (1993) Differential attraction of calyptrate flies (Diptera) to faeces. *Journal of Natural History* 27, 645–655.

Bonn, S. and Poschlod, P. (1998) *Ausbreitungsbiologie der Pflanzen Mitteleuropas.* Quelle & Meyer, Wiesbaden.

Brown, V.K. and Gange, A.C. (1989) Herbivory by soil-dwelling insects depresses plant species richness. *Functional Ecology* 3, 667–672.

de Bruijn (1979) Feeding ecology of the barn owl, *Tyto alba,* in the Netherlands. *Limosa* 52, 91–154.

Buddenhagen, I.W. (1977) Resistance and vulnerability of tropical crops in relation to their evolution and breeding. *Annals of the New York Academy of Sciences* 287, 309–326.

Burdon, J.J. (1987) *Diseases and Plant Population Biology.* Cambridge University Press, Cambridge.

Burel, F. (1992) Effect of landscape structure and dynamics on species diversity in hedgerow networks. *Landscape Ecology* 6, 161–174.

Cabo Alonso, A. (1978) Antecedentes históricos de las dehesas salmantinas. In: Consejo Superior de Investigaciones Científicas (ed.) *Estudio Integrado y Multidisciplinario de la Dehesa Salmantina.* Salamanca-Jaca, pp. 63–98.

Campos, P. (1995) Dehesa economy and conservation in the Iberian Peninsula. In: McCracken, D.I., Bignal, E. and Wenlock, S.E. (eds) *Farming on the Edge: the Nature of Traditional Farmland in Europe.* JNCC, Peterborough, pp. 112–117.

Carrascal, L.M., Bautista, L.M. and Lázaro, E. (1993) Geographical variation in the density of the white stork *Ciconia ciconia* in Spain: influence of habitat structure and climate. *Biological Conservation* 65, 83–87.

Charrier, S., Petit, S. and Burel, F. (1997) Movements of *Abax parallelipedus* (Coleoptera, Carabidae) in woody habitats of a hedgerow network landscape: a radio-tracing study. *Agriculture, Ecosystems and Environment* 61, 133–144.

Coclanis, P.A. (1995) The poetics of American agriculture: the United States rice industry in international perspective. *Agricultural History* 69, 140–162.

Conway, G. (1997) *The Doubly Green Revolution: Food for All in the Twenty-first Century.* Penguin Books, London.

Dennis, P. and Fry, G.L.A. (1992) Field margins: can they enhance natural enemy population densities and general arthropod diversity on farmland? *Agriculture,*

Ecosystems and Environment 40, 95–115.

Díaz, M., González, E., Muñoz-Pulido, R. and Naveso, M.A. (1996) Habitat selection patterns of common cranes *Grus grus* wintering in holm oak *Quercus ilex* dehesas of central Spain: effects of human management. *Biological Conservation* 75, 119–123.

Ellenberg, H. (1988) *Vegetation Ecology of Central Europe*. Cambridge University Press, Cambridge.

Evenson, R.E., Herdt, R.W. and Hossain, M. (1996) (eds) *Rice Research in Asia: Progress and Priorities*. CAB International, Wallingford, UK.

Firbank, L.G. and Watkinson, A.R. (1985) On the analysis of competition within two-species mixtures of plants. *Journal of Applied Ecology* 22, 503–517.

Foster, D.R. (1992) Land-use history (1730–1990) and vegetation dynamics in central New England, USA. *Journal of Ecology* 80, 753–772.

Gliessman, S.R. (1989) Integrating trees into agriculture: the home garden agro-ecosystem as an example of agroforestry in the tropics. In: Gliessman, S.R. (ed.) *Agroecology: Researching the Ecological Basis for Sustainable Agriculture*. Springer-Verlag, Berlin, pp. 160–168.

Gould, G. (1990) Ecological genetics and integrated pest management. In: Carroll, C.R., Vandermeer, J.H. and Rosset, P.M. (eds) *Agroecology*. McGraw-Hill, New York, pp. 441–458.

Greenslade, P.J.M. (1983) Adversity selection and the habitat templet. *American Naturalist* 122, 352–365.

Grime, J.P. (1977) Evidence for the existence of three primary strategies in plants and its relevance to ecological and evolutionary theory. *American Naturalist* 111, 1169–1194.

Grime, J.P. (1979) *Plant Strategies and Vegetation Processes*. John Wiley & Sons, Chichester.

Grime, J.P. (1997) Biodiversity and ecosystem function: the debate deepens. *Science* 277, 1260–1261.

Grime, J.P. and Hunt, R. (1975) Relative growth-rate: its range and adaptive significance in a local flora. *Journal of Ecology* 63, 393–422.

Halley, J.M. and Lawton, J.H. (1996) The JAEP ecology of farmland modelling initiative: spatial models for farmland ecology. *Journal of Applied Ecology* 33, 435–438.

Halley, J.M., Thomas, C.F.G. and Jepson, P.C. (1996) A model for the spatial dynamics of linyphiid spiders in farmland. *Journal of Applied Ecology* 33, 471–492.

Harper, J.L. (1977) *Population Biology of Plants*. Academic Press, London.

Harper, J.L. (1982) After description. In: Newman, E.I. (ed.) *The Plant Community as a Working Mechanism*. Blackwell Scientific Publications, Oxford, pp. 11–25.

Harrington, L. (1997) Diversity by design. *CGIAR News* 4, 5–8.

Harris, D., Natarajan, M. and Willey, R.W. (1987) Physiological basis for yield advantage in a sorghum/groundnut intercrop exposed to drought: 1. Dry matter production, yield, and light interception. *Field Crops Research* 17, 259–272.

Heckman, C.W. (1979) *Rice Field Ecology in Northeastern Thailand. The Effect of Wet and Dry Seasons on a Cultivated Aquatic Ecosystem*. Dr Junk, The Hague.

Heong, K.L., Teng, P.S. and Moody, K. (1995) Manipulating rice pests with less chemicals. *Geojournal* 35, 337–349.

Hildrew, A.G. and Townsend, C.R. (1987) Organization of freshwater benthic communities. In: Gee, J.H.R. and Giller, P.S. (eds) *Organisation of Communities*

Past and Present. Blackwell Scientific Publications, Oxford, pp. 347–372.

Hodgson, J.G. (1987) Why do so few species exploit productive habitats? An investigation into cytology, plant strategies and abundance within a local flora. *Functional Ecology* 1, 243–250.

Hodgson, J.G. and Grime, J.P. (1990) The role of dispersal mechanisms, regenerative strategies and seed banks in the vegetation dynamics of the British landscape. In: Bunce, R.G.H. and Howard, D.C. (eds) *Species Dispersal in Agricultural Habitats.* Belhaven Press, London, pp. 65–81.

Hokkanen, H.M.T. (1991) Trap-cropping in pest management. *Annual Review of Entomology* 36, 119–138.

Janzen, D.H. (1973) Tropical agroecosystems. *Science* 182, 1212–1219.

Jiménez-Díaz, L., Arévalo Vicente, G. and Prieto Guijarro, A. (1978) Aspectos téchnico-económicos de un grupo de dehesas salmantinas. Consejo Superior de Investigaciones Científicas (ed.) *Estudio Integrado y Multidisciplinario de la Dehesa Salmantina.* Salamanca-Jaca, pp. 149–203.

Knuth, B. (1989) Agrarstruktur und agrarregionale Differenzierung der Extremadura/Spanien. *Marburger Geographische Schriften* 113, 1–186.

Krysan, J.L., Foster, D.E., Branson, T.F., Ostlie, K.R. and Cranshaw, W.S. (1986) Two years before the hatch: rootworms adapt to crop rotation. *Bulletin of the Entomological Society of America* 32, 250–255.

Landauer, K. and Brazil, M. (eds) (1990) *Tropical Home Gardens. Selected Papers from an International Workshop, Bandung, Indonesia, December, 1985.* United Nations University Press.

Law, R., Bradshaw, A.D. and Putwain, P.D. (1977) Life history variation in *Poa annua. Evolution* 31, 233–246.

Lawton, J.H. (1994) What do species do in ecosystems? *Oikos* 71, 367–374.

Lele, U. and Balu Bumb [n.d.] *The Evolving Role of the World Bank: South Asia's Food Crisis, the Case of India.* World Bank, Washington, DC.

Liebman, M. (1995) Polyculture cropping systems. Agroecology. In: Altieri, M.A. (ed.) *The Science of Sustainable Agriculture.* Westview Press, Boulder, Colorado, pp. 205–218.

Mahn, E.G. (1988) Changes in the structure of weed communities affected by agrochemicals – what role does nitrogen play? *Ecological Bulletins* 39, 71–73.

Marshall, E.J.P. and Hopkins, A. (1990) Plant species composition and dispersal in agricultural land. In: Bunce, R.G.H. and Howard, D.C. (eds) *Species Dispersal in Agricultural Habitats.* Belhaven Press, London, pp. 98–116.

Matson, P.A., Parton, W.J., Power, A.G. and Swift, M.J. (1997) Agricultural intensification and ecosystem properties. *Science* 277, 504–509.

Matzke-Hajek, G. (1997) Zur Evolution und Ausbreitung apomiktischer *Rubus*-Arten (Rosaceae) in Offenland-Ökosystemen. *Bulletin of the Geobotanical Institute ETH* 63, 33–44.

Nair, P.K.R. (1989) Classification of agroforestry systems. In: Nair, P.K.R. (ed.) *Agroforestry Systems in the Tropics.* Kluwer, Boston, Massachusetts, pp. 39–52.

Oberdorfer, E. (1994) *Pflanzensoziologische Exkursionsflora.* Ulmer, Stuttgart.

Ouborg, N.J. and van Treuren, R.V. (1994) The significance of genetic erosion in the process of extinction. IV. Inbreeding load and heterosis in relation to population size in the mint *Salvia pratensis. Evolution* 48, 996–1008.

Pagiola, S. and Kellenberg, J. (1997) *Maintaining Biodiversity in Agricultural Development: Towards Good Practice. Global Diversity Program.* World Bank

Environment Paper No. 15.

Pain, D.J. and Pienkowski, M.W. (1997) *Farming and Birds in Europe: the Common Agricultural Policy and its Implications for Bird Conservation*. Academic Press, San Diego.

Pickett, S.T.A. and White, P.S. (1985) *The Ecology of Natural Disturbance and Patch Dynamics*. Academic Press, Orlando, Florida.

Pimentel, D.A., Stachow, U., Takacs, D.A., Brubaker, H.W., Dumas, A.R., Meaney, J.J., O'Neil, J.A.S., Onsi, D.E. and Corzilius, D.B. (1992) Conserving biological diversity in agricultural and forestry systems. *BioScience* 42, 354–364.

Plucknett, D.L. and Winklemann, D.L. (1995) Technology for sustainable development. *Scientific American* Sept. 1995, 182–186.

Poorter, H. and Remkes, C. (1990) Leaf area ratio and net assimilation rate of 24 wild species differing in relative growth rate. *Oecologia* 83, 553–559.

Potts, G.R. (1986) *The Partridge: Pesticides, Predation and Conservation*. Collins, London.

Pulido, F. and Escribano, M. (1995) The dehesa system: economy and environment. Analysis of typical dehesas of south west of Badajoz province (Spain). In: Albisu, L.M. and Romero, C. (eds) *Environmental and Land Use Issues: an Economic Perspective*. Wissenschaftsverlag Vauk Kiel, pp. 463–474.

Radosevich, S., Holt, J. and Ghersa, C. (1997) *Weed Ecology: Implications for Management*, 2nd edn. John Wiley & Sons, New York.

Ramakrishnan, P.S. (1992) *Shifting Agriculture and Sustainable Development: An Interdisciplinary Study from North-Eastern India*. UNESCO-MAB Series, Paris; Parthenon, Carnforth, UK.

Remmert, H. (1991) The mosaic-cycle concept of ecosystems – an overview. In: Remmert, H. (ed.) *The Mosaic-cycle Concept of Ecosystems*. Ecological Studies 85, Springer-Verlag, Berlin, pp. 1–21.

Rodwell, J.S. (ed.) (1992) *British Plant Communities*, Vol. 3, *Grasslands and Montane Communities*. Cambridge University Press, Cambridge.

Shajaat Ali, A.M. (1987) Intensive paddy agriculture in Shyampur, Bangladesh. In: Turner, B.L., II and Brush, S.B. (eds) *Comparative Farming Systems*. The Guilford Press, New York, pp. 276–305.

Sibly, R. and Calow, P. (1985) Classification of habitats by selection pressures: a synthesis of life-cycle and r/K theory. In: Sibly, R.M. and Smith, R.H. (eds) *Behavioural Ecology. Ecological Consequences of Adaptive Behaviour*. Blackwell Scientific Publications, Oxford, pp. 75–90.

Smithson, J.B. and Lenné, J.M. (1996) Varietal mixtures: a viable strategy for sustainable productivity in subsistence agriculture. *Annals of Applied Biology* 128, 127–158.

Southwood, T.R.E. (1977) Habitat, the templet for ecological strategies. *Journal of Animal Ecology* 46, 337–365.

Southwood, T.R.E. (1988) Tactics, strategies and templets. *Oikos* 52, 3–18.

Srivastava, J.P., Smith, N.J.H. and Forno, D.A. (1996) *Biodiversity and Agricultural Intensification. Partners for Development and Conservation*. The World Bank, Washington, DC.

Strysktra, R.J., Verweij, G.L. and Bakker, J.P. (1997) Seed dispersal by mowing machinery in a Dutch brook valley system. *Acta Botanica Neerlandica* 46, 387–401.

Surryanarayana, M.H. (1995) PDS: beyond implicit subsidy and urban bias – the Indian experience. *Food Policy* 20, 259–278.

Sutcliffe, O.L. and Thomas, C.D. (1996) Open corridors appear to facilitate dispersal by ringlet butterflies (*Aphantopus hyperantus*) between woodland clearings. *Conservation Biology* 10, 1359–1365.

Swift, M.J. and Anderson, J.M. (1996) Biodiversity and agroecosystem function in agricultural systems. In: Schulze, E.D. and Mooney, H.A. (eds) *Biodiversity and Ecosystem Function*. Springer-Verlag, Berlin, pp. 15–41.

Swift, M.J., Vandermeer, J., Ramakrishnan, P.S., Anderson, J.M., Ong, C.K. and Hawkins, B.A. (1996) Biodiversity and agroecosystem function. In: Mooney, H.A., Cushman, J.H., Medina, E., Sala, O.E. and Schulze, E.-D. (eds) *Functional Roles of Biodiversity*. John Wiley & Sons, Chichester, pp. 261–298.

Thurston, H.D. (1992) *Sustainable Practices for Plant Disease Management in Traditional Farming Systems*. Westview Press, Boulder, Colorado.

Tilman, D., Wedin, D. and Knops, J. (1996) Productivity and sustainability influenced by biodiversity in grassland ecosystems. *Nature* 379, 718–720.

Trenbath, B.R. (1976) Plant interactions in mixed crop communities. In: Papendick, R.I., Sanchez, P.A. and Tripplelt, G.B. (eds) *Multiple Cropping*. American Society of Agronomy, Madison, Wisconsin, pp. 129–170.

Tucker, G. (1997) Priorities for bird conservation in Europe: the importance of the farmed landscape. In: Pain, D.J. and Pienkowski, M.W. (eds) *Farming and Birds in Europe: the Common Agricultural Policy and its Implications for Bird Conservation*. Academic Press, San Diego, pp. 79–149.

USDA (1998) *Agricultural Statistics 1997*. United States Department of Agriculture, Washington, DC.

Voelcker, J.A. (1894) *Report of the Improvement of Indian Agriculture*. Eyre & Spottiswoode, London.

Wainhouse, D. and Howell, R.S. (1983) Intraspecific variation in beech scale populations and in susceptibility of their host *Fagus sylvatica*. *Ecological Entomology* 8, 351–359.

de Wet, J.M.J. and Harlan, J.R. (1975) Weeds and domesticates: evolution in the man-made habitat. *Economic Botany* 29, 99–107.

Williams, J.T. and Harper, J.L. (1965) Seed polymorphism and germination. 1. The influence of nitrates and low temperatures on the germination of *Chenopodium album*. *Weed Research* 5, 141–150.

Wolfe, M.S. (1985) The current status and prospects of multiline cultivars and variety mixtures for disease resistance. *Annual Review of Phytopathology* 23, 251–273.

Wratten, S.D. and Thomas, C.F.G. (1990) Farm-scale spatial dynamics of predators and parasitoids in agricultural landscapes. In: Bunce, R.G.H. and Howard, D.C. (eds) *Species Dispersal in Agricultural Habitats*. Belhaven Press, London, pp. 219–237.

Traditional Management of Agrobiodiversity

9

H.D. Thurston[1], J. Salick[2], M.E. Smith[3], P. Trutmann[4], J.-L. Pham[5] and R. McDowell[6]

[1]Department of Plant Pathology, Cornell University, Ithaca, NY 14850, USA; [2]Department of Botany, Porter Hall, Ohio University, Athens, OH 45701, USA; [3]Department of Plant Breeding, Cornell University, Ithaca, NY 14850, USA; [4]Institut für Agrarwirtschaft, ETH-Zentrum, 8092 Zürich, Switzerland; [5]Genetic Resources Centre, International Rice Research Institute, PO Box 933, 1099 Manila, Philippines; [6]Department of Animal Science, North Carolina State University, Raleigh, NC 27695, USA

Introduction

Vigorous efforts are under way worldwide to preserve the world's plant and animal biodiversity which is disappearing at an unprecedented rate. Nevertheless, it is apparent that these efforts may not be sufficient to preserve many plants and animals or to maintain many of the world's valuable ecosystems. Not only are plants, animals and irreplaceable ecosystems disappearing, but traditional cultures are also rapidly disappearing in much of the world. Agricultural practices of traditional cultures have had and continue to have a strong influence on the future of both wild and cultivated species. At present, insufficient efforts are being undertaken to preserve, or even to understand, the impact of the 10,000 years of traditional agriculture on biodiversity.

One expression of the high value that many traditional farmers place on biodiversity is shown by their actions in assembling collections of diverse crops and animals. Some crops, such as cassava and yams, are still central to indigenous religion and identity in some areas of Africa and South America (Boster, 1984; Thurston, 1992; Salick *et al.*, 1997). Some traditional agroecosystems have been sustainable for millennia, such as the paddy rice system (Thurston, 1992); others have failed and we should learn from this. However, agriculture may have had more negative effects on biodiversity – through restructuring of landscapes and, more recently, by the use of agrochemicals – than positive effects.

There is continuing debate on how the plant and animal materials maintained by traditional farmers over centuries should be preserved, studied and
made available. Wood and Lenné (1993) noted that efforts in the past have
tended towards two extremes: conservation in protected areas or conservation in genebanks. Wood and Lenné argue that the management of agricultural diversity by farm communities, an intermediate solution, can be an
effective mechanism for conserving these valuable resources (see Engels and
Wood, Chapter 14, and Smale and Bellon, Chapter 15, this volume).

The term 'traditional' is generally associated with primitive agroecosystems or pre-industrial, peasant agriculture. Traditional farming is
usually based on agriculture that has been practised for many generations.
Indigenous farmers are peoples native to an area who continue to use the
accumulated knowledge of traditional agriculture transmitted by their
ancestors. Most traditional societies had low population densities and were
based on agricultural strategies and social activities that assured their survival
for millennia. However, the standard of living obtained from traditional
farming systems was low in comparison with that obtained in modern
societies. Increased communication and the expanding knowledge of a better
life is influencing the rapid disappearance of traditional farming worldwide
(Giampietro, 1994).

Anthropologists, archaeologists, ethnobotanists and geographers have
been in the forefront of those trying to understand traditional agriculture. To
a lesser degree, economists, systematists and sociologists have also tried.
Unfortunately, those in the so-called 'hard agricultural sciences', such as
agronomy, plant pathology, soil science, entomology and ecology have paid
limited attention to traditional agriculture. Thus, the knowledge of and
respect for traditional agriculture among such disciplines is generally low.
Likewise, the disciplines that have contributed most to our knowledge of
wild biodiversity (ecology, systematics, conservation biology, economics and
others) have often had little exposure to the literature on traditional agriculture. The support for work on understanding agrobiodiversity and traditional agriculture has been minuscule in comparison with that for understanding of wild biodiversity (see Wood and Lenné, Chapter 17, this volume).

Traditional Agroecosystems

There are many different, diverse traditional farming systems (Thurston,
1992), and it is difficult to generalize about them (see Plates 9–12). Stable and
often highly productive systems are found, including the intensive and
strikingly productive systems of mainland China (Wittwer *et al.*, 1987), the
sophisticated systems of South and Southeast Asia (Marten, 1986), and the
durable Latin American crop production systems derived from ancient
Indian civilizations (Wilken, 1987) (Fig. 9.1). Through the use of different
species, varietal mixtures and highly diverse landraces, genetic diversity is a

Fig. 9.1. A maize, beans and squash garden in Mexico (courtesy of H. David Thurston, Cornell University).

characteristic of these systems (Fig. 9.2). Few have been carefully studied; thus, we still know little of the mechanisms which have kept the best of these systems stable, sustainable and productive.

Fig. 9.2. Maize, banana, cassava and yams in a slash–burn cropping system in Nigeria (courtesy of H. David Thurston, Cornell University).

The traditional farmer knowledge of the Hanunóo, a mountain tribe of Mindoro in the Philippines, is often cited as an example of the encyclopaedic knowledge of some traditional farmers. Conklin (1954) noted that on some aspects their agricultural knowledge is amazingly wide and practical. The Hanunóo distinguished numerous soil and mineral categories and understood the suitability of each for various crops, as well as the effects of soil erosion, exposure and over-farming. Regarding diversity, they distinguished over 1500 useful plant types, including 430 cultigens, and discerned minute differences in vegetative structure.

Traditional farming systems, especially in the tropics, resemble natural ecosystems in their diversity. An indication of the importance of diversity to traditional farmers is the classification systems some groups have developed for plants. One of the best examples is the comprehensive plant classification system of the Mayan (Tzeltal) Indians in Mexico. Berlin *et al.* (1974) stated that '471 mutually exclusive generic taxa were established as legitimate Tzeltal plant groupings.' There is no disagreement on whether diversity is a useful trait for crop species and traditional agroecosystems.

Traditional farming systems often show a high degree of stability, reliability, resilience and efficiency although yields are usually lower and labour demands higher than in modern farming systems. Seed is often saved from one season to the next in traditional systems; seed is usually purchased each year in modern farming systems (see Wright and Turner, Chapter 13, this volume). The common principles underlying crop protection in traditional systems are high degrees of both intra- and interspecific genetic diversity (Clawson, 1985; Thurston, 1992); a high degree of flexibility in farmer decision-making (Trutmann *et al.,* 1993); diverse cultural and biological control practices (Thurston, 1992); limited use of external inputs; and sustainable practices which often have multiple uses (Trutmann *et al.,* 1993). In contrast, modern farming systems rely on external inputs such as high-yielding, uniform varieties, often with multiple resistances to diseases, pests and stresses, and inorganic chemicals for both pest control and soil fertility. In spite of these inputs, serious disease and pest epidemics still occur in modern farming systems. Although history has recorded devastating plant disease epidemics in traditional farming systems (e.g. wheat rust of biblical times; the Irish potato famine; maize rust in Africa; the Bengal rice famine; see Allen *et al.*, Chapter 6, Polaszek *et al.*, Chapter 11, this volume), such occurrences are often associated with unusual climatic conditions (Large, 1940; Carefoot and Sprott, 1969).

Landraces

The variability within and between landraces developed in response to, and was maintained by, the tremendous diversity of cultural and environmental regimes present in traditional farming systems (Marshall, 1977). Buddenhagen (1981) noted that today millions of hectares of many crops are traditional cultivars or landraces selected by traditional farmers in prehistory or at least before agricultural science was developed (Fig. 9.3). He wrote:

> This is so for the *Dioscorea* yams, for most of the rice and cowpeas in West Africa, and for most of the maize and beans in Latin America. It is still true for several million acres of rice in Asia, and much of the potato crop of the Andes in South America. Sorghum and millet in tropical Africa are largely old landraces, as is much of the forage grass acreage of the world.

The role of landraces or folk crop varieties has received considerable attention in recent decades (Harlan, 1975, 1976, 1992; Plucknett *et al.*, 1987; National Research Council, 1993; Cleveland *et al.*, 1994). Few better descriptions of landraces are found than that of Harlan (1975):

> Landraces have a certain genetic integrity. They are recognizable morphologically; farmers have names for them and different landraces are understood to differ in adaptation to soil type, time of seeding, date of maturity, height, nutritive value, use and other properties. Most important, they are genetically diverse. Such balanced populations – variable, in equilibrium with both environment and pathogens, and genetically dynamic, are our heritage from past generations and cultivators.

The selection of landraces by farmers was also well described by Harlan (1976):

Fig. 9.3. Seventeen different types (species and landraces) of *Phaseolus* spp. beans grown by one farmer in Mexico (courtesy of H. David Thurston, Cornell University).

> The techniques of subsistence agriculture select continuously for a quality the French call '*rusticité*' and for which there is no appropriate English word. Cultivars or landraces with *rusticité* yield something despite flood, drought, disease, insects, nematodes, birds, monkeys, or witchweed. In basic subsistence agriculture there is little pressure for high yields, but a crop failure means famine and starvation. Dependability is essential, a matter of life or death to the cultivator. Crops developed under these conditions are hardly ever high yielding, but they have excellent general fitness for local conditions and good wide-spectrum resistance to the diseases of the region.

Landraces are dependable and stable and, although not necessarily high yielding, yield some harvest under all but the worst conditions. Yet, it is unlikely that traditional farmers did not also select for productivity within the pools of variability available to them (Frankel *et al.*, 1995). Landraces are often grown under severe physical and biological stresses with low inputs and in low densities. They are the basic building blocks of traditional farming systems.

A major impetus to scientific breeding of plants for disease resistance began after the potato late blight epidemic in Ireland during 1845–1846 when an estimated one million Irish died in the resulting famine (see Allen *et al.*, Chapter 6, this volume). However, traditional farmers have been selecting superior components from landraces, including for disease resistance, for thousands of years. Jennings (1976), who developed the variety 'IR8', which began the Green Revolution in rice, stated: 'The breeding methods devised by Neolithic Man remained standard until the 20th century, although in recent decades they were applied more systematically and with more sophistication. The technique is called pure line selection.' Leppik (1970), Harlan (1976) and Buddenhagen (1983) noted that the greatest number of genes for disease resistance is usually found in landraces, where the host and pathogen have coexisted for long periods of time. Browning (1975) suggested that in centres of origin there is more diversity, not only in crops, but also in their pathogens. He recommended that 'living-gene parks' be established to maintain biodiversity in crops, their ancestors and their pathogens (see Allen *et al.*, Chapter 6, this volume). Altieri and Merrick (1987) also suggested that crop genetic resources might be saved through maintenance of traditional agroecosystems.

The biodiversity found within the landraces selected and maintained by traditional farmers during millennia constitutes an impressive contribution to the human race. However, have the traditional farmers who conserved these landraces benefited? Dr M.S. Swaminathan, former President of the World Conservation Union and winner of the 1987 World Food Prize, suggested that despite the provision of the United Nations Convention on Biological Diversity, ratified by 170 nations after the Earth Summit in Rio de Janeiro in 1992, nothing protects the rights of indigenous farmers who maintain landraces. Furthermore, 'commercial plant breeders are more likely to benefit legally and financially from the sale of useful varieties than those who have conserved plant stocks for centuries' (Shore, 1998). There is considerable debate and discussion on these issues, and this threatens to reduce the valuable exchange of landraces between nations and, ultimately, jeopardize the future of the agroecosystems in which landraces survive.

Management Practices that Affect Agrobiodiversity in Traditional Systems

Many management practices which affect agrobiodiversity are used by traditional farmers (Table 9.1). In addition to direct management of plant diversity, farmers manage the environment, especially the soil, through practices such as site selection, fallowing and rotation, use of organic amendments, flooding, burning, mulching, use of raised beds and manipulation of shade. Thurston (1992) gives many more examples. Many of these practices have well-demonstrated positive effects on agrobiodiversity and are sustainable in the long term (see also Polaszek *et al.*, Chapter 11, this volume). Perhaps the most enlightened way to examine effects of management practices on agrobiodiversity is through specific studies of various crops. The following case studies look in more detail at management of diversity of cassava, maize, common bean, rice and domesticated animals by farmers in traditional systems in different continents.

Table 9.1. Effects of selected traditional agricultural practices on agrobiodiversity. Main source: Thurston (1992).

Traditional practice	Benefits for agrobiodiversity	References
Multiple cropping (intercropping; polycropping; home gardens)	Maintains biodiversity as interdependent crop variability Often reduces damage from pests and diseases favouring maintenance of agrobiodiversity	Dalrymple (1971), Okigbo and Greenland (1976), Minchon *et al.* (1986), Francis (1986, 1988), Landauer and Brazil (1990), Thurston (1992)
Varietal mixtures	Maintains intraspecific crop variability Often reduces damage from pests and diseases favouring maintenance of agrobiodiversity	Clawson (1985); Thurston (1992), Smithson and Lenné (1996), see case studies this chapter
Fallowing and rotation	Maintains soil biodiversity Manages soil pathogens and pests (through interruption of life cycles) favouring crop health and maintenance of agrobiodiversity	Cato (1934), Lewis (1941), Curl (1963), Palti (1981), Thurston (1992), Kennedy (1996)
Organic amendments	Soil enrichment favours soil biodiversity Development of suppressive soils Manages soil pathogens and pests favouring crop health and maintenance of agrobiodiversity	Palti (1981), Cook and Baker (1983), Youtai (1987), Thurston (1992, 1997)
Flooding	Nutrient enrichment favours soil biodiversity Reduces damage from weeds, pests and diseases favouring crop health (especially in paddy rice) and maintenance of agrobiodiversity	Kellman and Cook (1977), Cook and Baker (1983), Thurston (1992), see rice case study this chapter
Burning	Slash/burn systems maintain considerable agrobiodiversity Contributes to pest and disease management, crop health and maintenance of agrobiodiversity	Hardison (1976), Thurston (1992)

Table 9.1. Continued

Traditional practice	Benefits for agrobiodiversity	References
Mulching	Lowers soil temperature, protects against erosion, improves soil texture, provides nutrients and organic matter, reduces weed problems, suppresses soilborne pests and pathogens contributing to crop health and maintenance of agrobiodiversity	Wilken (1987), Thurston (1992)
Raised beds	Improves drainage, fertilization, frost control and irrigation, supports management of soilborne pathogens and pests contributing to crop health and maintenance of agrobiodiversity	Denevan and Turner (1974), Thurston (1992)
Site selection	Avoids diseases, pests and weeds associated with previous crops, matches soil fertility and drainage to crop and variety contributing to crop health and maintenance of agrobiodiversity	Thurston (1992)
Manipulating shade	Maintains biodiversity as interdependent multiple crop variability, e.g. in coffee and cocoa systems	Willey (1975), Thurston (1992)
	Manages pathogens and pests favouring crop health and maintenance of agrobiodiversity	

Traditional management of cassava in the upper Amazon – Jan Salick

In her contribution, Jan Salick notes the importance of various influences on cassava biodiversity in the upper Amazon. These include multiple culinary uses and cultivation in a variety of farming systems on a variety of soils; indigenous people value diversity and seek it. Gender plays an important role in traditional management of cassava germplasm. Exceptional individuals, such as shamans, maintain, collect and generate extraordinary numbers of cassava cultivars. Salick also discusses the role of indigenous beliefs, religion and rituals in the traditional management of cassava diversity.

Cassava (*Manihot esculenta*), as traditionally grown by indigenous people, has extremely high germplasm diversity, comparable to other traditional staples like rice (Conklin, 1957), sweet potato (Yen, 1968), potato (Brush *et al.*, 1981), common bean (Sperling *et al.*, 1994) and maize (Bellon and Brush, 1994). The upper Amazon is a prime example of an area with exceptional cassava diversity. Anthropologists (e.g. Kensinger, 1975) first drew it to our attention and then studied it directly through the perceptions of the indigenous people ('perceptual distinctiveness'; Boster, 1984).

Cassava in the upper Amazon is grown by indigenous people in a variety of farming systems, on a variety of soils, in various rotations (Salick, 1989; Salick and Lundberg, 1990) and in mixtures (Boster, 1984; Salick *et al.*, 1997; Table 9.1). These systems are typical of lowland South American swidden agriculture (Leeds, 1961; Carneiro, 1964; Denevan, 1971; Kensinger, 1975). Although any one field is diverse in cassava varieties itself, it is the diversity of types of fields and stages of fields over time which account for much of the diversity. The multiple uses of cassava may also account for diversity (Boster, 1984; but not Salick *et al.*, 1997) as cassava is boiled, roasted, processed into flour and meal (*farina*), brewed into beer (*masato*), or fermented into various other local delicacies.

Not only do indigenous people recognize distinct cassava varieties, but they also value and seek diversity (Boster, 1984; Salick *et al.*, 1997). People may maintain old varieties that no longer produce well, because they remember them fondly from the past. Varieties are traded, brought back from travels as living mementos, and even stolen from unfriendly tribes or neighbours. When asked why so many varieties of cassava are kept, local people often reply quite mechanistically, 'If I didn't plant them, I would lose those varieties.' Boster (1984) suggested that 'The Aguaruna maintain diversity for its own sake without the need for clearly articulated, pragmatic reasons for each variety maintained.' However, some old, unproductive varieties are purged. Cassava germplasm maintenance is dynamic: there is constant turnover, with new varieties being cultivated and old ones being abandoned (Boster, 1984; Salick *et al.*, 1997).

Although we have a good record of existing cassava diversity, we have very little understanding of rate of germplasm turnover under traditional management. The evolution of germplasm *in situ* is a subject that demands much more attention and research (see Maxted *et al.*, 1997; Engels and Wood, Chapter 14, and Smale and Bellon, Chapter 15, this volume). Conservation of cassava diversity must go beyond the relatively simple goal of saving existing biodiversity to develop strategies to deal with evolutionary processes. This ongoing process of evolution has been documented for other crops as well (for maize see Johannessen *et al.*, 1970; for potato see Johns and Keen, 1986).

Gender plays a very important role in traditional management of cassava germplasm. For many indigenous people of the upper Amazon, cassava is a women's crop (e.g. Barclay, 1985). Although strict gender roles are rarely adhered to in contemporary indigenous societies, the role of women is undoubtedly dominant except where market production is concerned. Indigenous women with a sense of tradition maintain many cassava varieties with associated myths, songs, names and indigenous production technologies (Salick *et al.*, 1997). Educating children in traditional practices, processes, production techniques and associated customs, is a most critical step in maintaining cassava diversity and its evolution into the future.

Exceptional individuals within an indigenous tribe may be a key factor in managing the cassava diversity of the tribe. A shaman (Salick *et al.*, 1997) maintains and collects extraordinary numbers of cassava varieties, far beyond any individual farmer of his tribe. He also breeds cassava by clonally propagating outcrossed, sexually reproduced, volunteer seedlings, assessing characteristics and productivity over several seasons. The importance of traditional crop breeding in the conscious generation of agricultural biodiversity is becoming increasingly recognized (e.g. Franquemont, 1988).

The role of the many indigenous beliefs and religion in traditional agricultural biodiversity has not been well addressed. Cassava is central to indigenous religion and identity in the upper Amazon. Smith (1977) analysed traditional cassava songs. Salick *et al.* (1997) recounted cassava myths and the meaning of sacred pedolyths: cassava began the universe, maintains the people and, in the end, the universe will revert to cassava again; cassava is mother.

The naming process is often important in the establishment of a new cassava variety among indigenous people. A traditional ritual of trance, coca and sleeplessness is held in which the name of a new variety is revealed by the cassava plant itself to the person undergoing the ritual (Salick *et al.*, 1997). Such formal ritual sanctifies varietal recognition and obligates varietal perpetuation. One belief of direct relevance to the generation of cassava diversity is the 'Gift of the Ancestors' (Salick *et al.*, 1997). In newly slashed-and-burned fields, long-dormant cassava seed will germinate from the soil seed-bank. These sexually reproduced, volunteer plants are believed to be gifts of ancestors which should be revered and clonally propagated for a second year to assess the quality of the gift. The process, perpetuated by belief, is akin to the traditional breeding of cassava, with the important difference that the seed may have originated decades before. Thus, bygone genetic material is not only recombined but recycled over long periods of time. This means that germplasm dynamics and evolution is not only a continuous process but may also loop through time, recycling past, successfully cultivated varieties. Again, this process underscores the urgent need for further research into the evolution of genetic diversity in cassava.

The role that traditional economic systems play in supporting biodiversity has not been well studied. However, the recent move to regional and national marketing of cassava is leading to decreased diversity in the field (Salick and Merrick, 1990). Market sellers and buyers are interested in marketable traits such as inner skin colour, root flesh colour and starchiness. There may also be other socio-economic variables that affect indigenous peoples' interest in or ability to maintain cassava diversity (Salick and Lundberg, 1990).

Smith (1996) states categorically that from a modern indigenous peoples' point of view, 'Biodiversity won't feed our children.' His research into the changing economies of the upper Amazon indicates that unless indigenous Amazonians can sustainably satisfy their growing demand for cash and market goods, they are not likely to give much priority to agrobiodiversity conservation. Smith also argues that the image of the indigenous Amazonian as a noble savage and steward of biodiversity may correspond to a past reality, before the introduction of industrial goods, modern school systems and the market economy. The careful manager of biodiversity is being replaced along with the indigenous gift economy ('Gift of the Ancestors'), which engendered a strong sense of value and obligation to share, to receive and to acknowledge debt as a socially binding force. Today, indigenous people are encouraged by the market economy to discount future returns on such nebulous factors as germplasm diversity in favour of ensuring satisfaction of today's perceived needs. Traditional management of cassava diversity is undergoing change (Salick and Lundberg, 1990).

Traditional management of maize in Mexico – Margaret E. Smith

Margaret Smith writes about maize in Mexico, home to both maize and its wild relatives. She notes the common practice of cultivating both modern varieties with intensive crop management techniques and traditional varieties with traditional methods of culture. Smith suggests that the high cost of labour and resulting labour migration in Mexico, rather than the introduction of improved varieties, will pose the greatest threat to the maintenance of maize diversity by Mexican farmers in the future.

Maize (*Zea mays*) and its wild teosinte (*Zea* spp.) relatives in Mexico provide an excellent case study of farmers' traditional approaches to management of agrobiodiversity. Mexico is part of the centre of origin of maize, and of the 40 races of maize identified in Mesoamerica, 31 are found in Mexico (Bretting and Goodman, 1989). Recent data show that 77% of the maize area in Mexico is planted with farmers' traditional or folk varieties (Lopez-Pereira and Morris, 1994). In certain regions, such as in Oaxaca, 93% of the maize area harvested is from folk varieties (Aragón-Cuevas, 1995). Mexico is also home to the closest wild relatives of maize: annual diploid teosintes, perennial tetraploid teosinte, and the only perennial diploid teosinte (*Z. diploperennis*). Many of these teosintes grow as weedy invaders in smallholder maize fields or in the areas around field margins. The Sierra de Manantlán Biosphere Reserve in Jalisco, Mexico, located where various species of teosinte grow (including the only known population of *Z. diploperennis*), is the first national *in situ* reserve established for the preservation of a crop wild relative (Benz, 1988). Thus, Mexico has unique genetic resources of maize and teosinte, and traditional farmers manage the environments in which many of these resources are located.

Recent studies of farmers' maize production in Mexico have shed light on traditional management of maize and teosinte biodiversity. Many small-holder farmers grow modern varieties or hybrids with intensive management techniques and folk or traditional varieties with traditional management techniques (Bellon, 1991). In a community in Chiapas, farmers kept 15 local varieties of maize (belonging to six different races and four race mixtures) because of their adaptation to specific soil conditions, their agronomic characteristics (like drought or lodging resistance), and their storage quality (Bellon and Brush, 1994; Table 9.1; Plate 4). Almost 85% of the farmers also planted an improved variety on about 45% of the total maize area in the community. In a community within the biosphere reserve in Jalisco, farmers grew six 'local' maize varieties (traditional within the region), 19 'foreign' varieties (including farmers' varieties from other regions and farmers' advanced generations of improved varieties), and one recent generation, improved variety (Louette *et al.*, 1997). Again, different varieties were grown to meet specific agronomic or end-use requirements. Similarly, community-level studies in Veracruz have shown that farmers maintain a mixture of modern and traditional varieties to fulfil diverse niches, both in terms of production (precipitation, temperature, altitude, slope, maturity) and utilization requirements (tortillas, specialty foods, feed, cash grain) (Rice *et al.*, 1997; Table 9.1). Thus, traditional Mexican maize farmers are maintaining a considerable number of traditional and modern varieties and adoption of modern varieties is not necessarily occurring at the expense of traditional ones.

Clearly the number of varieties grown by farmers may not be an accurate reflection of the diversity present among those varieties as different varieties may be closely related genetically (Morris and Heisey, 1997). Measures of diversity are complicated by issues such as the level of assessment (genes, blocks of genes, whole genomes, pedigree complexity, agronomic performance) and the appropriate scope of the assessment (*ex situ* collections, *in situ*, across farms within a community, over a larger geographical region) (Smale *et al.*, 1996; see Cox and Wood, Chapter 3, and Witcombe, Chapter 10, this volume). However, examining farmers' seed systems can help to shed light on changes in genetic diversity, since the process of choosing seed for the next planting determines how many and which genetic variants are included in that crop (see Wright and Turner, Chapter 13, this volume).

Farmer seed selection in traditional varieties appears to be a multi-step process, involving selection at some or all of the following stages: in the field, at harvest, as the crop is being used (often by women), and as selected ears are being shelled for planting (Rice *et al.*, 1997; see Wright and Turner, Chapter 13, and Smale and Bellon, Chapter 15, this volume). In Veracruz, modern varieties, which are susceptible to damage by storage insects due to their thin husk cover, tended to be selected at or around harvest time, while traditional varieties that resist storage insect damage were selected from stored seed during food preparation (Rice et al., 1997). Studies in Veracruz and Oaxaca both suggest that farmers view varietal traits as fixed, and thus use selection not to improve varieties, but rather to maintain their characteristics and to ensure good seed quality (Cleveland and Soleri, 1997; Rice et al., 1997).

It is not uncommon for farm families to suffer seed loss, due to weather-related crop failures, insect damage to stored ears, or insufficient grain to meet family needs (Louette and Smale, 1996; Rice et al., 1997). Additionally, farmers in one community in Jalisco believed that they had to obtain new seed on a regular basis to preserve the productivity of their varieties (Louette and Smale, 1996). Lastly, many farmers introduce new varieties for experimentation, and then may save their own seed of these varieties (Lamola and Bertram, 1994). For all of these reasons, introductions of seed obtained from family members, neighbours, other communities, and seed markets are common (Louette et al., 1997; Rice et al., 1997). Introduced seed may be assigned a variety name based on its phenotypic similarity to known varieties, and may be mixed – intentionally or coincidentally – with existing seed of those varieties (Bellon and Brush, 1994; Anonymous, 1997; Louette et al., 1997). This has also been noted for traditional management of common bean in Tanzania (Briand et al., 1998). In addition, since maize is an outcrossing crop, genetic mixing occurs readily between varieties planted in adjacent fields if their flowering dates are similar (Louette, 1997), making the concept of an 'improved open-pollinated variety' quite arbitrary (Morris and Heisey, 1997). These investigations indicate that the nature of farmer varieties of maize in Mexico is very dynamic, and suggest that the impacts of farmer seed selection and introduction of improved varieties on maize varietal biodiversity are not easy to predict.

Diversity at the gene (or allele) level could be lost from open-pollinated maize varieties when small numbers of ears (e.g. 40–50) are the seed source for the next generation (Crossa, 1989). As traditional farmers often plant only a small area with certain varieties, they require only a few ears to provide the necessary seed. One could expect loss of diversity and inbreeding depression to occur. However, in a community in Jalisco, although 32% of the seed lots contained fewer than 40 parent ears, no difference in genetic diversity (based on isozyme analysis) was detected between varieties from these seed lots and varieties routinely planted with large seed lots (Louette et al., 1997). This suggests that genetic introgression from mixing of new seed into existing varieties and from cross-pollination between adjacently planted varieties can maintain relatively high levels of diversity in Mexican farmers' maize varieties. Further support for this idea is provided by yield comparisons of collections made 20–30 years ago (obtained from germplasm banks) and the same varieties collected from the same areas today. Examples from the highlands of Mexico and from Medellín in Colombia (cited by Castillo González and Goodman, 1997), both showed better yield of recent collections individually compared to their older counterparts. This suggests that these varieties may have benefited from introgression of desirable genetic variation from other farmer varieties or from introduced modern varieties, and certainly indicates that they have not suffered from loss of heterozygosity. Thus, diversity at the genetic level appears to benefit from traditional farmers' management of introduced seed.

Maize diversity also could be affected by the presence of teosinte relatives, which can cross readily with maize. Studies of teosinte in Mexico and Guatemala revealed progeny from crosses with maize in all known teosinte populations (Wilkes, 1972). Subsequent work found only anecdotal claims for introgression of useful traits from teosinte to maize (Wilkes, 1977; Benz *et al.*, 1990) concluding that, if gene flow occurs, it is most likely from maize to teosinte. Wood and Lenné (1997) reviewed the considerable literature on maize–teosinte gene flow and concluded that there is little evidence for movement of genes from teosinte to maize in traditional farming systems, and that there is no proof that maize has acquired valuable traits due to gene exchange with teosinte. Although teosinte may not contribute genetic diversity to farmer-managed maize varieties, it is a potentially useful source of diversity for maize breeders (Paintner *et al.*, 1997) and thus is an important part of the agrobiodiversity that farmers manage in Mexico.

In most cases, teosinte occurs as a weed on field margins and in uncultivated lands (Wilkes, 1977). Wilkes concluded that without habitat conservation, the pressures of growing populations demand for land and urbanization would result in elimination of virtually all of Mexico's presently unprotected teosinte populations within 35 years. In the Sierra de Manantlán Biosphere Reserve in Jalisco, however, perennial teosinte – locally known as 'milpilla' – occurs as an integral component of the traditional maize-based agroecosystem, serving as an important source of fodder (Benz, 1988; Benz *et al.*, 1990). Its persistence depends on the continued farmer management of the cropping system. Mexico's traditional farmers, however, will probably play a limited role in conserving annual teosinte biodiversity. This reinforces the need for establishing good *ex situ* teosinte collections.

Although traditional farmers' activities in Mexico clearly provide benefits for maintenance of maize agrobiodiversity, there are definite threats to these processes. Advances in our knowledge and ability to manipulate plant genetics may result in varieties that are dramatically better in certain respects than those available to farmers now. In this case, adoption of the best performing variety over a broad area is possible, leading to what Morris and Heisey (1997) refer to as a 'social trap': what is best for the individual farmer is not best for farmers collectively or for society in general. Although possible, this scenario seems unlikely given the diversity of micro-environments and environmental constraints that confront traditional Mexican maize farmers. It would be difficult to develop a variety that is better than most or all others across such a range of conditions.

Probably the greater threat to maintenance of maize agrobiodiversity in

Mexico results from the opportunity cost of labour. In Veracruz, seed selection has been abandoned by farmers living in a coffee-growing community due to the conflicting demand for labour between this activity and coffee production (Rice *et al.*, 1997). A study in Oaxaca showed that fewer than 30% of people between the ages of 20 and 45 actually reside in their community of birth, and noted that this labour migration represented a major threat to diversity management (García-Barrios and García-Barrios, 1990). On visits to rural Mexican communities, one is often hard pressed to find young people between the ages of 15 and 30, and almost every family has a relative who is in the USA earning money. It seems likely that the opportunity cost of labour and resulting labour migration, rather than replacement by improved varieties, will pose the greatest threat to maintenance of agrobiodiversity by Mexico's maize farmers in the future (see Smale and Bellon, Chapter 15, this volume).

Traditional management of common bean in eastern Africa – Peter Trutmann

The close working relationship between Peter Trutmann, a plant pathologist, and several anthropologists for a number of years in traditional common bean farming systems in eastern Africa helped to develop a deeper understanding, not only of farmer perceptions regarding bean diversity and culture, but also of farmer maintenance and improvement of common bean, especially in varietal mixtures. Trutmann noted that women farmers play a dominant role in most aspects of bean culture as was also noted by Salick for cassava.

Common bean, *Phaseolus vulgaris*, was introduced into eastern Africa from the Americas almost five centuries ago by the Portuguese. Today, eastern Africa is regarded as an important secondary centre of bean diversity. Historically, beans in highland Africa were grown as varietal mixtures (Voss, 1992), and today most farmers in countries like Rwanda still prefer to grow varietal mixtures (Table 9.1). Descriptions of the cropping systems in eastern Africa from late 19th to early 20th century European explorers, for example, note that local farmers in Burundi cultivated at least 90 varieties of *P. vulgaris*, including heat-tolerant and cold-resistant varieties of various growth types (bush, sprawling and climbing beans) (Meyer, 1984). The grain types were diverse and multi-coloured (Plate 1). Beans were grown extensively in association with other crops including plantains, or sole-cropped in small fields. The tasks involved in agricultural food production were the role of women then, and remain so today (Voss, 1992).

In Rwanda, Burundi, Zaire, Malawi and southwestern Uganda, recent changes in bean production have not been dramatic, and the systems are fundamentally similar to those described early in the century. In contrast, in Kenya, Tanzania and the lowlands of Uganda, which have better infrastructure and easier access to markets, colonial policies discouraged traditional practices and promoted the use of uniform varieties. (C. Robertson, personal communication, in Voss, 1992). Clear regional differences now exist in bean agrobiodiversity within eastern Africa.

Research over the last 15 years has consistently shown that farmer's use and management of bean genetic diversity is sophisticated and based on traditional knowledge and understanding of bean diversity. For example, the knowledge of traditional bean farmers on control of bean diseases is often impressive. Traditional farmers in the highlands of eastern Africa manipulate the planting density of beans in damp and fertile conditions to ensure that plants do not touch each other (Trutmann *et al.*, 1996). In drier conditions, beans are planted more densely to ensure adequate yields. Rate of sowing is also altered depending on soil fertility, weed pressure and seed viability. Farmer manipulations also include training climbing beans on stakes so that plants do not touch each other. Avoiding plant contact and reducing humidity by regulation of plant density reduces disease incidence (Table 9.1). Such manipulations undoubtedly have a strong influence on the bean selections made by farmers in these systems.

Scheidegger (cited in Sperling *et al.*, 1994) found 550 local varieties in use in Rwanda alone. In a postal survey of 29 districts in Uganda, only 3% of the farmers surveyed grew mixtures (David, 1994); however, in southwest Uganda, 22% of farmers grew four or more varieties in mixtures. This survey also indicated the existence of at least 135 landraces in Uganda (Grisley and Sengoba, 1993). In Rwanda, Lamb and Hardman (1985) and Voss (1992) found a mean of 11 and a range of 6 to 29 visually different 'types' or varieties per mixture (Table 9.1). In Malawi, Fergusen and Sprecher (1987) noted a mean of 13 varieties per mixture while in the southern highlands of Tanzania, Teverson *et al.* (1994) studied mixtures with from 6 to 15 different varieties. In Zaire, mixtures varied from very phenotypically diverse to those containing only a few varieties that are sorted into colours for markets such as those in Kinshasa.

The primary aim of farmer mixture management is production stability. By using small fields with distinct combinations of diverse genotypes, farmers cope with the highly diverse and fluctuating conditions found in eastern Africa. Typically, mixtures are selected for different conditions (Table 9.1). Small-seeded mixtures are used in poor soils, while larger-seeded mixtures are used in fertile soils. Upright varieties are interplanted with bananas. Highly competitive varieties are selected against in fertile soils but promoted in poorer soils (Trutmann *et al.*, 1996). Farmers appear to manage the proportion of the disease-resistant varieties in mixtures depending on the disease pressure in a particular environment. This may be related to differences in altitude and rainfall. For example, in higher, wetter elevations, farmers' mixtures comprise a greater proportion of varieties resistant to anthracnose (*Colletotrichum lindemuthianum*) than mixtures from lower, drier areas with less disease pressure (Trutmann *et al.*, 1993). The evolution of resistance is probably a combination of farmer and natural selection.

A better understanding of farmer manipulation of mixtures for disease management is needed by both breeders and pathologists.

Most farmers change the composition of their mixtures from time to time. Farmers obtain bean seed from neighbours, their local markets or use their own seed (see Wright and Turner, Chapter 13, this volume). In Rwanda, Burundi and Zaire, more than 70% of farmers obtain new seed from neighbours (usually the man's parents). In Uganda and Tanzania, most farmers use their own seed. Local seed is regarded as more adapted (Sperling *et al.*, 1994). In some regions of eastern Africa, communities recognize 'good seed' farmers (Sperling, 1994), whereas in other areas such farmers were not found (David, 1994). In the Great Lakes Region, farmers moving to new areas will collect seed of available varieties to add to their mixtures (Trutmann *et al.*, 1993). However, in established areas, farmers' mixtures are generally finely tuned and most farmers are reluctant to add new varieties to their mixtures until they have been tested in separate plots (Trutmann *et al.*, 1993). New varieties are added if they are compatible with existing components, if they continue to perform well when combined, and if their addition does not threaten the vitality of existing components (Sperling *et al.*, 1994). Farmers recognize the importance of improving their mixtures and will pay premium prices for new varieties.

By active selection for desired and against undesirable seed types, women prevent single or a few varieties from dominating their mixtures, as would be the case under natural selection (Voss, 1992). Although the quality of local seed is relatively good (David, 1994), there is active selection against blemished seed (Fairhead, 1990). Significant yield increases are possible by cleaning out blemished seed which often harbours seedborne pathogens (Trutmann and Kaytare, 1991). However, the availability of seed greatly influences the quality of seed farmers can use.

Maintenance of varietal mixtures by farmers, even if preferred, is not guaranteed. Population pressure drives the need for intensification of food production. However, the common increase in diseases in intensified systems requires farmers to maintain flexibility in their bean production systems. When a very good variety is introduced, or insufficient economic options are available to farmers, displacement of varietal mixtures will occur with associated loss of diversity. For example, in Rwanda, the high-yielding, climbing variety 'Urubano' (G2333), with multiple disease resistance, was extensively and rapidly adopted by farmers after introduction in response to production pressures and, in some areas, susceptibility of traditional bush bean mixtures to root rots. The extensive displacement of mixtures with 'Urubano' and, the resulting loss of varietal diversity, was of concern during the initial introduction phase. It became a greater concern later when 'Urubano' was also found to be susceptible to root rots (Sperling *et al.*, 1994). Efforts are now being made to provide a greater choice of varieties and to increase the participation of farmers in their selection. A diversity of options is of critical importance to continued farmer management of bean diversity in traditional systems as they build new technologies onto existing knowledge to increase productivity (Trutmann *et al.*, 1993).

Traditional management of rice diversity in Southeast Asia – Jean-Louis Pham

Rice is the major food for 60% of humanity and the paddy rice system may be the world's oldest sustainable agroecosystem. In the following section, Jean-Louis Pham provides a deeper understanding of the changes occurring in traditional management of rice diversity in Southeast Asia, where rice originated. The striking loss of diversity in rice in Asia since the introduction of high-yielding semi-dwarf varieties and the effect of modern rice varieties on traditional management of rice diversity are also considered. As with cassava, in traditional systems rice has religious and social functions as well as being a staple food.

The history of traditional management of rice diversity in Southeast Asia began a few thousand years ago, when farmers domesticated *Oryza sativa* from wild populations. The most likely hypothesis is that Southeast Asia, along with South Asia, was part of the 'non-centre' of domestication (*sensu* Harlan, 1992) of the *japonica* subspecies of *Oryza sativa* (Second, 1982). History suddenly accelerated in 1966 with the release of 'IR8', the first high-yielding, semi-dwarf rice variety. Since then, the agroecological and human environments of rice farming have dramatically changed. In many areas of Asia, only two or three modern varieties are recommended by local extension agencies and seed is typically renewed every 3 years. There is little traditional management of rice diversity: the process by which farmers choose, grow and manage rice varieties each year from a diverse set of available varieties and seed is saved from season to season. This will be the main issue developed in this case study.

Some fascinating examples of traditional rice cultivation still exist, such as the Banaue terraces in the Philippines. Conklin (1986) mentioned that Ifugao farmers distinguish more than 50 parts of the rice plant, 70 uses of rice, 125 techniques to cultivate, store and process postharvest, and more than 250 varietal characteristics. He listed 42 Ifugao varietal names (Conklin, 1982). However, although useful information can be gathered from the study of such traditional systems (e.g. use of the rice–sweet potato rotation to incorporate organic matter into rice terraces – see Yen, 1974), these systems are so unique that they can be considered remnant. They are not representative of current traditional rice cultivation.

In comparison with the irrigated ecosystem, the impact of modern plant breeding has been much lower in rainfed ecosystems (rainfed-upland, rainfed-lowland, and flood-prone) which are generally more heterogeneous and risk-prone and where there is little or no water control. Studies on traditional management of diversity have consequently focused on these ecosystems where diversity is still present. It does not mean, however, that rainfed ecosystems are free from any changes in diversity management or threats to loss of genetic diversity. Firstly, the upland systems have undergone dramatic changes as farmers have shifted from rice to cash crop cultivation (Pandey, 1996). Secondly, institutional plant breeding has decentralized to increase its efficiency and impact in rainfed ecosystems. Studies of traditional management of rice diversity therefore should not be restricted to farmer's management of traditional varieties only. Whether the varieties are traditional or modern does not seem to matter to farmers. In a recent survey, several Vietnamese farmers identified modern, semi-dwarf varieties as being traditional (Bellon, Tuyen, Huong and Pham, Vietnam, 1996, unpublished data); 'modern' varieties from a breeder's perspective were no longer considered 'modern' by farmers. As far as we know, there are no cases reported in the literature of farmers growing or discarding traditional varieties just because they are 'traditional'.

For a farming household, planting different varieties of one crop is a way of attaining goals and solving problems (Bellon, 1996). Table 9.2 gives examples of how rice farmers use diversity to address various goals and problems which vary in space and time. The relative importance of farmers' goals and concerns varies among rice ecosystems. In a comparative study of farmers' perceptions in three rice ecosystems in the Cagayan Valley in the Philippines (Bellon *et al.*, 1997a), farmers in the uplands and rainfed-lowlands valued consumption quality as the main positive trait of their varieties, whereas in the irrigated systems, good yield was the principal factor mentioned by farmers (see Witcombe, Chapter 10, this volume). Susceptibility to biotic stresses and crop management traits were important in all three ecosystems, but were more frequently mentioned in the two rainfed ecosystems.

While some concerns are location-specific, traits related to growth cycle characteristics can be identified among the general key factors in farmer's management of diversity. Whenever significant variation for these traits is present, the trait becomes a predominant part of a farmer's classification system. This was previously reported by Dumont in 1935 (see Dumont, 1995, for a revised edition) for the varieties of the Red River Delta in northern Vietnam. Farmers clustered varieties into '5th month' and '10th month' varieties depending on the month of sowing. Duration is also a key trait for categorizing varieties in the rainfed-lowland ecosystem in the Cagayan Valley (S. Morin, Philippines, 1997, personal communication). Classifications based on duration facilitate the management of rice diversity in relation to available water resources. Cambodian farmers divide rainfed-lowland varieties into early, medium and late maturing according to flowering and harvest dates (Lando and Mak, 1994). The varieties are planted according to the topography of the fields. Early-maturing varieties are usually planted in highest fields (the most drought-prone), medium- and some late-maturing varieties in middle fields, and late-maturing varieties in lowest fields (the least drought-prone).

Table 9.2. Examples of the functions of rice infraspecific diversity for farmers (adapted from Bellon *et al.*, 1998).

Functions	Examples
To farm in a variety of environments characterized by different soil qualities, temperature and rainfall regimes, topographies, etc.	Many rice farmers match different varieties to different field levels, that in turn reflect different regimes of water availability (Lambert, 1985; Lando and Mak, 1994)
To make harvest easier	In Lao PDR, Bolikhan farmers grow varieties with long peduncle panicles so that they can easily cut them and tie them into bundles (Appa Rao *et al.*, 1997)
To provide variety in a monotonous diet	Some farmers in the Philippines mix glutinous and non-glutinous varieties to improve texture
To provide rice for particular uses	Glutinous rice is the basic ingredient in sweets prepared for weddings and festivals, especially the Khmer lunar New Year in April (Lando and Mak, 1994) Farmers in Vietnam prepare special cakes with glutinous rice for the Tet festival
To provide rice with high volume expansion grain	Cambodian farmers mentioned that a small amount of milled rice feeds a large family (Lando and Mak, 1994)
To fulfil rituals, generate prestige, forge social ties	The Iban of Sarawak, an ethnic group in Malaysia, locate a special ritual segment in the middle of a field where a special rice variety is planted (Sutlive, 1978). Balinese farmers grow black, red and white grain varieties to produce rice corresponding to the four Orients (the fourth colour, yellow, being obtained from turmeric) (Friedberg, 1996)

In less favourable environments, when modern varieties are perceived to perform as well as or better than traditional varieties, the management of diversity is affected (Morin *et al.*, 1998). In the Cagayan Valley, traditional varieties are planted almost exclusively in the wet season, when there is no water limitation. When the need for traditional varieties to address specific water availability regimes decreases, then the tension between modern and traditional varieties increases. Crop duration becomes a major issue in terms of the trade-off between traditional and modern varieties. When asked why they discarded traditional varieties, farmers in the rainfed-lowland eco-system in Cagayan Valley mentioned duration as a major factor (Bellon *et al.*, 1997b). For long-duration varieties, only one rice crop per year can be grown versus two crops for the short-duration varieties. There is therefore a clear threat to long-duration varieties, still grown because of their consumption quality and their high market price, if the balance in farmers' priorities changes.

Bellon et al. (1997a) reviewed data on the number of varieties or landraces maintained by households in 14 villages or clusters of villages in different rice ecosystems, mainly in Cambodia and Thailand. Individual rice farmers maintain on average between one and three varieties; at the village level, more diversity is maintained, with 10.8 varieties on average, with the exception of a village in the swamp ecosystem with 32 varieties (Plate 3). These figures are consistent with recent observations in the Philippines and Vietnam (IRRI and the Hue University of Agriculture and Forestry, Vietnam, unpublished data). The number of varietal names present at the regional level can be very high. In the Cagayan Valley, 152 different names were collected in 12 villages in three ecosystems. Genetic analyses demonstrated that some landraces include very different genotypes while other landraces are specific, well-defined genetic entities (Pham et al., 1997). The analyses also showed slight genetic differences among collections with the same modern variety name. Comparison of the farmers' varieties with those derived from breeder seeds also indicated a divergence between the two samples. These results indicate a high degree of outcrossing among farmers' varieties and/or the misnaming of several varieties (Sebastian et al., 1998). It appears that farmers' management of rice diversity is a dynamic process with associated genetic changes. It suggests that in some locations, even after thousands of years of cultivation, the contribution of traditional management of rice diversity to the evolution of the crop is still considerable.

Traditional management of domesticated animal diversity — Robert McDowell

The knowledge of many traditional farmers regarding animals is as broad and practical as their knowledge of cultivated plants. Robert McDowell draws on his decades of experience with farm animals in the developing world in this last case study. Farmer selection for superior animal characteristics has resulted in a tremendous diversity of animal breeds (see Plates 5–8). The wider introduction of improved, modern breeds and rapid global environmental and economic changes are considerable threats to the maintenance of domesticated animal biodiversity (see Steane, Chapter 4, this volume).

With no knowledge of the formal principles of genetics and animal breeding, traditional farmers have long been aware of the relative merits of their animals and the concept that 'like begets like' (Wilson, 1994). They have exerted selection in favour of superior draught power, milk yield, meat production, and colour pattern (ILCA, 1995). For example, the Bateng cattle (Bos sondaicus), considered the most 'primitive' type of cattle which contributed to present breeds, have many of the phenotypic characteristics of deer: short tail, sloping rump, small bones, and branching horns. Small groups of Bateng continue to live in isolated forested areas of Southeast Asia (e.g. Burma, Cambodia and Indonesia). Bateng move rapidly and are therefore able to avoid would-be predators more readily than modern cattle breeds. They are inferior, however, in preferred production traits (McDowell, 1972).

Environmental conditions and agriculture systems have had a high impact on the diversity of breeds of livestock. In India and Pakistan, with the largest reservoir of 'pure Zebu' type cattle, varied climatic conditions led to the development of 37 breeds, highly variable in phenotypic characteristics. In the Great India Desert of northwest India, feed is limited and cattle were selected for milk production, while on the plains, different breeds were developed for ploughing heavy and light soils. Also in India, breeds were developed for cartage, entertainment (including a breed for racing – *Hallikar*), and a small breed for forested areas in the northeast. Although livestock improvement is important in India and Pakistan, care is being taken to maintain native breeds.

Cattle from India have been widely dispersed globally (Gupta, 1992). Around the 11th century, Indian traders sailed along the coasts of the Arabian Sea and Africa and into the Indian Ocean. Their cattle were developed into the breeds used by pastoralists and indigenous agriculturalists throughout Africa today. Indian breeds have also been widely used in tropical Latin America to strengthen the heat tolerance of local breeds (Pratt and Gwynne, 1977), while the Brahma cattle of the USA came directly from India. In China, a country of highly variable topography, primitive cattle from Mongolia were developed into more than 100 breeds, adapted to many local systems for draught purposes, cartage, food, skin and fibre production.

The yak, a long-haired, wild ox native to the Tibetan highlands at about 3000 m, has long external hair, up to 20 cm, and a dense woollen undercoat for winter. Yaks must be able to dig through 2 m of snow to obtain feed during winter. The long and short hair is used for clothing and yaks are also important for cartage. Males, especially, can tolerate altitudinal changes while hauling loads. Yaks are, however, maintained as a pure breed at very high altitudes only (Nivsarkar *et al.*, 1997). Elsewhere, they are crossed with cattle to obtain larger, faster animals with higher milk production. These crosses are known *yakow* or *chouri*. The first-generation male is sterile but female progeny are fertile (McDowell, 1992).

Buffaloes have been domesticated for thousands of years and remain highly valued in 30 countries. The two main types of domesticated buffalo (*Bubalina* spp.) share some unique features of management. The riverine type originated in India and is used for milk in Egypt, Iran, Pakistan and India. Males are useful for ploughing heavy soils. The swamp type or *carabao* from South Asia is adapted to the paddy rice growing areas (Cockrill, 1974) (Plate 6). Both males and females can be worked and are quite docile. Often small farms will have a single animal which essentially becomes a member of the family.

Herders in the Gobi Desert of Mongolia and China selected horses, rather than cattle, for milk production because horses could more easily escape predators. During the grass growing season, the mares produce 300–400 litres of milk which is processed and stored for food. In this same high plains region, the bi-humped camel (*Camel bactrianus*) was selected for milk production and work. Pastoralists may follow their camel herd for up to 1000 miles (1600 km) per year over the range area. Milk is the basic food and animals are bartered from time to time to obtain other family needs (Harris, 1974).

One unique pool of domesticated animal germplasm diversity is found in the sheep maintained by sub-tribes of Kurdish people. During summer, the flocks are grazed in the Pontic Mountains of Turkey, the Elburz Mountains of Iran, and the highlands of Afghanistan. They are moved to lower elevations to graze stubble or straw in winter. Germplasm from the Kurd flocks has been used in most of the world's sheep breeds (Mason, 1988). About 100 distinct types have emerged from this pool including long-haired for carpet-making, long wool, medium wool, short wool, and short-haired for meat. Milk is obtained and cheese made under most primitive conditions. Nearly all breeds are fat-tailed (storing pure fat for retrieval during poor feeding times). Some types have double fat-tails to further aid in survival. The flock owners are especially selective for breeding rams. In late summer, tribal groups from a region will gather and arrange marriages of their sons and daughters. Rams are exchanged among flocks. An outstanding ram of desired phenotype can, over his life, be used in nearly all regions from west to east (about 3000 km distance).

Pigs are classed as a warm-climate species as they originated in southern China (Plate 7). During early domestication, pigs were reared to utilize human food wastes. Approximately 80% of the world pig population is found on small farms and there remains a large reservoir of traditional types ranging from dwarf to (potentially) 450 kg, and sows capable of producing over 20 piglets per litter. Seldom do small farms have problems of in-breeding, as there is generally a 'community boar' in small villages that meets the approval of most farmers (McDowell, 1992).

The two main types of poultry, improved breeds and native types, descended from the jungle fowl of China. The latter are excellent scavengers and produce tasty meat but the egg production is low. They are tempera-mental and include fighting breeds. There are thousands of native types on small farms throughout the developing world. These birds are better than ducks, pigs and guinea fowl in their use of household wastes and insects. Native types readily cross with improved breeds. In rural areas of Africa, Asia and Latin America, the wide diversity of native types should continue to survive since they are well adapted to local environments.

Comprehensive coverage of the great genetic diversity of domesticated animal species which are key features of small farms in developing countries around the world is not possible here. Others include the alpaca, capybara, donkey or ass, goat, rabbit, guinea pig, duck, goose, guinea fowl and ostrich. Today sub-Saharan Africa is probably the richest region for animal bio-diversity, including their pests and parasites (ILCA, 1995). Conservation of existing biodiversity among and within animal species is needed as a defence against threats posed by pests, diseases and environmental hazards (Wilson, 1994; see Steane, Chapter 4, this volume). There are, however, serious economic threats to these indigenous systems as supermarkets are becoming the main source of food for purchasers. Producers and sellers of native breeds are going out of business. For example, the small eggs from a native chicken, or the pork and lamb chops from native pigs and sheep, will be rejected by buyers when the same products from modern breeds are available. Overall, the best conservers of existing diversity are small farmers in developing countries. Further traditional management of domesticated animal bio-diversity depends on the continued existence of these farmers and their farming systems (see Steane, Chapter 4, this volume).

Summary

Traditional farmers have made and could continue to make invaluable contributions to maintaining, managing and improving the biodiversity of the plants and animals that feed us. Traditional farmers maintain agrodiversity for diversity of diet and income source, stability of production, reduced insect and pest damage, intensification of production with limited resources, optimization of natural resource use, indigenous beliefs and rituals, minimization of risk and the lack of suitable modern varieties (Clawson, 1985; Thurston, 1992; Table 9.1). Traditional management of agrobiodiversity shows remarkable parallels across crops, cultures and continents (Wood and Lenné, 1993) and this is clearly shown by the case studies on cassava, maize, common bean, rice and domesticated animals.

Serious obstacles are also highlighted in these studies to the continued management of the agrobiodiversity of traditional crop varieties and animal breeds. Market forces, as mentioned for cassava, maize, rice and farm animals, are a wide-ranging and increasing threat to continued management (see Smale and Bellon, Chapter 15, this volume). Traditional animal breeds and crop varieties often cannot compete with modern equivalents as they are less productive and less profitable (see Steane, Chapter 4, and Witcombe, Chapter 10, this volume). The increasingly high cost of labour and the labour-intensive nature of traditional farming is another reason why agrobiodiversity may decrease, as illustrated by maize. Both human-induced and natural environmental changes will also affect future maintenance of many traditional crops and animals, as numerous types have been selected and managed by farmers to fill the specific ecological niches of traditional agroecosystems. As farming becomes 'homogeneous', these niches will disappear together with their traditional varieties. Some traditional farmers have demonstrated considerable skill in utilizing a mix of modern and traditional varieties to their advantage, as illustrated here by both maize and cassava, and also by potato (Brush *et al.*, 1981; see Wright and Turner, Chapter 13, this volume). This approach may be a more sustainable alternative for traditional farming systems in the future.

Both *in situ* and *ex situ* collections of crops and farm animals will be useful to future generations (see Engels and Wood, Chapter 14, this volume). However, as Wood and Lenné (1993) argue, the management of agricultural diversity by farm communities may be the most effective mechanism for conserving the valuable resources that traditional farmers have produced and maintained in the last 10,000 years. A major challenge is to ensure that farmer management of this important agrobiodiversity continues as an ongoing process. It deserves the support of mankind if there is interest in the needs of future generations.

To close, Weatherwax (1954) quotes a wonderful taxonomic statement on biodiversity that came from an illiterate Indian woman in a market in Mexico when asked how many kinds of maize her people had. She answered as follows:

> We have four months maize and six months maize; then there is white maize and black maize and yellow maize; and there is large maize and small maize. But they are all the same thing. There is just one name for all of them. They are all maize.

References

Altieri, M.A. and Merrick, L.C. (1987) *In situ* conservation of crop genetic resources through maintenance of traditional farming systems. *Economic Botany* 41, 86–96.

Anonymous (1997) Conservation of genetic diversity and improvement of crop production in Mexico: a farmer-based approach. *Project MILPA – Annual Report to the McKnight Foundation*, Universidad Nacional Autónoma de México and University of California, Davis.

Appa Rao, S., Bounphanouxay, C., Phetpaseut, V., Schiller, J.M., Phannourath, V. and Jackson, M.T. (1997) Collection and preservation of rice germplasm from southern and central regions of the Lao PDR. *Lao Journal of Agriculture and Forestry* 1, 43–56.

Aragón-Cuevas, F. (1995) *La Producción de Maíz en Oaxaca, Situación Actual y Perspectivas Futuras.* Instituto Nacional de Investigaciones Forestales y Agropecuarias (INIFAP), Centro de Investigación Regional del Pacífico Sur, Mexico.

Barclay, F. (1985) *Analisis de la Division del Trabajo y de la Economía Domestica entre los Amuesha de la Selva Central.* PEPP/USAID, Lima, Peru.

Bellon, M.R. (1991) The ethnoecology of maize variety management: a case study from Mexico. *Human Ecology* 19, 389–418.

Bellon, M.R. (1996) The dynamics of crop infraspecific diversity: a conceptual framework at the farmer level. *Economic Botany* 50, 26–39.

Bellon, M.R. and Brush, S.B. (1994) Keepers of maize in Chiapas, Mexico. *Economic Botany* 48, 196–209.

Bellon, M.R., Pham, J.L. and Jackson, M.T. (1997a) Genetic conservation: a role for rice farmers. In: Maxted, N., Ford-Lloyd, B.V. and Hawkes. J.G. (eds) *Plant Conservation: the In Situ Approach*. Chapman & Hall, London, pp. 263–289.

Bellon, M.R., Pham J.L., Sebastian, L.S., Erasga, D., Abrigo, G., Sanchez, P., Calibo, M., Quilloy, S. and Francisco, S.R. (1997b) Farmers' perceptions and variety selection: implications for on-farm conservation of rice. In: *Proceedings of an International Conference Building the Basis for Economic Analysis of Genetic Resources in Crop Plants*, CIMMYT and Stanford University, 17–19 August 1997, Palo Alto, California, (in press).

Bellon, M.R., Brar, D., Lu, B.R. and Pham, J.L. (1998). Rice genetic resources. In: Dowling, N.G., Greenfield, S.M. and Fischer, K.S. (eds) *Sustainability of Rice in the Global Food System*. The Pacific Basin Study Center, Davis, California, and IRRI, Philippines, pp. 251–283.

Benz, B.F. (1988) *In situ* conservation of the genus *Zea* in the Sierra de Manantlán Biosphere Reserve. In: *Recent Advances in the Conservation and Utilization of Genetic Resources, Proceedings of the Global Maize Germplasm Workshop*, CIMMYT, Mexico, pp. 59–69.

Benz, B.F., Sánchez-Velásquez, L.R. and Santana Michel, F.J. (1990) Ecology and ethnobotany of *Zea diploperennis*: preliminary investigations. *Maydica* 35, 85–98.

Berlin, B.D., Breedlove, E. and Raven, P.H. (1974) *Principles of Tzeltal Plant Classification. An Introduction to the Botanical Ethnography of a Mayan-speaking People of Highland Chiapas*. Academic Press, New York.

Boster, J.S. (1984) Classification, cultivation and selection of Aguaruna cultivars of *Manihot esculenta* (Euphorbiaceae). *Advances in Economic Botany* 1, 34–47.

Bretting, P.K. and Goodman, M.M. (1989) Karyotypic variation in Mesoamerican races of maize and its systematic significance. *Economic Botany* 43, 107–124.

Briand, L., Brown, A.E., Lenné, J.M. and Teverson, D.M. (1998) Random amplified polymorphic DNA variation within and among bean landrace mixtures (*Phaseolus vulgaris* L.) from Tanzania. *Euphytica* 102, 371–377.

Browning, J.A. (1975) Relevance of knowledge about natural ecosystems to development of pest management programs for agroecosystems. *Proceedings of the American Phytopathological Society* 1, 191–194.

Brush, S.B., Carney, H.J. and Huaman, Z. (1981) Dynamics of Andean potato culture. *Economic Botany* 35, 70–88.

Buddenhagen, I.W. (1981) Conceptual and practical considerations when breeding for tolerance or resistance. In: Staples, R.C. and Toenniessen, G.H. (eds) *Plant Disease Control. Resistance and Susceptibility*. John Wiley & Sons, New York, pp. 221–234.

Buddenhagen, I.W. (1983) Breeding strategies for stress and disease resistance in developing countries. *Annual Review of Phytopathology* 21, 385–409.

Carefoot, G.L. and Sprott, E.R. (1969) *Famine on the Wind: Plant Diseases and Human History*. Angus & Robertson, London.

Carneiro, R. (1964) Shifting cultivation among the Amahuaca of eastern Peru. *Völkerkundliche Abhandlungen* 1, 9–18.

Castillo González, F. and Goodman, M.M. (1997) Research on gene flow between improved maize and landraces. In: *Proceedings of Gene Flow Among Maize Landraces, Improved Maize Varieties, and Teosinte: Implications for Transgenic Maize*. CIMMYT, Mexico, DF, pp. 67–72.

Cato, M.P. (1934) *On Agriculture.* Translated by Hooper, W.D. and Ash, H.B., Harvard University Press, Cambridge, Massachusetts.

Clawson, D.L. (1985) Harvest security and intraspecific diversity in traditional tropical agriculture. *Economic Botany* 39, 56–67.

Cleveland, D.A. and Soleri, D. (1997) Genetic diversity, genotype-by-environment interaction and plant selection. In: *Proceedings of an International Conference on Building the Basis for the Economic Analysis of Genetic Resources in Crop Plants,* CIMMYT and Stanford University, 17–19 August 1997, Palo Alto, California.

Cleveland, D.R., Soleri, D. and Smith, S.E. (1994) Do folk crop varieties have a role in sustainable agriculture? *BioScience* 44, 740–751.

Cockrill, W.R. (ed.) (1974) *The Husbandry and Health of the Domestic Buffalo.* FAO, Rome.

Conklin, H.C. (1954) An ethnoecological approach to shifting agriculture. *Transactions of the New York Academy of Science* 17, 133–142.

Conklin, H.C. (1957) Hanunóo agriculture: a report on an integral system of shifting cultivation in the Philippines. *FAO Forestry Development Paper* No. 12, FAO, Rome.

Conklin, H.C. (1982) Ifugao rice varieties. Paper presented at the IRRI Special Seminar, 10 August 1982.

Conklin, H.C. (1986) Des orientements, de vents, de riz . . . pour une étude lexicologique des savoirs traditionnels. *Journal d'Agriculture Traditionnelle et de Botanique Appliquée* 33, 3–10.

Cook, R.J. and Baker, K.F. (1983) *The Nature and Practice of Biological Control of Plant Pathogens.* The American Phytopathological Society, St Paul, Minnesota.

Crossa, J. (1989) Methodologies for estimating the sample size required for genetic conservation of outbreeding crops. *Theoretical and Applied Genetics* 77, 153–161.

Curl, E.A. (1963) Control of plant disease by crop rotation. *Botanical Review* 29, 413–479.

Dalrymple, D.G. (1971) Survey of multiple cropping in less developed nations. *Foreign Economics Development Report* No. 12, Foreign Economic Research Service, USDA/USAID, Washington, DC.

David, S. (1994) Local bean seed systems in Uganda: preliminary results from surveys in two districts. In: *Proceedings of a Working Group Meeting,* Kampala, Uganda, 10–13 October 1994. CIAT African Workshop Series No. 32, pp. 16–27.

Denevan, W.M. (1971) Campa subsistence in the Gran Pajonal, eastern Peru. *Geographical Review* 61, 496–518.

Denevan, W.M. and Turner, B.L., II (1974) Forms, function and associations of raised fields in the old world tropics. *Journal of Tropical Geography* 39, 24–33.

Dumont, R. (1995) *La Culture du Riz dans le Delta du Tonkin. Etude et Propositions d'Amélioration des Techniques Traditionnelles de Riziculture Tropicale.* Edition de 1935 revue, corrigée et augmentée. Grand Sud. Hommes et Sociétés d'Asie du Sud-Est, Série 'Classiques', 6. Prince of Songkla University.

Fairhead, J. (1990) Fields of struggle: towards a social history of farming knowledge and practice in a Bishira community, Kivu, Zaire. PhD Thesis, School of Oriental and Asian Studies, University of London.

Fergusen, A. and Sprecher, S. (1997) *Women and Plant Genetic Diversity: the Case of Beans in the Central Region of Malawi.* Bean Cowpea CRSP, Michigan State University Press, East Lansing, Michigan.

Francis, C.A. (ed.) (1986) *Multiple Cropping Systems*. Macmillan, New York.

Francis, C.A. (1988) Biological efficiencies in multiple-cropping systems. *Advances in Agronomy* 42, 1–42.

Frankel, O.H., Brown, A.D.H. and Burdon, J.J. (1995) *The Conservation of Plant Biodiversity*. Cambridge University Press, Cambridge.

Franquemont, C. (1988) *Proceedings of a Workshop on Traditional Cultivars*. Latin American Studies Center, Cornell University, Ithaca, New York.

Friedberg, C. (1998) Diversité, ordre et unité du vivant dans les savoirs populaires. *Natures Sciences Sociétés* 5, 5–17.

García-Barrios, R. and García-Barrios, L. (1990) Environmental and technological degradation in peasant agriculture: a consequence of development in Mexico. *World Development* 18, 1569–1585.

Giampietro, M. (1994) Sustainability and technological development in agriculture. *BioScience* 44, 677–689.

Grisley, W. and Sengoba, T. (1993) Bean cultivars sown by farmers in Uganda during 1989–90: results of a survey of district agricultural officers. In: Grisley, W. (ed.) *A Collection of Papers on the Economics of Beans in Uganda and Eastern and Southern Africa*. CIAT Regional Bean Program, Uganda.

Gupta, R.P. (ed.) (1992) *Cattle Breeds of India*, 4th edn. Priyadarshini Vihar, New Delhi.

Hardison, J.R. (1976) Fire and flame for plant disease control. *Annual Review of Phytopathology* 14, 355-379.

Harlan, J.R. (1975) Our vanishing genetic resources. *Science* 188, 618–621.

Harlan, J.R. (1976) The plants and animals that nourish man. *Scientific American* 235, 89–97.

Harlan, J.R. (1992) *Crops and Man*, 2nd edn. American Society of Agronomy, Madison, Wisconsin.

Harris, M. (1974) *Cows, Pigs, Wars and Witches: the Riddles of Culture*. Random House, New York.

ILCA (1995) *Characterizing and Conserving Livestock Genetic Resources*. International Livestock Centre for Africa, Addis Ababa.

Jennings, P. (1976) The amplification of agricultural production. *Scientific American* 235, 180–194.

Johannessen, C.L., Wilson, M.R. and Davenport, W.A. (1970) The domestication of maize: process or event. *Geographical Review* 60, 393–413.

Johns, T. and Keen, S.L. (1986) Ongoing evolution of the potato on the altiplano of western Bolivia. *Economic Botany* 40, 409–424.

Kellman, A. and Cook, R.J. (1977) Plant pathology in the People's Republic of China. *Annual Review of Phytopathology* 15, 409–429.

Kennedy, A.C. (1996) Soil microbial diversity in agricultural systems. In: Olson, R., Francis, C. and Kaffka, S. (eds) *Exploring the Role of Diversity in Sustainable Agriculture*. American Society of Agronomy, Madison, Wisconsin, pp. 35–54.

Kensinger, K.M. (1975) Studying the Cashinahua. In: Dwyer, J.P. (ed.) *The Cashinahua of Eastern Peru. Studies in Anthropology and Material Culture*, Vol. 1. Haffenreffer Museum of Anthropology, Brown University, Providence, Rhode Island, pp. 9–25.

Lamb, E.M. and Hardman, L.L. (1985) *Final Report of the Survey of Bean Varieties Grown in Rwanda*, January 1984 – June 1985. AID Rwanda Local Crop Storage Co-operative Research, Projet Agricole de Karama.

Lambert, D.H. (1985) *Swamp Rice Farming: the Indigenous Pahang Malay Agricultural System*. Westview Press, Boulder, Colorado.

Lamola, L.M. and Bertram, R.B. (1994) Experts gather in Mexico to seek new strategies in preserving agrobiodiversity. *Diversity* 10, 15–17.

Landauer, K, and Brazil, M. (eds) (1990) *Tropical Home Gardens*. Selected papers from an International Workshop, Bandung, Indonesia, December 1985. United Nations University Press.

Lando, R.P. and Mak, S. (1994) *Cambodian Farmers' Decision-Making in the Choice of Traditional Rainfed Lowland Rice Varieties*. IRRI Research Paper Series 154.

Large, E.C. (1940) *The Advance of the Fungi*. Jonathan Cape, London.

Leeds, A. (1961) Yaruro incipient tropical forest horticulture: possibilities and limits. In: Wilbert, J. (ed.) *The Evolution of Horticultural Systems in Native South America, Causes and Consequences*. Antropologica Supplementary Publications No. 2, Caracas, Venezuela, pp. 13–46.

Leppik, E.E. (1970) Gene centers of plants as sources of disease resistance. *Annual Review of Phytopathology* 8, 323–344.

Lewis, C.D. (1941) *The Georgics of Virgil: a New Translation. Book 1*. Jonathan Cape, London.

Lopez-Pereira, M.A. and Morris, M.L. (1994) *Impacts of International Maize Breeding Research in the Developing World, 1966–90*. CIMMYT, Mexico, DF.

Louette, D. and Smale, M. (1996) *Genetic Diversity and Maize Seed Management in a Traditional Mexican Community: Implications for in Situ Conservation of Maize*. CIMMYT, Mexico, DF, NRG 96–03.

Louette, D., Charrier, A. and Berthaud, J. (1997) *In situ* conservation of maize in Mexico: genetic diversity and maize seed management in a traditional community. *Economic Botany* 51, 20–38.

Marshall, D.R. (1977) The advantages and hazards of genetic homogeneity. *Annals of the New York Academy of Sciences* 287, 1–20.

Marten, G.G. (ed.) (1986) *Traditional Agriculture in Southeast Asia*. Westview Press, Boulder, Colorado.

Mason, I.L. (1988) *World Dictionary of Livestock Breeds*. CAB International, Wallingford.

Maxted, N., Ford-Lloyd, B.V. and Hawkes, J.G. (1997) *Plant Genetic Resource Conservation: the In Situ Approach*. Chapman & Hall, London.

McDowell, R.E. (1972) *Improvement of Livestock Production in Warm Climates*. W.H. Freeman, San Francisco.

McDowell, R.E. (1992) *Choosing Animal Species for Warm Climate Regions*. Department of Animal Science, North Carolina State University, Raleigh, Mimeo No. 101.

Meyer, H. (1984) *Les Barundi: une Etude Ethnologique en Afrique Orientale*. Société Française d'Histoire d'Outre-mer, Paris. (Translated by R. Willmann.)

Minchon, G., Mary, F. and Bompard, J. (1986) Multistoried agroforestry garden system in West Sumatra. *Agroforestry Systems* 4, 315–338.

Morin, S.R., Pham, J.L., Sebastian, L.S., Abrigo, G., Erasga, D., Bellon, M.R., Calibo, M. and Sanchez, P. (1998) Integrating indigenous technical knowledge and *in situ* conservation: collaborative research in the Cagayan Valley. *Philippines Symposium/Workshop on Indigenous Knowledge on Biodiversity Conservation and Management*, NUFFIC-PCARRD, Cebu City (Philippines), 4–6 March, 1998.

Morris, M.L. and Heisey, P.W. (1997) Achieving desirable levels of crop genetic diversity in farmers' fields: factors affecting the production and use of improved seed. In: *Proceedings of an International Conference Building the Basis for Economic Analysis of Genetic Resources in Crop Plants.* CIMMYT and Stanford University, 17-19 August 1997, Palo Alto, California.

National Research Council (1993) *Managing Global Genetic Resources: Agricultural Crop Issues and Policies.* National Academy Press, Washington, DC.

Nivsarkar, A.E., Gupta, S.G. and Gupta, N. (1997) *Yak Production.* Indian Council of Agricultural Research, Pusa, New Delhi.

Okigbo, B.N. and Greenland, D.J. (1976) Intercropping systems in tropical Africa. In: Papendick, R.I., Sanchez, P.A. and Triplett, G.B. (eds) *Multiple Cropping.* American Society of Agronomy Special Publication 27, Madison, Wisconsin, pp. 63–101.

Paintner, S., McCouch, S.R., Smith, M.E. and Tanksley, S.D. (1997) Discovery of desirable QTL alleles from teosinte. In: *Proceedings of the 52nd Annual Northeastern Corn Improvement Conference,* Ottawa pp. 37–38.

Palti, J. (1981) *Cultural Practices and Infectious Crop Diseases.* Springer-Verlag, Berlin.

Pandey, S. (1996) Socioeconomic context and priorities for strategic research on Asian upland rice ecosystems. In: Piggin, C. (ed.) *Upland Rice Research in Partnership: Proceedings of the Upland Rice Consortium Workshop,* 4–13 January 1996, Padang, Indonesia, International Rice Research Institute Discussion Paper No. 16, pp. 103–124.

Pham, J.L., Sebastian, L.S., Sanchez, P., Calibo, M., Quilloy, S., Bellon, M.R., Francisco, S.R., Erasga, D. and Abrigo, G. (1997) Genetic data and analysis in the Philippines on-farm. *2nd Participants Meeting for the Project: Strengthening the Scientific Basis of In Situ Conservation of Agricultural Biodiversity.* IPGRI–FAO, Rome, 25–29, August 1997.

Plucknett, D.L., Smith, N.J.H., Williams, J.T. and Anishetty, N.M. (1987) *Gene Banks and the World's Food.* Princeton University Press, Princeton, New Jersey.

Pratt, D.J. and Gwynne, M.D. (eds) (1977) *Rangeland Management and Ecology in East Africa.* Hodder & Stoughton, London.

Rice, E., Smale, M. and Blanco, J.L. (1997) Adoption and impact of improved seed selection practices in Mexican maize: evidence and issues from the Sierra Santa Marta. In: *Proceedings of an International Conference Building the Basis for Economic Analysis of Genetic Resources in Crop Plants.* CIMMYT and Stanford University, 17–19 August 1997, Palo Alto, California.

Salick, J. (1989) Ecological basis of Amuesha agriculture. *Advances in Economic Botany* 7, 189–212.

Salick, J. and Lundberg, A. (1990) Variation and change in Amuesha indigenous agricultural systems. *Advances in Economic Botany* 8, 199–223.

Salick, J. and Merrick, L. (1990) Use and maintenance of genetic resources: crops and their wild relatives. In: Carroll, C.R., Vandermeer, J.H. and Rosset, P.H. (eds) *Agroecology.* McGraw-Hill, New York, pp. 517–548.

Salick, J., Cellinese, N. and Knapp. S. (1997) Indigenous diversity of cassava: generation maintenance, use and loss among the Amuesha, Peruvian upper Amazon. *Economic Botany* 51, 6–19.

Sebastian, L.S., Garcia, J., Quilloy, S.M., Hipolito, L., Sanchez, P.L., Calibo, M.C. and Pham, J.L. (1998) Assessment of diversity and identity of farmers' rice varieties using microsatellite markers. Poster presented at the Plant and Animal Genome VI Conference, 18–22 January 1998, San Diego, California.

Second, G. (1982) Origin of the genetic diversity of cultivated rice (*Oryza* spp.): study of the polymorphism scored at 40 isozyme loci. *Japanese Journal of Genetics* 57, 25–57.

Shore, K. (1998) *Protecting Biodiversity: Toward the Fair and Equitable Sharing of Genetic Resources.* IDRC 1997–1998, Ottawa, pp. 39–40.

Smale, M., Aquino, P., Crossa, J. and del Toro, E. (1996) *Understanding Gobal Trends in the Use of Wheat Diversity and International Flows of Wheat Genetic Resources.* Economics Working Paper, CIMMYT, Mexico, DF, 96–02.

Smith, R.C. (1977) Deliverance from chaos for a song: preliminary discussion of Amuesha music. PhD dissertation, Cornell University, Ithaca, New York.

Smith, R.C. (1996) Biodiversity won't feed our children. In: Redford, K.H. and Mansour, J.A. (eds) *Traditional Peoples and Biodiversity Conservation in Large Tropical Landscapes.* The Nature Conservancy, Arlington, Virginia, pp. 197–218.

Smithson, J.B. and Lenné, J.M. (1996) Varietal mixtures: a viable strategy for sustainable productivity in subsistence agriculture. *Annals of Applied Biology* 128, 127–158.

Sperling, L. (1994) *Analysis of Bean Seed Channels in the Great Lakes Region: South Kivu, Zaire, Southern Rwanda and Selected Bean-Growing Zones of Burundi.* CIAT Occasional Publications Series, No. 13, CIAT/RESAPAC, Butare, Rwanda.

Sperling, L., Scheidegger, U. and Buruchara, R. (1994) *Enhancing Small Farm Seed Systems: Principles Derived from Bean Research in the Great Lakes Region.* CIAT Network on Bean Research in Africa Occasional Publication Series, No. 15.

Sutlive, V.H. (1978) *The Iban of Sarawak.* AHM Publishing Co., Arlington Heights, Virginia.

Teverson, D.M., Taylor, J.D. and Lenné, J.M. (1994) Functional diversity for disease resistance in *Phaseolus vulgaris* seed mixtures in Tanzania. *Aspects of Applied Biology* 39, 163–172.

Thurston, H.D. (1992) *Sustainable Practices for Plant Disease Management in Traditional Farming Systems.* Westview Press, Boulder, Colorado.

Thurston, H.D. (1997) *Slash/Mulch Systems. Sustainable Methods for Tropical Agriculture.* Westview Press, Boulder, Colorado.

Trutmann. P. and Kaytare, E. (1991) Disease control and small farmers multiplication plots to improve quality and small farmer dry bean yields in Central Africa. *Journal of Applied Seed Production* 9, 36–40.

Trutmann, P., Voss, J. and Fairhead, J. (1993) Management of common bean diseases by farmers' in the Central African Highlands. *Tropical Pest Management* 39, 334–342.

Trutmann, P., Voss, J. and Fairhead, J. (1996) Local knowledge and farmer perceptions of bean diseases in the Central African Highlands. *Agriculture and Human Values* 13, 64–70.

Voss, J. (1992) Conserving and increasing on-farm genetic diversity: farmer management of varietal bean mixtures in Central Africa. In: Moock, J.L. and Rhoades, R.E. (eds) *Diversity, Farmer Knowledge and Sustainability.* Cornell University Press, Ithaca, New York, pp. 34–51.

Weatherwax, P. (1954) *Indian Corn in Old America.* Macmillan, New York.

Wilken, G.C. (1987) *Good Farmers. Traditional Agricultural Resource Management in Mexico and Central America*. University of California Press, Berkeley, California.

Wilkes, H.G. (1972) Maize and its wild relatives. *Science* 177, 1071–1077.

Wilkes, H.G. (1977) Hybridization of maize and teosinte in Mexico and Guatemala and the improvement of maize. *Economic Botany* 31, 254–293.

Wilkes, H.G. (1997) Teosinte in Mexico: personal retrospective and assessment. In: *Proceedings of Gene Flow Among Maize Landraces, Improved Maize Varieties, and Teosinte: Implications for Transgenic Maize*. CIMMYT, Mexico, DF, pp. 10–17.

Willey, R.W. (1975) The use of shade in coffee, cocoa and tea. *Horticultural Abstracts* 45, 791–798.

Wilson, R.T. (1994) *Conserving the World's Genetic Resources*. Centre for Domestic Animal Diversity, FAO, Rome.

Wittwer, S., Youtai, Y., Han, S. and Lianzheng, W. (eds) (1987) *Feeding a Billion: Frontiers of Chinese Agriculture*. Michigan State University Press, East Lansing, Michigan.

Wood, D. and Lenné, J.M. (1993) Dynamic management of domesticated biodiversity by farming communities. In: *Proceedings of the Norway/UNEP Expert Conference on Biodiversity*, Trondheim, Norway, pp. 84–98.

Wood, D. and Lenné, J.M. (1997) The conservation of agrobiodiversity on-farm: questioning the emerging paradigm. *Biodiversity and Conservation* 6, 109–129.

Yen, D.E. (1968) Natural and human selection in the Pacific sweet potato. In: Drake, E.T. (ed.) *Evolution and Environment*. Yale University Press, New Haven, Connecticut, pp. 387–412.

Yen, D.E. (1974) *The Sweet Potato and Oceania. An Essay in Ethnobiology*. Bishop Museum Press, Honolulu, Hawaii.

Youtai, Y. (1987) Agricultural history over seven thousand years. In: Wittwer, S., Youtai, Y., Han, S. and Lianzheng, W. (eds) *Feeding a Billion: Frontiers of Chinese Agriculture*. Michigan State University Press, East Lansing, Michigan, pp. 19–33.

Does Plant Breeding Lead to a Loss of Genetic Diversity?

10

J.R. Witcombe

Centre for Arid Zone Studies, University of Wales, Bangor, Gwynedd LL57 2UW, UK

Introduction

In 1995 the International Development Research Centre (IDRC) organized a conference in Delhi on using agricultural diversity on-farm. One school of thought, of the 'grass roots' non-governmental organizations (NGOs), reflected some of the emerging assumptions for on-farm conservation of genetic diversity that were later criticized by Wood and Lenné (1997). In particular, the assumptions, implicit or explicit, were that modern varieties were responsible for an overall loss of traditional varieties, and that traditional varieties are 'locally adapted', and therefore of greater value to farmers than modern varieties (MVs). Given these assumptions, the proposed solutions by grass-root NGOs were the use of seed-banks to preserve traditional cultivars, the adding of economic value to traditional varieties by promoting their qualities, and a need to prevent or hinder the introduction of MVs into the farming system (Sateesh, 1996; Magnifico, 1996; Maheshwari, 1996). A second school of thought, presented by, for example, Weltzein *et al.* (1996), Joshi *et al.* (1996), Witcombe and Joshi (1996) and Sthapit *et al.* (1996a), was that MVs had an essential role to play in the agricultural system, and more participatory methods would help to preserve biodiversity in cropping systems that involve MVs. This approach was accepted by many, but not all, to be more realistic, as preventing farmers' access to new varieties is difficult and ethically dubious as it denies economic benefits, often to those in greatest need. Once the essential role of MVs in the farming system is accepted then it must also be accepted that they will have an impact on biodiversity.

© 1999 CAB *International. Agrobiodiversity*
(D. Wood and J.M. Lenné)

This chapter first critically reviews how biodiversity is measured in agricultural systems. It then examines how genetic diversity is affected by genetic improvement. The factors that most influence how genetic improvement affects biodiversity are the production system (marginal or high potential), the degree of farmer participation employed in plant breeding (centralized or decentralized participatory systems) and the breeding methods employed (broad or narrow genetic bases). The paper also examines *de novo* genetic variation and the potential impact of modern breeding techniques – molecular marker-assisted selection and genetic transformation – on agrobiodiversity.

Methods of Estimating Diversity

Witcombe *et al.* (1996) briefly reviewed the impact of participatory methods of varietal selection and plant breeding on biodiversity in crops. They discussed the importance for maintaining or increasing biodiversity, of having a high genetic dissimilarity between existing varieties and those that they replace, and of having a large number of cultivars that are adopted to equal extents. Biodiversity over time (temporal diversity), rather than only over space (spatial diversity), and the possible pattern of adoption of a new cultivar from a few or many foci were also considered. Souza *et al.* (1994) have considered these issues in a quantitative manner using the number of cultivars, their genetic relatedness and the proportion of area they occupy to determine indices for both spatial and temporal diversity (Box 10.1; and see Cox and Wood, Chapter 3, this volume). First average diversity is determined and then weighted for proportion of area, but there are still many other measures of diversity that are relevant from a functional viewpoint (Box 10.1).

Box 10.1. Diversities defined.

Average diversity is the diversity among cultivars growing in any specified region unweighted by the cultivated areas they occupy. It is equal to the dissimilarity between all possible pairs of varieties for coefficients of parentage (CP), including the CP of a cultivar with itself, subtracted from 1 so that high values indicate high diversity (Fig. 10.1 and Box 10.2). A weakness of this measure is that it does not necessarily give higher values when there are a greater number of genotypes present. Indeed, Frankel *et al.* (1995), in discussing the measurement of genetic variation, define *richness* as the total number of genotypes present and point out it has a smaller role in determining diversity than evenness (see below).

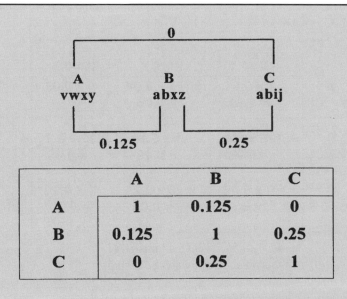

$$\text{Average diversity} = 1 - \text{mean CP}$$
$$= 1 - 0.42 = 0.58$$

Fig. 10.1. The coefficients of parentage of three cultivars, A, B and C, derived from nine grandparental genotypes, a, b, i, j, v, w, x, y and z (top). It is assumed that parents contribute equally to offspring so that the degree of relatedness of full sibs (that have both parents in common) is 0.5. B and C have half their parents in common so the degree of relatedness is 0.25, A and C have no common parents, and A and B have a quarter of their parents in common, so the degree of relatedness is 0.125. Once the CPs are calculated, a matrix of the CPs is calculated (bottom).

Weighted diversity is the average diversity where the CP between each pair of cultivars is weighted by the percentage area they occupy (Fig. 10.2). This gives a more functional measure of diversity deployed in farmers' fields, as it accounts for reduced diversity caused by one, or a few, varieties occupying a large proportion of the cultivated area of the crop. The more unequal the adoption of cultivars, the greater the tendency for weighted diversity to be lower than average diversity (Fig 10.3). Weighted diversity can be higher than average diversity, but only when smaller areas are occupied by the most related varieties and unrelated varieties occupy greater and about equal areas. In a related concept, Frankel *et al.* (1995) discuss genetic diversity in terms of *evenness* in the frequency of different types in a population or sample. In estimates of *gene diversity* they point out that the measure is largely, but not solely, dependent on the evenness of allelic frequencies.

Continued

Box 10.1. Continued

	A W =0.1	B W =0.4	C W =0.5
A W =0.1	1 x 0.01 = 0.01	0.125 x 0.04 = 0.05	0 x 0.05 = 0
B W =0.4	0.125 x .04 = 0.05	1 x 0.16 = 0.16	0.25 x 0.2 =0.05
C W =0.5	0 x 0.05 = 0	0.25 x 0.2 = 0.05	1 x 0.25 = 0.25

W = **proportion of land occupied.**
Weighted diversity = 1- 0.53 = 0.47

Fig. 10.2. The coefficients of parentage (taken from Fig. 10.1) of three cultivars, A, B and C, weighted by the areas they occupy to derive the weighted diversity.

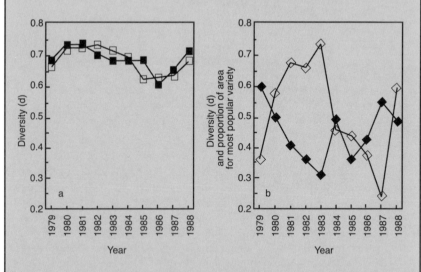

Fig. 10.3. Diversity in spring wheat, Yaqui Valley, Mexico, replotted from Souza *et al.* (1994). (a) The recommended (solid squares) and average diversity (open squares). (b) The weighted diversity (solid diamonds) and the proportion occupied by the most popular wheat cultivar (open diamonds). Note how the weighted diversity differs dramatically from recommended and average diversity, and how the weighted diversity is influenced by the area occupied by the most commonly grown variety.

Recommended diversity is the average diversity of cultivars recommended by extension services for a given region. It will differ significantly from the average diversity when farmers refuse to adopt recommended varieties, continue to grow varieties that have been withdrawn from the recommended list, or adopt varieties that have never been recommended. When participatory methods of varietal selection are employed, the recommended diversity is invariably found to be lower than the average diversity of the cultivars farmers find desirable.

Temporal diversity is a measure of change in diversity over time. It is measured as the CP between varieties at time 1 and time 2 and can also be unweighted or weighted by area. The weighted temporal diversity can be determined using the CPs and the areas occupied at two given times (Fig. 10.4). High temporal diversity occurs when there are high replacement rates of cultivars by cultivars that are unrelated to them. High temporal diversity is desirable to prevent epidemics, but often results from epidemics that force farmers to replace susceptible varieties.

Latent diversity is the underlying genomic diversity that only becomes apparent when the crop is exposed to the appropriate biotic or abiotic stress. The coefficient of parentage is an indicator of latent genetic diversity as it measures the probability that two cultivars are identical by descent for any

T1 T2	A $W=0$	B $W=0.6$	C $W=0.4$
A $W=0.2$	1 x 0 = 0	0.125 x 0.12 = 0.15	0 x 0.08 = 0
B $W=0.8$	0.125 x 0 = 0	1 x 0.48 = 0.48	0.25 x 0.32 = 0.16
C $W=0$	0 x 0 = 0	0.5 x 0 = 0	1 x 0 = 0

Weights (proportion of land occupied) are derived for time 1 (T1) and time 2 (T2)
Weighted temporal diversity = 1 - 0.58 = 0.42

Fig. 10.4. The coefficients of parentage (taken from Fig. 10.1) of three cultivars, A, B and C, weighted by the areas they occupy at T1 and T2 to determine the temporal diversity from T1 to T2.

Continued

Box 10.1. Continued

character. Low latent diversity is best illustrated by the use of male-sterile cytoplasm in hybrid breeding. The low latent diversity in the male-sterile cytoplasm of maize was revealed by the epidemic of southern corn leaf blight (*Cochliobolus heterostrophus*) of maize. All maize hybrids were based on a single male-sterile cytoplasm, Texas (T), and once a pathogen evolved to which the T cytoplasm was susceptible a disastrous epidemic ensued. Low latent diversity in cytoplasm is still found in crops such as pearl millet (almost complete reliance on A_1 cytoplasm) and in rice (where there is an equally high reliance on wild abortive cytoplasm). Hybrids in these crops can thus be considered to be vulnerable. Clearly, certain modern methods of plant breeding result in very low latent diversity, although it must be said that the use of male-sterile cytoplasm in the production of hybrid cultivars is the most extreme example.

Apparent diversity is manifest in the phenotype and is particularly obvious when genes are deployed with specific resistances to biotypes of pests and pathogens. It is exemplified by the use of multilines (Browning and Frey, 1981). Apparent diversity for disease and pest resistance can be increased with the use of molecular marker-assisted selection. Apparent genetic diversity, as well as latent genetic diversity, will often be related to average diversity.

Reserve diversity is the diversity available in the short term to replace existing varieties should they become obsolete because of, for example, disease susceptibility (see Cox and Wood, Chapter 3, this volume). When the T cytoplasm of maize became susceptible there was no reserve diversity because no alternative cytoplasms were available in advanced lines. Replacement hybrids were rapidly produced, not by utilizing reserve cytoplasmic diversity, but by resorting to detasseling of the female parent. In contrast, in pearl millet an alternative to the currently widely used A_1 cytoplasm exists and could be deployed in the near term. The major sources of reserve diversity are varieties in formal breeders' trials, and breeders' advanced lines. Duvick (1984) defines first-line reserves as those already on-farm. The second line of genetic reserves are cultivars in advanced yield trials and are always more numerous than the released cultivars. He defines third-line reserves as cultivars and hybrids in preliminary trials.

Resource diversity could be considered to be the total diversity available as genetic resources to plant breeders, *ex situ* or *in situ*, and in primary, secondary and tertiary gene pools (Harlan and de Wet, 1971). Richness is more important than evenness for resource diversity. The *genetic base* of the crop is often used in a narrower sense to refer to the total deployed genetic variability rather than the total variability available to breeders (see Simmonds, 1979).

Box 10.2. Measuring degrees of genetic relatedness using coefficients of parentage.

Coefficient of parentage is determined using ancestries that commence with 'founder' parents for which no ancestry is known. There is assumed to be no coefficient of parentage between these parents, i.e. $r = 0$. Ancestry of the cultivars is then traced to determine the CP (Fig. 10.1). This figure is a much simplified model of three cultivars derived from nine founder grandparents ab, ij and vwxy. In practice, the number of parents in the ancestry of a cultivar is usually much higher, and many of the parents will have degrees of relatedness that makes the calculations of CP more demanding. Once the CP is known between all of the cultivars (including the cultivar upon itself) it can be used to calculate average diversity or weighted diversity if the areas devoted to the cultivers are known (Fig. 10.2).

Instead of the use of the CP, diversity can be estimated using dissimilarity between molecular markers. There are many examples in the literature of diversity analysis using molecular markers and some are briefly reviewed in Table 10.1. Some of these studies arrive at surprising conclusions. For example, Chen *et al.* (1994) did not find agreement between the wheat cultivar groups (*sphaerococcum, spelta, compactum* and *aestivum*) or geographical origin and molecular marker data, and this result may have been due to inadequate sample size. None the less, considering the errors involved in estimating diversity using molecular markers and in estimating CP, a number of studies do show significant, albeit often low, correlations between them (Table 10.1). The highest correlations occur when the number of polymorphic bands studied is high (e.g. Cao and Oard, 1997) or when the genetic material examined is highly diverse and well defined (Prabhu *et al.*, 1997).

More recent developments in molecular marker technology make it more discriminating and cost-effective. For example, Senior *et al.* (1998) have used simple sequence repeats (SSR) as an alternative to RFLPs and isozymes. The genetic divergence of 94 elite maize inbreds revealed by using 70 SSR marker loci were consistent with known pedigrees, heterotic groups and market classes. As few as five SSR loci provided unique genetic fingerprints for each of the inbred lines.

The advent of less expensive molecular marker techniques has also made it possible to examine genetic diversity using mapped markers (Zhu *et al.*, 1999). Zhu *et al.* have been able, in rice, to examine diversity using a large number of AFLP markers that have been mapped. For map-based diversity analysis each chromosome was divided into smaller segments or blocks, the size and boundary of which depend on the availability of markers in any region. The blocks are termed fingerprint linkage blocks

Table 10.1. A summary of some molecular marker diversity studies.

Crop[a]	Marker used	Sample size	Comment	Reference
US rice cultivars recommended in Louisiana (20)	RAPD Pedigree data	92 polymorphic bands	Mean genetic distances by pedigree data were significantly larger. Grouping by both methods was related to maturity group and grain type. Correlation coefficient between genetic distances calculated by RAPD and pedigree data was 0.72***	Cao and Oard (1997)
Soybean (10)	DAF RFLP RFLP Pedigree data	18 12 clones 18 fragments 53 clones	Extraordinarily good agreement between methods but probably dependent on the genotypes used (pair of near isogenic lines, cultivar and backcross product, exotic introduction, five North American cultivars) and the low number of them (10)	Prabhu et al. (1997)
North American oat cultivars (83)	RFLP Coefficients of parentage	48 clones 205 bands	Clearly separated spring- and autumn-sown cultivars. Relationship between RFLP data and CP was significant but low, $r = -0.32$. Genetic distances were higher for CP than RFLP	O'Donoughue et al. (1994)
Durum wheat cultivars and landraces (113)	RFLP Morphophysiological traits CP	36 probes 165 bands 16 traits on 76 entries	Low correlation coefficient between genetic distances from CP and from RFLP, $r = 0.23$, because it had to be assumed unknown ancestors were unrelated when they were not, and that reselections from landraces had CP of 0.75 with landraces. $r = 0.47$ between RFLP data and morphophysiological traits. The overall mean genetic distances by CP and RFLP were similar.	Autrique et al. (1996)

Sweet corn inbreds (46)	RFLP Isozyme Morphological traits Coefficients of relationships (COR)	71 probes ? bands 31 isozymes 60 alleles 24 traits	Only RFLPs accurately matched known pedigree relationships. RFLP genetic distances and COR correlated well ($r =0.54**$) despite the fact that a quarter of the inbreds studied had unknown or partially known backgrounds	Gerdes and Tracy (1994)
Hard red spring wheat (45)	STS PCR markers	31 primer enzyme combinations	Marker data did not agree either with cultivar group (*sphaerococcum, spelta, compactum* or *aestivum*) or geographical origin	Chen *et al.* (1994)
European barley germplasm (48)	RFLP	62 clone enzyme combinations (127 bands?)	Division into winter and spring types clear	Melchinger *et al.* (1994)
Old US soybean cultivar groups (106)	RFLP Morphological	37–50 RFLP loci	In eight instances where there was no RFLP diversity morphological markers distinguished them in five cases	Lorenzen and Shoemaker (1996)
Cultivated races of sorghum (190)	RAPD	53 primers 162 bands	Only 13% of total genetic variation was shown to be between regions. No separation of accessions into racial or geographic groups	Menkir *et al.* (1997)
Common bean landraces from Chile (95)	RAPD Morphological data	25 primers 105 bands 7 quantitative 13 quantitative	RAPDs discriminated between meso American and Andean accessions reliably, even with a low sample size. Morphological data differences were less marked	Johns *et al.* (1997)

$**P <0.01$; $***P <0.001$.

[a]Numbers in parentheses refer to the number of cultivars, landraces, inbreds, etc.

(FLBs). The power of the diversity analysis is enormous (Fig. 10.5). For example, although 'IR64' and '63-83' are the most genetically divergent they are closely related at FLB 1A, and although 'IR64' and 'IR15324' are genetically close they are highly dissimilar at FLB 4A. 'IR64' differs from the other varieties most on FLB 6, whereas differences are the smallest for blocks such as 2A. The authors suggest that this technique offers a more precise view of genetic differentiation, a more powerful method of defining and utilizing core collections of germplasm, and a new era of genomic breeding in which the whole genome is traced and selected by markers.

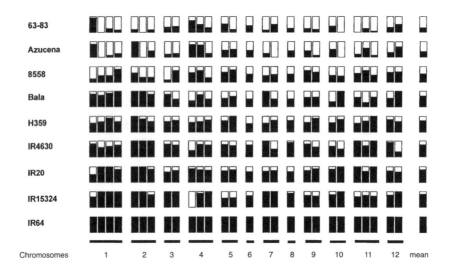

Fig. 10.5. A comparison of eight varieties with IR64, with the similarity to IR64 in each genotype linkage block indicated by the black shading. From Zhu et al. (1999).

Sources of error in estimating diversity

All methods of estimating diversity are subject to errors. The assumptions made in determining coefficients of parentage can be incorrect (Box 10.3). The results of molecular marker analysis need to be treated with caution if a small sample size of markers is employed. If 50 bands are scored on a dissimilarity index (0% no bands dissimilar, 100% no bands alike) then the standard errors of the dissimilarity coefficients are high, particularly for intermediate values. For example, with $n = 50$ then a similarity coefficient of 50% has a SE of ± 7, and produces an estimate that is expected to range between 43 and 57 % (Fig. 10.6). Given the importance of the sample size it is surprising that several of the studies summarized in Table 10.1 fail to report the number of polymorphic bands evaluated. Those that do, have an adequate sample size of between 92 and 205.

Box 10.3. Sources of error when measuring coefficients of parentage.

The records of ancestry of some cultivars may be incorrect and are often incomplete. The assumption that parents contribute to offspring in accordance with average expectations will sometimes be very inaccurate as large divergences can be expected in a low proportion of cases. The assumption is made that a cultivar derived from a cross between two parents will have half its genetic material from each parent. Sampling theory indicates that the proportion inherited from any parent will follow a binomial distribution. St Martin (1982), in support of the use of the CP, calculates that for soybean with 20 chromosomes, 88% of the time the value will lie between 0.4 and 0.6. However, this means that 12% of the time the deviation is appreciably large. Research using molecular markers can obtain real estimates of the deviation. For example, Howell *et al.* (1996) analysed the parental contribution to the gametes produced from the F₁ in a *Brassica napus* cross that has 19 chromosomes. In only about 63% of the cases did the value lie between 0.4 and 0.6. Moreover, the proportions contributed by one of the two original parents varied from < 25% to > 70% in the population of 200 plants that were analysed.

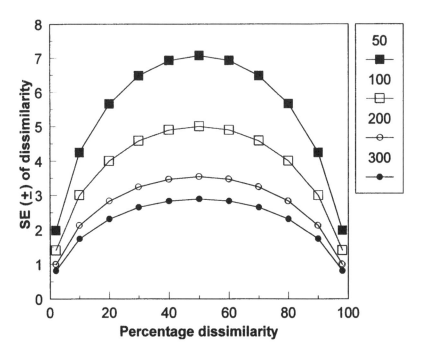

Fig. 10.6. Standard errors (SE ±) of percentage dissimilarity with varying sample sizes of *n* = 50, 100, 200 or 300.

Genetic Base of Cultivars

It is often argued that the genetic base of modern varieties is highly restricted, and that there is an increasing trend for it to narrow. The evidence from studies on the parentage of MVs does not fully support this pessimistic view.

Tanksley and McCouch (1997) quote Duvick (1977) who argued that nearly all modern North American soybean cultures could be traced back to a dozen strains from a small area in northeastern China. A still narrow, but broader genetic base than was reported by Duvick (1977) was found in a more recent study by Gizlice *et al.* (1994), who analysed the pedigrees of 258 cultivars of modern North American soybean cultivars released between 1947 and 1988. They found that the genetic base was fully defined by 80 ancestors. Their contribution was unequal and ranged from practically zero to 12%. More than half the genetic base was constituted by only six ancestors. The contribution of 17 ancestors accounted for 84% of the total genetic base of soybean and there were no discernible changes in this genetic base with time.

Tanksley and McCouch (1997) quote Harlan (1987) who reported that the majority of hard red winter wheat varieties in the USA originated from just two lines imported from Poland and Russia. A more recent study by van Beuningen and Busch (1997) gives a more optimistic picture. They analysed the pedigrees of spring wheat from Canada (47), the USA (133) and Mexico (90). In Canada, it was argued that diversity has declined, probably due to strict quality regulations limiting the diversity of released varieties. In contrast, the diversity in wheats in the USA and Mexico increased over time. They conclude that, in the latter two cases, the introduction of novel alleles from wild ancestors or other sources has contributed to the broadening of the genetic base and has reduced the genetic vulnerability of the gene pools.

Evans (1997) examined the number of landraces in the backgrounds of three cereals:

- 11 International Maize and Wheat Improvement Center (CIMMYT) spring wheat varieties;
- 37 International Rice Research Institute (IRRI) rice varieties; and
- 38 Pioneer maize hybrids adapted to central Iowa.

In all cases there has been a significant increase over time in the number of landraces that are used. The most dramatic example is for CIMMYT wheats, when the number of landraces used increased from less than ten in 1950 to over 60 in 1990. The information for IRRI rice is presented in Table 10.2 (data kindly supplied by Dr Gurdev Khush, IRRI). See also Cox and Wood, Chapter 3, this volume.

Smale *et al.* (1996) carried out a similar analysis for CIMMYT wheat cultivars. The average number of different pedigrees in wheat varieties released in developing countries increased over time from about 20 in the 1960s to about 50 in releases in 1990. Since the 1980s, national programmes

Table 10.2. Number of landraces in the parentage of selected IRRI rice cultivars.

Year	Varieties	Number of landraces in parentage	Mean
1966	IR8	4	4
1967	IR5	4	4
1969	IR20, IR22	11, 5	8
1971	IR24	22	22
1973	IR36	35	35
1974	IR28, IR30	31, 28	30
1975	IR32, IR34	18, 31	25
1976	IR36	35	35
1977	IR42	35	35
1978	IR43, IR44, IR45, IR46	24, 31, 29, 28	28
1979	IR48, IR50	31, 37	34
1980	IR52, IR54	38, 38	38
1982	IR56	39	39
1983	IR58, IR60	38, 39	39
1984	IR62	37	37
1985	IR65	39	39
1987	IR66	42	42
1988	IR68, IR70, IR72, IR74	42, 40, 46, 43	43
1991	PSBRC 2, PSBRC 4	40, 45	43
1992	PSBRC 10	47	47
1994	PSBRC 18, PBSRC 20	44, 47	46

have released varieties produced from crosses in which an average of one new landrace appears each year.

Although these data appear convincing and are presented by Evans (1997) as an indication of the broadening of the genetic base of the crop, this may not be the case. Early varieties in a breeding programme inevitably have only a few landraces as parents. These have no known pedigree although they must be the result of hybridization between many landraces. Hence, at an early stage the number of landrace parents is underestimated. As a breeding programme advances, crosses are made between elite cultivars with different landrace parentages so the pedigrees inevitably increase in complexity. It would be surprising, with this increasing complexity, if there was no increase in the number of landraces because it could only happen if breeders repeatedly used the same small set of landraces. What is important is the rate of increase, and an average of one new landrace a year indicates that plant breeders are expanding the range of landraces used at a respectable rate. What may be even more important from the view-point of biodiversity is the total number of landraces in the pedigrees of varieties released at about the same time. How many landraces do contemporaneous varieties have in common? However, analysed data are currently unavailable.

IRRI initiated a study in 1975 to measure the impact of the early semi-dwarf rices on genetic diversity in Asian rice breeding programmes (Hargrove, 1979). It traced the diffusion of parents from 1965–67 until 1974–75 at 14 agricultural research centres in seven Asian nations (Table 10.3). The most popular single gene source in the 1965–67 sample of crosses was the Taiwanese cultivar 'Taichung native 1' (TN1), probably the first semi-dwarf rice developed through hybridization (Athwal, 1971), and it had the DGWG gene for dwarfing.

Although the original stiff-strawed cultivars such as 'TN1' and 'IR8' were virtually phased out of breeding programmes as direct parents by the mid-1970s, almost all the semi-dwarf parents that replaced them were their progeny. The diffusion of semi-dwarf genetic materials into breeding programmes has reduced the genetic diversity of current and future cultivars. More than 80% of the 1974–75 crosses carry the DGWG gene for dwarfism, and most new semi-dwarfs were *indica* rices. These largely replaced *japonica*, *ponlai* and other races in the breeding programmes.

Participatory Approaches to Plant Breeding and Varietal Selection

Recent work on increased farmer participation in the plant breeding process has discussed the impact on biodiversity (e.g. Witcombe *et al.*, 1996). Two methods of farmer participation in the plant breeding process can be employed. The first is participatory varietal selection (PVS), where farmers are given

Table 10.3. Percentage of rices of different plant height used as parents in crosses. From 819 parents used in 355 randomly selected crosses at 14 agricultural experiment stations and universities in seven Asian nations, 1965–1975.

Plant height	1965–67[a]	1970–71[b]	1974–75[c]
Rices used in crosses			
Tall	74	57	45
Intermediate	51	39	35
Semi-dwarf	61	86	84
Floating or deepwater	2	1	2
Rices used as individual parents			
Tall	40	30	24
Intermediate	31	22	17
Semi-dwarf	28	48	58
Floating or deepwater	1	–	1

[a]277 rice cultivars and lines used in 119 crosses.
[b]351 rices used in 147 crosses.
[c]191 rices used in 89 crosses.

varieties (the finished products from plant breeding) to try in their own fields. After a successful PVS programme, the varieties found to be liked by farmers can be used as parents in a breeding programme in which farmers participate as consultants or collaborators. This involves breeding and selection to create new varieties and is termed participatory plant breeding (PPB) by Witcombe *et al.* (1996). Others use the term PPB more broadly to include PVS.

A PVS programme has the following stages:

- participatory surveys to discover what varieties farmers are growing;
- a search and procurement process for suitable new varieties;
- experimentation by farmers, in their own fields, with the new varieties; and
- wider dissemination of the identified improved varieties.

The assumption is made that varieties already exist that are better than those currently grown, but farmers have simply never had the opportunity to try them. This assumption is almost invariably correct. Work the author has been involved in with colleagues in India and Nepal in many crops and two production systems has usually successfully identified existing varieties that farmers greatly prefer but had never before had the opportunity to try (Table 10.4).

Using participatory techniques, varieties can rapidly spread from farmer to farmer. A study on 'Kalinga III' rice variety has shown an extremely rapid spread with more than 120 villages receiving seed by 1997 from three case study villages that first received seed in 1994 (Witcombe *et al.*, 1999). Clearly, in areas of high biodiversity the rapid spread of an introduced variety such as 'Kalinga III' can have a major impact on average diversity and an overwhelming impact on weighted diversity. In high-potential production systems, PVS in rice, wheat and green gram is resulting in an increase in the number of varieties farmers

Table 10.4. Varieties preferred by farmers in PVS programmes in two production systems and two countries.

Country	Production system	Crop	Preferred varieties identified	References
India	Rainfed	Rice	1	Joshi and Witcombe
		Maize	1	(1996, 1998)
		Chickpea	3	
		Black gram	2	
		Pigeonpea	1	
	Irrigated	Wheat	>5	Virk *et al.* (1998)
		Rice	>4	
		Green gram	3	
Nepal	Irrigated	Rice	>8	Joshi *et al.* (1998)
		Wheat	2	

wish to adopt, thus increasing average diversity. It is too soon to say what the longer-term impact will be on weighted diversity, but it is almost certain to be favourable. These contrasting scenarios, from rice in marginal areas to rice and wheat in high potential production systems, indicate the manifold ways in which the varieties farmers are growing can be replaced by others to change levels of agrobiodiversity (see Thurston *et al.*, Chapter 9, this volume). The greatest increase in biodiversity will occur when:

- *existing biodiversity* in farmers' varieties is *low;*
- these varieties are *partially* replaced by a new variety or varieties;
- there are *many new varieties;*
- when the new varieties have a *high genetic dissimilarity* with the existing varieties; and
- when all varieties occupy *similar areas* (weighted diversity tends to be less when one variety occupies a large area).

This is illustrated by a number of examples (Table 10.5).

In general, participatory approaches in varietal selection in marginal areas often lead to an increase in biodiversity, but sometimes result in a reduction. In high-potential production systems, where MVs are already grown, then PVS has increased biodiversity. It would be expected that this would be by far the most common situation because:

- under the almost universal transfer of technology extension systems used in developing countries there is extremely low recommended diversity and farmers grow only a few varieties; and

Table 10.5. Examples of impact on biodiversity resulting from PVS.

Example	Replacement	Impact on biodiversity	Genetic relatedness of replacements to existing	References
Chickpea, India	Partial	Increase	Low	Joshi and Witcombe (1996)
Rice, India	Partial	Increase[a]	Low	Joshi and Witcombe (1996)
Beans, Rwanda	Partial	Increase	Not reported	Sperling *et al.* (1993)
Rice, Gambia	Complete	Reduction	Not reported	Cromwell and Wiggins (1993)
Rice, India	Partial	Increase	Moderate (all dwarf MVs)	Virk *et al.* (1998)
Rice, Nepal	Partial	Increase	Low	Joshi *et al.* (1998)

[a]Case of 'Kalinga III' rice. Increase in average diversity. Weighted diversity may have declined as it occupies an increasingly high proportion of the area devoted to upland rice in unbunded fields.

- it is simpler, when compared to marginal areas, to find a good range of suitable cultivars for high-potential production systems since the greatest number of varietal releases are for such systems.

For example, Witcombe *et al.* (1995) examined the number of rice cultivars released in India for different production systems. Cumulative releases from the 1940s to 1993 were examined. There were about 250 releases for irrigated and transplanted conditions, and about 100 releases for upland conditions. Even for developing countries, with small breeding programmes, the choice is large. Experience shows that introducing varieties from one country to another is effective as long as the original agroecological conditions in which the variety is popular match those of the area into which it is introduced.

This suggested approach of plant introduction can differ somewhat from the strategies of the International Agricultural Research Centres (IARCs) that have had extensive programmes of distributing germplasm between countries. This germplasm has not always been appropriately targeted and often the same genotypes are sent to many countries with the risk of reducing interregional diversity. The products from local breeding efforts can significantly outperform entries in IARC international nurseries (Sthapit *et al.,* 1996b). More recently, there has been a trend towards more decentralized breeding methods that should help promote interregional biodiversity. For example, Ceccarelli *et al.* (1994) describe how IARC breeding programmes can make crosses targeted at specific countries by using locally adapted landraces and varieties as parents.

Biotechnology

Genetic engineering allows the transfer of cloned genes 'transgenes' between and within species to produce transgenic crops. The area under transgenic crops has increased from nothing in the recent past to 31.5 Mha in 1997 in six different countries, namely (in descending order), the USA, China, Argentina, Canada, Australia and Mexico (James, 1997).

With the introduction of transgenic crops, the average diversity has increased because there is a completely new allele in the transgenic variety and the genetic base of the crop has been extended. However, the weighted diversity in the USA has declined in cotton and maize because genetically transformed *Bt* cotton and herbicide-resistant maize varieties have occupied a large proportion of the cultivated area. Such reductions in weighted diversity are to be expected when transgenic crops are first introduced that have advantages that are appreciated greatly by farmers over non-transgenic crops. Hence, in the shorter term weighted biodiversity will decrease, although average biodiversity in farmers' fields may increase as some farmers choose to grow non-transgenic crops. However, this scenario is based on the first transgenic crops that have been introduced by private sector companies using transgenes protected by intellectual property rights (IPR). In the longer term, as IPR protection expires, transgenes will enter into the public domain and can then be used to transform any variety. If the value of such transgenes is high, and if cost-effective, genotype-independent transformation systems are available, as, for example, in rice and potato, there is no reason to presuppose a marked reduction in the biodiversity. Under these circumstances farmers can be offered many background genotypes into which the transgenes have been introduced.

In the even longer term, transgenes will be available in parental material used in the breeders' crossing block. The transgene, as long as it remains useful, will tend to become universal in the crop and its impact, both negative and positive, on the biodiversity of the cultivated crop will be neutral (but positive in terms of the genetic resources of the crop). When transgenes eventually become common, varieties may differ in the number and type of transgenes they have, and in the different alleles (constructs) of the same transgene. The overall impact will then be an increase in biodiversity, that is, in the richness of the resource diversity.

Overall, the impact of transgenic technology is difficult to predict because it is affected by many factors, and is dependent on time (summarized in Box 10.4).

Box 10.4. Changes over time in the use of transgenics: possible scenarios.

1. Private sector domination of transgenics protected by IPR. Deployment of one or two transgenes in few varieties, only in crops of private sector interest. A few transgenic varieties become popular because of their advantages resulting in large reductions in weighted diversity. Number of transgenes and transgenic varieties deployed increases over time.
2. Transgenes in public sector introduced (a possible first stage in crops such as rice). The number of transgenes and varieties deployed increases over time.
3. Varieties are deployed that have both private and public sector transgenes. Transgenic technologies become cheaper and more genotype independent. Between-cultivar diversity for transgenes increases.
4. Some transgenes become universal in breeding material.
5. IPR on private sector transgenes expires. Choice of transgenes increases. Transgenes useful for particular environments and situations emerge. Conventional breeding, using marker-assisted selection (MAS), increasingly deploys transgenes. Overall biodiversity increased because the genetic base has expanded.

Impact of Molecular Marker Technologies on Biodiversity

The utilization of wild relatives

Tanksley and McCouch (1997) argue that recent applications of genome mapping suggest that the genetic diversity stored in germplasm banks can be utilized with a much higher level of efficiency than previously imagined.

They argue that the usual method of utilizing genetic resources has been to screen genebank accessions for phenotypically expressed characters, usually those with a high heritability that can be easily evaluated. The utilization of genetic resources for quantitative characteristics of low heritability is more difficult and breeders have usually concentrated on elite by elite crosses to improve such traits. However, it is unlikely that all desirable alleles have been accumulated in elite germplasm, but it is difficult to identify germplasm having these desirable alleles. Prior to the advent of molecular marker technology, exotic improved parents were chosen for breeding programmes on their provenance and morphology. Recent developments in molecular marker technology to map quantitative trait loci (QTL) have made this process more predictable. It is now known that many complex traits are controlled by very few loci and that parents that are inferior for any trait such as disease resistance or yield have QTL that contribute positively for the trait. For example:

- A downy mildew susceptible parent of pearl millet had some QTLs that contributed to resistance (Jones *et al.*, 1995).
- *Lycopersicum pimpinellifolium,* a small-fruited wild tomato species, contributed QTL for large fruit size in cultivated tomato (Tanksley *et al.* 1996).
- Tomato lines have been created that have specific QTLs from the wild species *L. hirsutum* that outperform the original elite variety by 48% for yield, 22% for soluble solids content and 33% for fruit colour (Tanksley and McCouch, 1997).
- An advanced backcross method was used to examine alleles from the wild rice species *O. rupifogon* in the genetic background of an elite Chinese hybrid. Two QTLs were identified that each increased crop yield by 17% compared with the original hybrid (personal communication cited in Tanksley and McCouch, 1997).

In the case of tomato, the breeding procedure was first to QTL map the target trait in the second backcross generation to the elite parent (the BC2 generation) and then to carry out MAS to incorporate the favourable QTL from the wild species into the elite germplasm. Without the use of QTL mapping and MAS such lines could only be attempted by conventional breeding with a very low probability of success.

Tanksley and McCouch (1997) ask how do we decide which accessions to sample in order to maximize the chance of finding new and useful genes? They suggest that the sampling of exotic germplasm should emphasize the genetic composition rather than appearance of exotic accessions by using DNA profiles.

The conclusion of Tanksley and McCouch (1997) is correct in that QTL mapping and MAS are an opportunity to increase the efficiency of germ-plasm utilization. However, they overestimate the difficulties of utilizing wild germplasm for quantitative traits of low heritability and underestimate the extent to which it has been used. For example, over 20 years ago Frey (1976) reported on the successful utilization of wild relatives to improve quantitative characters in cultivated oats. He described the contribution of genes from the wild species *Avena sterilis* into *A. sativa*, using backcross breeding. In all of the five backcross generations in eight crosses between wild and cultivated oats there were transgressive segregants for higher yield. The yield in the cultivated oats increased by 25–30%, an increase twice as large as the 14% accomplished with all oat breeding resources in the midwestern USA between 1905 and 1960.

Multiline varieties

The use of MAS makes possible techniques such as gene pyramiding, the incorporation of multiple loci governing the same resistance trait. It is difficult to predict the impact of this upon biodiversity. However, rather than pyramiding all of the resistance loci into a single genotype, multiline varieties can be produced. Multiline varieties consist of a number of lines that differ only for the resistance genes they have at different loci against a target pest or disease. The farmer cultivates what appears to be a uniform cultivar, but it is heterogeneous for the resistance loci. This intracultivar diversity for resistance imparts improved stability of resistance. To produce multilines, resistance genes at different loci, or different alleles at the same locus, are introduced singly, or in multiples, into the same genetic background by backcross breeding. Multiline varieties have been bred and deployed (Browning and Frey, 1981). However, the cost of doing so with conventional breeding has prevented their widespread use. MAS can transform the economics of multiline breeding and can make complex strategies possible that would not be considered using conventional breeding (Witcombe and Hash, 1999).

Genetic Diversity from *de Novo* Variation and Elevated Epistasis

Rasmusson and Phillips (1997), in a thought-provoking paper, argue that narrow gene pools do not prevent the occurrence of useful genetic variation that allows genetic response to selection pressure. They suggest that elite gene pools have inherent mechanisms to provide a continuing source of new genetic variability. They point out that modest levels of genetic diversity, based on pedigree information, have permitted sizeable genetic gains. The barley cultivar 'Stander' is quoted as an example because its resistance to lodging is superior to any of its parental genotypes including 'Bumper', the most lodging resistant.

They discuss two neglected sources of variation. One source is *de novo*-generated variation: genetic change, other than that due to hybridization and recombination, that produces alleles with modified effects. A second and complementary source of variation could come from interaction or epistatic effects that are larger than quantitative theory suggests (Moreno, 1994), involving *de novo* diversity as well as original genetic diversity. Individual alleles can have a dramatic or elevated effect on phenotype via gene inter-action when placed in a new genetic environment.

Selection experiments performed over many generations on closed populations lend weight to the hypothesis of continuing sources of genetic variation. They have been performed on the flour beetle, *Tribolium*, where pupa weight responded to selection for more than 130 generations (Enfield, 1980; Enfield and Braskerud, 1989), the fruit fly *Drosophila melanogaster* (Mather and Harrison, 1949), and maize (Dudley and Lambert, 1992). The University of Illinois long-term selection studies for modified oil and protein in maize (Dudley and Lambert, 1992) are some of the most well known. These lines have been selected for more than 90 generations with variability still present – enough to achieve progress from selection. The Illinois High Protein, Reverse High Protein, Reverse Low Protein, and Low Protein lines have vastly different numbers of ribosomal RNA genes (Phillips, 1978). Gene numbers in the Reverse High Protein (RHP) lines differed more than twofold during the years of selection.

Rasmusson and Phillips (1997) argue that original variation, including that released by recombination augmented by *de novo* variation, can lead to epistatic relationships that are substantially more important than previously recognized. They give as an example Doebley *et al.* (1995), who analysed two QTL controlling differences in plant and inflorescence architecture between maize and teosinte (*Zea mays* subsp. *mexicana*). When brought together in one genotype, the two QTL substantially transformed both plant and inflor-escence architecture. Also, in studying pairs of QTL in soybean, Lark *et al.* (1995) concluded that interactions between QTL are frequent and control large effects. In their investigation, an allele at one locus that explained little or no height variation by itself had a large effect on height via interaction with other loci. Geiger (1988) concluded that long-term selection accumulates coadapted gene arrangements, which can be disrupted in sub-sequent generations.

More than 30 years ago, Sprague *et al.* (1960) and Russell *et al.* (1963) noted that doubled haploid lines of maize soon accumulated considerable variation in agronomic traits. With the advent of molecular biology, geneticists have documented an ever-increasing number of ways to alter the expression of genes. These mutation and mutation-like events include single allele changes, intragenic recombination, unequal crossing over, element transpositions, DNA methylation, paramutation, gene amplification and single-base substitutions, insertions, deletions and duplications (Rasmusson and Phillips, 1997).

Intragenic recombination

Recombination frequency within the limits of a gene can be considerably higher than in the rest of the genome. In maize, for example, Dooner (1986) examined recombination at the *bronze* (*bz1*) locus and Brown and Sundaresen (1991) at the colourless aleurone (*a1*) locus. Recombination rates, measured in centiMorgans (cM), can be compared to the physical length of DNA. At these two loci 1 cM had a physical DNA length 30–100 times smaller than was found on average over the whole of the genome, that is, recombination was 30–100 times more likely to occur than expected from the physical length of the DNA.

Transposable elements

It is difficult to dispute that transposable elements could be a constant source of variation. The number of elements in a genome also can be impressively high (Walbot, 1992).

DNA methylation

The pattern of methylation can be modified in progeny of self-pollinations and crosses (Chandler and Walbot, 1986).

Paramutation

Another means of generating new variation is through the interaction of alleles in a heterozygote. Certain alleles (paramutagenic alleles) have the capacity to alter the expression of other alleles (paramutable alleles) at a locus in the heterozygote. The altered state of the paramutable alleles is heritable and represents a heritable change in transcription (Chandler, 1995).

Gene amplification

Selection for methotrexate resistance in insects (Shotkoski and Fallon, 1993) resulted in tetraploid cells plus an extra chromosome, of which 50% was dihydrofolate reductase genes.

Conclusions

Whether genetic improvement leads to a loss of agrobiodiversity can only be judged using objective measures of agrobiodiversity. This chapter reviews methods used by Souza *et al.* (1994) to measure spatial and temporal diversity. It is simple to compute them, but the need for complex data on: (i) the ancestry of cultivars and (ii) their spread is onerous. Although the need for data on ancestry can be replaced by a diversity analysis using molecular markers, greater statistical rigour than is often found in published papers is required. When an adequate sample size of molecular markers is used this is still resource demanding, although developments in marker technology, such as microsatellites and AFLPs, now offer more cost-effective solutions. Apart from spatial and temporal diversity, there are many other measures of biodiversity relevant to agriculture. These are defined and discussed.

In areas that already grow modern cultivars, continued genetic improvement does not necessarily lead to loss of genetic diversity. Indeed, in many cases it leads to an increase in diversity, particularly when participatory methods and more innovative plant breeding strategies are employed. In areas that are not already growing modern cultivars, genetic improvement will often reduce biodiversity. The use of participatory plant breeding will limit the rate of loss, and put a ceiling on its reduction. Modern developments in plant breeding, such as genetic transformation and marker-assisted selection, will affect biodiversity. Molecular marker techniques will help in broadening the genetic base of the crop and facilitate the deployment of multiline varieties. The impact of genetic transformation is difficult to predict since it depends on socio-economic as well as technical factors. There is convincing evidence that, even with a narrow genetic base, continuing genetic progress can be made. This may be because of the creation of *de novo* genetic variation and the effects of epistasis.

References

Athwal, D.S. (1971) Semidwarf rice and wheat in global food needs. *Quarterly Review of Biology* 46, 1–34.

Autrique, E., Nachit, M.M., Monneveux, P., Tanksley, S.D. and Sorrells, M.E. (1996) Genetic diversity in durum wheat based RFLPs, morphophysiological traits and coefficient of parentage. *Crop Science* 36, 735–742.

van Beuningen, L.T. and Busch, R.H. (1997) Genetic diversity among North American spring wheat cultivars: I. Analysis of the coefficient of parentage matrix. *Crop Science* 37, 570–579.

Brown, J. and Sundaresen, V. (1991) A recombination hotspot in the maize *Al* intragenic region. *Theoretical and Applied Genetics* 81, 185–188.

Browning, J.A. and Frey, K.J. (1981) The multiline concept in theory and practice. In: Jenkyn, J.F. and Plumb, R.T. (eds) *Strategies for the Control of Cereal Disease*. Blackwell Scientific Publications, Oxford, pp. 37–46.

Cao, D. and Oard, J.H. (1997) Pedigree and RAPD-based DNA analysis of commercial U.S. rice cultivars. *Crop Science* 37, 1630–1635.

Ceccarelli, S., Erskine, W., Hamblin, J. and Grando, S. (1994) Genotype environment interaction and international breeding programmes. *Experimental Agriculture* 30, 177–187.

Chandler, V.L. (1995) A review of paramutation at *b*: an allelic interaction that causes heritable changes in transcription. In: Oono, K. and Takaiwa, F. (eds) *Modification of Gene Expression and Non-Mendelian Inheritance*. National Institute of Agrobiological Resources, Tsukuba, Japan, pp. 109–118.

Chandler, V.L. and Walbot, V. (1986) DNA modification of a maize transposable element correlates with loss of activity. *Proceedings of the National Academy of Sciences USA* 83, 1767–1771.

Chen, H.B., Martin, J.M., Lavin, M. and Talbert, L.E. (1994) Genetic diversity in hard red spring wheat based on sequence-tagged-site PCR markers. *Crop Science* 34, 1628–1632.

Cromwell, E. and Wiggins, S. (1993) Performance assessment. In: Cromwell, E., Wiggins, S. and Wentzel, S. (eds) *Sowing Beyond the State. NGOs and Seed Supply in Developing Countries*. Overseas Development Institute, London, pp. 81–89.

Doebley, J., Stec, A. and Gustus, C. (1995) *Teosinte branched1* and the origin of maize: evidence for epistasis and the evolution of dominance. *Genetics* 141, 333–346.

Dooner, H.K. (1986) Genetic fine structure of the *bronze* locus in maize. *Genetics* 113, 1021–1036.

Dudley, J.W. and Lambert, R.J. (1992) Ninety generations of selection for oil and protein in maize. *Maydica* 37, 81–87.

Duvick, D.N. (1977) Major United States crops in 1976. *Annals of the New York Academy of Sciences* 287, 86–96.

Duvick, D.N. (1984) Genetic diversity in major farm crops on the farm and in reserve. *Economic Botany* 38, 161–178.

Enfield, F.D. (1980) Long-term effects of selection: the limits to response. In: Robertson, A. (ed.) *Proceedings of the Symposia of Selection Experiments in Laboratory and Domestic Animals*. Harrogate, UK, 21–22 July 1979. Commonwealth Agricultural Bureaux, Farnham Royal, pp. 69–86.

Enfield, F.D. and Braskerud, O. (1989) Mutational variance for pupae weight in *Tribolium castaneum*. *Theoretical and Applied Genetics* 77, 416–420.

Evans, L.T. (1997) Adapting and improving crops. The endless task. *Philosophical Transactions of the Royal Society of London* 352, 901–906.

Frankel, O.H., Brown, A.D.H. and Burdon, J.J. (1995) *The Conservation of Plant Diversity*. Cambridge University Press, Cambridge.

Frey, K.J. (1976) Plant breeding in the seventies: useful genes from wild plant species. *Egyptian Journal of Genetics and Cytology* 5, 460–482.

Geiger, H.H. (1988) Epistasis and heterosis. In: Wier, B.S., Eisen, E.J., Goodman, M.M. and Namkoong, G. (eds) *Proceedings of the 2nd International Conference on Quantitative Genetics*. Sinauer, Sunderland, Massachusetts, pp. 395–399.

Gerdes, J.T. and Tracy, W.F. (1994) Diversity of historically important sweet corn inbreds as estimated by RFLPs, morphology, isozymes, and pedigree. *Crop Science* 34, 26–33.

Gizlice, Z., Carter, T.E. and Burton, J.W. (1994) Genetic base for North American soybean cultivars released between 1947 and 1988. *Crop Science* 34, 1143–1151.

Hargrove, T.R. (1979) Diffusion and adoption of semidwarf rice cultivars as parents in Asian rice breeding programmes. *Crop Science* 19, 571–573.

Harlan, J.R. (1987) Gene centers and gene utilization in American agriculture. In: Yeatman, C.W., Kafton, D. and Wilkes, G. (eds) *Plant Genetic Resources: a Conservation Imperative.* AAAS selected Symposium 87. Boulder, Colorado, pp. 111–129.

Harlan, J.R. and de Wet, J.M.J. (1971) Towards a rational classification of cultivated plants. *Taxon* 20, 509–517.

Howell, P.M., Marshall, D.F. and Lydiate, D.J. (1996) Towards developing intra-varietal substitution lines in *Brassica napus* using marker-assisted selection. *Genome* 39, 348–358.

James, C. (1997) *Global Status of Transgenic Crops* 1997. ISAAA (International Service for the Acquisition of Agri-Biotech Applications). Brief No. 5.

Johns, M.A., Skroch, P.W., Nienhuis, J., Hinrichsen, P., Bascur, G. and Munoz-Schick, C. (1997) Gene pool classification of common bean landraces from Chile based on RAPD and morphological data. *Crop Science* 37, 605–613.

Jones, E.S., Liu, C.J., Gale, M.D., Hash, C.T. and Witcombe, J.R. (1995) Mapping quantitative trait loci for downy mildew resistance in pearl millet. *Theoretical and Applied Genetics* 91, 448–456.

Joshi, A. and Witcombe, J.R. (1996) Farmer participatory crop improvement. II. Farmer participatory varietal selection in India. *Experimental Agriculture* 32, 461–477.

Joshi, A. and Witcombe, J.R. (1998) Farmer participatory approaches for varietal improvement. In: Witcombe, J.R., Virk, D.S. and Farrington, J. (eds) *Seeds of Choice. Making the Most of New Varieties for Small Farmers.* Published for CAZS and ODI by Oxford IBH, New Delhi, and Intermediate Technology Publications, London, pp. 172–190.

Joshi, K.D., Rana, R.B., Subedi, M., Kadayat, K.B. and Sthapit, B.R. (1996) Addressing diversity through farmer participatory variety testing and dissemination approach: a case study of *chaite* rice in the western hills of Nepal. In: Sperling, L. and Loevinsohn, M.L. (eds) *Using Diversity. Proceedings of Conference on Using Diversity and Maintaining Genetic Resources on Farm.* New Delhi, June 1995. International Development Research Centre, New Delhi, pp. 158–176.

Joshi, K.D., Rana, R.B., Gadal, B. and Witcombe, J.R. (1998) The success of participatory varietal selection for *Chaite* rice in high potential production systems in the Nepal *Terai.* Paper presented at the *International Conference on Food Security and Crop Science,* 3-6 November 1998. Haryana Agricultural University, Hisar.

Lark, K.G., Chase, K., Alder, F., Mansur, L.M. and Orf, J.H. (1995) Interactions between quantitative trait loci in soybean in which trait variation at one locus is conditional upon a specific allele at another. *Proceedings of the National Academy of Sciences USA* 92, 4656–4660.

Lorenzen, L.L. and Shoemaker, R.C. (1996) Genetic relationships within old U.S. soybean cultivar groups. *Crop Science* 36, 743–752.

Magnifico, F.A. (1996) Community-based resource management: CONSERVE (Philippines) experience. In: Sperling, L. and Loevinsohn, M.L. (eds) *Using Diversity. Proceedings of Conference on Using Diversity and Maintaining*

Genetic Resources on Farm. New Delhi, June 1995. International Development Research Centre, New Delhi, pp. 289–302.

Maheshwari, J.K. (1996) Maintenance and conservation of 'heirloom' varieties in Indian agro-ecosystems. In: Sperling, L. and Loevinsohn, M.L. (eds) *Using Diversity. Proceedings of Conference on Using Diversity and Maintaining Genetic Resources on Farm*. New Delhi, June 1995. International Development Research Centre, New Delhi, pp. 309–318.

Mather, K. and Harrison, B.J. (1949) The manifold effect of selection. *Heredity* 3, 1–52.

Melchinger, A.E., Graner, A., Singh, M. and Messmer, M.M. (1994) Relationships among European barley germplasm: I. Genetic diversity among winter and spring cultivars revealed by RFLPs. *Crop Science* 34, 1191–1199.

Menkir, A., Goldsbrough, P. and Ejeta, G. (1997) RAPD based assessment of genetic diversity in cultivated races of sorghum. *Crop Science* 37, 564–569.

Moreno, G. (1994) Genetic architecture, genetic behaviour, and character evolution. *Annual Review of Ecological Systems* 25, 31–44.

O'Donoughue, L.S., Souza, E., Tanksley, S.D. and Sorrells, M.E. (1994) Relationships among north America oat cultivars based on restriction fragment length polymorphisms. *Crop Science* 34, 1251–1258.

Phillips, R.L. (1978) Molecular cytogenetics of the nucleolus organizer region. In: Walden, D.B. (ed.) *Maize Breeding and Genetics*. John Wiley & Sons, New York, pp. 711–741.

Prabhu, R.R., Webb, D., Jessen, H., Luk, S., Smith, S. and Gresshoff, P.M. (1997) Genetic relatedness among soybean genotypes using DNA amplification fingerprinting (DAF), RFLP and pedigree. *Crop Science* 37, 1590–1595.

Rasmusson, D.C. and Phillips, R.L. (1997) Plant breeding progress and genetic diversity from *de novo* variation and elevated epistasis. *Crop Science* 37, 303–310.

Russell, W.A., Sprague, G.F. and Penny, L.H. (1963) Mutations affecting quantitative characters in long-term inbred lines of maize. *Crop Science* 3, 175–178.

Sateesh, P.V. (1996) Genes, gender and biodiversity: Deccan Development Society's community genebanks. In: Sperling, L. and Loevinsohn, M.L. (eds) *Using Diversity. Proceedings of Conference on Using Diversity and Maintaining Genetic Resources on Farm*. New Delhi, June 1995. International Development Research Centre, New Delhi, pp. 268–275.

Senior, M.L., Murphy, J.P., Goodman, M.M., and Stuber, C.W. (1998) Utility of SSRs for determining genetic similarities and relationships in maize using an agarose gel system. *Crop Science* 38, 1088–1098.

Shotkoski, F.A. and Fallon, A.M. (1993) An amplified mosquito dihydrofolate reductase gene: amplicon size and chromosomal distribution. *Insect Molecular Biology* 2, 155–161.

Simmonds, N.W. (1979) Epidemics, populations and the genetic base. In: Simmonds, N.W. (ed.) *Principles of Crop Improvement*. Longman, Harlow, pp. 262–269.

Smale, M., Aquino, P., Crossa, J., del Toro, E., Dubin, J., Fischer, T., Fox, P., Khairallah, M., Mujeeb-Kazi, A., Nightingale, K.J., Ortiz-Monasteria, I., Rajaram, S., Sing, R., Skovmand, B., van Ginkel, M., Varughese, G. and Ward, R. (1996) *Understanding Global Trends in the Use of Wheat Diversity and International Flows of Wheat Genetic Resources*. International Maize and Wheat Improvement Centre (CIMMYT), Mexico, DF.

Souza, E., Fox, P.N., Byerlee, D. and Skovmand, B. (1994) Spring wheat diversity in irrigated areas of two developing countries. *Crop Science* 34, 774–783.

Sperling, L., Loevinsohn, M.E. and Ntabomvra, B. (1993) Rethinking the farmer's role in plant breeding: local bean experts and on-station selection in Rwanda. *Experimental Agriculture* 29, 509–519.

Sprague, G.F., Russell, W.A. and Penny, L.H. (1960) Mutations affecting quantitative traits in selfed progeny of doubled monoploid maize stocks. *Genetics* 45, 855–865.

Sthapit, B.R., Joshi, K.D. and Witcombe, J.R. (1996a) Farmers' participatory high altitude rice breeding in Nepal: providing choice and utilizing farmers' expertise. In: Sperling, L. and Loevinsohn, M.L. (eds) *Using Diversity. Proceedings of Conference on Using Diversity and Maintaining Genetic Resources on Farm.* New Delhi, June 1995. International Development Research Centre, New Delhi, pp. 186–205.

Sthapit, B.R., Joshi, K.D. and Witcombe, J.R. (1996b) Farmer participatory crop improvement. III. Farmer participatory plant breeding in Nepal. *Experimental Agriculture* 32, 479–496.

St Martin, S.K. (1982) Effective population size for the soybean improvement program in maturity groups 00 to IV. *Crop Science* 22, 151–152.

Tanksley, S.D. and McCouch, S.R. (1997) Seed banks and molecular maps: unlocking genetic potential from the wild. *Science* 277, 1063–1066.

Tanksley, S.D., Grandillo, S., Fulton, T.M., Zamir, D., Eshed, Y., Petiard, V., Lopez, J. and Beckbunn, T. (1996) Advanced backcross QTL analysis in a cross between an elite processing line of tomato and its wild relative *L. pimpinellifolium. Theoretical and Applied Genetics* 92, 213–224.

Virk, D.S., Raghuvanshi, B.S., Sodhi, P.S. and Witcombe, J.R. (1998) Participatory crop improvement in high potential production systems. Paper presented at the International Conference on Food Security and Crop Science. 3–6 November 1998. Haryana Agricultural University, Hisar.

Walbot, V. (1992) Strategies for mutagenesis and gene cloning using transposon tagging and T-DNA insertional mutagenesis. *Annual Review of Plant Physiology and Plant Molecular Biology* 43, 49–82.

Weltzien, E., Whitaker M.L. and Dhamotharan, M. (1996) Diagnostic methods for breeding pearl millet with farmers in Rajasthan. In: Sperling, L. and Loevinsohn, M.L. (eds) *Using Diversity. Proceedings of Conference on Using Diversity and Maintaining Genetic Resources on Farm.* New Delhi, June 1995. International Development Research Centre, New Delhi, pp. 127–140.

Witcombe, J.R. and Hash, C.T. (1999) Resistance gene deployment strategies in cereal hybrids using marker-assisted selection: gene pyramiding, three-way hybrids, and synthetic parent populations. *Euphytica* (in press).

Witcombe, J.R. and Joshi, A. (1996) The impact of farmer participatory research on biodiversity of crops. In: Sperling, L. and Loevinsohn, M.L. (eds) *Using Diversity. Proceedings of Conference on Using Diversity and Maintaining Genetic Resources on Farm.* New Delhi, June 1995. International Development Research Centre, New Delhi, pp. 87–101.

Witcombe, J.R., Joshi, A., Joshi, K.D. and Sthapit, B.R. (1996) Farmer participatory crop improvement. I: Varietal selection and breeding methods and their impact on biodiversity. *Experimental Agriculture* 32, 445–460.

Witcombe, J.R., Raj, A.G.B., Packwood, A.J. and Virk, D.S. (1995) *Small Farmer*

Seed Supply. Reforming Regulatory Frameworks for Testing Release and Dissemination. Final Technical Report. Centre for Arid Zone Studies, University of Wales, Bangor, pp. 48–50.

Witcombe, J.R., Petre, R., Jones, S. and Joshi, A. (1999) Farmer participatory crop improvement. IV: The spread and impact of Kalinga III – a rice variety identified by participatory varietal selection. *Experimental Agriculture* (in press).

Wood, D. and Lenné, J.M. (1997) The conservation of agrobiodiversity on-farm: questioning the emerging paradigm. *Biodiversity and Conservation* 6, 109–129.

Zhu, J., Stephensen, P., Laurie, D., Li, W., Tang, D. and Gale, M.D. (1999) Towards rice genome scanning by map-based AFLP fingerprinting. *Molecular and General Genetics* 261, 184–195.

The Effects of Pest Management Strategies on Biodiversity in Agroecosystems

11

A. Polaszek[1], C. Riches[2] and J.M. Lenné[3]

[1] CABI Bioscience UK Centre, Silwood Park, Ascot, Berkshire SL5 7PY, UK; [2] Natural Resources Institute, University of Greenwich, Central Avenue, Chatham Maritime, Kent ME4 4TB, UK; [3] Agrobiodiversity International, 13 Herons Quay, Sandside, Milnthorpe, Cumbria LA7 7HN, UK

Introduction

Pests are critically important components of agroecosystems globally, for both economic and biological reasons. On the one hand, damaging pests may reduce crop yield and diversity – the annual preharvest yield losses due to pests in major crops have been estimated at 20–30% globally (Pimentel *et al.*, 1989; Terry, 1996). On the other hand, pests play a key role in creating and fostering biodiversity through evolutionary processes (Ridley, 1930; Gillet, 1962; Janzen, 1973; Burdon, 1987; see Allen *et al.*, Chapter 6, this volume).

Most farmers implement management strategies to minimize pest losses. Such management interventions may have direct or indirect effects on biodiversity in agroecosystems. For example, the application of pesticides may reduce pest populations but also affect beneficial insect populations, soil biota and bird diversity (see page 276; also Edwards *et al.*, Chapter 12, this volume). Similarly, the introduction of a rust-resistant wheat variety may lead to an overall increase or decrease in regional wheat diversity depending on the existing varietal diversity, the diversity in the new variety and its extent of adoption by farmers (Heisey *et al.*, 1997). Clearly the pest management strategies deployed by farmers in agroecosystems may have profound effects on both planned and unplanned biodiversity.

In this chapter, we concentrate on three important groups of pests – insects, weeds and pathogens – mostly above ground. The nature and importance of harmful pathogen and beneficial insect biodiversity have already been considered in Chapters 6 and 7, respectively, and Chapters 5 and 12 cover the nature, function and management of soil biodiversity in agroecosystems. Here, we will briefly look at the importance of pests in agroecosystems and then consider the effects of management strategies on pest diversity and the interrelationships between pests, crops and the surrounding vegetation. The overall effects of pest management strategies on the productivity and stability of agroecosystems is discussed in Lenné and Wood, Chapter 18, this volume.

Pests in Agroecosystems

Insect pests

Insect pests are components of the biodiversity of many, if not most, agroecosystems. Many important pests are notable for having invaded and colonized agroecosystems geographically remote from their original areas of distribution; others are the direct products of their immediate agroecosystems. In either case, the populations of certain insects, or groups of insects, are often favoured when they encounter an agroecosytem with abundant food sources for exploitation. This effect leads us to regard such favoured insect groups as crop pests.

Insect pests often multiply and cause damage when an ecosystem is simplified (Altieri, 1994) by replacement with a dominant crop and reduction of biodiversity (Kogan and Lattin, 1993), although there are exceptions. For example, irrigated rice is an ancient, simplified cropping system where predator diversity is maintained by the paddy system (see Settle et al., 1996, and discussed later). Many crop monocultures contain a lower diversity of organisms, including natural enemies of possible pest species, and present a more easily encountered food source. However, other factors, such as weather patterns (Dent, 1991) and host resistance breakdown, may have major synergistic effects in generating insect pest problems in monocultures. Where insecticides are over-used, development of resistance by the insect, sometimes coupled with emergent inadequate host plant resistance and non-target effects, generate new insect pest problems. For example, in rice in tropical Asia, a large number of pest outbreaks have been associated more with over-use of pesticides than with the breakdown of host resistance and high cropping intensity (Kenmore et al., 1984; Schoenly et al., 1996). The genetic variability of certain insect populations which enables them to assume pest status has been usefully reviewed by Kim and McPheron (1993).

Weeds

Weeds form integral components of agroecosystems. Because of their detrimental impact on crop yields due to competition, and/or because they provide reservoirs of insect pests and pathogens, they have traditionally been regarded as undesirable. Indeed, weed control has been referred to as 'the oldest profession' (Froud-Williams, 1991), while government officials in China recognized its importance as long ago as 6000 BC (Li, 1987). Despite the application of a range of weed management technologies, with recurrent investment of financial resources and labour, weeds remain a serious constraint to the productivity of farming systems in both industrialized and less developed countries.

All cropping systems are managed by farmers to reduce the impact of these most ubiquitous of pests. Worldwide losses due to weeds are roughly 10–15% of attainable production of the principle food and cash crops (Zimdahl, 1980; Oerke *et al.*, 1994), with developing countries being burdened by greater loss than the industrialized world. Some 18–20% of cotton, rice and maize is estimated to be lost due to weeds in developing countries of Africa, Asia and the Americas, compared with 9–11% in the industrialized economies of these areas (Terry, 1996). In the USA it has been estimated that more than $6 billion is spent annually on herbicides, tillage and cultivation to control weeds (Chandler, 1991), but crop losses due to infestation currently exceed $4 billion (Bridges and Anderson, 1992). In low-input, smallholder agricultural systems of the developing world, hand-weeding is a major burden; in semi-arid Zimbabwe, for example, weed control accounts for up to 60% of the labour used in maize production (Riches *et al.*, 1997).

Pathogens

Plant pathogens are an extremely heterogeneous group and include fungi, bacteria, phytoplasmas, viruses and nematodes, with diverse ecological requirements and modes of parasitism, reproduction, dispersal and survival (Burdon, 1993; see Allen *et al.*, Chapter 6, this volume). In most agroecosystems, pathogens interact with a wide variety of organisms and play a crucial role in defining the diversity for disease resistance in associated economic plants (Lenné, 1998). It is somewhat incongruous that continued generation of biodiversity for disease resistance for future crop improvement may depend on active, dynamic pathogen populations which, in this process, reduce food production (Wills, 1996).

Throughout history, pathogens have had serious effects on crops in agroecosystems with often far-reaching consequences for society. Examples include the Irish potato famine of 1845–1849 caused by late blight (*Phytophthora infestans*); the Indian wheat rust epidemics of 1850–1950 (*Puccinia* spp.); the Sri Lankan coffee rust epidemic of the 1870s (*Hemileia vastatrix*); the great Bengal rice famine of 1942–1943 associated with a devastating epidemic of brown spot (*Biploris oryzae*); 15 serious epidemics of groundnut rosette in sub-Saharan Africa since the early 1900s; and the recent epidemic of sorghum ergot in Latin America (Large, 1940; Carefoot and Sprott, 1969; Padmanabhan, 1973; Safeeula, 1977; Bandyopadhyay *et al.*, 1998; Naidu *et al.*, 1998). Some pathogens have continued to evolve in areas of primary and secondary diversity causing frequent epidemics; others have newly encountered crops which have been moved intercontinentally; while other pathogens have re-encountered their original hosts after a period of separation and have resulted in serious epidemics (Buddenhagen, 1977; Allen, 1983; see Allen *et al.*, Chapter 6, this volume). It is not unexpected that pathogen diversity is seen as harmful in agroecosystems.

Pest Management Strategies in Agroecosystems: Some Examples

Management of insect pests

Available management strategies for insect pests can be broadly categorized as chemical, cultural and biological, and also combined as components of integrated pest management (IPM). Host plant resistance, transgenic plants, 'interference methods' (Dent, 1991), microbial insecticides and biopesticides are usually treated either separately, or as additional IPM components. In practice, however, IPM may mean different things in different parts of the world (Morse and Buhler, 1997). In particular, shifting the balance between the chemical component and other components will have significantly different effects on local agrobiodiversity.

Chemical practices which affect diversity

INSECTICIDES: NON-TARGET LETHAL AND SUBLETHAL EFFECTS The overall effect of using insecticides is a reduction in local agrobiodiversity. This effect may be carried on through the food chain, resulting in further losses in other components of biodiversity (e.g. birds). One of the best-documented direct consequences of insecticide use is the reduction or elimination of the natural enemies of the targeted pest, resulting in a consequent increase in the pest population (Jepson, 1989). In fact, insecticides have been used for experimental purposes to deliberately increase pest populations by killing natural enemies, for example *Aphytis* spp. (Hymenoptera: Aphelinidae), parasitoids of California red scale, *Aonidiella aurantii* (Hemiptera: Diaspididae) (DeBach and Rosen, 1991). In general, agricultural pests of which the stage targeted by the insecticide is sessile and protected by wax or thick chitin are more likely to resist contact insecticides. Such pests include scale insects and whiteflies (Hemiptera: Coccoidea and Aleyrodidae), groups which also have a rich natural enemy fauna, especially parasitoids, whose beneficial effects on the crop are often disrupted by insecticide use.

Indeed, there are numerous examples of the exacerbation of scale insect and whitefly problems due to insecticide use, on various crops. Attempts to control the *Bemisia tabaci* (Hemiptera: Aleyrodidae) whitefly complex illustrate both non-target effects of pesticide use and the development of insecticide-resistant populations or species, so-called 'biotypes'. *Bemisia tabaci* is now a global pest although it had only been a sporadic pest in Africa and Asia earlier in the 20th century. Resurgence of *B. tabaci* following insecticide use has now been documented for cotton in Sudan (Abdelrahman, 1986; A.A. Abdelrahman, Sudan, 1993, personal communication) and in one case in India (Rajak and Diwakar, 1987). Gupta and Katiyar (1991), however, noted that the use of a pyrethroid mixture on cotton in India did not cause resurgence of *B. tabaci*. Clearly circumstances and applications may differ markedly, with differing results.

In general, indiscriminate use of pesticides will generate pest problems rather than alleviate them. Other documented cases of resurgence of *B. tabaci* following pesticide usage may actually be attributable to what is now known to be widespread displacement of 'original' *B. tabaci* populations by the so-called 'B-biotype', also known as *B. argentifolii* (Bedford *et al.*, 1992, 1994; Bellows *et al.*, 1994). *Bemisia argentifolii* has been blamed for US$500 million in crop losses in the USA in a single year (Perring *et al.*, 1993).

As well as the direct, lethal effects of chemicals on natural enemies, especially parasitoids, the beneficial actions of natural enemies on pest populations may be further reduced by sublethal effects. These include reductions in fecundity, altered or diminished host-searching efficiency, disruption of sexual communication and reduced longevity (Elzen, 1989).

INSECTICIDES: DEVELOPMENT OF RESISTANCE IN KEY PESTS Insecticide use is an extremely potent selective force on insect populations, inevitably leading to targeted pests becoming resistant (May, 1985). There are currently well over 500 documented cases of the development of insecticide resistance (Georghiou and Lagunes-Tejeda, 1991). Resistance to a range of insecticides has been demonstrated in a number of regional populations of the *B. tabaci* complex, and this has contributed largely to its increased pest status (Cahill *et al.*, 1995; de Barro, 1995), as noted above. The actual mechanisms of resistance development vary greatly, and may be due to simple multiplication of single genes, as shown for *Myzus persicae* (Hemiptera: Aphididae) (Devonshire and Sawicki, 1979) or the actions of transposable elements, as claimed for *Culex pipiens* (Diptera: Culicidae) (Salvado *et al.*, 1994).

As well as responding to insecticides, resistance has arisen in pests in response to juvenile hormone mimics (Shemshidini and Wilson, 1990) and *Bacillus thuringiensis* (*Bt*) toxins (Heckel, 1994; Tabashnik, 1994). In the latter case, it has been postulated that resistance is less likely to evolve if several *Bt* toxins are used simultaneously, the theory being that independent mutations are required to counter each toxin. This is referred to as the multiple-toxin approach (Gould, 1991). Recently, however, a single autosomal recessive gene conferring extremely high resistance to four different

Bt toxins has been found in field populations of the diamond-back moth, *Plutella xylostella* (Lepidoptera: Plutellidae) (Tabashnik *et al.*, 1997).

According to Gould *et al.* (1991), the combination of partial host plant resistance with insecticide use will speed up the evolution of insecticidal resistance in a population. Conversely, replacing insecticides with natural enemies will slow down that process of adaptation. The complex and varying relationships between pests, host plant resistance and natural enemies have been reviewed by Thomas and Waage (1996) and briefly by Verkerk *et al.* (1998). It would seem that the spectrum of relationships between host plant resistance and natural enemies ranges from synergistic to antagonistic (Verkerk *et al.*, 1998), and each situation needs to be examined individually. In many cases, while a particular pest is being controlled by pesticides, host plant resistance to the pest is less important. With emergent insecticide-resistant pest populations, the problem of inadequate host plant resistance emerges. As Way and Heong (1994) have stated, 'the dramatic losses of cultivar resistances to virulent populations of several rice plant hoppers and gall midges all occurred in circumstances of widespread and intense insecticide treatment'.

In summary, the use of insecticides, microbials, juvenile hormone mimics or *Bt* toxins forces a rapid evolutionary response in the pest. The longer-term effects of transgenic insecticidal cultivars (TICs) on resistance development in insects, however, have been little studied. To date, no instances of resistance to TICs developing in insects in the field have been recorded (F. Gould, Raleigh, USA, 1998, personal communication). Although insects are less likely to develop rapid resistance to TICs than with the direct application of toxins, it is believed that the potential is there and management strategies involving TICs must be developed with care (Gould, 1998). Increased use of TICs could lead to reduced pesticide use, which will have positive effects on beneficial insects (Krattiger, 1998).

Cultural practices which reduce or increase agrobiodiversity

MANAGEMENT OF CROP DIVERSITY There is some contention over the extent to which crop monocultures, with reduced plant biodiversity, are inherently unstable and invite insect pest problems (Southwood and Way, 1970; Altieri, 1994; Way and Heong, 1994). There is no simple answer, as conditions in the field vary considerably. In relation to insect pests, increasing biodiversity at the first trophic level, throughout the agroecosystem, will have both positive and negative effects on the pest population. The overall effect will depend on so many factors, including the reactions of other herbivores and the response of natural enemies. Increasing crop diversity by intercropping or polyculture, practices particularly suitable for small-scale farmers, generally reduces insect pest attack (McLaughlin and Mineau, 1995), but adoption by farmers may be problematic (van den Berg *et al.*, 1998).

MANAGEMENT OF SURROUNDING VEGETATION Interactions between crops, their pests, surrounding vegetation and natural enemies of pests are complex and often unpredictable. For example, providing alternative host plants for cereal stem borers in land surrounding maize fields has several effects. The plants may function as trap crops, enticing moths away from the maize, or they may also act as reservoirs for the borers after the maize has been harvested, allowing the pest to persist through the off-season. Often natural enemies will be preferentially attracted to borers on natural host plants, with which they have coevolved. Wild grasses, commonly growing near farmers' fields, provide refugia for natural enemies of stem borers, providing additional sources of stem borer hosts, such as tortricids and cossids, which are not generally found on the crop itself (Polaszek and Khan, 1998; van den Berg *et al.*, 1998). Repellent plants may also be used, as part of a 'push–pull' strategy. For example, in a recent study in Kenya, Khan *et al.* (1997) demonstrated that molasses grass (*Melinis minutiflora*), repellent to stem borers, when intercropped with maize or sorghum, significantly decreased levels of infestation by stem borers and also increased larval parasitism of stem borers by the braconid, *Cotesia sesamiae* (Khan *et al.*, 1997). Other specific examples of the effects of managing the vegetation surrounding the crop on insect pests are given by Thresh (1981), Altieri (1994) and Verkerk *et al.* (1998); see also LaSalle, Chapter 7, this volume.

OTHER OPTIONS In a pest response which parallels the development of pesticide resistance, Krysan *et al.* (1986) showed that certain populations of the northern corn rootworm (*Diabrotica barberi*, Coleoptera: Chrysomelidae) appear to be capable of prolonging their normal annual diapause to a 2-year cycle in order to avoid the rotational year when non-host crops are grown. Table 11.1 lists several other options for cultural management of insect pests and diseases and their possible effects on biodiversity (see also Thurston *et al.*, Chapter 9, and Edwards *et al.*, Chapter 12, this volume).

Biocontrol practices that reduce or increase agrobiodiversity

The preservation, enhancement and exploitation of agrobiodiversity are central to successful biological control in its broadest sense (see LaSalle, Chapter 7, this volume). The concept of biodiversity as a source of control agents has also been extensively reviewed relatively recently (LaSalle, 1991; Waage, 1991). Here, we focus on the effects that various biocontrol practices have on existing agrobiodiversity in several cropping systems.

In the classical biological control example of an invading species becoming a pest, often on an introduced crop, we are frequently faced with the basic problem of 'filling in the missing biodiversity': the natural enemy or enemies. Perhaps the simplest example is the successful biocontrol of the cassava mealybug, *Phenacoccus manihoti* (Herren and Neuenschwander, 1991), which resulted in enormous savings of a staple food across the cassava belt of Africa (see LaSalle, Chapter 7, this volume). The biocontrol literature

Table 11.1. Effect of selected cultural practices on management of insect pests and pathogens and agrobiodiversity.

Cultural practice	Effect on agrobiodiversity	References
Tillage	Reduced tillage favours some insect pests and pathogens but may also favour some beneficial organisms; overall effects on agrobiodiversity variable; no-tillage systems have other benefits (e.g. water conservation and reduced erosion)	Walters (1975), Thurston (1992), McLaughlin and Mineau (1995), see also Edwards *et al.*, Chapter 12, this volume
Grazing crop residues	Minimal effect on insect pest incidence but may reduce pathogen inoculum; general increase in grazer-associated diversity but overall effect may be a reduction in organismal diversity	Madsen *et al.* (1990), Lenné (1989), van den Berg *et al.* (1998)
Burning stubble	May reduce carry-over of insect pests and pathogens but more likely to substantially reduce crop-associated beneficial biodiversity, especially in subsistence conditions	Thurston (1992), van den Berg *et al.* (1998)
Planting density	Decreasing plant density generally decreases insect pest and pathogen damage (some viruses are an exception) and may increase agrobiodiversity, but effects are variable and the practicality (in terms of food production) is questionable	Thurston (1992), van den Berg *et al.* (1998)
Crop rotation	Decreases insect pest and pathogen damage and increases agrobiodiversity, but insects have potential to adapt life cycles to crop rotation patterns	Krysan *et al.* (1986), Dent (1991), Thurston (1992)

is full of examples of the successful regulation of an individual insect pest by a single parasitoid, multiple parasitoid releases adding little to the outcome (Hawkins, 1993). In theory at least, it is of paramount importance to locate the geographical origin of the introduced pest and those effective natural controlling agents for importation into the pest's new area of occurrence, effectively restoring the balance. Reality is, of course, rarely quite as simple as the theory, and the particular situation is often complicated by the actions of local natural enemies, hyperparasitoids and climate.

Conversely, invading pests may bring invading biocontrol agents with them. This has happened with the spread of *B. argentifolii*, whose increase in range coincides with the increasing range of several of its natural enemies (Polaszek *et al.*, 1992). Similarly, the so-called spiralling whitefly, *Aleurodicus dispersus,* has brought two natural enemies with it into most of its newly colonized distribution, but not everywhere (A. Polaszek, 1998, unpublished results). In parts of its new distribution, *A. dispersus* is kept under fortuitous biological control by these parasitoids. Fortuitous biocontrol can also emerge through indigenous natural enemies moving on to the invading pest (see LaSalle, Chapter 7, this volume).

Non-target insect hosts can be affected by imported biological control agents. Host range testing of prospective biocontrol agents is rarely undertaken thoroughly, for reasons of feasibility and practicality. In particular, there is the difficulty of equating field with laboratory conditions. Where the alternative hosts are 'generally undesirable' insects, such as aphids, scale insects or whiteflies, there is usually very little concern about non-target effects. More recently, concern has been raised, particularly where predators have been involved, but there is a general belief that parasitoid Hymenoptera are host-specific. While this may often be true at family, genus, species (and even population) levels, there are many exceptions. In particular, family or subfamily host specificity appears to be the rule among many whitefly parasitoids.

LaSalle (see Chapter 7, this volume) has correctly drawn attention to the fact that even where concerns exist about non-target effects of introduced biocontrol agents, they may often overlook the third trophic level. For example, in whitefly classical biocontrol, the introduced parasitoid species may not only affect endemic whiteflies, but may lead both directly and indirectly to the extinction of endemic parasitoids. Indirect extinction of these species could be brought about by competition with the introduced species for hosts. Also, because many *Encarsia* whitefly parasitoids exhibit heteronomous host relationships (reviewed by Williams and Polaszek, 1996), in which the males develop as hyperparasitoids of their own or other species, certain introduced parasitoids could utilize endemic parasitoids as hosts for the development of males. Accidental or deliberate introduction of parasitoids could therefore lead to the direct extinction of these endemics through parasitization, as well as indirect extinction through competition. Island endemics by their very nature are particularly vulnerable to extinction (Williamson, 1996), and it is hoped that any further introductions will be carefully planned. Harris (1990), discussing non-target effects of biocontrol agents, states that some rare herbivores may increase in abundance as a result of the decline in the targeted pest population. However, where the rare species is host- and niche-specific, this is less likely.

Local and global extinctions of insects due to pest management
Occasionally the impact of a particular pest management strategy leads to the extinction of the pest, either directly or indirectly. Examples are scarce, but do serve to illustrate what can happen. They include the global extinction of the 'typical' form of the blackfly *Simulium soubrense* (Diptera: Simulidae); eradication of *S. neavei* from Kenya; and of the *djodji*-form of *S. sanctipauli* from the mountainous area on the Togo–Ghana border. In each of these cases, extinction resulted directly from the use of chemical insecticides (Davies, 1994; R.J. Post, London, 1998, personal communication). Using a method less drastic, but equally effective, the tsetse fly *Glossina palpalis* was completely eradicated from the island of Principe in West Africa by using back-carried sticky traps (Da Costa *et al.*, 1916). Before the 20th century, the Rocky Mountain grasshopper (*Melanoplus spretus*, Orthoptera: Acrididae) was the most serious agricultural pest in the western USA and Canada (Lockwood and DeBrey, 1990). Apparently, a combination of various agricultural practices, including tillage, irrigation and introduction of grazing, coupled with a natural population crash in the late 1880s, led to its complete extinction.

Management of weeds

Management strategies can affect diversity in weed floras at two levels. Shifts in species composition and abundance can result from the adaptation of species to the opportunities provided by the continuous use of a particular tillage, rotation or herbicide regime. Individual species also have the capacity to evolve resistance to, or tolerance of, management practices because of the considerable genetic variability which exists within and among weed populations (Barrett, 1988; Liebman and Gallandt, 1997). Changes in weed management practices must therefore be expected to be associated with dynamic responses in weeds which allow their continued evolution, adaptation and persistence (see Edwards *et al.*, Chapter 12, this volume).

Effects of non-chemical farming practices
The species composition at a particular site reflects antecedent tillage and weed management practice (e.g. Hume *et al.*, 1991; McLaughlin and Mineau, 1995; Liebman and Gallandt, 1997; Shield and Bacon, 1998). The weed species shifts which occur at various stages of a bush–fallow system of shifting cultivation and the temporal effects of both swidden agriculture and long-term cultivation on vegetation biodiversity are well known (de Rouw, 1995; de Jong, 1997). A bush–fallow has the effect of initially reducing population and species diversity of non-woody weeds, particularly grasses, which

later return to predominate over broadleaf species 2–3 years into the next cropping phase (Sanchez and Nurena, 1970; Cancian, 1972; Moody 1975). With reduced fallow periods associated with increased demand for food due to population growth, rotational fallow becomes less effective in suppressing both monocotyledonous and dicotyledonous weeds, probably due to longevity of the weed seed-bank (Moody, 1975).

Indeed, where upland rice was cultivated in cycles of less than 10 years fallow in Sierra Leone, rice production became uneconomic due to weeds (Nyoka, 1982). Repeated cutting and burning may destroy the seed-bank of forest plants leading to the replacement of the forest fallow by a thicket (de Rouw, 1995). Degraded upland rice systems in West Africa and northeast India have been extensively invaded by the tropical American exotic *Chromolaena odorata* (Kushwahja, *et al.*, 1981; Slaats, 1992). The weed flora is less diverse in invaded areas, and in West Africa it is becoming apparent that several problematic weed species are less important in crops following *C. odorata* short fallow (Anon., 1995). Cropping systems are therefore evolving to utilize the weed suppression and soil fertility potential of the invaded fallow (de Rouw, 1995; Autfray and Gabaka Tchetche, 1997).

The economics of agriculture in developing countries rarely allows farmers access to herbicides so weed control continues to depend on hand cultivation, the provision of a clean seedbed, or inter-row cultivation by animal-drawn implements (Terry, 1996; Riches *et al.*, 1997). Such systems may lead to the dominance of a particular species. In semi-arid southern Africa, for example, the rhizome system of the grass *Cynodon dactylon* is adapted to the annual cycle of disturbance and re-establishment associated with shallow mouldboard ploughing (Phillips, 1993).

While the majority of annual species in the tropics are relatively easy to control with a hoe or mechanical weeder, many species are a problem because they produce abundant seed and emerge in large numbers with the onset of the rains and the period of crop establishment. Common examples in Malawi are *Ageratum* spp., *Gallinsoga*, *Ocimum* and *Nicandra physalodes* (Hillocks *et al.*, 1996).

Deployment of resistance to parasitic weeds

Parasitic weeds in the genus *Orobanche* (*Orobanchaceae*) are important constraints in sunflower and grain legume production in Eastern Europe, the Iberian Peninsula, the Middle East and North Africa, while *Alectra* and *Striga* (*Scrophulariaceae*) are devastating pests in sub-Saharan Africa (Parker and Riches, 1993; Plate 13). Inter- and intrapopulation genetic variability provides a number of problems for the development and sustained deployment of resistant cultivars (Riches, 1994).

The existence of geographical races of *Orobanche cumana* with differential virulence to sunflower cultivars led to the 'breakdown' of resistance when lines developed in Russia were planted in Eastern Europe (Cubero, 1986). The variability in virulence of *Striga gesnerioides* from different countries in West Africa on different cowpea lines also complicated the development of resistant cultivars (Parker and Polniaszek, 1990), although sources of resistance to parasite strains identified so far can be combined to provide multiple strain resistance (Lane *et al.*, 1997).

The selection pressure exerted by the host on the parasitic weed population is an important factor which may influence the stability of resistance and hence the utility of resistance genes. The mechanism of resistance plays an important role in determining the level of selection pressure. A number of sorghum cultivars developed for use in *Striga*-infested areas produce low levels of the root exudate necessary to stimulate parasite germination (Parker and Riches, 1993) but they are only partially resistant. Recent work with cowpea (Shawe and Ingrouille, 1993) suggests that individual cultivars exert a strong selection pressure on *S. gesnerioides* populations, so as to diversify and select virulent pathotypes. There is therefore a risk of selection and build-up of virulent sub-populations of *Striga* species in the field.

Use of herbicides

Reduced-tillage systems which rely on herbicides have been considered an integral component of sustainable high-input agriculture in many areas such as the Canadian prairies, particularly as these allow soil and moisture conservation (see Edwards *et al.*, Chapter 12, this volume). Shifts in annual weed dominance and increased perennial weed populations have been observed within 3–4 years in herbicide-treated, no-tillage maize fields (Triplett and Lytle, 1972). In a 5-year period in Maryland, USA, weed dominance shifted from the grass *Setaria faberi* to the broadleaf *Amaranthus hybridus* (Coffman and Frank, 1991). An increased association of wind dispersed weed species with reduced tillage or zero tillage in North America and temperate regions of Europe is well known, probably due to the lack of weed distribution by cultivation equipment (Froud-Williams *et al.*, 1981; Derksen *et al.*, 1993).

The progressive switch from mechanical methods to chemical weed control in industrialized economies was one of the primary reasons for the fivefold increase in farmer productivity achieved between 1920 and 1980 (Klingsman and Ashton, 1982). Herbicides are now used on 85–100% of all major crops in the industrialized world (Caseley, 1996). As with tillage, repeated use of a herbicide itself can influence weed succession. The use of glyphosate, for example, in some tropical plantation crops has resulted in excellent grass control but, with the reduced competition, aggressive broadleaf weeds, such as *Asystasia gangetica* and *Mikania micrantha*, have become important (Lam *et al.*, 1993).

A more widespread problem is the evolution of weed populations which are resistant to herbicides. This has occurred in situations where repeated applications have been made of herbicides with a single mode of action (Moss and Rubin, 1993). By 1995/96, 183 herbicide-resistant biotypes had been recorded in 42 countries, with 124 weed species that have evolved resistance to one or more herbicide active ingredients (Heap, 1997). Notable examples are the widespread occurrence of propanil-resistant jungle rice (*Echinochloa colona*) in rice in Central America (Garita *et al.*, 1995); of isoproturon-resistant *Phalaris minor* in wheat in India (Malik and Malik, 1994); and of populations of annual rye-grass (*Lolium rigidum*) with multiple resistance mechanisms conferring cross-resistance to a number of herbicide modes of action in Australia (Gill and Holmes, 1997). Common mechanisms of resistance which have evolved are found in biotypes which have either elevated levels of enzymes responsible for herbicide metabolism and detoxification, or altered binding sites for herbicidally active molecules on usually susceptible enzymes. The loss of reliable weed control from products which have been standard treatments for many years has been one of the motivations behind the move, in high-input systems, to a more sustainable system of integrated weed management (Swanton and Weise, 1991).

A combination of methods is therefore now seen as essential for the management of existing resistant populations and the prevention of further resistances. This may include pre-plant weed control by tillage or herbicide, integrated with in-crop control using rotation or sequences of herbicides with different modes of action and cultural practices including grazing of crop residues, increased crop density, or capture and destruction of weed seeds (e.g. Jutsum and Graham, 1995; Valverde, 1996; Gill and Holmes, 1997).

Weed problems associated with intensification

The intensification of production within high-input systems is dependent on reduced tillage and selective herbicides, which tend to ignore cultural methods of weed control, and has encouraged volunteer crop plants as weeds (see Edwards *et al.*, Chapter 8, this volume). In northern Europe these include volunteer oilseed rape or field beans in winter cereals, weed beet in sugar beet, and volunteer potatoes (Orson, 1993). As well as competing with young crops, volunteers provide reservoirs of many crop pathogens (Yardam and Gladders, 1993). The threat of volunteer crops as weeds can influence the choice and timing of stubble management methods and primary tillage. Now that straw burning is banned in many areas, farmers may need to plough to control volunteers, and this has associated effects on biodiversity. Ploughing, for example, kills more earthworms and the predators of aphids (which transmit barley yellow dwarf virus) than reduced tillage methods (Glen and Jordan, 1992) (see Edwards *et al.*, Chapter 12, this volume).

Consequences of herbicide-tolerant crops

Perhaps the greatest revolution in weed control since the introduction of the first selective herbicide, 2,4-D in 1946, will be provided by the introduction of genetically modified (GM) (or transgenic) herbicide-tolerant crops. Since tolerance to the broad-spectrum herbicide glyphosate produced by transgenic means was first announced in 1985, the development and introduction of GM cultivars has been rapid. Some 800,000 ha of glyphosate-tolerant cotton were planted in the USA in 1997; industry sources predicted a 1.2

Mha market for glyphosate-tolerant maize in USA in 1999; while the potential market for herbicide-tolerant soybean in North and South America has been estimated at 100 Mha (Agrow, 1996, 1998a,b). GM oilseed rape cultivars, tolerant to glufosinate or glyphosate, and non-transgenic imidazolinone-tolerant cultivars are now widely grown in Canada. Conversely, delays in the necessary legislation make it unlikely that commercial GM crops of herbicide-tolerant rapeseed, maize, sugar beet and eventually wheat will be grown in Europe much before 2000 (Marshall, 1998).

Although herbicide tolerance provides the opportunity for new strategies and greater flexibility in weed management (Wilcut et al., 1996), as well as probable reduction in herbicide use (Krattiger, 1998), there are a number of concerns about the commercialization of the technology. These include effects on the future diversity of the weed flora, the development of hard-to-control volunteer weeds from resistant crop plants, and the risk of resistance genes moving to sexually compatible weeds (Cole, 1994). Volunteer crops, particularly potatoes and oilseed rape, already cause considerable weed control problems in Europe and North America. The herbicide/crop combination used in a particular area must therefore be considered carefully. Crawley et al. (1993) concluded that glufosinate-tolerant rapeseed was no more invasive, or persistent, than conventional cultivars. Although glyphosate-tolerant volunteers could increase in follow-on crops in Europe or in Canada when pre-planting applications of glyphosate are made to stubble, the susceptibility of GM rapeseed to existing alternative herbicides (sulphonylureas or 2,4-D), would solve the problem (Marshall, 1998). Glyphosate is, however, a valuable treatment for the control of volunteer potatoes, so the introduction of glyphosate-tolerant cultivars may well compromise volunteer control.

The risk of movement of transgenes in pollen is highly dependent on geography (Cole, 1994). For example, GM herbicide-tolerant sunflower poses a higher risk of gene 'escape' in the USA (because of the presence of weedy relatives of sunflower) than do herbicide-resistant maize, soybean, cotton and wheat (which have no wild relatives in the USA). In Canada, gene transfer from the three major oilseed rape crops, Brassica napus, B. campestris or B. juncea, to the competitive weedy relative Sinapsis avensis was not achieved even under the most favourable conditions (Downey et al., 1991). Studies in Europe, however, have confirmed that fertile, transgenic weed-like plants can occur after just two generations of hybridization and backcrossing between rapeseed and the weedy relative B. campestris (Jorgensen and Anderson 1994; Mikkelsen et al., 1996). In the context of UK agriculture, the risk of gene flow to wild species has been assessed as minimal in potato, wheat, maize, tomato, cucumber and grain legumes, but high in carrots, sugar beet, cabbage and forage grasses (Raybold and Gray, 1993).

Herbicide-tolerant rice is considered to be high risk because rice is often associated with weedy relatives, for example, red rice and various weedy Oryza spp. (Tally, 1996). Based on the results of hybridization studies conducted under controlled conditions, there appears to be a distinct possibility of transfer and expression of glufosinate resistance in F_1 hybrids between rice and red rice (Sankula *et al.*, 1998), although natural introgression of engineered genes from rice has not yet been observed.

Much of the work on gene flow completed to date has been undertaken on small plots and may not be readily extrapolated to field or landscape scale (Timmons *et al.*, 1995). However, when the probability and consequences of gene transfer to wild relatives are considered as 'potential environmental impact', some cases do appear to carry a risk (Ahl Goy and Duesing, 1996), leading to suggestions that cultural methods, including adequate isolation distances, will be needed to minimize the opportunity of the flowering period of the crop and wild relatives coinciding.

In practice, the containment of herbicide-tolerant transgenes is probably quite impractical (Marshall, 1998). Use of integrated weed management programmes may become important for the control of herbicide-resistant weeds. Other strategies, such as the design of self-destruct mechanisms in gene-constructs with either inactive tolerance genes or so-called terminator genes which prevent the germination of hybrids, may become important for environmental protection as well as the protection of the intellectual property of biotechnology companies (Gressel, 1993).

Management of pathogens

Common management strategies deployed in agroecosystems to control pathogens include chemicals such as fungicides, host plant resistance, cultural methods and biological control. Several examples of the effects of these strategies on various components of biodiversity in selected agroecosystems are discussed in this section.

Effects of fungicides on pathogens and associated diversity

The fungal genus *Colletotrichum* consists of many plant pathogenic and saprophytic species which frequently occur together and may interact with each other and other microorganisms on host surfaces (Sherriff *et al.*, 1999). Overuse of fungicides to control pathogenic *Colletotrichum* species may result in the development of resistant strains which must then be controlled by alternative fungicides, host plant resistance or biological control (Bailey and Jeger, 1992). For example, fungicide-resistant and fungicide-sensitive strains of *Colletotrichum truncatum*, the causal fungus of brown blotch of cowpea, responded differently to *Penicillium* sp., *Phoma* sp. and *Pestalotiopsis* sp. from the cowpea phyllosphere. The three fungi strongly inhibited conidial germination of the fungicide-resistant strains of *C. truncatum* compared with the fungicide-sensitive strains (Ravi and Anilkumar, 1991), indicating some potential for combining fungicide and biological control in an integrated management strategy.

Colletotrichum kahawae, the causal agent of coffee berry disease (CBD) in eastern and southern Africa, exists in intimate association on coffee with *C. gloeosporioides* and *C. acutatum*, two common saprophytic species (Masaba and Waller, 1992). Under appropriate conditions, these species and other phylloplane microorganisms are capable of inhibiting mycelial growth, conidial germination and appressorial formation of *C. kahawae* on coffee. However, when fungicides are used to control CBD, disease severity often increases. This is due both to the development of fungicide resistance in *C. kahawae* (Ramos and Kamidi, 1982) and to changes in the microflora on coffee surfaces, especially a reduction in populations of antagonistic *C. gloeosporioides* and *C. acutatum* (Masaba and Waller, 1992). It may be possible to manipulate chemical control to favour the antagonistic microflora and improve CBD control on coffee (Masaba and Waller, 1992). Similar associations between pathogenic *Colletotrichum* spp. and saprophytic species have been found for other diseases including anthracnose of the tropical pasture legume *Stylosanthes* spp. in Peru (Lenné and Brown, 1991), postbloom fruit drop of citrus in Florida (Agostini and Timmer, 1994) and fruit anthracnoses in Israel (Freeman *et al.*, 1998), and may be more widespread (Sherriff *et al.*, 1999). There is therefore potential for developing management strategies which stimulate microbial diversity on perennial hosts to improve control of important pathogens.

Increasing diversity through host plant resistance

Host plant resistance that provides comprehensive protection from pathogens is perhaps the most valuable contribution that crop breeding can make to the hundreds of millions of small, poor farmers in developing countries (Allen and Lenné, 1998). The wealth of pest management strategies developed by farmers since the beginnings of agriculture (see Thurston, 1992, and Thurston *et al.*, Chapter 9, this volume for pathogens) strongly suggests that early varieties (landraces) lacked sufficient pathogen resistance. This realization stimulated plant breeders to place considerable effort in increasing genetic resistance to major diseases.

The widespread adoption of modern, disease-resistant varieties of staple food crops such as rice and wheat by millions of small, poor farmers, however, has been strongly criticized as one of the major causes of genetic erosion and crop vulnerability to diverse pathogens (see Marshall, 1977; Brown, 1983). The incorporation of reliable resistances into acceptable, higher yielding varieties, or varietal mixtures (see below), need not reduce local crop diversity and, in fact, in many cases will increase diversity (Smithson and Lenné, 1996; Tripp, 1996; see Witcombe, Chapter 10, this volume), especially for disease resistance.

Some modern varieties of rice and wheat have very comprehensive pedigrees and can be highly genetically diverse (Boyle and Lenné, 1997). For example, the rice 'IR66' has 42 landraces in its parentage while the wheat 'Kauze' has 49 landraces including multiple disease and pest resistances (Hargrove *et al.*, 1988; Skovmand *et al.*, 1992). The number of landraces used in popular CIMMYT wheat varieties and IRRI rice varieties has increased tenfold over the past 30 years (Frankel *et al.*, 1995; see Witcombe, Chapter 10, this volume). In addition, farmers often grow modern varieties on part of their land, continuing to grow traditional landraces on the rest of their land for sound socio-economic reasons (Brush *et al.*, 1981; Wood and Lenné, 1997).

Breeding for single gene resistances is generally easier than breeding for multiple gene resistance and more widely practised (Marshall, 1977; Frankel *et al.*, 1995), but it may reduce the ability of a crop to combat a variable pathogen. Some single gene resistances, however, have proved remarkably stable (e.g. the *I* gene for resistance to BCMV, the *Are* gene for resistance to bean anthracnose (Allen and Lenné, 1998), and single gene resistances to several cereal diseases (Brown, 1995)); others have proved transient (e.g. single genes for resistance to rice blast (Zeigler *et al.*, 1994) and wheat rusts (Simmonds and Rajaram, 1988) in both modern and traditional crop varieties). It should be remembered that it is not lack of diversity that is the problem, but homogeneity for the one or a few genes that condition susceptibility to a particular pest or pathogen (Buddenhagen, 1977).

Multiple single gene resistances to one or several pathogens will increase diversity for disease resistance and, potentially, its durability (Brown, 1995; Allen and Lenné, 1998) and the use of resistance genes from exotic sources such as wild species (Lenné and Wood, 1991; Frankel *et al.*, 1995), and through biotechnological methods including transgenes (Tanksley and McCouch, 1997) will also increase diversity (see Witcombe, Chapter 10, this volume).

Use of transgenics

Genetic transformation for disease resistance in food crops is an increasingly important technology (Krattiger, 1998; Witcombe and Harris, 1998). Techniques are well advanced for developing transgenic varieties of major crops for resistance to fungal, bacterial, viral and nematode diseases. For example, the deployment of transgenic rice resistant to rice yellow mottle virus will provide a control option for many small poor African farmers where nothing presently exists (Witcombe and Harris, 1998). It will allow increasing production on existing land, taking pressure off marginal land and reducing threats to natural biodiversity.

The role of quarantine

Quarantine has played a very effective role in restricting the movement of pathogens (Khan, 1977), especially during the past 50 years – the most active period of plant germplasm movement. This has reduced the potential for aggressive pathogens to reduce crop plant diversity in agroecosystems through devastating disease epidemics such as the late blight epidemic which caused the Irish potato famine (see Allen *et al.*, Chapter 6, this volume) and has also reduced the potential for evolution of pathogen diversity in new environments. The effectiveness of quarantine, however, has varied among crops and pathogens. Many legume pathogens have spread widely because the hard testa and often large cotyledons provide a safe haven, especially for bacterial and viral pathogens (Allen and Lenné, 1998). In spite of effective quarantine systems in countries such as the USA and Australia, serious pathogens or new variants of existing pathogens are still occasionally introduced with often considerable effects on plant and pathogen diversity (e.g. for potato late blight see Fry *et al.*, 1993; for sorghum ergot see Bandyopadhyay *et al.*, 1998; see also Allen *et al.*, Chapter 6, this volume).

Management of cropping systems

Management of pathogens should address the cropping system as a whole, if full advantage is to be taken of control measures (Allen and Lenné, 1998), including those based on manipulation of crop-associated biodiversity (see Wood and Lenné, Chapter 2, this volume). Such an approach allows greater use of diversity as a management tool in the agroecosystem. The history of the use of integrated disease management strategies in traditional farming systems has been extensively reviewed by Thurston (1992). Practices include a very wide range of cropping and cultural practices.

Cropping practices (e.g. multilines, varietal mixtures, intercropping and multiple cropping) which increase plant diversity usually decrease pathogen damage (Thurston, 1992; Smithson and Lenné, 1996; Finckh and Wolfe, 1998; see Allen *et al.*, Chapter 6, and Thurston *et al.*, Chapter 9, this volume). Several mechanisms contribute to reduced damage including increased distance between susceptible plants; resistant plants acting as barriers to pathogen spread; reduced inoculum production; and various host–pathogen interactions such as cross-protection. It is somewhat paradoxical that plant diversity is most used in those agroecosystems where limited research has been done (i.e. in developing countries), and least used where considerable research effort has been carried out (i.e. in developed countries, especially Europe) (Smithson and Lenné, 1996).

Cultural practices (e.g. adjustment of sowing date, use of cultivars of different duration, fallow, rotation, tillage, site selection, management of the surrounding vegetation, flooding, cultivation and landform pattern such as raised beds and terraces, plant population and spacing patterns, depth of planting, mulching, suppressive soils, trap crops, fire and other sanitation methods such as clean seed) often effectively manage pathogens above and below ground by manipulating the diversity of certain components of the agroecosystem including crop-associated biodiversity (Thresh, 1981; Thurston, 1992; see Thurston *et al.*, Chapter 9, and Edwards *et al.*, Chapter 12, this volume; Tables 9.1 and 11.1). Many cultural practices, however, unfortunately remain the domain of traditional agricultural systems.

Recent investigations of farmers' traditional management of common bean systems in the Great Lakes region of Africa found that local strategies were based on microclimate regulation, genetic diversity and sanitation (Trutmann *et al.*, 1993). All of these practices were found to have positive effects in reducing bean diseases. It was concluded that enhanced disease management should be possible in these systems through improved host plant resistance and seed health while maintaining plant diversity. The importance of maintaining existing flexibility in management technologies was emphasized. Traditional systems such as these offer models for disease management in modern systems through the manipulation of both crop and crop-associated biodiversity (e.g. competitors, antagonists, endophytes, mycorrhizae, etc.). The ultimate aim should be the development of safe, economic and durable management strategies through a combination of cultural practices, host plant resistance, cropping practices and, in some systems, chemicals (Allen and Lenné, 1998). It is unrealistic to assume that the manipulation of biodiversity will ever solve all pest problems all of the time without the need for pesticides (Way and Heong, 1994).

Interrelationships Between Pests and Crops in Agroecosystems

The complexity of the interrelationships between pests and crops in agroecosystems has generated quite different views on pest management. Janzen (1973) suggested that the agricultural potential of many parts of the tropics, especially the seasonally dry tropics, might well be improved by the destruction of the natural vegetation. This would destroy the tremendous numbers of pests and diseases which are harboured by this vegetation. In contrast, Altieri (1994) stressed the importance of using the full potential of pest controls in natural vegetation by designing diverse, pest-stable, self-sustained, environmentally sound agroecosystems through the long-term maintenance of vegetational and landscape diversity.

Vegetational diversity in agroecosystems, however, is a mixed blessing to farmers. The benefits depend on whether the vegetation (weeds and wild plants) interspersed among crop fields harbours more or fewer pests and more or fewer parasitoids, predators and antagonists. The outcome is generally site-specific and may be either beneficial or detrimental to the crop. The distribution of cultivated land in a matrix of natural and semi-natural vegetation gives abundant opportunities for the movement of pests into crops (Thresh, 1981). Striking parallels can be drawn among a range of crops in temperate and tropical areas regarding the detrimental effects of surrounding vegetation which allows invasion of pathogenic fungi, nematodes and viruses into crops (see Thresh, 1981, for many examples). The spread of virulent rust strains from wild relatives of wheat in the Himalayan foothills to cultivated wheat in India and Pakistan with resultant epidemics (Joshi, 1986) reinforces the role of natural vegetation in supporting pathogen evolution and the emergence of new races capable of overcoming crop resistances (see Allen et al., Chapter 6, this volume).

For insect pests, however, the role of the surrounding vegetation is unpredictable (Thresh, 1981; Andow, 1991). On the one hand, surrounding vegetation can be a source of beneficial predators and parasites (Altieri, 1994; Polaszek and Khan, 1998) but, on the other hand, it may harbour serious crop pests (see van Emden and Williams, 1974; Thresh, 1981). Locust plagues provide a striking example of the way in which serious problems can arise in agriculture when insect pests invade crops from natural vegetation (Thresh, 1981). Positive results from the manipulation of surrounding vegetation depend on the composition of the vegetation and its ability to host predators and parasitoids of crop pests (van Emden, 1981; Polaszek and Khan, 1998). In European agriculture various forms of field margin management are now promoted in order to enhance arthropod diversity, including the use of grass strips in place of weed-free hedge bottoms or fence lines, and modified herbicide regimes along 'conservation headlands' where a diversified weed flora is maintained as a source of predatory arthropods which migrate through the crop canopy (Dennis and Fry, 1992; Hassall et al., 1992; Kromp and Steinberger, 1992; Moreby et al., 1994). As Andow (1991) stated 'While some of the major twists in the Gordian knot of vegetational diversity can be perceived, we are a long way from unravelling its complexity' and effects on pest populations and crop productivity.

Weeds greatly influence the organization and functioning of agroecosystems. Weed–pest or weed–pathogen interactions are important at two levels. On the one hand weeds may be alternate hosts to crop pests, diseases and nematodes (e.g. Bendixon, 1983; Terry, 1991; Hillocks et al., 1996). This can be a particular problem when weedy relatives of crops are represented in the flora and provide overwintering sites for the survival of pathogens, for example rice yellow mottle virus on wild Oryza species (Terry, 1991) and many sorghum diseases on Sorghum halepense (Warrick and Black, 1983). However, the presence of weeds in a crop can also increase the activity of pest predators by providing shelter, modifying crop microclimates and altering crop background to enhance predator colonization (William, 1981; Altieri and Letourneau, 1982; Andow, 1988; Afun et al., 1995). There is increasing interest in managing the weed diversity available in a crop from both the conservation standpoint, linked to the maintenance of invertebrate flora as a food source for wildlife and game birds (Hassall et al., 1992), and to enhance predation of pest species (Altieri et al., 1977; Altieri, 1994).

Whatever effects within- and between-field diversity might have on pests in agroecosystems, it should be remembered that there are some very successful monocultures in the tropics, for example, irrigated rice whose stability is based on characteristics inherent in the paddy system. In particular, a generalist predator population is supported in the early stages of the rice crop by feeding on abundant populations of detritus-feeding and plankton-feeding insects in the paddy (Settle et al., 1996). This process leads to strong suppression of rice pest populations as the generalist predators have a 'head start' on the pest populations. The decoupling of predator populations from strict dependence on rice pests lends stability to irrigated rice ecosystems. Increased diversity *per se* cannot be assumed to reduce insect pest (Way, 1979) or pathogen (Thresh, 1981) damage. The devastating dieback epidemics (caused by *Phytophthora cinnamomi*) in species-rich forests in southern Australia (more than 13 botanical families were killed – see Newhook and Podger, 1972, and Allen *et al.*, Chapter 6, this volume) is sufficient reminder that it is the quality, not the quantity, of diversity that matters.

Acknowledgements

AP would like to acknowledge the advice of Fred Gould on resistance development to TICs, and Martin Hall and Rory Post (The Natural History Museum, London) for information on insect extinctions.

References

Abdelrahman, A.A. (1986) The potential of natural enemies of the cotton whitefly in Sudan Gezira. *Insect Science and its Application* 7, 69–73.

Afun, J.V.K., Heinrichs, E.A., Johnson, D.E. and Russell-Smith, A. (1995) The influence of weeds in the natural control of upland rice insect pests in Côte d'Ivoire. *Proceedings of the Brighton Crop Protection Conference – Weeds*, pp. 181–186.

Agostini, J.P. and Timmer, L.W. (1994) Population dynamics and survival of strains of *Colletorichum gloeosporioides* on citrus in Florida. *Phytopathology* 84, 420–425.

Agrow (1996) Biotechnology revolution hits US agriculture. *Agrow* 256, 14–15.

Agrow (1998a) D & PL pulls roundup-ready cotton. *Agrow* 300, 3.

Agrow (1998b) DeKalb to expand R-R maize. *Agrow* 307, 20.

Ahl Goy, P. and Duesing, J.H. (1996) Assessing the environmental impact of gene transfers to wild relatives. *Biotechnology* 14, 39–40.

Allen, D.J. (1983) *The Pathology of Tropical Food Legumes: Disease Resistance in Crop Improvement.* John Wiley & Sons, Chichester.

Allen, D.J. and Lenné, J.M. (1998) *The Pathology of Food and Pasture Legumes.* CAB International, Wallingford.

Altieri, M.A. (1988) The impact, uses and ecological role of weeds in agroecosystems. In: Altieri, M.A. and Liebman, M. (eds.) *Weed Management in Agroecosystems: Ecological Approaches.* CRC Press, Boca Raton, Florida, pp. 1–6.

Altieri, M.A. (1994) *Biodiversity and Pest Management in Agroecosystems.* Food Products Press, New York.

Altieri, M.A. and Letourneau, D.K. (1982) Vegetation management and biological control in agroecosystems. *Crop Protection* 1, 405–430.

Altieri, M.A., van Schoonhoven, A. and Doll, J. (1977) The ecological role of weeds in insect pest management systems: a review illustrated by bean (*Phaseolus vulgaris*) cropping systems. *PANS* 23, 195–205.

Andow, D.A. (1988) Management of weeds for insect manipulation in agroecosystems. In: Altieri, M.A. and Liebman, M. (eds) *Weed Management in Agroecosystems: Ecological Approaches.* CRC Press, Boca Raton, Florida. pp. 266–301.

Andow, D.A. (1991) Vegetational diversity and arthropod population response. *Annual Review of Entomology* 36, 561–586.

Anon. (1995) Women and weeds in a changing agro-ecosystem. *International Institute of Tropical Agriculture, Annual Report,* Ibadan, Nigeria, pp. 24–26.

Autfray, P. and Gabaka Tchetche, H. (1997) L'utilisation de *Chromolaena oderata* pour fixer l'agriculture en zone forestiere de Côte d'Ivoire. *Agriculture et Developpement* 13, 3–12.

Bailey, J.A. and Jeger, M.J. (eds) (1992) *Colletotrichum: Biology, Pathology and Control.* CAB International, Wallingford.

Bandyopadhyay, R., Frederickson, D.E., Mclaren, N.W., Odvody, G.N. and Ryley, M.J. (1998) Ergot, a new disease threat to sorghum in the Americas and Australia. *Plant Disease* 82, 356–367.

de Barro, P.J. (1995) *Bemisia* biotype B: a review of its biology, distribution and control. *CSIRO Division of Entomology Technical Paper* 33, 1–57.

Barrett, S.C.H. (1988) Genetics and evolution of agricultural weeds. In: Altieri, M.A. and Liebman, M. (eds) *Weed Management in Agroecosystems: Ecological Approaches.* CRC Press, Boca Raton, Florida, pp. 57–75.

Bedford, I.D., Briddon, R.W., Markham, P.G., Brown, J.K. and Rosell, R.C. (1992) *Bemisia tabaci* – biotype characterisation and the threat of this whitefly species to agriculture. *Proceedings of the Brighton Crop Protection Conference – Pests and Diseases,* pp. 1235–1240.

Bedford, I.D., Briddon, R.W., Brown, J.K., Rosell, R.C. and Markham, P.G. (1994) Geminivirus transmission and biological characterisation of *Bemisia tabaci* (Gennadius) biotypes from different geographic regions. *Annals of Applied Biology* 125, 311–325.

Bellows, T.S, Jr, Perring, T.M., Gill, R.J. and Headrick, D.H. (1994) Description of a species of *Bemisia* (Homoptera: Aleyrodidae). *Annals of the Entomological Society of America* 87, 195–206.

Bendixen, L.E. (1983) Another criterion in defining the world's worst weeds. *Proceedings of the 9th Asian–Pacific Weed Science Society Conference,* pp.176–180.

van den Berg, J., Nur, A.F. and Polaszek, A. (1998) Cultural control. In: Polaszek, A. (ed.) *African Cereal Stem Borers; Economic Importance, Taxonomy, Natural Enemies and Control.* CAB International, Wallingford, pp. 333–347.

Bridges, D.C and Anderson, R.L. (1992) Crop losses due to weeds in the United States by state. In: Bridges, D.C. (ed.) *Crop Losses Due to Weeds in the United States – 1992.* Weed Science Society of America, Champaign, Illinois, pp. 1–60.

Boyle, T. and Lenné, J.M. (1996) Defining and meeting needs for information: agriculture and forestry perspective. In: Hawksworth, D.S., Kirk, P.M. and Dextre-Clarke, S. (eds) *Biodiversity Information: Needs and Options.* CAB International, Wallingford, pp. 31–54.

Brown, J.K.M. (1995) Pathogens' responses to the management of disease resistance genes. *Advances in Plant Pathology* 11, 75–102.

Brown, W.L. (1983) Genetic diversity and genetic vulnerability: an appraisal. *Economic Botany* 37, 4–12.

Brush, S.B., Carney, H.J. and Huaman, Z. (1981) Dynamics of Andean potato agriculture. *Economic Botany* 35, 70–88.

Buddenhagen, I.W. (1977) Resistance and vulnerability of tropical crops in relation to their evolution and breeding. *Annals of the New York Academy of Sciences* 287, 309–326.

Burdon, J.J. (1987) *Diseases and Plant Population Biology.* Cambridge University Press, Cambridge.

Burdon, J.J. (1993) The structure of pathogen populations in natural plant communities. *Annual Review of Phytopathology* 31, 305–323.

Cahill, M., Byrne, F.J., Gorman, K., Denholm, I. and Devonshire. A.L. (1995) Pyrethroid and organophosphate resistance in the tobacco whitefly *Bemisia tabaci* (Homoptera: Aleyrodidae). *Bulletin of Entomological Research* 85, 181–187.

Cancion, F. (1972) *Change and Uncertainty in a Peasant Economy.* Stanford University Press, Stanford.

Carefoot, G.L. and Sprott, E.R. (1969) *Famine on the Wind: Plant Diseases and Human History.* Angus & Robertson, London.

Caseley, J.C. (1996) The progress and development of herbicides for weed management in the tropics. *The Planter, Kuala Lumpur* 72, 323–346.

Chandler, J.M. (1991) Estimated losses of crops to weeds. In: Pimentel, D. (ed.) *CRC Handbook of Pest Management in Agriculture,* Vol. 1. CRC Press, Boca Raton, Florida, pp. 53–65.

Coffman, C.B. and Frank, J.R. (1991) Weed-crop responses to management systems in conservation tillage corn (*Zea mays*). *Weed Technology* 5, 67–81.

Cole, D.J. (1994) Introduction of herbicide-resistant crops. *Pesticide Outlook* June 1994, 32–36.

Crawley, M.J., Hails, R.S., Rees, M., Kohn, D. and Buxton, J. (1993) Ecology of transgenic oilseed rape in natural habitats. *Nature* 363, 620–623.

Cronk, C.B. and Fuller, J.L. (1995) *Plant Invaders.* Chapman & Hall, London.

Cubero, J.I. (1986) Breeding for resistance to *Orobanche* and *Striga*: a review. In: ter Borg, S.J. (ed.) *Biology and Control of Orobanche.* Wageningen, pp. 127–139.

Da Costa, B.F.B., Sant'ana, J.F., Santos, A.C. and Alvares, M.G.A. (1916) *Sleeping Sickness. A Record of Four Years War Against it in Principe, Portuguese West Africa.* Balliere, Tindall & Cox, London.

Davies, J.B. (1994) Sixty years of *Onchocerciasis* vector control: a chronological summary with comments on eradication, reinvasion, and insecticide resistance. *Annual Review of Entomology* 39, 23–45.

DeBach P. and Rosen, D. (1991) *Biological Control by Natural Enemies.* Cambridge University Press, Cambridge.

Dennis, P. and Fry, G.L.A. (1992) Field margins: can they enhance natural enemy population densities and general arthropod diversity on farm land? *Agriculture, Ecosystems and Environment* 40, 95–115.

Dent, D. (1991) *Insect Pest Management.* CAB International, Wallingford.

Derksen, D.A., Lafond, G.P., Thomas, G., Loeppky, A. and Swanton, C.J. (1993) Impact of agronomic practices on weed communities: tillage systems. *Weed Science* 41, 409–417.

Devonshire, A.L. and Sawicki, R.M. (1979) Insecticide-resistant *Myzus persicae* as an example of evolution by gene duplication. *Nature* 280, 140–141.

Downey, R.K., Bing, D.J. and Rakow, G.F.W. (1991) Potential of gene transfer among oilseed *Brassica* and their weedy relatives. In: *Proceedings 8th International Rapeseed Congress,* Saskatoon, Canada, pp. 1022–1027.

Elzen, G. (1989) Sublethal effects of pesticides on beneficial parasitoids. In: Jepson, P.C. (ed.) *Pesticides and Non-Target Invertebrates*. Intercept Limited, Andover, pp. 129–150.

van Emden, H.F. and Williams, G.C. (1974) Insect stability and diversity in agro-ecosystems. *Annual Review of Entomology* 19, 455–475.

Finckh, M.R. and Wolfe, M.S. (1998) The use of biodiversity to restrict plant diseases and some consequences for farmers and society. In: Jackson, L.E. (ed.) *Agricultural Ecology*. Academic Press, New York.

Frankel, O.H., Brown, A.H.D. and Burdon, J.J. (1995) *The Conservation of Plant Biodiversity*. Cambridge University Press, Cambridge.

Freeman, S., Katan, T. and Shabi, E. (1998) Characterization of *Colletotrichum* species responsible for anthracnose diseases of various fruit. *Plant Disease* 82, 596–605.

Froud-Williams, R.J. (1991) Novel approaches to weed control: new tricks for the oldest profession. In: *Proceedings of the Brighton Crop Protection Conference – Weeds*, pp. 143–154.

Froud-Williams, R.J., Chancellor, R.J. and Drennan, D.H.S. (1981) Potential changes in weed floras associated with reduced cultivation systems for cereal production in temperate regions. *Weed Research* 21, 99–101.

Fry, W.E., Goodwin, S.B., Dyer A.T., Matussak, J.M., Drenth, A., Tooley, P.W., Sujkowski, L.S., Koh, Y.J., Cohen, B.A., Spielman, L.J., Deahl, K.L., Inglis, D.A. and Sandlan, K.P. (1993) Historical and recent migrations of *Phytophthora infestans*: chronology, pathways and implications. *Plant Disease* 77, 653–661.

Garita, I., Valverde, B.E., Vargas, E., Chacon, L.A., de la Cruz, R., Riches, C.R. and Caseley, J.C. (1995) Occurrence of propanil resistance in *Echinochloa colona* in Central America. In: *Proceedings of the Brighton Crop Protection Conference – Weeds*, pp.193–196.

Georghiou, G.P. and Lagunes-Tejeda, A. (1991) *The Occurence of Resistance to Pesticides in Arthropods*. FAO, Rome.

Gill, G.S. and Holmes, J.E. (1997) Efficacy of cultural control methods for combating herbicide-resistant *Lolium rigidum*. *Pesticide Science* 51, 352–358.

Gillett, J.B. (1962) Pest pressure, an underestimated factor in evolution. *Systematics Association Publications* 4, 37–46.

Glen, D.M. and Jordan, V.W.L. (1992) The implications for pests and diseases of practices to reduce nitrate leaching. *Aspects of Applied Biology* 30, 343–350.

Gould, F. (1991) The evolutionary potential of crop pests. *American Scientist* 79, 496–507.

Gould, F. (1998) Sustainability of transgenic insecticidal cultivars: integrating pest genetics and ecology. *Annual Review of Entomology* 43, 701–726.

Gould, F., Kennedy, G.G. and Johnson, M.T. (1991) Effects of natural enemies on the rate of herbivore adaptation to resistant host plants. *Entomologia Experimentalis et Applicata* 58, 1–14.

Gressel, J. (1993) Advances in achieving the needs for biotechnically-derived herbicide resistance crops. *Plant Breeding Review* 11, 155–198.

Gupta, G.P. and Katiyar, K.N. (1991) Bioefficacy of tank-mix insecticides for control of bollworm complex (*Earias* spp. and *Pectinophora gossypiella*) and impact on whitefly (*Bemisia tabaci*) in upland cotton (*Gossypium hirsutum*). *Indian Journal of Agricultural Sciences* 61, 531–534.

Hargrove, T.R., Cabanilla, V.L. and Coffman, W.R. (1988) Twenty years of rice breeding: The role of semidwarf varieties in rice breeding for Asian farmers and the effects on cytoplasmic diversity. *BioScience* 38, 675–681.

Harris, P. (1990) Environmental impact of introduced biological control agents. In: Mackauer, M., Ehler, L.E. and Roland, J. (eds) *Critical Issues in Biological Control.* Intercept, Andover, pp. 289–300.

Hassall, M., Hawthorne, A., Maudsley, M., White, P. and Cardwell, C. (1992) Effects of headland management on invertebrate communities in cereal fields. *Agriculture, Ecosystems and Environment* 40, 155–178.

Hawkins, B.A. (1993) Parasitoid species richness, host mortality and biocontrol. *American Naturalist* 141, 634–641.

Heap, I.M. (1997) The occurrence of herbicide-resistant weeds worldwide. *Pesticide Science* 51, 235–243.

Heckel, D.G. (1994) The complex genetic basis of resistance to *Bacillus thuringiensis* toxin in insects. *Biocontrol Science and Technology* 4, 405–417.

Heisey, P.W., Smale, M., Byerlee, D. and Souza, E. (1997) Wheat rusts and the costs of genetic diversity in the Punjab of Pakistan. *American Journal of Agricultural Economics* 79, 726–737.

Herren, H.R. and Neuenschwander, P. (1991) Biological control of cassava pests in Africa. *Annual Review of Entomology* 36, 257–283.

Hillocks, R.J., Logan, J.W.M., Riches, C., Russell-Smith, A. and Shaxson, L.J. (1996) Soil pests in traditional farming systems in sub-Saharan Africa – a review. Part 1. Problems. *Tropical Pest Management* 42, 241–251.

Hume, L., Tessier, S. and Dyck, F.B. (1991) Tillage and rotation influences on weed community composition in wheat (*Triticum aestivum* L.) in southwestern Saskatchewan. *Canadian Journal of Plant Science* 71, 783–789.

Janzen, D. (1973) Tropical Agroecosystems. *Science* 182, 1212–1219.

Jepson, P.C. (1989) *Pesticides and Non-target Invertebrates.* Intercept Limited, Andover.

de Jong, W. (1997) Developing swidden agriculture and the threat to biodiversity loss. *Agriculture, Ecosystems and Environment* 62, 187–197.

Jorgensen, R. and Anderson, B. (1994) Spontaneous hybridisation between oilseed rape (*Brassica napus*) and weedy *B. campestris* (Brassicaceae): a risk of growing genetically modified oilseed rape. *American Journal of Botany* 81, 1620–1626.

Joshi, L.M. (1986) Perpetuation and dissemination of wheat rusts in India. In: Joshi, L.M., Singh, D.V. and Srivastava, K.D. (eds) *Problems and Progress of Wheat Pathology in South Asia.* Malhotra Publishing House, Delhi, pp. 41–68.

Jutsum A.R. and Graham, J.C. (1995) Managing weed resistance: the role of the agrochemical industry. In: *Proceedings of the Brighton Crop Protection Conference – Weeds*, pp. 557–566.

Kenmore, P.E., Cariño, F.O., Perez, C.A., Dyck, V.A. and Gutierrez, A.P. (1984) Population regulation of the rice brown planthopper within rice fields in the Philippines. *Philippine Journal of Plant Protection in the Tropics* 1, 19–37.

Khan, R.P. (1977) Plant quarantine: principles, methodology, and suggested approaches. In: Hewitt, W.B. and Chiarappa, L. (eds) *Plant Health and Quarantine and the International Transfer of Genetic Resources.* CRC Press, Cleveland, Ohio.

Khan, Z.R., Ampong-Nyarko, K., Chiliswa, P., Hassanali, A., Kimani, S., Lwande, W. Overholt, W.A., Pickett, J.A., Smart, L.E., Wadhams, L.J. and Woodcock, C.M. (1997) Intercropping increases parasitism of pests. *Nature* 388, 631–632.

Kim, K.C. and McPheron, B.A. (eds) (1993) *Evolution of Insect Pests: Patterns of Variation.* John Wiley & Sons, New York.

Klingsman, G.C. and Ashton, F.M. (1982) *Weed Science: Principles and Practices.* John Wiley & Sons, New York.

Kogan, M. and Lattin, J.D. (1993) Insect conservation and pest management. *Biodiversity and Conservation* 2, 242–257.

Krattiger, A. (1998) *The Importance of Ag-biotech to Global Prosperity.* ISAAA Briefs No. 6, 10 pp.

Kromp, B. and Steinberger, K.-H. (1992) Grassy field margins and arthropod diversity: a case study on ground beetles and spiders in eastern Austria (Coleoptera: Carabidae; Arachnida: Aranei, Opiliones). *Agriculture, Ecosystems and Environment* 40, 71–93.

Krysan, J.L., Foster, D.E., Branson, T.F., Ostlie, K.R. and Cranshaw, W.S. (1986) Two years before the hatch: rootworms adapt to crop rotation. *Bulletin of the Entomological Society of America* 32, 250–253.

Kushwaha, S.P.S., Ramakrishanan, P.S. and Tripathi, R.S. (1981) Population dynamics of *Eupatorium oderatum* in successional environments following slash and burn agriculture. *Journal of Applied Ecology* 18, 529–535.

Lam, C.H., Lim, J.K. and Jantan, B. (1993) Comparative studies of a paraquat mixture and of glyphosate and/or its mixtures on weed succession in plantation crops. *The Planter, Kuala Lumpur* 69, 525–533.

Lane, J.A., Child, D.V., Reiss, G.C., Entcheva, V. and Bailey, J.A. (1997) Crop resistance to parasitic plants. In: Crute, I.R., Holub, E.B. and Burdon, J.J. (eds) *The Gene-for-Gene Relationship in Plant–Parasite Interactions.* CAB International, Wallingford, pp. 81–98.

Large, E.C. (1940) *The Advance of the Fungi.* Jonathan Cape, London.

LaSalle, J. (1991) Parasitic Hymenoptera, biological control and biodiversity. In: LaSalle, J. and Gauld, I.D. (eds) *Hymenoptera and Biodiversity.* CAB International, Wallingford, and The Natural History Museum, London, pp. 197–215.

Lenné, J.M. (1989) Evaluation of biotic factors affecting grassland production – history and prospects. Section Introduction Paper. In: *Proceedings of the XVI International Grasslands Congress,* Nice, October 1989, pp. 1811–1815.

Lenné, J.M. (1998) The biodiversity and conservation of crops for disease resistance. *Seventh International Congress of Plant Pathology,* Edinburgh, abstract 4.1.1S.

Lenné, J.M. and Brown, A.E. (1991) Factors affecting the development of pathogenic and weakly-pathogenic isolates of *Colletotrichum gloeosporioides* on leaf surfaces of *Stylosanthes guianensis. Mycological Research* 95, 227–232.

Lenné, J.M. and Wood, D. (1991) Plant diseases and the use of wild germplasm. *Annual Review of Phytopathology* 29, 35–63.

Li, Yank-Han (1987) The development and future trend of weed science in Mainland China. *Weed Technology* 1, 259–264.

Liebman, M. and Gallandt, E.R. (1997) Many little hammers: ecological management of crop–weed interactions. In: Jackson, L.E. (ed.) *Ecology in Agriculture.* Academic Press, San Diego, pp. 291–343.

Lockwood, J.A. and DeBrey, L.D. (1990) A solution for the sudden and unexplained extinction of the Rocky Mountain grasshopper (Orthoptera: Acrididae). *Environmental Entomology* 19, 1194–1205.

Madsen, M., Overgaard Nielsen, B., Holter, P., Pedersen, O.C., Brøchner Jespersen,

J., Vagn Jensen, K.M., Nansen, P. and Grønvold, J. (1990) Treating cattle with Ivermectin: effects on the fauna and decomposition of dung pats. *Journal of Applied Ecology* 27, 1–15.

Malik, R.K. and Malik, Y.S. (1994) Development of herbicide resistance in India. In: Sastroutomo, S.S. and Auld, B.A. (eds) *Appropriate Weed Control in Southeast Asia.* CAB International, Wallingford, pp. 64–82.

Marshall, D.R. (1977) The advantages and hazards of genetic homogeneity. *Annals of the New York Academy of Sciences* 287, 1–20.

Marshall, G. (1998) Herbicide-tolerant crops – real farmer opportunity or potential environmental problem. *Pesticide Science* 52, 394–402.

Masaba, D. and Waller, J.M. (1992) Coffee berry disease: the current status. In: Bailey, J.A. and Jeger, M.J. (eds) *Colletotrichum: Biology, Pathology and Control.* CAB International, Wallingford, pp. 237–249.

May, R.M. (1985) Evolution of pesticide resistance. *Nature* 315, 12–13.

McLaughlin, A. and Mineau, P. (1995) The impact of agricultural practices on biodiversity. *Agriculture, Ecosystems and Environment* 55, 201–211.

Mikkelsen, T.T., Anderson, B. and Jorgensen, R.B. (1996) The risk of crop transgene spread. *Nature* 380, 31.

Moody, K. (1975) Weeds in shifting cultivation. *PANS* 21, 188–194.

Moreby, S.J., Aebischer, N.J., Southway, S.E. and Sotherton, N.W. (1994) A comparison of the flora and arthropod fauna of organically and conventially grown winter wheat in southern England. *Annals of Applied Biology* 125, 13–27.

Morse, S. and Buhler, W. (1997) *Integrated Pest Management: Ideals and Realities in Developing Countries.* Lynne Rienner Publishers, Boulder, Colorado.

Moss, S.R. and Rubin, B. (1993) Herbicide resistant weeds: a worldwide perspective. *Journal of Agricultural Science, Cambridge* 120, 141–148.

Naidu, R.A., Bottenberg, H., Subrahmanyam, P., Kimmins, F.M., Robinson, D.J. and Thresh, J.M. (1998) Epidemiology of groundnut rosette disease: current status and future research needs. *Annals of Applied Biology* 132, 525–548.

Newhook, F.J. and Podger, F.D. (1972) The role of *Phytophthora cinnamomi* in Australian and New Zealand forests. *Annual Review of Phytopathology* 10, 299–326.

Nyoka, G.C. (1982) The influence of fallow period on weed vegetation and rice yields in Sierra Leone. *Tropical Pest Management* 28, 1–9.

Oereke, E.C., Dehne, H.W., Schonbeck, F. and Weber, A. (1994) *Crop Production and Crop Protection: Estimated Losses in Major Food and Cash Crops.* Elsevier, New York.

Orson, J.H. (1993) The penalties of volunteer crops as weeds. *Aspects of Applied Biology* 35, 1–7.

Padmanabhan, S.Y. (1973) The great Bengal rice famine. *Annual Review of Phytopathology* 11, 11–26.

Parker, C. and Polniaszek, T.I. (1990) Parasitism of cowpea by *Striga gesnerioides*: variation in virulence and discovery of a new source of host resistance. *Annals of Applied Biology* 37, 305–311.

Parker, C. and Riches, C.R. (1993) *Parasitic Weeds of the World: Biology and Control.* CAB International, Wallingford.

Perring, T.M., Cooper, A.D., Rodriguez, R.J., Farrar, C.A. and Bellows, T.S., Jr (1993) Identification of a whitefly species by genomic and behavioral studies. *Science* 259, 74–77.

Phillips, M.C. (1992) A survey of the arable weeds of Botswana. *Tropical Pest Management* 36, 13–21.

Phillips, M.C. (1993) Use of tillage to control *Cynodon dactylon* under small-scale farming conditions. *Crop Protection* 12, 267–272.

Pimentel, D., McLaughlin, L., Zepp, A., Lakitan, B., Kraus, T., Kleinman, P., Vancini, F., Roach, W., Graap, E., Keeton, W. and Selig, G. (1989) Environmental and economic impacts of reducing US agricultural pesticide use. *Handbook of Pest Management in Agriculture* 4, 223–278.

Polaszek, A. and Khan, Z.R. (1998) Host plants. In: Polaszek, A. (ed.) *African Cereal Stem Borers: Economic Importance, Taxonomy, Natural Enemies and Control.* CAB International, Wallingford, pp. 3–10.

Polaszek, A., Evans, G.A. and Bennett, F.D. (1992) *Encarsia* parasitoids of *Bemisia tabaci* (Hymenoptera: Aphelinidae, Homoptera: Aleyrodidae): a preliminary guide to identification. *Bulletin of Entomological Research* 82, 375–392.

Rajak, R.L. and Diwakar, M.C. (1987) Resurgence of cotton whitefly in India and its integrated management. *Plant Protection Bulletin, India* 39, 13–14.

Ramos, A.H. and Kamidi, R.E. (1982) Determination and significance of the mutation rate of *Colletotrichum coffeanum* from benomyl sensitivity to benomyl tolerance. *Phytopathology* 72, 181–185.

Ravi, K. and Anilkumar, T.B. (1991) Effect of cowpea phylloplane fungi on fungicide resistant strains of *Colletotrichum truncatum* (Schw.) Andrus & Moore. *Zentralblatt für Mikrobiologie* 146, 209–212.

Raybold, A.F. and Gray, A.J. (1993) Genetically modified crops and hybridization with wild relatives: a UK perspective. *Journal of Applied Ecology* 30, 199–219.

Riches, C.R. (1994) Variability of parasitic weeds and their hosts: implications for witchweed management in Africa. *Aspects of Applied Biology* 39, 145–154.

Riches, C.R., Twomlow, S.J. and Dhliwayo, H. (1997) Low-input weed management and conservation tillage in semi-arid Zimbabwe. *Experimental Agriculture* 33, 173–187.

Ridley, H.N. (1930) *The Dispersal of Plants throughout the World.* L. Reeve, Ashford.

de Rouw, A. (1995) The fallow period as a weed-break in shifting cultivation (tropical weed forests). *Agriculture, Ecosystems and Environment* 54, 31–43.

Safeeula, S. (1977) Genetic vulnerability: the basis of epidemics in India. *Annals of the New York Academy of Sciences* 287, 72–85.

Salvado, J.C., Bensaadi-Merchermek, N. and Mouches, C. (1994) Transposable elements in mosquitoes and other insect species. *Comparative Biochemistry and Physiology B, Biochemistry and Molecular Biology* 109, 531–544.

Sanchez, P.A. and Nurena, M.A. (1970) *Upland Rice Improvement Under Shifting Cultivation Systems in the Amazon Basin of Peru.* North Carolina Experiment Station, North Carolina (mimeo).

Sankula, S., Braverman, M.P. and Oard, J.H. (1998) Genetic analysis of glufosinate resistance in crosses between transformed rice (*Oryza sativa*) and red rice (*Oryza sativa*). *Weed Technology* 12, 209–214.

Schoenly, K.D., Cohen, J.E., Heong, K.L., Arida, G., Barrion, A.T. and Litsinger, J.A. (1996) Quantifying the impact of insecticides on food web structure of rice-arthropod populations in Philippine farmers' irrigated fields: a case study. In: Polis, G. and Winemiller, K. (eds) *Food Webs: Integration of Patterns and Dynamics.* Chapman & Hall, New York, pp. 343–351.

Settle, W.H., Ariawan, H., Astuti, E.T., Cahayana, W., Hakim, A.L., Hindayana, D., Sri Lestari, A. and Pajarningsih (1996) Managing tropical rice pests through conservation of generalist natural enemies and alternative prey. *Ecology* 77, 1975–1988.

Shawe, K.G. and Ingrouille, M.J. (1993) Isoenzyme analysis demonstrates host selection of parasite pathotypes in the association between cowpea and *S. gesnerioides*. *Proceedings of the Brighton Crop Protection Conference –Weeds*, pp. 919–924.

Shemshedini, L. and Wilson, T.G. (1990) Resistance to juvenile hormone and an insect growth regulator in *Drosophila* is associated with an altered cytosolic juvenile hormone-binding protein. *Proceedings of the National Academy of Sciences USA* 87, 2072–2076.

Sherriff, C., Lenné, J.M. and Julian, A. (1999) The importance of pre-penetration structures and infection events of *Colletotrichum* on aerial plant surfaces, and the potential for novel management strategies. *Plant Pathology* (in press).

Shield, I.F. and Bacon, E.T.G. (1998) A comparison of management regimes for one-year rotational set-aside with a sequence of winter wheat crops, and of growing wheat without interruption. 2. Effects on vegetation and weed control. *Journal of Agricultural Science, Cambridge* 130, 389–397.

Simmonds, N.W and Rajaram S. (eds) (1988) *Breeding Strategies for Rusts of Wheat.* CIMMYT, Mexico, DF.

Skovmand, B., Varughese, G. and Hettel, G.P. (1992) *Wheat Genetic Resources at CIMMYT: Their Preservation, Enrichment and Distribution.* CIMMYT, Mexico, DF.

Slaats, J.J.P. (1992) Transformation de system de culture itinerant en system de culture permanents. In: Vooren, A.P., Schork, W., Blokhuis, W.A. and Spiukerman, A.J.C. (eds) *Compte Rendu Seminaire l'Amenagement Integre des Forets Denses Humides et des Zones Agricoles Peripheriques, Tropenbos Series* 1. Tropenbos, Wageningen, pp. 123–132.

Smithson, J.B. and Lenné, J.M. (1996) Varietal mixtures: a viable strategy for sustainable productivity in subsistence agriculture. *Annals of Applied Biology* 128, 127–158.

Southwood, T.R.E. and Way, M.J. (1970) Ecological background to pest management. In: Rabb, R.L. and Guthrie, F.E. (eds) *Concepts of Pest Management.* North Carolina University Press, Raleigh, pp. 6–28.

Swanton, C.J. and Weise, S.F. (1991) Integrated weed management: the rationale and the approach. *Weed Technology* 5, 657–663.

Tabshnik, B.E. (1994) Evolution of resistance to *Bacillus thuringiensis*. *Annual Review of Entomology* 39, 47–79.

Tabashnik, B.E., Biao, L.Y., Finson, N., Masson, L. and Heckel, D.G. (1997) One gene in diamondback moth confers resistance to four *Bacillus thuringiensis* toxins. *Proceedings of the National Academy of Sciences USA* 94, 1640–1644.

Tally, S. (1996) The red rice riddle. *Purdue Agriculture* 1995–6, 9.

Tanksley, S.D. and McCouch, S.R. (1997) Seed banks and molecular maps: unlocking genetic potential from the wild. *Science* 277, 1063–1066.

Terry, P.J. (1991) Grassy weeds – a general overview. In: Baker, F.W.G. and Terry, P.J. (eds) *Tropical Grassy Weeds.* CAB International, Wallingford, pp. 5–38.

Terry, P.J. (1996) The use of herbicides in the agriculture of developing countries. *Proceedings of Second International Weed Control Congress*, Copenhagen, pp. 601–609.

Thomas, M. and Waage, J. (1996) *Integrating Biological Control and Host Plant Resistance Breeding*. CTA, Wageningen.

Thresh, J.M. (ed.) (1981) *Pests, Pathogens and Vegetation*. The Pitman Press, Bath.

Thurston, H.D. (1992) *Sustainable Practices for Plant Disease Management in Traditional Systems*. Westview Press, Boulder, Colorado.

Timmons, A.M., O'Brian, E.T., Charters, Y.M., Dubbels, S.J. and Wilkinson, M.J. (1995) Assessing the risks of wind pollination from fields of genetically modified *Brassica napus* spp. *oleifera*. *Euphytica* 85, 417–423.

Triplett, G.B. and Lytle, G.D. (1972) Control and ecology of weeds in continuous corn grown without tillage. *Weed Science* 20, 453–457.

Tripp, R. (1996) Biodiversity and modern crop varieties: sharpening the debate. *Agriculture and Human Values* 13, 48–63.

Trutmann, P., Voss, J. and Fairhead, J. (1993) Management of common bean diseases by farmers in the central African highlands. *International Journal of Pest Management* 39, 334–342.

Valverde, B.E. (1996) Management of herbicide resistant weeds in Latin America: the case of propanil-resistant *Echinochloa colona* in rice. In: *Proceedings of Second International Weed Control Congress,* Copenhagen, pp. 415–420.

Verkerk, R.H.J., Leather, S.R. and Wright, D.J. (1998) The potential for manipulating crop-pest–natural enemy interactions for improved insect pest management. *Bulletin of Entomological Research* 88, 493–501.

Vernon, R. (1979) Weed control in Zambia's maize. In: *Proceedings of the 7th East Africa Weed Science Society Conference*, Nairobi, Kenya.

Waage, J.K. (1991) Biodiversity as a resource for biological control. In: Hawksworth, D.L. (ed.) *The Biodiversity of Microorganisms and Invertebrates: Its Role in Sustainable Agriculture.* CAB International, Wallingford, pp. 149–163.

Walters, M.C. (1975) Evolution in tillage techniques and impact on entomological research, with special reference to the maize stalk borer, *Busseola fusca* (Fuller). In: *Proceedings of the First Congress of the Entomological Society of Southern Africa*, pp. 235–244.

Warwick, S.I. and Black, L.D. (1983) The biology of Canadian weeds. 61. *Sorghum halepense* (L.) Pers. *Canadian Journal of Plant Science* 62, 997–1014.

Way, M.J. (1979) Significance of diversity in agroecosystems. In: *Proceedings of the Opening Session and Plenary Session Symposium, IXth International Congress of Plant Protection*, Washington, DC, pp. 9–12.

Way, M.J. and Heong, K.L. (1994) The role of biodiversity in the dynamics and management of insect pests in tropical irrigated rice – a review. *Bulletin of Entomological Research* 84, 567–587.

Wilcut, J.W., Coble, H.D., York, A.C. and Monks, D.W. (1996) The niche for herbicide resistant crops in US agriculture. In: Duke, S.O. (ed.) *Herbicide Resistant Crops.* Lewis Publishers, Boca Raton, Florida, pp. 213–230.

William, R.D. (1981) Complementary interactions between weeds, weed control practices, and pests in horticultural cropping systems. *HortScience* 16, 508–513.

Williams, T. and Polaszek, A. (1996) A re-examination of host relations in the Aphelinidae (Hymenoptera: Chalcidoidea). *Biological Journal of the Linnean Society* 57, 35–45.

Williamson, M. (1996) *Biological Invasions*. Chapman & Hall, London.

Wills, C. (1996) Safety in diversity. *New Scientist* 2002, 38–42.

Wilson, A.K. (1981) *Euphorbia heterophylla*: a review of distribution, importance and control. *Tropical Pest Management* 27, 32–38.

Witcombe, J.R. and Harris, D. (1998) The impact of transgenic food crops and the environment – some issues for the developing world. In: *Seventh International Congress of Plant Pathology*, Edinburgh, abstract 4.9.2S.

Wood, D. and Lenné, J.M. (1997) The conservation of agrobiodiversity on-farm: questioning the emerging paradigm. *Biodiversity and Conservation* 6, 109–129.

Yardam, D.J. and Gladders, P. (1993) Effect of volunteer plants on crop diseases. *Aspects of Applied Biology* 35, 75–82.

Zeigler, R.S., Leong, S.A. and Teng, P.S. (1994) *Rice Blast Disease*. CAB International, Wallingford.

Zimdahl, R.L. (1980) *Weed–Crop Competition: a Review*. Oregon State University, Corvallis, Oregon.

The Effects of Alternative Tillage **12** Systems on Biodiversity in Agroecosystems

P.J. Edwards[1], C. Abivardi[1] and W. Richner[2]

[1]*Geobotanisches Institut, ETH, Zürichbergstrasse 38, 8044 Zürich, Switzerland;* [2]*Institute of Plant Sciences, ETH-Zentrum, 8092 Zürich, Switzerland*

Introduction

A revolution is occurring in the way we think about agricultural systems. For most of the 20th century, agricultural research has primarily been concerned with maximizing production, and agricultural systems have been viewed, first and foremost, as production systems. We can characterize this view of agriculture, in which the principal focus of interest is the quantity of food produced (i.e the output) in relation to the quantity of inputs (measured either in financial terms or as the amounts of agrochemicals used), as the *production paradigm*. Many of the concepts used to characterize agricultural systems, and even the language used to describe them, have been borrowed from economics and manufacturing industry. One of the features of this paradigm is that the criteria for judging the effectiveness of a particular agricultural practice are relatively narrow and short term, and concerned chiefly with the relationship between inputs and outputs. Furthermore, research into different aspects of agricultural production has often been relatively separate. For example, plant breeders were principally concerned with increasing yields, but paid little attention to the consequences of new varieties for insect pest problems or fertilizer use. These problems were taken up subsequently by other specialists, often working in different research institutions (see Polaszek *et al.*, Chapter 11, this volume).

The revolution has been to regard agricultural systems as ecosystems, in which internal processes such as nutrient recycling, maintenance of soil fertility and the regulation of populations of insect pest by natural enemies are important for the sustainability of the system. Obviously, the production of food continues to be the major objective of agriculture, but within what may be called the *ecosystem paradigm*, research focuses not only on how management practices affect inputs and outputs, but also the structure and functioning of the system as a whole. This new paradigm emphasizes the interconnectedness of the various components of an agroecosystem – for example the crop, the soil subsystem, and populations of antagonists and beneficial organisms (see Wood and Lenné, Chapters 2 and 18, this volume). The concern to reduce inputs lies not simply in short-term economic advantage, but also in the realization that these inputs may be damaging for the sustainability of the system. Indeed, it was the accumulating adverse consequences of production-focused approaches that forced the shift to an ecosystem paradigm for agriculture. For example, the use of pesticides has had consequences for many compartments of the agroecosystem. It has led not only to the evolution of pesticide-resistant pests (Lemon, 1994), pathogens (Russell, 1995) and weeds (Powles *et al.*, 1997), but also to a drastic reduction in populations of natural enemies (Pimentel and Andow, 1984; Pimentel *et al.*, 1992, and Polaszek *et al.*, Chapter 11, this volume). It can also affect the decomposer system and hence soil conditions and nutrient turnover (Simon-Sylvestre and Fournier, 1979; Pimentel and Andow, 1984; see Wardle *et al.*, Chapter 5, this volume).

Increasingly, agricultural management is viewed from an ecosystem perspective; even when certain cultural practices are introduced to deal with a specific problem (e.g. soil compaction or weed control), there is a need to understand their consequences for the functioning of the system as a whole. An excellent example of the value of an ecosystem approach is in the use of alternative cultivation techniques such as conservation tillage in place of conventional tillage by ploughing. In this chapter we examine in detail the effects of using alternative tillage systems for wheat before taking a briefer look at experiences with maize and rice. In each case, we examine the perceived problems of conventional tillage which were the motivation for introducing alternative methods before considering the consequences of these practices for various aspects of the agroecosystem. In the final section we attempt to draw some general lessons from the effects of alternative methods of tillage about the functioning of agroecosystems (see Wardle *et al.*, Chapter 5, this volume).

Wheat

Conservation tillage systems for wheat

One of the major problems associated with the intensified wheat production of recent years is deterioration of soil structure. Although conventional tillage with ploughing creates a favourable environment for germination and crop establishment, which is the primary objective of tillage, it often leads to soil compaction, a decrease in organic matter content, and soil losses due to erosion (Hakansson, 1994). Furthermore, a very fine seedbed tends to reduce soil aeration, especially under wet conditions, and may thus reduce early wheat growth and the tolerance of plants to abiotic and biotic stresses.

Losses of nitrogen are also a major problem in some intensive wheat areas. Although wheat is not commonly thought of as a crop with a high potential for nitrate leaching, a considerable amount of leaching may occur during winter when the demand of the wheat crop is low but conditions for nitrogen mineralization may be favourable (moist, not too cold soil). Winter leaching losses of 50 kg N ha^{-1} are common (Jenkinson, 1986), but much higher losses have been reported (Whitmore and Addiscott, 1986). The main source of nitrogen leached at this time appears to be residual soil nitrogen from previous crops (e.g. potatoes) in excess of what can be utilized by the small wheat seedlings. Furthermore, intensive tillage for seedbed preparation may enhance mineralization and thus leaching potential. Repeated stimulation of mineralization as a result of ploughing tends to reduce the soil organic content in the ploughed layer in the long term, especially if straw and other harvest residues are regularly removed from the field.

An important advance in agricultural technology, encouraged by the severe wind erosion of soil in the North American Great Plains in the 1930s (Blevins and Frye, 1993) and the energy crisis in the early 1970s (Ilnicki and Enache, 1992), was the development of reduced tillage practices as an alternative to conventional cultivation by ploughing. The terminology used to describe these practices is a little confusing. As defined by OECD (1995), conservation tillage comprises systems of minimum tillage or no-tillage, in which the seed is placed directly in the soil with little or no preparatory cultivation. In such practices the mouldboard plough is replaced by non-inverting tillage implements and the number of tillage operations is reduced. The most extreme form of conservation tillage is no-tillage or direct drilling, characterized by the absence of any soil tillage between the harvest of the preceding crop and new seeding (Baeumer and Bakermans, 1973). According to Baker *et al.* (1996), conservation tillage is the collective 'umbrella term' commonly given to no-tillage, reduced-drilling, minimum-tillage and/or ridge tillage, and denotes that the practice has a conservation goal of some nature. The retention of at least 30% surface cover by residues from the previous crop at the time of planting or during the non-crop period is often regarded as the minimum criterion for conservation-tillage (CTIC, 1992). In fact this is a somewhat misleading definition since the goal of minimum tillage may not be the conservation of residues or topsoil, but rather reduction of labour, fuel or machinery costs.

Conservation tillage techniques have been increasingly used for wheat cultivation since the 1970s (Cannell, 1984; Davies, 1988; Christian and Ball, 1994). In such tillage systems, the mouldboard plough is most often replaced by chisel ploughs or rotary hoes for primary tillage. Secondary tillage, i.e. the preparation of the seedbed, is done with rotary hoes or different harrow

forms, either simultaneously with primary tillage using combined cultiva-
tors or in one or more additional tillage operations. No-tillage, the most
extreme form of minimum tillage, does not involve any tillage operations for
seedbed preparation at all. Instead, the wheat seed is delivered into a slot
which is formed by a disc or tine.

If properly done, conservation tillage systems improve the soil structure
(Lal, 1989) and alleviate some of the negative side-effects of conventional
tillage and repeated heavy traffic, especially by reducing soil compaction and
erosion and by conserving soil moisture (Unger and Cassel, 1991). The
practice is of particular benefit on medium- and fine-textured soils. However, the effects of minimal tillage are not solely upon the physical properties
of the soil. In the following sections we demonstrate how the form of tillage
affects the structure and functioning of all compartments of the agroecosystem.

The crop

Many studies have investigated the effect of conservation tillage upon yield.
In several long-term experiments in Switzerland, only small effects of tillage
on wheat yields were found (Vez, 1977; Maillard and Vez, 1988; Anken et al.,
1997), but on a heavy soil, yield reductions were reported for conservation
tillage systems (Anken et al., 1997). Similarly, in a 10-year experiment in
Belgium, Frankinet et al. (1979) found only minor differences in winter
wheat yields between deep mouldboard tillage (30 cm), shallow mouldboard tillage (15 cm) and direct drilling. In the United Kingdom, long-term
average yields of winter cereals in conservation tillage systems rarely
differed from those with conventional tillage systems with mouldboard
ploughing, whereas soil compaction, weed infestation and difficulty of
sowing into residues caused problems in no-tillage systems (Ball, 1990).
Christian and Bacon (1990) reported that, in England, yields of winter
wheat in plots that were either mouldboard ploughed to 20 cm, tine
cultivated to 7 cm or non-tilled did not differ when stubble was burned
prior to wheat sowing. On the other hand, minimum tillage lowered wheat
yields when the straw was not removed. Nuttall et al. (1986) found no
significant effect of tillage (mouldboard ploughing, chisel ploughing,
one-way and double disc cultivation) on spring wheat yields in a 24-year
monoculture experiment in the Canadian prairie (Saskatchewan). In
experiments in Prince Edward Island (Canada), Carter (1991) found similar
spring wheat yields with mouldboard ploughing, shallow tillage with rotary
harrow or disc harrow, and no-tillage in years with normal amounts of
rainfalls. Kirkegaard (1995), reviewing medium- and long-term tillage
experiments in Australia, found only small differences in wheat yield
between direct drilling and conventional tillage. However, effects of stubble
management on wheat yields were larger than tillage impacts. In all regions,

stubble retention reduced yields compared with burning. In contrast, in a 9-year trial at two sites in New Zealand, wheat yields tended to be greater under no-tillage compared with conventional tillage (Francis and Knight, 1993). From all these studies we can conclude that minimum tillage does not have a consistent effect on wheat grain yields, although there may be significant year-to-year variation in the effects of tillage and residue management on wheat yield (Kirkegaard, 1995).

In cases where the direct sowing of wheat does lead to yield reductions compared with conventional tillage, the reason is usually reduced early growth. Reduced tillage is associated with generally lower soil temperatures (especially in combination with surface plant residues) and reduced soil penetrability, both important soil factors that tend to reduce the germination and emergence of seedlings (Addae *et al.*, 1991). Residue management appears to be the key factor in the yield response of minimum tillage systems in humid temperate regions (Carter, 1994) as it strongly affects germination and early growth of crops. Large amounts of crop residues on the soil surface in minimum tillage systems may impede seeding operations, delay drying and warming of the soil after winter, release allelopathic substances that decrease plant growth, increase incidence of crop pests and diseases, and reduce the efficacy of fertilizers (especially nitrogen) and pesticides (e.g. soil applied herbicides) (Carter, 1994). Consequently, wheat yield reductions are mainly observed with high levels of crop residues; for this reason, removal of crop residues is often seen as essential in humid, temperate areas (Christian and Ball, 1994). However, in subtropical and tropical areas, a lowering of soil temperatures by minimum tillage may even be beneficial.

If there are yield reductions after conversion to reduced tillage, they commonly occur in the first few years but later decline or completely disappear (Gooding and Davies, 1997). This is attributed to positive changes in soil characteristics that compensate for the negative impacts on crop growth of increased bulk density and greater resistance to penetration (soil strength). Very important in this respect are increasing numbers of vertically oriented, continuous biopores that are created mainly by plant roots and earthworms. As these biopores are not annually disrupted by ploughing in no-tillage systems, exploration of the soil by plant roots and water infiltration and percolation improve steadily when non-inversion tillage is introduced (Pearson *et al.*, 1991). A gradual increase in soil organic matter with time in no-tillage soils (Ellis and Howse, 1980; Ball *et al.*, 1989) may also compensate for possibly lower N mineralization rates in non-tilled soils (Dowdell and Cannell, 1975), so that nitrogen mineralization rates are similar to those in regularly tilled soils. As a result, any initial decline in N availability to crops after conversion to no-tillage (Dowdell and Crees, 1980; Vaidyanathan and Davies, 1980) may be of short duration.

Weeds

The adoption of no-tillage practices has major consequences for the population dynamics of some arable weeds, and for the composition of the weed community in general (see Polaszek *et al.*, Chapter 11, this volume). Since one of the reasons for ploughing is to control weeds, it is not surprising that weed problems tend to be greater under conservation tillage systems. Indeed, ineffective weed control (e.g. of downy brome *Bromus tectorum*; see Rasmussen, 1995) has been one of the dominant impediments to the adoption of conservation tillage by wheat growers in many wheat-growing areas of the world (e.g. Young *et al.*, 1996).

A persistent seed-bank is a feature of many agricultural weed species, and under conventional tillage most of these seeds are buried well below the soil surface; it is only when they are returned to the surface by subsequent ploughing that they find conditions suitable for germination. While ploughing tends to produce a uniform vertical distribution of seeds, conservation tillage concentrates seeds at the soil surface (e.g. Macchia *et al.*, 1996). For example, in a study in Argentina Feldman *et al.* (1997) found the smallest seed-banks under conventional ploughing, but with no difference in seed density or composition at 0–5 cm and 5–10 cm depth. In contrast, under various forms of conservation tillage there were larger seed-banks, which tended to be concentrated in the upper part of the soil profile. Diversity of seed-banks increased from mouldboard plough, to disc, to chisel, to no-tillage, supporting the hypothesis that the systems causing less disturbance allow the build-up of a larger and more diverse soil seed-bank (see Feldman *et al.*, 1997).

Long-term experiments have shown that grass weed populations tend to increase under conservation tillage, especially in Europe (Cannell and Hawes, 1994). Among the species that increase in Europe, *Bromus* spp. and *Alopecurus myosuroides* are of particular importance; they are more effectively controlled by conventional tillage because their small seeds do not survive for many years after being ploughed in (Gooding and Davies, 1997). Similarly, most perennial weeds are unable to persist in fields which are regularly ploughed, but can become a problem under conservation tillage. For example, an over 20-year-old no-tillage experiment with winter wheat in rotation with winter rape and maize (at Changins, Switzerland) revealed that the reduction in depth and intensity of tillage in the unploughed plots increased populations of perennial weeds such as *Polygonum convolvulus, Cirsium arvense, Agropyron repens* and *Equisetum arvensis* (Mayor and Maillard, 1995).

Pests

In contrast to the bare soil environment of conventional tillage, the lack of soil disturbance and the presence of surface crop residues in conservation tillage provide a potential habitat for pests. Slugs are often the most serious problem. In a long-term no-tillage experiment at the Federal Agricultural Research Station at Changins, Switzerland, Maillard and Vez (1993) report increased slug damage to winter wheat, rape and maize, though damage was economically important only in crops with a low plant density such as maize and rape.

Residues may be particularly important for the survival of pests through the winter. For example, Morrill *et al.* (1993) found that mortality of over-wintering larvae of the wheat stem sawfly (*Cephus cinctus*, Hymenoptera: Cephidae) at low temperatures in Montana (USA) was over 92% in wheat stubble exposed on the soil surface by tillage while there was 91–100% survival of larvae in untilled stubble. Slower seedling emergence as a result of lower soil temperatures in spring increases the time of exposure to pest attack (All and Musick, 1986). An indirect effect of conservation tillage in the northern USA is that some pests have become more serious because the increased use of herbicides has eliminated their preferred hosts (All and Musick, 1986).

Antagonists

If conservation tillage benefits insect pests, then it may also benefit the invertebrates which prey on them. One important group are large carabids, which can play a significant role in preventing outbreaks of pests in crops under conservation tillage but are vulnerable to conventional tillage (All and Musick, 1986; Brust *et al.*, 1986). Hance and Wibo (1987) found that carabid beetles which breed in the spring are very sensitive to autumn ploughing and bare soil in winter. Other work has shown that various carabid species react differently to the cultivation techniques according to the crop concerned (winter wheat, winter barley, maize and sugar beet) and whether they are spring or autumn breeders (Stassart and Wibo, 1983). Direct drilling increases the populations of some species, especially by leaving in place a thick layer of organic material on the soil that protects the developing insects both from winter cold and from insecticides. Since conventionally tilled fields do not provide suitable overwintering habitats for these beetles, they move into the crop from adjacent non-agricultural habitats in the spring. This takes time, during which young crop seedlings are more vulnerable to pest infestation.

Carabids are also predators of weed seeds, though the extent to which they can affect the population dynamics of plants is uncertain. Brust (1994) observed that seed-predators such as carabids and field crickets (known to be enhanced by no-tillage) preferred to feed on the seeds of broadleaved species, and suggested that this might be a reason for their reduced abundance in some low-input no-tillage systems. It was found that approximately 2.3 times more seeds were consumed in no-tillage systems than with conventional (Brust and House, 1988). However, in greenhouse experiments, it was found that decreasing seed-predators by 50% had no effect on the yields of broadleaved weeds compared with controls without seed-predators.

In practice, although populations of carabids and other beneficial insects may be favoured by no-tillage, this tendency is counteracted because higher applications of pesticides are used (Roger and Gupta, 1995). Outbreaks of slugs in these systems also dictate frequent application of molluscicides (Brust, 1990; Samsoe-Petersen *et al.*, 1992). In field trials, the total herbicides glyphosate and paraquat, which are frequently used in no-tillage systems, significantly reduced carabid populations. Brust (1990) found that large carabids (> 10 mm length) did not return to paraquat- and glyphosate-treated field areas until approximately 28 days after application. In greenhouse studies he also showed that simazine and atrazine had a repellent effect on carabids.

Plant diseases

The influence of tillage system upon the incidence of plant diseases is complex. Although there is abundant evidence that conservation tillage practices do affect the incidence of crop pathogens, it is not possible to generalize about what kind of response occurs under particular forms of tillage.

The complexity is well illustrated by the case of take-all on wheat, caused by *Gaeumannomyces graminis* var. *tritici*. Two reports from Canada (Cook and Haglund, 1991; Cook, 1992) demonstrated that the poor performance of wheat planted directly into wheat or barley stubble was the result of take-all, together with rhizoctonia and pythium root rots (caused by *Rhizoctonia solani* AG8 and *Pythium* spp.). These diseases were apparently favoured by the combination of no crop rotation together with surface residues that keep the soil cool and moist. Cooler soil temperatures are known to increase take-all incidence on winter wheat (Bockus *et al.*, 1994) and, in addition, the period when seedlings are vulnerable to attack is extended because of their slower growth in spring (All and Musick, 1986). In contrast, in a study conducted in Australia, Cotterill and Sivasithamparam (1988) found that inoculum of the take-all fungus was most severe in deep-cultivated plots and least in direct-drilled plots, even though the postharvest propagule numbers were greatest in the direct-drilled and deep-cultivated treatments and least in shallow-cultivated plots. Many factors interacting with the form of tillage are probably responsible for such contrasting results. One of them is crop succession. Sattin (1994) found that following the lucerne/wheat (winter wheat) sequence, chiselling conserved a high surface inoculum (73% infected roots), whereas ploughing preserved a low inoculum (3–13%) throughout the soil horizon. On the other hand, following wheat/wheat sequence, chiselling preserved a very high surface inoculum (100%) but the inoculum level was low at > 20 cm depth (3%), whereas ploughing distributed a high inoculum level throughout the soil layers (19–38%). In the field it was shown that the percentage of infected plants is increased with host frequency in the crop succession. Where the inoculum level was high because the previous crop was also a host, disease was increased by chiselling and decreased by ploughing.

Other diseases of wheat are also affected – some positively, some negatively – by the form of tillage. For example, Sutton and Vyn (1990) found that tan spot (caused by *Pyrenophora tritici-repentis*) and blotch (caused by *Phaeospheria nodorum*) were increased, while septoria tritici blotch (caused by *Mycosphaerella graminicola*) was suppressed under minimum or zero tillage. The opposite occurred when wheat followed other crops in all tillage treatments (i.e. barley, oats, maize, soybean or lucerne) or followed wheat under conventional tillage. Similarly, while eye spot (*Pseudocercosporella herpotrichoides*) may decrease in minimum tillage systems (Maillard and Vez, 1993), rust (*Puccinia recondita*) and blotch (*P. nodorum*) (Bailey and Duczek, 1996; Stover *et al.*, 1996) tend to increase with stubble retention, particularly where the next crop is susceptible to the same disease as the previous crop.

While large amounts of organic residues on the soil surface may harbour pathogens (Boosalis *et al.*, 1986), they may also harbour antagonists (Pfender, 1988; Pfender *et al.*, 1993). For example, under laboratory conditions, a fast-growing basidiomycete, *Limonomyces roseipellis*, present in the microbial community of wheat straw was able to reduce ascocarp and ascospore production by *Pyrenophora tritici-repentis*, a stubble-borne pathogen of winter wheat (Pfender, 1988). Pfender *et al.* (1993) also showed that *L. roseipellis* could significantly reduce residue-borne primary inoculum of *P. tritici-repentis*. Nevertheless, although a 60–80% reduction of inoculum was achieved, greater reductions would be necessary for acceptable biological control of this disease (Pfender *et al.*, 1993).

The soil subsystem

The influence of tillage upon populations of earthworms is particularly important for the structure and functioning of the soil subsystem, since these organisms function as ecological engineers (Jones *et al.*, 1994; see Wardle *et al.*, Chapter 5, this volume). Numerous studies reveal the positive effect of conservation tillage on the population dynamics of earthworms (e.g. Wyss and Glasstetter, 1992; Bligh, 1994; Fraser, 1994). In one case, the positive effect was greatest for *Lumbricus rubellus* and *L. terrestris*, whose populations increased by 66 and 9 times, respectively (Andersen, 1987). The contrasting effects of tillage on different earthworm species appear to be related to their burrowing behaviour (Sochtig and Larink, 1992; Wyss and Glasstetter, 1992). Conservation tillage systems have a positive effect on vertically burrowing (*anecic*) species (Wyss and Glasstetter, 1992). An increased abundance of these species improves soil quality and crop growth by increasing the biological activity and nutrient availability, and facilitating deep root penetration (Ehlers *et al.*, 1980, 1983) and water infiltration (Ehlers, 1975; Willoughby *et al.*, 1997). Vertically oriented burrows of *Lumbricus terrestris* in no-tillage fields have been shown to provide preferential flow paths for water and agrochemicals (e.g. atrazine), especially during intense summer storms (Edwards *et al.*, 1992, 1993). Furthermore, earthworm activity is reported to induce substantial changes in the soil characteristics of the burrow walls (namely, pH, clay content, total- and water-soluble organic C, and alkaline phosphatase activity) that may affect soil fertility and pesticide sorption (Stehouwer *et al.*, 1993).

Tillage also has profound effects on the microbial composition of the soil. Its influence can be both direct, through damage and disturbance, and indirect, through altering the physical and chemical properties of the soil. While tillage favours organisms in the bacterial-based compartment of soil food webs, no-tillage systems favour the fungal-based compartment (see Wardle *et al.*, Chapter 5, this volume). Thus, fungal-based food webs tend to be more important for decomposition in no-tillage systems, while bacterial-based food webs predominate in conventional tillage systems. Within a given soil, there is considerable variation with depth in the composition of the microbial community. In no-till soils, microbial activities differ drastically with depth, with the greatest microbial activity occuring near the surface; in tilled systems, activities are more evenly distributed throughout the plough layer (Doran, 1980; Kennedy and Smith, 1995). Nevertheless, neither the time course nor the outcome of changing management practices are well understood. Only when diversity studies of the total community are conducted will we be able to understand the type and magnitude of the ecological disturbance and the means to return the soil to its near-original state (Kennedy and Smith, 1995).

Ecosystem function

The wide range of effects of tillage upon the biota of agroecosystems are reflected at the ecosystem level in the patterns of nutrient turnover. For environmental reasons, the influence on nitrogen balance has received most attention. Doran and Power (1983) found that differences in microbial populations between minimum and conventionally cultivated maize and wheat in the USA were related to changes in soil moisture content and soil aeration. Reduced soil aeration in conservation tillage resulted in less nitrate-N and increased denitrification potential. Jacinthe and Dick (1997) also found that the seasonal N_2O loss from chisel-till plots were significantly higher than from no-till (NT) or ridge-till plots. Studies of Rodriguez and Giambiagi (1995), investigating factors involved in denitrification in tillage and no-tillage pampean soils under wheat cultivation, also revealed that under NT treatment N_2O losses increased from the 0–10 cm sampling depth through the impact of NT on mineralizable carbon and water field pore space.

Studies of Wilson *et al.* (1995) in western Tennessee and Kentucky (USA) on the effect of tillage and cover crop on nitrate leaching (in a soybean/wheat/maize (SWM) rotation), revealed that leaching was greatest during the winter and spring when nitrate concentrations were low. While the tillage systems did not significantly affect the annual nitrate leaching, cropping system and rainfall timing had pronounced effects. Wheat managed as a grain crop was the most susceptible to nitrate losses in the SWM rotation. Meek *et al.* (1995), studying the effects of crop sequence and tillage on nitrate leaching under furrow irrigation, report that the soil nitrate-N in the 1.35–3.3 m depth after 2 years was 21 kg ha^{-1} greater for the maize/wheat rotation under conventional tillage than under the no-till system. In addition, the groundwater NO_3-N concentration in a no-tillage system (NT: 4 year rotation of maize/winter wheat/soybean/soybean) was found to be much lower under a low-input sustainable agriculture system (LISA system) than under no-tillage and at a depth of 180 cm the mineral content under NT was almost double that observed under the LISA system (see Craig and Weil, 1993).

Maize

Both the motivation for using alternative forms of tillage with maize and the consequences for the ecosystem are generally similar to those with wheat. Soils under maize are particularly vulnerable to erosion because of the wide spacing of the rows and because in some production systems the soil remains unprotected for up to 8 months of the year. Conservation tillage systems (Plate 14) are often recommended to decrease soil degradation and erosion associated with intensive row cropping. Recent studies of Meyer *et al.* (1997) on the effect of different crop production systems to control erosion and reduce runoff show how no-tillage can reduce soil loss under maize by >80% compared with a conventionally tilled system. Surface runoff is also greatly reduced by no-till. In one experiment, a combination of no-till maize and a vetch cover reduced runoff to levels <75% of those in a conventional system.

As with wheat, eliminating tillage causes major shifts in weed populations (Triplett and Lytle, 1972). In untilled fields in the USA, perennial species such as poison ivy (*Rhus radicans*), horsenettle (*Solanum carolinense*), trumpet creeper (*Campsis radicans*), and tree seedlings can become especially troublesome.

As conservation tillage practices have gained acceptance in the US corn belt, slug problems have also increased (Hammond *et al.*, 1996). However, no-tillage systems may also enhance populations of at least some antagonists. The effect of different tillage methods upon carabid beetle assemblages proves to be species dependent (Tonhasca, 1993). For example, while *Harpalus pensylvanicus* and *Cyclotrachelus sodalis* were more abundant in no-till plots, *Pterostichus chalcites* and *Scarites substriatus* were more common with conventional tillage. Similarly, Bigler *et al.* (1995) found no impact of cropping system on populations of Chrysopidae and Anthocoridae in maize, whereas the number of carabids caught in pitfall traps was highest in one of the no-tillage treatments (no-tillage, rye sown in autumn, rye mulched in spring before drilling maize in rotavated bands, and herbicides applied in drill-bands). Predation on insect larvae and pupae placed as baits on the soil and on maize plants increased with decreasing tillage and with increasing green or dead cover on the soil. Predation under no-tillage treatments was up to four times as high as in treatments with conventional ploughing

Surface residues associated with minimum tillage may also favour certain fungal and bacterial pathogens of maize. For example, residues act as the primary source of initial inoculum of *Exserohilum turcicum* (the causal agent of the northern leaf blight) and support high populations of *Colletotrichum graminicola* (which causes anthracnose) and *Phyllosticta maydis* (the causal agent of yellow leaf blight) (Boosalis *et al.*, 1986). Furthermore, surface residues which harbour *Pseudomonas syringae* (the causal agent of holcus leaf spot of maize) play a key role in the survival and spread of this bacterial pathogen.

Changes in the soil subsystem under conservation tillage, and the great reduction in surface runoff have important consequences for various ecosystem processes. For example, rates of denitrification may be twice as high as under conventional tillage (Palma *et al.*, 1997), apparently because of changes in soil moisture content and soil aeration (Doran and Power, 1983). The movement and fate of pesticides and other agrochemicals is also greatly altered. By means of long-term lysimeter studies, Gish *et al.* (1995) showed that under no-tillage all of the applied agrochemicals (including atrazine, alachlor and nitrate) initially moved deeper into the soil profile. In the soil matrix, however, the agrochemicals disappeared more rapidly under no-till than with conventional tillage, possibly due to enhanced biological activity. Pesticides and nitrate concentrations at about 160–190 cm were always less on no-till fields. Similarly, Isensee and Sadeghi (1995) found that the amounts of three herbicides (atrazine, alachlor and cyanazine) recovered in the surface 10 cm of no-till soil were half that under conventional tillage. This was partly because the crop residues intercepted much of the applied herbicide, but also because of differences in the patterns of herbicide leaching; these are complex and vary according to the chemical in question (Sadeghi and Isensee, 1997).

Paddy Rice

The conventional tillage for paddy farming consists of dry tillage and puddling (i.e. a tillage under flooded conditions). For example, in Japan dry tillage is conducted in autumn or winter and once again in spring just before puddling; both procedures are usually conducted with a rotary tiller fitted to a tractor. The main objective of puddling is to minimize seepage loss and percolation rate by destroying the soil structure (Koenigs, 1963). Fine soil particles are dispersed as the clods are destroyed and fill soil cracks and macropores. In addition, both compression by the wheels and the scraping action of the rotary tiller play a role in eliminating macropores. Other objectives of puddling are to level the soil surface, to control weeds, and to mix residual plants and manure into the soil.

In several countries and for a variety of reasons, alternatives to conventional tillage have been introduced or at least tested (e.g. Korea, Japan, Pakistan, Nigeria). In Japan, minimum tillage has been investigated as a way of reducing labour costs and as a solution to particular site problems (Kaneta, 1997). For example, it may facilitate seedling establishment in heavy clay soils where tillage produces large clods, it may improve trafficability in ill-drained fields, and it may enable a paddy–upland rotation by avoiding soil damage under wet conditions. However, the proportion of paddy fields in Japan where minimum tillage is practised is very limited, amounting to only 500 ha in 1995.

Minimum tillage has major effects upon the soil subsystem. Soil hardness tends to increase (Naganoma *et al.*, 1991; Takahashi, 1993; Choi-Chung *et al.*, 1996) and the soil becomes more structured. The soil also becomes more aerobic (Naganoma *et al.*, 1991), partly because straw is not mixed into it and partly as a result of cracks and biopores left by the roots of previous crops (Sato, 1992). The morphology and continuity of the network of biopores strongly influence water permeability of soil. Non-tilled soils are more permeable and, as a result, gleying tends to be suppressed.

These changes in the redox conditions of the soil have important implications for microbial processes. In conventional tillage systems these are dominated by bacterial reduction processes such as denitrification (e.g. by *Pseudomonas aeruginosa* and *P. fluorescens*; Li *et al.*, 1992), methanogenesis (Min *et al.*, 1997), and sulphate reduction (e.g. by *Desulfovibrio desulfuricans*; Dalsgaard and Bak, 1994). In particular, paddy fields are recognized as an important source of atmospheric methane (Kagatsume, 1991; Yagi and Minami, 1991; Minami and Neue, 1994). Rates of methane emission vary widely according to season, nutrient conditions and the amount of organic matter incorporated, but values as high as 880 mg m^{-1} day^{-1} have been reported in China (Mitsch and Wu, 1995). There is evidence that minimum tillage reduces methane emissions from paddy soil (Kaneta *et al.*, 1992; Sass *et al.*, 1992; Ito *et al.*, 1995). Similarly, denitrification rates are lower under minimal tillage.

Although the rate of bacterial reductive processes generally tends to be higher in conventional tillage systems, the tillage process itself promotes nutrient mineralization. For example, both puddling and the air-drying of soil which follows dry tillage enhance the mineralization of nitrogen. In no-tillage fields, the rate of nitrogen mineralization is slow, which affects plant growth, particularly at younger stages. Furthermore, nutrients tend to be unevenly distributed in minimum tillage fields because organic matter derived from rice plants and weeds accumulate on the soil surface. The contents of carbon, nitrogen, phosphorus and silica near the soil surface (0–5 cm depth) are higher under minimum tillage treatments (Kaneta, 1997).

These differences in the distribution and mineralization of nutrients have an influence upon the timing of plant growth. The initially lower nitrogen supply means that early seedling growth is slower than under conventional tillage (Hamada *et al.*, 1994). For example, Yoshii *et al.* (1996) found that no-tillage decreased rice seedling growth at early stages of development but that growth became vigorous in middle stages and finally greater than that of traditional tillage systems. Certainly, there are no consistent effects of tillage upon harvestable yield.

As for wheat and maize, one of the most troublesome problems of minimum tillage in paddy fields is that weeds, particularly perennial species, are unchecked. Choi-Chung *et al.* (1996) found that the most problematic weed species in no-tillage rice fields in Korea were *Echinochloa crus-galli*, *Leersia japonica* and *Aneilema keisak*. Indeed, *A. keisak* alone could reduce yields by 50%, and high levels of herbicide were needed to control it (Kwon *et al.*, 1996). Populations of some pest species may also be affected. For example, in Pakistan survival of the hibernating larvae of several rice stem borers, such as *Tryporyza* (*Scirophaga*) and *Sesamia* species, was highest in untilled fields, whereas rotavation caused the complete destruction of stubble and left no hibernating larvae for the next season (Abdul *et al.*, 1997). There is also reason to suppose that the more aerobic conditions produced by minimum tillage may enhance populations of root-feeding nematodes, though the evidence is not clear-cut. Certainly, populations of nematodes tend to be low under anaerobic conditions; thus numbers of plant parasitic nematodes are usually much lower in paddy fields than in upland rice fields (Van den Berg and De-Waele, 1989). However, Stirling *et al.* (1989) found that treatment of soils infested with the needle nematode (*Paralongidorus australis*) by adding straw and various sulphur-containing compounds and flooding for 6–12 weeks failed to reduce nematode numbers in the subsequent rice crop, indicating that products of anaerobic decomposition did not control the nematode. None the less, there was a marked reduction in the percentage of root tips damaged by the nematode in the straw + sulphur treatment.

There is very little published information about the influence of tillage upon beneficial invertebrates. However, there are reports that populations of spiders, which are thought to be important biological control agents in paddy fields (Ye and Wang, 1987), are enhanced under minimum tillage (S. Akita, Tokyo, 1997, personal communication).

In general, reduced tillage tends to provoke problems of disease control. For example, Ghaffar (1987) found that under maximum tillage sclerotia of *Sclerotium oryzae* (*Magnaporthe salvinii*, causal agent of stem rot) completely lost viability and rice yields were 59% higher compared with minimum or no-tillage treatments. Furthermore, under certain conditions, the microbial decomposition of crop residues in or on the soil can result in the release of compounds that are toxic to crop plants and also predispose plant roots to attack by soilborne pathogens. It is now known that phytotoxin development in soil is favoured by cool, wet or anaerobic soils, conditions such as often occur early in the growing season in temperate regions, particularly under conservation tillage (see Doran and Linn, 1993).

Generalizing About the Effect of Tillage on Agroecosystems

Tillage as a form of ecological disturbance

Disturbance is a topic of central importance in ecology. When an ecosystem is disturbed the established order – in terms of the interactions between organisms which structure a community – is disrupted and a process of community change or succession is initiated. When the disturbance is not total – that is, when something remains of the previous ecosystem in the form of living organisms, their propagules, or their organic residues – the process of ecosystem change is referred to as secondary succession. A great deal has been written about the ecological characteristics of organisms at different stages in succession (Bazzaz, 1979; Brown and Southwood, 1987). For example, plants of early secondary successions are characterized by a rapid growth rate, a high reproductive allocation and a short life span (*r*-selected characteristics). These characteristics equip an organism to succeed in short-lived but otherwise favourable habitats, but not to persist later in a succession when other pressures, particularly competition by other organisms for the same resources, become increasingly important. Later successional species show so-called *K*-selected characteristics such as longer life spans, slower growth rates, and more allocation to vegetative structures (see Allen *et al.*, Chapter 6, and Wood and Lenné, Chapter 17, this volume). As a broad generalization, one can say that in early successional communities population sizes are determined by the influence of abiotic conditions upon the growth of organisms while, at later successional stages, interactions (competition, predation, symbiosis, etc.) become much more important. For example, in early succession, nutrient cycles are typically open and the availability of minerals for growth is not dependent upon biological recycling. In contrast, in later successional stages, the processes of decomposition which convert organically bound mineral elements into an available form become increasingly important.

To an ecologist, tillage is not just a way of preparing the soil for the next crop; far more, it is a major disturbance which is likely to affect all components of the ecosystem and how they function. Thus conventional tillage leads to massive mortality, not only of weeds but also of invertebrates such as earthworms, slugs, and beneficial and harmful insects. Fungal mycelia nets are broken and mycorrhizal associations are disrupted. Even those organisms not directly affected by tillage are indirectly affected, for example by changes in microclimatic or redox conditions, or by release from competition or predation. There are probably no organisms in an arable ecosystem whose activities are unaffected by tillage. Many of the generalizations just described about the structure and function of early successional ecosystems apply to an arable field. Most arable weeds are short-lived and have a very high relative growth rate; nutrient cycles are open and dependent upon inputs in the form of fertilizer, populations of insect herbivores are often not effectively regulated by natural enemies and can increase rapidly unless controlled by artificial means.

Methods of conservation tillage systems involve a much less drastic disturbance of the ecosystem than conventional ploughing. Obviously there is still disturbance – the crop is harvested, seeds are drilled, fertilizers and pesticides are still applied – but important components of the ecosystem remain relatively undisrupted. We can hypothesize that these components will tend to show the characteristics of later successional communities. For example, we could predict that organisms with longer life cycles may succeed, the biological regulation of pest populations becomes more important, and there will be greater recycling of nutrients.

To what extent are these predictions evident in practice? Certainly some of them are correct. For example, in all three crops considered here, perennial weeds, whose populations are kept in check under normal tillage, become a problem under forms of conservation tillage. Similarly, there is evidence that the biological regulation of insect populations and fungal diseases is more effective. The biological recycling of some nutrients also becomes more significant under reduced tillage. For example, in terrestrial systems with wheat and maize, leaching losses of nitrates are lower in conservation tillage systems and decomposition processes play a more important role in supplying plants with their nutrients. In this sense, the nutrient cycle becomes more closed.

Implications for sustainability

How important are biological processes of pest control and nutrient cycling for the sustainability of agroecosystems? In the case of wheat, there are serious concerns that the rapid decline of biodiversity which has been evident in many areas in the past few decades, and the losses of the ecosystem function which this diversity provides, are harmful. In contrast, there is a general view that paddy field rice is an extremely stable and sustainable production system. In the Mekong valley, there is archaeological evidence of rice production as far back as 3500 years ago, representing some of the longest continuous agricultural production anywhere in the world (see Thurston et al., Chapter 9, this volume). The conclusion seems to be that biological diversity does not play an important role in the stability of rice production (Abivardi *et al.*, 1998; but see Settle *et al.*, 1996).

This evidence seems to suggest that the loss of biodiversity, particularly in the soil biota, may be more significant for the stability of terrestrial systems such as wheat and maize than for paddy rice. If this apparent difference is real, an important reason for it may lie in the strongly contrasting nature of the two ecosystems. In many areas, paddy rice represents a semi-aquatic system with a characteristically open nutrient cycle. As in many natural wetland ecosystems (e.g. reed beds) growth is maintained through an input of nutrients in the water supply. The stability of the system therefore depends upon the stability of the hydrological conditions, rather than upon the recycling of nutrients within the ecosystem. This implies that, at least as far as nutrient supply is concerned, paddy fields can be maintained sustainably as long as an appropriate water regime is maintained. In contrast, dryland crops such as wheat are terrestrial systems. Natural terrestrial ecosystems usually have a more or less closed nutrient cycle in which nutrients for plant growth become available through the processes of decomposition.

As we have seen, not all of the diverse effects of conservation tillage are desirable from an agricultural point of view. For paddy rice in Japan, it was concluded that because of the associated problems such as weed infestation, minimum tillage only has a role to play in special circumstances, for example when the land is very heavy and trafficability is difficult (Abivardi *et al.*, 1998). However, for wheat and maize the advantages of conservation tillage for soil protection are increasingly appreciated. None the less, new methods must be found to combat some of the less desirable effects of conservation tillage systems. Many pest and disease problems associated with conservation tillage can be minimized in a variety of ways that do not significantly increase production costs. Crop rotation is especially valuable for controlling diseases associated with surface tillage (Boosalis *et al.*, 1986); for instance, a rotation sequence of 2 or 3 years induces sufficient deterioration of residues to reduce pathogen populations. Many annual broadleaf weeds can be suppressed if mulches, especially grain cover crops, are left on the soil surface (e.g. Shilling *et al.*, 1985). This beneficial effect, largely due to allelopathic interactions, can help to suppress difficult-to-control annual broadleaf weeds in many broadleaf crops and possibly reduce the need for post-emergence herbicide applications with surface- or no-tillage. Progress of this kind illustrates how the negative effects of alternative tillage systems can be minimized while obtaining the benefits in terms of improved soil structure and maintenance of fertility (see Griffith *et al.*, 1986).

References

Abdul, R., Sabir, B.A. and Zafar, M.A. (1997) Control of insect pests on rice crop using tillage practices. *Agricultural Mechanization in Asia, Africa and Latin America* 28, 29–30.

Abivardi, C., Akita, S., Edwards, P., Nakamoto, T., Richner, W., Stamp, P. and Yamaji, A. (1998) A comparison of intensive rice production in Japan and wheat production in Europe: the contribution of minimal tillage to sustainability. *Plant Production Science*, Japan 1, 149–164.

Addae, P.C., Collis-George, N. and Pearson C.J. (1991) Overriding effects of temperature and soil strength on wheat seedlings under minimum and conventional tillage. *Field Crops Research* 28, 103–116.

All, J.N. and Musick, G.J. (1986) Management of vertebrate and invertebrate pests. In: Sprague, M.A. and Triplett, G.B. (eds) *No-Tillage and Surface-Tillage Agriculture: the Tillage Revolution*. John Wiley & Sons, New York, pp. 347–387.

Andersen, A. (1987) Effects of direct drilling and ploughing on populations of earthworms. *Tidsskrift for Planteavlforsog* 91, 3–14.

Anken, T., Heusser, J., Weisskopf, P., Zihlmann, U., Forrer, H.R., Högger, Ch., Scherrer, C., Mozafar, A. and Sturny, W.G. (1997) Bodenbearbeitungssysteme – Direktsaat stellt höchste Anforderungen. *FAT-Berichte* Nr. 501, Switzerland, 14 pp.

Baeumer, K. and Bakermans, W.A.P. (1973) Zero-tillage. *Advances in Agronomy* 25, 77–123.

Bailey, K.L. and Duczek, L.J. (1996) Managing cereal diseases under reduced tillage. *Canadian Journal of Plant Pathology* 18, 159–167.

Baker, C.J., Saxton, K.E. and Ritchie, W.R. (1996) *No-tillage Seeding: Science and Practice*. CAB International, Wallingford.

Ball, B.C. (1990) Reduced tillage for energy and cost savings with cereals: practical and research experience. *Agricultural Engineer* 45, 2–6.

Ball, B.C., Lang, R.W., O'Sullivan, M.F. and Franklin, M.F. (1989) Cultivation and nitrogen requirements for continuous winter barley on a gleysol and a cambisol. *Soil and Tillage Research* 13, 333–352.

Bazzaz, F.A. (1979) The physiological ecology of plant succession. *Annual Review of Ecology and Systematics* 10, 351–371.

Bigler, F., Waldburger, M. and Frei. G. (1995) Important insect and spiders as natural enemies in maize. *Agrarforschung* (Switzerland) 2, 383–386.

Blevins, R.L. and Frye, W.W. (1993) Conservation tillage: an ecological approach to soil management. *Advances in Agronomy* 51, 33–78.

Bligh, K. (1994) No-tillage sowing decreases water erosion on loamy soils and increases earthworm activity. *Journal of Agriculture* 35, 47–50.

Bockus, W.W., Davis, M.A. and Norman, B.L. (1994) Effect of soil shading by surface residues during summer fallow on take-all of winter wheat. *Plant Disease* 76, 50–54.

Boosalis, M.G., Doupnik, B.L. and Watkins, J.E. (1986) Conservation tillage in relation to plant diseases. In: Sprague, M.A. and Triplett, G.B. (eds) *No-Tillage and Surface-Tillage Agriculture: the Tillage Revolution*. John Wiley & Sons, New York, pp. 389–408.

Brown, V.K. and Southwood, T.R.E. (1987) Secondary succession: patterns and strategies. In: Gray, A.J., Crawley, M.J. and Edwards, P.J. (eds) *Colonization, Succession and Stability*. Symposia of the British Ecological Society, 25, Blackwell Scientific Publications, Oxford, pp. 315–337.

Brust, G.E. (1990) Direct and indirect effects of four herbicides on the activity of carabid beetles (Coleoptera: Carabidae). *Pesticide Science* 30, 309–320.

Brust, G.E. (1994) Seed-predators reduce broadleaf weed growth and competitive ability. *Agriculture, Ecosystems and Environment* 48, 27–34.

Brust, G.E. and House, G.J. (1988) Weed seed destruction by arthropods and rodents in low-input soybean agroecosystems. *American Journal of Alternative Agriculture* 3, 19–25.

Brust, G.E., Stinner, B.R. and McCartney, D.A. (1986) Predator activity and predation in corn agroecosystems. *Environmental Entomology* 15, 1017–1021.

Cannell, R.Q. (1984) Mitigation of soil constraints to cereal production. In: Gallagher, E.J. (ed.) *Cereal Production.* Butterworth, London, pp. 193–210.

Cannell, R.Q. and Hawes, J.D. (1994) Trends in tillage practices in relation to sustainable crop production with special reference to temperate climates. *Soil and Tillage Research* 30, 245–282.

Carter, M.R. (1991) Evaluation of shallow tillage for spring cereals on a fine sandy loam. 1. Growth and yield components, N accumulation and tillage economics. *Soil and Tillage Research* 21, 23–35.

Carter, M.R. (1994) A review of conservation tillage strategies for humid temperate regions. *Soil and Tillage Research* 31, 289–301.

Choi, C.D., Won, J.G., Lee, W.H., Kim, C.R. and Choi, B.S. (1996) Weed occurrence and its effective control measures in no-tillage paddy field. *RDA Journal of Agricultural Science* (Crop-Protection) 38, 414–420.

Christian, D.G. and Bacon, E.T.G. (1990) A long-term comparison of ploughing, tine cultivation and direct drilling on the growth and yield of winter cereals and oilseed rape on clayey and silty soils. *Soil and Tillage Research* 18, 311–331.

Christian, D.G. and Ball, B.C. (1994) Reduced cultivation and direct drilling for cereals in Great Britain. In: Carter, M.R. (ed.) *Conservation Tillage in Temperate Agroecosystems.* Lewis Publishers Inc., Boca Raton, Florida, pp. 117–140.

Cook, R.J. (1992) Wheat root health management and environmental concern. *Canadian Journal of Plant Pathology* 14, 76–85.

Cook, R.J. and Haglund, W.A. (1991) Wheat yield depression associated with conservation tillage caused by root pathogens in the soil not phytotoxins from the straw. *Soil Biology and Biochemistry* 23, 1125–1132.

Cotterill, P.J. and Sivasithamparam, K. (1988) The effect of tillage practices on distribution, size, infectivity and propagule number of the take-all fungus (*Gaeumannomyces graminis* var. *tritici*). *Soil and Tillage Research* 11, 183–195.

Craig, J.P. and Weil, R. (1993) Nitrate leaching to a shallow mid-atlantic coastal plain aquifer as influenced by conventional no-till and low-input sustainable grain production systems. *Water Science and Technology* 28, 691–700.

CTIC (Conservation Technology Information Center) (1992) *National Survey of Conservation Tillage Practices.* CTIC, West Lafayette, Indiana.

Dalsgaard, T. and Bak, F. (1994) Nitrate reduction in a sulfate reducing bacterium, *Desulfovibrio desulfuricans,* isolated from rice paddy soil: sulfide inhibition, kinetics, and regulation. *Applied and Environmental Microbiology* 60, 291–297.

Davies, B. (1988) *Reduced Cultivation for Cereals.* Research Review, Home Grown Cereals Authority, UK, No. 5, 71 pp.

Doran, J.W. (1980) Soil microbial and biochemical changes associated with reduced tillage. *Journal of the Soil Science Society of America* 44, 765–771.

Doran, J.W. and Linn, D.M. (1993) Microbial ecology of conservation management systems. In: Hatfield, J.L. and Stewart, B.A. (eds) *Soil Biology: Effects on Soil Quality.* Lewis Publishers, London, pp. 1–27.

Doran, J.W. and Power, J.F. (1983) *The Effects of Tillage on the Nitrogen Cycle in*

Corn and Wheat Production. Special Publication, College of Agriculture Experiment Stations, University of Georgia 23, 441–455.

Dowdell, R.J. and Cannell, R.Q. (1975) Effect of plowing and direct drilling on soil nitrate content. *Journal of Soil Science* 26, 53–61.

Dowdell, R.J. and Crees, R. (1980) The uptake of ^{15}N labeled fertilizer by winter wheat and its immobilization in a clay soil after direct drilling or ploughing. *Journal of the Science of Food and Agriculture* 31, 992–996.

Edwards, W.M., Shipitalo, M.J., Traina, S.J., Edwards, C.A. and Owens, L.B. (1992) Role of *Lumbricus terrestris* (L.) burrows on quality of infiltrating water. *Soil Biology and Biochemistry* 24, 1555–1561.

Edwards, W.M., Shipitalo, M.J., Owens, L.B. and Dick, W.A. (1993) Factors affecting preferential flow of water and atrazine through earthworm burrows under continuous no-till corn. *Journal of Environmental Quality* 22, 453–457.

Ehlers, W. (1975) Observations on earthworm channels and infiltration on tilled and untilled loess soil. *Soil Science* 119, 242–249.

Ehlers, W. and Claupein, W. (1994) Approaches toward conservation tillage in Germany. In: Carter, M.R. (ed.) *Conservation Tillage in Temperate Agroecosystems.* Lewis Publishers, Boca Raton, Florida, pp. 141–165.

Ehlers, W., Khosla, B.K., Kopke, U., Stulpnagel, R., Bohm, W. and Baeumer, K. (1980) Tillage effects on root development, water uptake and growth of oats. *Soil and Tillage Research* 1, 19–34.

Ehlers, W., Kopke, U., Hesse, F. and Bohm, W. (1983) Penetration resistance and root growth of oats in tilled and untilled loess soil *Avena sativa. Soil and Tillage Research* 3, 261–275.

Ellis, F.B. and Howse, K.R. (1980) Effects of cultivation on the distribution of nutrients in the soil and the uptake of nitrogen by spring barley and winter wheat on three soil types. *Soil and Tillage Research* 1, 35–46.

Feldman, S.R., Alzugaray, C., Torres, P.S. and Lewis, P. (1997) The effect of different tillage systems on the composition of the seedbank. *Weed Research* 37, 71–76.

Francis, G.S. and Knight, T.L. (1993) Long-term effects of conventional and no-tillage on selected soil properties and crop yields in Canterbury, New Zealand. *Soil and Tillage Research* 26, 193–210.

Frankinet, M., Rixhon, L., Crohain, A. and Grevy, L. (1979) Labour, demi-labour ou semis direct en continu: conséquences phytotechniques. *Bulletin des Recherches Agronomiques de Gembloux* 14, 35–95.

Fraser, P.M. (1994) The impact of soil and crop management practices on soil macrofauna. In: Pankhurst, C.E., Doube, B.M., Gupta, V.V.S.R. and Grace, P.R. (eds) *Soil Biota: Management in Sustainable Farming Systems.* CAB International, Wallingford, pp. 125–132.

Ghaffar, A. (1987) Soil management practices for the control of sclerotial fungi. *Final Research Report 1st April 1984–31st October 1987.* University of Karachi, Karachi.

Gish, T.J., Shirmohammadi, A., Wienhold, B.J. and Vyravipillai, R. (1995) Chemical transport below the root zone as influenced by tillage practice. In: *Proceedings Clean Water, Clean Environment, 21st Century Team Agriculture, Working to Protect Water Resources,* 5–8 March 1995, Kansas City, Missouri, and St Joseph, Michigan, American Society of Agricultural Engineers, volume 3, pp. 109–112.

Gooding M.J. and Davies, W.P. (1997) *Wheat Production and Utilization: Systems, Quality and the Environment.* CAB International, Wallingford.

Griffith, D.R., Mannering, J.V. and Box, J.E. (1986) Soil and moisture management with reduced tillage. In: Sprague, M.A. and Triplett, G.B. (eds) *The Tillage Revolution*. John Wiley & Sons, New York, pp. 19–57.

Hakansson, I. (1994) Soil tillage for crop production and for protection of soil and environmental quality – a Scandinavian viewpoint. *Soil and Tillage Research* 30, 109–124.

Hamada, Y., Nakajima, Y., Seki, M., Izawa, T., Sawada, Y. and Ibuka, T. (1994) Studies on no-tillage direct sowing culture of rice in well-drained paddy fields in the Anjo Region. 1. Cultivation method with the Aichi–Nohsohshi-type planter. *Report of the Tokai Branch of the Crop Science Society of Japan*, No. 117, 15–19 (in Japanese).

Hammond, R.B., Smith, J.A. and Beck, T. (1996) Timing of molluscicide applications for reliable control in no-tillage field crops. *Journal of Economic Entomology* 89, 1028–1232.

Hance, T. and Wibo, G.C. (1987) Effect of agricultural practices on carabid populations. *Acta Phytopathologica et Entomologica Hungarica* 22, 147–160.

Ilnicki, R.D. and Enache, A.J. (1992) Subterranean clover living mulch: an alternative method of weed control. *Agriculture, Ecosystems and Environment* 40, 249–264.

Isensee, A.R. and Sadeghi, A.M. (1995) Impact of tillage on herbicide leaching to groundwater. In: *Proceedings Clean Water, Clean Environment, 21st Century Team Agriculture, Working to Protect Water Resources*, 5-8 March 1995, Kansas City, Missouri, and St Joseph, Michigan, American Society of Agricultural Engineers, volume 1, pp. 97–100.

Ito, C., Kaneta, Y. and Iizuka, F. (1995) The effect of non-tillage cultivation on suppression of methane emission from paddy fields of strong-gley soils. *Tohoku Agricultural Research* 48, 103–104 (in Japanese).

Jacinthe, P. and Dick, W.A. (1997) Soil management and nitrous oxide emissions from cultivated fields in southern Ohio. *Soil and Tillage Research* 41, 221–235.

Jenkinson, D.S. (1986) Nitrogen in the UK arable agriculture. *Journal of the Royal Agricultural Society of England* 147, 178–189.

Jones, C.L., Lawton, J.H. and Shackak, M. (1994) Organisms as ecological engineers. *Oikos* 69, 373–386.

Kagatsume, M. (1991) Global warming and the emission of methane from rice growing: the international movement and prediction. *Farm Accounting Studies* 24, 133–142 (in Japanese).

Kaneta, Y. (1997) Progress and prospect of the research on paddy soil management under various rice growing system. 3. Soil management and fertilization method for non-tillage rice culture. *Soil Science and Plant Nutrition* 68, 69–74 (in Japanese).

Kaneta, Y., Awasaki, H. and Yamaya, S. (1992) Soil management of heavy subsoil for the rotational use of paddy fields. 4. Effects of straw application onto soil surface on production of methane and growth of rice in the non tillage paddy fields. *Tohoku Agricultural Research* 45, 77–78 (in Japanese).

Kennedy, A.C. and Smith, K.L. (1995) Soil microbial diversity and the sustainability of agricultural soils. In: Collins, H.P., Robertson, G.P. and Klug, M.J. (eds) *The Significance and Regulation of Soil Biodiversity*. Kluwer Academic, Dordrecht, pp. 75–86.

Kirkegaard, J.A. (1995) A review of trends in wheat yield responses to conservation cropping in Australia. *Australian Journal of Experimental Agriculture* 35, 835–848.

Koenigs, F.F.R. (1963) The puddling of clay soils. *Netherlands Journal of Agricultural Science* 11, 145–156.

Kwon, O.D., Shin, H.R., Park, T.D., Guh, J.O. and Lim, J.S. (1996) Control of spiderwort (*Aneilema keisak* Hassk) in no-tillage rice. *Korean Journal of Weed Science* 16, 100–107.

Lal, R. (1989) Conservation tillage for sustainable agriculture: tropics versus temperate environments. *Advances in Agronomy* 42, 85–197.

Lemon, R.W. (1994) Insecticide resistance. *Journal of Agricultural Science* 122, 329–333.

Li, L.M., Li, Z.G., Pan, Y.H. and Wu, S.C. (1992) Denitrifying bacteria and denitrification process in rice rhizosphere. In: *Proceedings International Symposium on Paddy Soils*, Nanjing, China, 15–19 September 1992, Academia Sinica (Beijing), pp. 90–98.

Macchia, M., Cozzani, A. and Bonari, E. (1996) Effects of soil tillage on weed seed bank structure and dynamics in a biennial winter wheat (*Triticum aestivum* L.)–soybean (*Glycine max* (L.) Merr.) rotation. *Rivista di Agronomia* 30, 136–141.

Maillard, A. and Vez, A. (1988) Influence du travail du sol et d'un engrais vert sur le blé d'automne et le mais grain en rotation depuis quinze ans. *Revue Suisse d'Agriculture* 20, 321–326.

Maillard, A. and Vez, A. (1993) Résultats d'un essai de culture sans labour depuis plus de 20 ans à Changins: I. Rendement des cultures, maladies et ravageurs. *Revue Suisse d'Agriculture* 25, 327–337.

Mayor, J.P. and Maillard, A. (1995) Results from an over-20-years-old ploughless tillage experiment at Changins. IV. Seed bank and weed control. *Revue Suisse d'Agriculture* 27, 229–236.

Meek, B.D., Carter, D.L., Westermann, D.T., Wright, J.L. and Peckenpaugh, R.E. (1995) Nitrate leaching under furrow irrigation as affected by crop sequence and tillage. *Journal of Soil Science Society of America* 59, 204–210.

Meyer, L.D., Dabney, S.M., Murphree, C.E., Harmon, W.C. and Grissinger, E.H. (1997) Crop production systems to control erosion and reduce runoff. In: *American Society of Agricultural Engineers Annual International Meeting*, Minneapolis, 10–24 August, 1997. Paper No. 97–2224, 19 pp.

Min, H., Zhao, Y.H., Chen, M.C. and Zhao, Y. (1997) Methanogens in paddy rice soil. *Nutrient Cycling in Agroecosystems* 49, 163–169.

Minami, K. and Neue, H.U. (1994) Rice paddies as methane source. *Climatic Change* 27, 13–26.

Mitsch, W.J. and Wu, X. (1995) Wetlands and global change. In: Lal, R., Kimble, J., Levine, E. and Stewart, B.A. (eds) *Soil Management and Greenhouse Effect*. CRC Press, London, pp. 205–230.

Morrill, W.L., Gabor, J.W. and Wichman, D. (1993) Mortality of the wheat stem sawfly (Hymenoptera: Cephidae) at low temperatures. *Environmental Entomology* 22, 1358–1361.

Naganoma, H., Kodama, T., Kaneta, Y. and Yamaya, S. (1991) Effect of tillage method on physical properties of heavy soil of rotational paddy field. *Soil Physical Condition and Plant Growth*, Japan 62, 43–52 (in Japanese).

Nuttall, W.F., Bowren, K.E. and Campbell, C.A. (1986) Crop residue management practices, and N and P fertilizer effects on crop response and on some physical and chemical properties of a black Chernozem over 25 years in a continuous wheat rotation. *Canadian Journal of Soil Science* 66, 159–171.

OECD (Organization for Economic Co-operation and Development) (1995) *Sustainable Agriculture: Concepts, Issues and Policies in OECD Countries.* OECD, Paris.

Palma, R.M., Rimolo, M., Saubidet, M.I. and Conti, M.E. (1997) Influence of tillage system on denitrification in maize-cropped soils. *Biology and Fertility of Soils* 25, 142–146.

Pearson, C.J., Mann, I.G. and Zhang, Z.H. (1991) Changes in root growth within successive wheat crops in a cropping cycle using minimum and conventional tillage. *Field Crops Research* 28, 117–133.

Pimentel, D. and Andow, D.A. (1984) Pest management and pesticide impacts. *Insect Science and its Application* 5, 141–149.

Pimentel, D., Acquay, H., Biltonen, M., Rice, P., Silva, M., Nelson, J., Lipner, V., Giordano, S., Horowitz, A. and D'Amore, M. (1992) Environmental and economic costs of pesticide use. *BioScience* 42, 750–760.

Pfender, W.F. (1988) Suppression of ascocarp formation of *Pyrenophora tritici-repentis* by *Limonomyces roseipellis*, a basidiomycete from reduced-tillage wheat straw. *Phytopathology* 78, 1254–1258.

Pfender, W.F., Zhang, W. and Nus, A. (1993) Biological control to reduce inoculum of the tan spot pathogen *Pyrenophora tritici-repentis* in surface-borne residues of wheat fields. *Phytopathology* 83, 371–375.

Powles, S.B., Preston, C., Bryan, I.B. and Jutsum, A.R. (1997) Herbicide resistance: impact and management. *Advances in Agronomy* 58, 57–93.

Rasmussen, P.E. (1995) Effects of fertilizer and stubble burning on downy brome competition in winter wheat. *Communications in Soil Science and Plant Analysis* 26, 7–8, 951–960.

Rodriguez, M.B. and Giambiagi, N. (1995) Denitrification in tillage and no tillage pampenean soils: relationships among soil water, available carbon, and nitrate and nitrous oxide production. *Communications in Soil Science and Plant Analysis* 26, 3205–3220.

Roger, M.M. and Gupta, V.V.S.R. (1995) Management practices and soil biota. *Australian Journal of Soil Research* 33, 321–339.

Russell, P.E. (1995) Fungicide resistance: occurrence and management. *Journal of Agricultural Science* 124, 317–323.

Sadeghi, A.M. and Isensee, A.R. (1997) Alachlor and cyanazine persistence in soil under different tillage and rainfall regimes. *Soil Science* 162, 430–438.

Samsoe-Petersen, L., Bieri, M. and Buchs, W. (1992) Interpretation of laboratory measured effects of slug pellets on soil dwelling invertebrates. *Aspects of Applied Biology* 31, 87–96.

Sass, R.L., Fisher, F.M., Wang, Y.B., Turner, F.T. and Jund, M.F. (1992) Methane emission from rice fields: the effect of floodwater management. *Global Biogeochemical Cycles* 6, 249–262.

Sato, T. (1992) Significance of soil macropores in the heavy clay soil of paddy field in Hachiro-gata Polder. *Journal of the Japanese Society of Irrigation, Drainage and Reclamation Engineering* 60, 25–30 (in Japanese).

Sattin, M. (1994) Influence of crop succession and soil tillage on wheat take-all (*Gaeumannomyces graminis* var. *tritici*). In: *Proceedings of the Third Congress of the European Society for Agronomy*, Abano-Padova, Italy, 18–22 September 1994, pp. 672–673.

Settle, W.H., Ariawan, H., Astuti, E.T., Cahayana, W., Hakim, A.L. and Hindayana, D. (1996) Managing tropical rice pests through conservation of generalist natural enemies and alternative prey. *Ecology* 77, 1975–1988.

Shilling, D.G., Liebl, R.A. and Worsham, D. (1985) *Rye (*Secale cereale *L.) and Wheat (*Triticum aestivum *L.) Mulch: the Suppression of Certain Broadleaved Weeds and the Isolation and Identification of Phytotoxins.* ACS-Symposium Series No. 268, American Chemical Society, Washington, DC, pp. 243–271.

Simon-Sylvestre, G. and Fournier, J.C. (1979) Effects of pesticides on the soil microflora. *Advances in Agronomy* 31, 1–92.

Sochtig, W. and Larink, O. (1992) Effect of soil compaction on activity and biomass of endogenic lumbricids in arable soils. *Soil Biology and Biochemistry* 24, 1595–1599.

Sprague, M.A. and Triplett, G.B. (1986) No-tillage and surface-tillage. In: Sprague, M.A. and Triplett, G.B. (eds) *The Tillage Revolution.* John Wiley & Sons, New York, pp. 21–37.

Stassart, P. and Wibo, G.C. (1983) Influence of soil tillage on carabid populations in large-scale fields: preliminary results. *Mededelingen van de Faculteit Landbouwwetenschappen Rijksuniversiteit Gent* 48, 465–474.

Stehouwer, R.C., Dick, W.A. and Trina, S.J. (1993) Characteristics of earthworm burrow lining affecting atrazine sorption. *Journal of Environmental Quality* 22, 181–185.

Stirling, G.R., Vawdrey, L.L. and Shannon, E.L. (1989) Options for controlling needle nematode (*Paralongidorus australis*) and preventing damage to rice in northern Queensland. *Australian Journal of Experimental Agriculture* 29, 223–232.

Stover, R.W., Francl, L.J. and Jordahl, J.G. (1996) Tillage and fungicide management of foliar diseases in a spring wheat monoculture. *Journal of Production Agriculture* 9, 261–265.

Sutton, J.C. and Vyn, T.J. (1990) Crop sequences and tillage practices in relation to diseases of winter wheat in Ontario (Canada). *Canadian Journal of Plant Pathology* 12, 358–368.

Takahashi, Y. (1993) Uptake ratio of paste fertilizer-N by side dressing and specificity of N-absorption on the non-tilled transplanting culture of paddy rice plant. *Soil Science and Plant Nutrition* 64, 681–684 (in Japanese).

Tonhasca, A. (1993) Carabid beetle assemblage under diversified agroecosystems. *Entomologia Experimentalis et Applicata* 68, 279–285.

Triplett, G.B. and Lytle, G.D. (1972) Control and ecology of weeds in continuous corn grown without tillage. *Weed Science* 20, 453–457.

Unger, P.W. and Cassel, D.K. (1991) Tillage implement disturbance effects on soil properties related to soil and water conservation: a literature review. *Soil and Tillage Research* 19, 363–382.

Vaidyanathan, L.V. and Davies, D.B. (1980) Response of winter wheat to fertilizer nitrogen in undisturbed and cultivated soil. *Journal of the Science of Food and Agriculture* 31, 414–415.

Van den Berg, E. and De-Waele, D. (1989) Further observations on nematodes associated with rice in South Africa. *Phytophylactica* 21, 125–130.

Vez, A. (1977) Dix ans d'expérience de cultures sans labour. *Revue Suisse d'Agriculture* 9, 59–70.

Whitmore, A.P. and Addiscott, T.M. (1986) Computer simulation of winter leaching losses of nitrate from soils cropped with winter wheat. *Soil Use and Management* 2, 26–30.

Willoughby, G.L., Kladivko, E.J. and Savabi, M.R. (1997) Seasonal variations in infiltration rate under no-till and conventional (disk) tillage systems as affected by *Lumbricus terrestris* activity. *Soil Biology and Biochemistry* 29, 481–484.

Wilson, G.V., Tyler, D.D., Logan, J., Thomas, G.W., Blevins, R.L., Dravillas, M.C. and Caldwell, W.E. (1995) Tillage and cover crop effects on nitrate leaching. In: *Proceedings of Clean Water – Clean Environment – 21st Century*, Vol. 2, *Nutrients*. Kansas City, Missouri, 5–8 March, American Society of Agricultural Engineers, St Joseph, pp. 251–254.

Wyss, E. and Glasstetter, M. (1992) Tillage treatments and earthworm distribution in a Swiss experimental corn field. *Soil Biology and Biochemistry* 24, 1635–1639.

Yagi, K. and Minami, K. (1991) Effect of organic matter application on methane emission from some Japanese paddy fields. *Soil Science and Plant Nutrition* 36, 599–610.

Ye, Z.X. and Wang, D.D. (1987) Compositions and dynamics of the spider fauna in paddies of Jiangxi Province. *Chinese Journal of Biological Control* 3, 11–14 (in Chinese).

Yoshii, M., Murakami, K. and Horiuchi, E. (1996) Growth and yield of rice (*Oryza sativa* L.) plants by non-tillage and transplant systems. *Bulletin of the Experimental Farm College of Agriculture*, Ehime-University No. 17, 53–58 (in Japanese).

Young, F.L., Ogg, A.G., Thill, D.C., Young, D.L. and Papendick, R.I. (1996) Weed management for crop production in the northwest wheat (*Triticum aestivum*) region. *Weed Science* 44, 429–436.

Seed Management Systems and Effects on Diversity

13

M. Wright[1] and M. Turner[2]

[1]*VSO Programme Office, PO Box 137, Belmopan, Belize;*
[2]*Seed Unit, ICARDA, PO Box 5466, Aleppo, Syria*

Introduction

The great majority of the world's food crops are annual species for which seed must be sown each season to establish a new crop. Consequently, seeds are the fundamental biological component of agricultural production. Regardless of their circumstances or location, farmers have a special concern about the quality of the seed they use and the time from sowing to establishment is of critical importance in nearly all cropping systems. Problems which appear at this time can seldom be fully rectified. This concern by farmers relates to the quality of seed in terms of its 'planting value', which is primarily a function of germination, health and purity status. However, seed also carries the genetic information which determines the characteristics and potential of the crop. In modern crop production systems these characteristics are defined in a 'variety', whereas in traditional production systems that concept is often less precise.

These two aspects of 'quality' embodied in the seed are largely independent and are influenced in very different ways. For example, the germination quality of seed may be poor in one season because of adverse weather conditions at harvest but the genetic quality may be unchanged in the plants that do establish and grow to maturity. On the other hand, in good production environments, physical seed quality may remain high through many cycles of multiplication but the genetic quality of the variety may decline rapidly due to contamination from different sources.

This is important when analysing seed management systems because it is always necessary to consider these two elements of seed quality. In modern seed systems, where a high level of management is applied throughout the production process, both quality attributes are normally ensured by adopting specific technical procedures. Seed quality has therefore become 'formalized' by technology, management and often by legislation also. However, in traditional seed systems, management practices by farmers are more variable and the priority given to different attributes of the seed may be less clear. The issue of 'diversity', which is the main concern of this book, derives from the variety aspects of seed management.

This chapter discusses the relationship between seed management systems and genetic diversity. The main focus is on farmers in developing countries, particularly small-scale farmers, with whom much diversity still resides. However, this is compared with the organized seed supply systems of countries with a developed, commercial agriculture so as to illustrate the result of two to three centuries of evolution away from traditional seed management systems. This comparison is important because of the debate about ways to increase food production. We can see a broad separation between traditional and commercial approaches to farming which lies at the heart of sustainability. Modern production systems rely on high levels of technology and are driven by short-term financial considerations, often strongly influenced by official policies and legislation. The current interest in organic agriculture in many countries is a conscious effort to reconstruct a more traditional approach in pursuit of greater sustainability and other more philosophical motives.

Despite these conservational moves in technologically advanced nations, much agricultural development activity in the 'Third World' has been directed at accelerating the move away from traditional agriculture in the cause of increased productivity, which is an economic necessity in many countries. If there is to be a continuing shift towards commercial production then it is important to understand the implications that could have for genetic diversity in those countries.

In recent years there has been a dramatic increase in the study of 'farming systems', especially the more complex systems characteristic of many developing countries. This has been prompted by the realization that technological innovations alone do not solve the problems of small resource-poor farmers. Seeds have often figured in this analysis because of the debate about the use of modern/high-yielding/hybrid varieties and monocultures in relation to sustainability and genetic diversity (see Witcombe, Chapter 10, this volume). Thus seeds have a clear place in the farming systems debate. There has been a corresponding increase in publications on this subject, with many well-documented site-specific examples, but it can be difficult and unwise to generalize from these.

By comparison, the routine technical activities (the seed management systems) of the commercial seed industry are not the subject of extensive literature. Seed companies were for many years rather small local businesses which attracted little attention and generated a modest income for their owners, who were typically family members. It is only in the past 20 years that the activities of seed companies have come into the spotlight, in the context of sensitive issues such as multinational ownership, property rights and the application of biotechnology.

This chapter is concerned primarily with true seeds rather than with vegetative crops such as roots and tubers since the principles of selection and the handling practices for those materials are very different.

Seed Production and Management Systems

The impact of crop breeding systems

Good quality seed is crucial to productive agriculture. The management of this material depends very much on the reproductive behaviour of the crop, which can be cross-pollinating, self-pollinating or asexual (vegetative). Relatively few major annual crops are cross-pollinating, the most notable examples being maize, sunflower, pearl millet and, to some extent, sorghum. There are more cross-pollinators among the vegetable crops, for example the brassicas, cucurbits and onions. Among grain crops, self-pollination is the principal breeding system in almost all of the 60–70 agriculturally important cereals and pulses (Zohary, 1984).

With self-pollinating crops, diversification in the local gene pool is primarily a function of farmer adoption or abandonment of varieties. Rapid change within varieties is only likely if a mutation has a strong competitive advantage or if it is actively selected out and multiplied by an alert farmer. This was the origin of some early 'named varieties' when formal crop improvement began in the 19th century. Varieties of these self-pollinated crops appear relatively distinct and are clearly recognized/identified by farmers. A name may be used over a wide area covering a general variety type, although of course this may contain a wide range of minor genetic variation. A good example of this is the landraces of barley found in Syria (Ceccarelli *et al.*, 1987). Since there is little intrinsic genetic deterioration, farmers can maintain their stocks for many years and may see little reason to renew or replenish from external sources. Even where farmers are using 'improved varieties', they may maintain them for many years on-farm, for example in rice, which is also relatively easy to store under adverse tropical conditions.

In cross-pollinating crops, genetic changes take place much more readily than with self-pollinators, as a result of cross-pollination with other local populations of the crop or with wild and weedy relatives. Such populations are heterozygous and genetic differences are expressed at an individual level, so that plants may vary in given characteristics. However, at a population level, the sum of genetic variability remains reasonably constant, unless a completely new source of genes is introduced or a new selection pressure is imposed by cultivation. Populations are therefore less well defined and less stable, but of course more adaptable.

Through conscious selection, farmers may take advantage of the adaptability of cross-pollinated crops to develop varieties more suited to their own needs and to the local environment while simultaneously removing unwanted characteristics. Natural selection will also act on the variation within and between varieties. The result is that the nature of the variety is constantly changing. This evolving population may be maintained indefinitely in traditional cropping systems or farmers may decide to replace their seed stock from an outside source to obtain the variety in its original form. Manipulating varieties to achieve minor improvements is much more likely in cross-pollinating crops but it is also more difficult to maintain the improvement because of the constant risk of contamination.

The importance of vegetative propagation is that, as with self-pollination, genetic changes can be readily fixed in the plant population (Zohary, 1984). In this way, uniform clones or varieties (even if not formally identified) may become widely distributed and be preserved for many years as distinct varieties without any genetic change. The crops which are normally propagated vegetatively by farmers include many perennials – root and tuber crops as well as the fruit trees – which can produce seed by cross-pollination and others, such as bananas, that may never produce seed.

Clones can be continually replanted without any changes in the varietal integrity becoming apparent, although some somatic mutations may occur leading to minor variations. Indeed somatic mutation is the only source of variation in crops such as banana, where the majority of varieties are sterile triploids, never producing seeds. For crops usually clonally propagated, but retaining the capacity to produce seed, significant genetic change remains a possibility. Seedlings would then normally be highly heterozygous, could show hybrid vigour, and could contain useful gene combinations which could be easily preserved by subsequent clonal propagation. This is the origin of very many novel varieties of perennial crops, including apple and avocado.

Traditional seed systems

Landraces and modern varieties
Modern (also called improved or high-yielding) varieties have been developed through systematic plant breeding and are typified by being genetically and morphologically distinct, uniform and stable. In contrast, landraces (also known as local, traditional, folk or farmer varieties) have been defined as a 'population of plants, typically genetically heterogeneous, commonly developed in traditional agriculture from many years of farmer-directed selection, and which is specifically adapted to local conditions' (Committee on Managing Global Genetic Resources, 1993; see also Thurston *et al.*, Chapter 9, this volume).

Modern and traditional varieties represent opposite ends of a spectrum. However, all intermediate stages exist, depending mainly on seed management and the breeding system of the crop. Many modern varieties are direct selections from landraces, and, under farm conditions, modern varieties may be managed exactly as landraces. Consequently, it is often difficult to identify true landraces on-farm. Instead, it is often more practical to denote as 'local' those varieties which are being farmer-managed via home-saving of seed, irrespective of original source of planting material. Two examples of traditional management of the cross-pollinated crop, maize, serve to illustrate this point. In Nicaragua and Honduras, the improved maize varieties 'Rocamex' and 'Sintético Tuxpeño' have been transformed into local varieties through farmers' selection (Almekinders *et al.*, 1994). In Honduras, farmers were seen to select the offspring of crosses between a commercial maize hybrid, which grew on the valley bottom, with local varieties which grew on the higher valley sides, to find maize types that were suitable for growing on the intermediate lower hill sides (de Bruijn *et al.*, 1994). There may even be varietal diversification by farmers of a modern variety. Variety No. 29 of soybean was introduced into East Java in 1924 from Taiwan. It was not officially promoted after its original release but has now been diversified by farmers into a large number of varieties with differing maturity periods and yield potentials (Linnemann and Siemonsma, 1989).

In commercial seed systems, the absolute relationship of a name with a genetic identity and a definitive seed stock lies at the heart of modern variety legislation and the distinctness, uniformity and stability (DUS) test. In traditional crop production systems, names may be used casually and may be loosely associated with geographical origin or a plant type. For example, in Syria two barley landraces are found across a very large area (Ceccarelli *et al.*, 1987). They are known by local farmers under two 'generic names', 'Arabi Aswadh' and 'Arabi Abiad', respectively black and white, but both contain much genetic variation and have been used as a pool for selecting improved varieties (van Leur *et al.*, 1989).

The circumstances of traditional variety management

Within a traditional farming system, diversity is a function of the number of crops grown and the number of varieties of each crop which are in use (see Thurston *et al.*, Chapter 9, this volume). Each crop and variety is valued for specific characteristics and farmers endeavour to retain those characters. Subsistence farmers are particularly sensitive to domestic attributes of varieties such as storability and cooking quality. Farmers will often grow both modern and traditional varieties in order to benefit from the perceived advantages of each. However, the abilities of farmers to obtain or renew desired varieties will depend on the local seed system. Also, the farmer's ability or desire to produce seed on-farm (and to thereby avoid reliance on external seed sources) will depend on many circumstances. These include:

- farm size, with a tendency for larger farms to grow modern varieties (Grisley and Shamambo, 1993; Persson, 1993; Almekinders *et al.*, 1994).
- Market integration, with commercial farmers growing modern varieties (Brush, 1986; Cleveland *et al.,* 1994).
- Range of consumption needs, with local varieties meeting a wider range than modern varieties (Schneider, 1995; Louette and Smale, 1996).
- Field diversity, with traditional varieties matched to specific micro-conditions (Brush, 1986; Andrade-Aguilar and Hernandez-Xolocotzi, 1991; Thomasson, 1991).
- Farmers' ability or need to experiment with unknown varieties (Johnson, 1972; Widodo, 1995; Bellon, 1996).
- Socio-economic changes external to the farms, including labour costs (Zimmerer, 1991), and production changes by neighbours increasing weed infestation (Cummings, 1978) or grazing damage (Zimmerer, 1991).

An important parameter for understanding seed systems is the rate of varietal turnover under traditional management. Andrade-Aguilar and Hernandez-Xolocotzi (1991) found that of the 145 bean samples collected in Mexico, 84% had been in use for only 15 years or less. Other examples of high varietal turnover include Almekinders *et al.* (1994) for maize and beans in Central America, Sperling and Loevinsohn (1993) for beans in Central Africa, and a review in Wood and Lenné (1997). If varietal turnover is high, this will preclude significant local adaptation of varieties. However, adoption of a new variety need not lead inevitably to abandonment of another. In many cases farmers simply incorporate the new varieties into their existing system (Cleveland *et al.*, 1994; Brush, 1995; Bellon, 1996; see Witcombe, Chapter 10, this volume).

Farmer seed management
Seed is rarely produced as a distinct crop by traditional farmers, although some reports do exist, for example, for soybean in Indonesia by Linnemann and Siemonsma (1989). Usually material intended for seed use is selected from within the main food (grain) crop. Farmers have adopted many strategies to ensure that the seed they select maintains its viability and its value for cultivation (Janssen *et al.*, 1992; Wright and Tyler, 1994; Wright *et al.*, 1994; Bandyopadhyay and Saha, 1996). Options available to the farming community are listed in Table 13.1.

Seed selection is often the responsibility of women and selection is based on visual criteria such as large, disease-free and undamaged seeds. Seed may be selected from the standing crop, when the quality of the parent plant may be taken into consideration in addition to the quality of the seed itself. Farmers have developed elaborate management systems in order to maintain the quality of their stored seed. Almekinders *et al.* (1994) report that Central American farmers produce kidney bean seed in the off-season when pest and disease pressures are lower, to obtain better quality seed. Farmers were also

Table 13.1. Options for seed selection.

Activity	Options
Seed selection	In the field prior to harvest
	At harvest time
	During storage
	Immediately before planting
Seed cleaning	As harvested or threshed/shelled, etc.
	Hand picking to remove insect-damaged seeds
	Drying in the sun or over the kitchen fire
Seed storage	Seeds stored together with grain or in a special separate container
	Mixing with chemicals or traditional products to control insects
	Storage in the house/roof

found to give their 'tired' bean varieties to farmers in cooler, more fertile areas, for regeneration (see Engels and Wood, Chapter 14, this volume).

This practice of 'seed rotation' has also been described by Andrade-Aguilar and Hernandez-Xolocotzi (1991) as part of the local domestication process. This is said to be a common practice under rainfed conditions in Mexico and consists of cultivating beans and maize on land of different quality, over a number of years, which leads to greater adaptability of the crops.

If the production zone of a vegetatively propagated crop such as potato has a relatively high incidence of virus, seed may be produced elsewhere, under low virus incidence. Andean farmers attempt to keep potatoes free from virus by producing the 'seed crop' in a zone 1000 m above the zone of crop production (Baumann, 1992). Brush (1986) notes another function of seed rotation. Traditional potato varieties in the Andes do not store well and so are maintained by rotating seed between different altitudinal zones. In some respects the farmers are using the fields as their potato store. These rotation systems are not generally applied to improved varieties, for which alternative sources of supply exist. There are many anecdotes about these and other aspects of farmer seed management and an urgent need for research to confirm the real value of farmer practices.

Wright *et al.* (1995) found that seed selection was frequently an *ad hoc* activity. Table 13.2 shows the number of farmers leaving seed selection until just prior to planting and taking seed from their general food stocks in Ghana, Malawi and Tanzania. Similarly, Griffiths (1994) reported that in Bangladesh, the smaller the holding the less likely is the farmer to store seeds. He also suggests that in the majority of cases, 'seed' is simply obtained from the residual grain stock which is available at the next sowing season.

In areas with an unpredictable climate, farmers will often store sufficient seed for two seasons (e.g. Sattar and Hossain, 1986, for rice in Bangladesh). This compensates for bad harvests but also allows for re-planting in the event of poor crop establishment or crop failure. However, farmers who

Table 13.2. The proportion of small-scale farmers leaving seed selection until just before planting in three African countries. Source: Wright *et al.* (1995).

Country	Crop	% farmers selecting just prior to planting	Number of farmers sampled
Ghana	Maize (main crop)	58	200
	Maize (minor crop)	38	100
	Cowpea (main crop)	51	150
	Cowpea (minor crop)	21	100
Malawi	Maize	100	150
	Cowpea	71	53
	Beans	99	150
Tanzania	Maize (first season)	36	160
	Maize (second season)	8	160
	Beans (first season)	41	160
	Beans (second season)	18	160

have little land or resources to adopt this strategy are highly vulnerable and may lose their seed stocks entirely in times of environmental stress. They may have an insufficient harvest to allow a portion to be reserved for seed or they may lose their seed as a result of unforeseen events such as flood or pest outbreak. Under these circumstances, farmers are obliged to obtain their seed from off-farm sources. Although this is primarily an issue of domestic seed security, there can be more serious implications for diversity if there is widespread famine or civil unrest in which the local seed stocks are largely consumed as grain.

Farmers usually keep their varieties separate during storage. At planting, the varieties may be kept discrete or intentionally mixed. A range of benefits from mixing has been suggested, including meeting prevailing weather conditions, matching family needs, taking advantage of micro-environmental differences in the field, or as a disease avoidance strategy (Smithson and Lenné, 1996). For example, bean farmers in Rwanda reconstitute planting mixtures each year from different varietal components (Voss, 1992; see Thurston *et al.*, Chapter 9, this volume).

It has been reported that some farmers will specifically look for, and collect, off-types when selecting seeds from crops, for example Richards (1989) for rice in West Africa. These are then planted separately from the main body of the crop and evaluated during the growing season for useful characteristics. These retained off-types become the basis for a new variety if shown to be promising. This is certainly not a general practice; Fujisaka *et al.* (1993) report that Filipino rice farmers specifically select seed from areas of the field where there are no off-types. The extent of farmer selection is no doubt influenced by the complexity of the local farming system and the relative importance of the subsistence or cash economies.

In general, in traditional systems, farmers use a wide range of techniques to ensure that they keep seed in good condition for planting the following season and they may impose on that an element of variety selection/ maintenance if they act before or at harvest time (Plate 15; Fig. 13.1).

The potential role of farmers in biodiversity maintenance

Agriculture is a dynamic process. Farmers' needs evolve as their socio-economic situations change and new varieties evolve which may, or may not, be taken up by the farming community. Clearly there is benefit in farmers using the best available genetic combinations (varieties), but who should take responsibility for ensuring that the full range of genetic diversity is available in some form so that useful genes may be accessed in the future? It has been argued that farmers may be ideally placed to act as custodians of this diversity as an extension of the activities which they already carry out (see Engels and Wood, Chapter 14, this volume). Whether farmers are suitable for this role revolves around two points:

- are farmers technically capable of maintaining good quality seed?
- are farmers willing to be custodians of agricultural diversity?

Farmers have often been portrayed as being conservative and technically

Fig. 13.1. Traditional onion seed production in Nigeria (courtesy of Michael Turner, ICARDA).

poor in their ability to store seed. However, several studies have shown that, on the contrary, farmers are able to maintain the quality of their seed to a standard that is acceptable for their purposes. The seed may not comply with national seed standards but, in terms of germination and purity, the seed quality is entirely appropriate to its immediate circumstances. In fact, for the majority of staple crops, traditional farmers have to be 'seed-wise' because, until recently, no alternative source of supply existed. When new crops are introduced, such as wheat in tropical countries, there may be problems because the grain is not well adapted to storage in those difficult environ-ments. However, that does not have any impact on diversity because, in these situations, farmers have access to only a small number of varieties, often introduced from international breeding programmes.

Wright *et al.* (1995), in an extensive survey of five crops in Ghana, Malawi and Tanzania, showed that most farmer-saved samples had germin-ation in excess of 75%. Tripp *et al.* (1998) found similar results for sorghum, cowpea and groundnut in Zambia, while Gore (1987) found germination figures of more than 94% among seed saved by maize farmers in Zimbabwe. Janssen *et al.* (1992) evaluated bean seed in Colombia, and found that normally there was no yield difference between farmers' seed and nominally clean seed produced by researchers. It was unclear whether this was because the seeds were of equal quality or whether environmental conditions did not allow differences to be expressed. Tsega (1994), working with Ethiopian farmers, found that, for a range of common crops, the decline in germination over a 6–7 month storage period was less than 10% in more than three-quarters of the samples.

A further important point regarding seed quality is that many farmers traditionally use high sowing rates as an insurance against poor germination. Although technically this is an 'opportunity cost' because the extra seed has a cash value, if seed/grain is not in short supply then the 'insurance value' of the extra seed sown may be seen as greater than its value as grain. Transplanted rice is a special safeguard against poor germination because it is possible to acquire seedlings from another source if germination is poor.

It therefore seems clear that, in general, farmers are capable of main-taining seed on-farm to a sufficiently high standard to be entrusted with bio-diversity conservation activities. Whether they are prepared to do so deserves further consideration (see Smale and Bellon, Chapter 15, this volume).

Farmers are often reported as being innovators, experimenters and plant breeders, through seed selection activities (de Boef *et al.*, 1995). While it is undoubtedly true that there are always individuals who do fit this de-scription, they represent a small minority of the farming community. The examples in the literature may be of interest simply because they are unusual. For example, Berg (1992), in his often quoted paper, highlights the abilities of the seed selectors in Tigray, but also concedes that many farmers no longer practise traditional seed selection. As noted elsewhere in this chapter, seed management, other than good storage practice, is often a very *ad hoc* process for the majority of farmers.

Farmers, particularly those living in marginal areas, need to be pragmatic and they cannot afford to take risks with their crop production. They do not maintain crop varieties purely for sentimental reasons but because they serve some important function. An interesting example is given by the work of the Academy of Development Sciences (ADS) in India. ADS is a non-governmental organization based in Maharashtra which works closely with local farmers to maintain local rice germplasm. ADS set up a community seed-bank housing seed of approximately 300 local varieties and offered these to farmers for reintroduction to the farming system. However, farmers only expressed interest in about 50 of the varieties available and an evaluation of the programme, carried out in 1993, showed that not all varieties tried by the farmers were subsequently continued. It was found that of the 300 farmers who had requested traditional varieties, 60% had discontinued growing them because they no longer met their current (mostly yield) needs. The remaining 40% tended to be the smaller farmers working marginal lands to whom the varieties were still relevant (Wright and Kameswara Rao, 1997).

There are serious dangers with generalizing about farmers and their role in biodiversity conservation. Farmers throughout the world work in very different agroecological conditions, with a wide range of crops, and under a wide range of socio-economic constraints. Nevertheless, some broad features of seed management systems give indications as to how best biodiversity can be maintained.

It appears that farmers have the ability to store and regenerate seeds successfully. They also have clear requirements that dictate their choice of varieties and these change over time as circumstances alter (see Smale and Bellon, Chapter 15, this volume). Farmers, it seems, will also conserve different varieties but only while it is in their best interests to do so. As Schneider (1995) points out 'it is difficult to imagine a role for farmers doing conservation *per se* except for a few individuals with the soul of a guardian of crops' (see Thurston *et al.*, Chapter 9, this volume).

It is suggested that the responsibility for genetic variation maintenance be divided among the interested parties in a way that capitalizes on the farming reality and institutional strengths. The formal sector should take responsibility for the maintenance, regeneration, and, in particular, characterization of varieties held in genebanks. With better characterization, including the use of molecular techniques where necessary, genebanks should be able to rationalize what is being maintained. It is not necessary (or realistic) to have collections of every single ecotype so long as the genetic variability available in the crop gene pool is broadly represented in the collections. Sufficient resources should be made available from the international community in order to allow these institutions to function satisfactorily (see Engels and Wood, Chapter 14, this volume).

Farmers should be left to conserve the varieties which they wish to use, thereby allowing continuing evolution of these plant types. New varieties, as well as abandoned varieties that may have renewed value, can be offered to them but, in general, the choice of varieties to be grown will continue to rest with the farmers themselves. However, it is unrealistic to expect farmers to maintain varieties which no longer serve their domestic needs (see Smale and Bellon, Chapter 15, this volume). This role should fall instead to the wider national and international communities. The sustainability of 'coercive conservation' would be highly questionable.

Non-governmental organizations, such as ADS in India, and other community-based organizations are then ideally placed to act as facilitators and provide a communication interface between the formal, *ex situ*, system of genebanks and the informal, *in situ*, farmer system (see Engels and Wood, Chapter 14, this volume). Their close links with farmers and farmer groups should allow them to:

- determine the farmers' changing needs and to solicit suitable varieties from the genebanks to meet these requirements; and
- be alert to varieties which are being abandoned so that they may be maintained elsewhere.

This model may not be applicable in all circumstances and, importantly, it does not address issues surrounding ownership of genetic resources or of farmers' rights which fall outside the remit of this chapter. However, it does encompass a pragmatic acceptance of what is reasonable and feasible.

Farmers have, over millennia, been instrumental in creating the great diversity of crop types that currently exist but which are under threat of being lost. They have a continuing role in the maintenance and further development of that diversity through their seed selection activities.

Modern seed systems

The seed supply system which serves commercial agriculture in developed countries differs in virtually all respects from that of traditional systems just described. There has been a gradual evolution of a 'seed-trade' over the last 200–300 years, with the most rapid change occurring in the past century, leading to the highly developed seed industry which we see today. In this process, seed production has become a specialized activity within the wider context of the 'agricultural supply trade', the commercial channels through which external inputs are provided to farmers. It is beyond the scope of this review to trace this story in detail, and indeed it is influenced strongly by the social structure of agriculture and the crops grown in individual countries. However, it is important to consider the main factors which prompted this development. As in the earlier discussion about quality attributes, both 'variety' and 'seed' elements can be identified in this.

Firstly, at the end of the 19th and in the early 20th centuries, following the work of Mendel, there was the emergence of genetics as a new branch of biology, and its application in systematic crop improvement. This gave rise to distinct 'improved varieties' from plant breeding programmes. Named varieties had existed before this as farmer's selections or landraces associated with particular regions but, with the advent of plant breeding, the number of varieties increased and their identity became more clear because they had originated from a systematic selection process to a state of reasonable uniformity.

Secondly, and concurrently, there was the recognition of the components of seed quality and the emergence of seed testing to define and confirm the quality of commercial seed lots. From a small beginning in 1876 these procedures developed rapidly between 1880 and 1920 with the establishment of seed testing stations in several European countries and the introduction of minimum seed standards embodied in law (Justice, 1972; Thomson, 1979). These standards covered the major aspects of seed quality such as purity and germination. However, laboratory seed testing has one major limitation in that the identity of the seed (i.e. the variety) cannot usually be established by visual inspection. This led to the development of 'certification schemes' involving the inspection of the seed crop in the field to confirm variety identity and purity. In the course of time, the scope of seed certification expanded to embrace the laboratory standards also so that today it is seen as a comprehensive quality assurance package ultimately represented by an official label fixed to the package in which seed is marketed.

These parallel developments of seed quality standards and the breeding of new varieties underpinned the development of the seed trade. Seed production became a distinct and organized activity, which is a fundamental difference from the traditional system in which seed is a relatively minor modification of the food grain system. International organizations were also established, such as the International Seed Testing Association (ISTA) and the International Seed Trade Federation (FIS), as seed production and commerce extended far beyond national boundaries (Thomson, 1974). The technology associated with seed quality testing has continued to advance during the 20th century, but the underlying concepts and practices have not changed greatly. Neither do they have major implications for diversity in the context of this review. However, developments on variety issues have had much more impact. On the one hand, there has been a steady trend towards uniformity, driven by bureaucratic and legal requirements to describe, register and protect new varieties. At the same time there have been expectations of agronomic uniformity from farmers who are concerned to meet the needs of mechanized production and more demanding markets.

The impact of the variety regulations through testing and registration procedures have been very pronounced. High standards of uniformity are required to meet distinctness, uniformity and stability criteria, while threshold levels of performance or quality (value for cultivation and use; VCU) effectively limit the number of varieties released. Collectively, these official regulations and commercial pressures have led to the use of very uniform varieties throughout advanced agriculture and a reduction in the number of varieties in use (see Gisselquist, Chapter 16, this volume). Commercial factors also tend to favour the reduction in the number of varieties, for example by simplifying production, processing and marketing operations of seed companies (Fig. 13.2). In these circumstances, there is now no opportunity or expectation of manipulating diversity at a farm level, except for the occasional selection of an off-type by an alert farmer. Consequently, the responsibility for 'diversity management' has shifted to those who breed varieties. With the decline of public sector breeding programmes in many countries, most breeders are now employed by private companies. This creates strong competitive pressure and incentive to use the diversity which is available to best commercial advantage although this may also militate against highly innovative breeding strategies which do not provide quick results.

The typical situation found for agricultural crops in many developed markets is that a few varieties occupy a large part of the market at any given moment while there is a frequent turnover of varieties as a result of new introductions from breeders. The market is therefore dynamic but is not characterized by great diversity at any one time. There is particular concern when a variety occupies a very large part of the market in case of 'loss of resistance' to disease resulting in sudden yield decline (Brennan and Byerlee, 1991). Indeed, even when there are several varieties in use, they may share the

Fig. 13.2. Modern seed processing plant (courtesy of Michael Turner, ICARDA).

same genetic resistance factors. However, modern varieties increasingly have multiple pest and disease resistance, and are less likely to be vulnerable than earlier generations of varieties (Byerlee, 1994; Tripp, 1996; and see Witcombe, Chapter 10, this volume). Where individual varieties do have limited resistances, there has been a recurrent interest in multilines and mixtures of varieties to reduce the pathogen pressure (Wolfe *et al.*, 1992). These have not been widely adopted because of the problems of harvesting and especially of marketing seed lots containing mixed varieties; Eastern Europe is an exception.

The desire to achieve uniformity in varieties does not present major technical problems in the self-pollinating crops. The main effort is required at the stage of initial variety purification to ensure that one specific plant type is selected as the basis of the variety after segregation has finished. However, in the few cross-pollinating crops which are of major agricultural importance (i.e. maize, sunflower and sorghum) there has been a strong incentive to produce F_1 hybrid varieties which exploit the cross-pollinating habit to overcome the 'problem' of population variability and provide a means of proprietary protection.

To summarize, one may say that the transition from traditional to commercial seed management systems has seen a physical separation of the seed activity from the main farming activities which it supplies, as represented in Fig. 13.3. This has enabled the seed industry to meet the high quality standards demanded by a competitive market but it has also separated the diversity generation into specialized facilities dedicated to plant breeding. Furthermore, the existence of official testing systems places a filter between the place of diversity generation and the farmer clients who use the products.

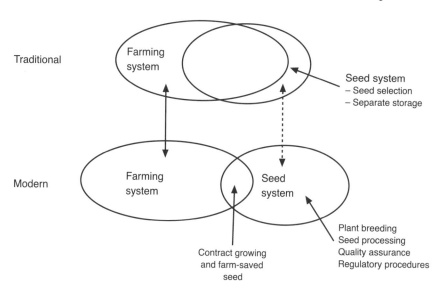

Fig. 13.3. The relationship of seed production activities to the farming system.

This filtering process is considered to be beneficial overall but it is certainly not perfect and it does also limit diversity in the production field. The incidence of 'farmer–breeders', once an alternative source of varieties, has decreased in recent years because of the substantial resources required and the difficulty of competing with large breeding programmes.

Finally, it should be noted that saving seed of proprietary varieties by farmers, which is still a widespread practice in self-pollinated crops in many developed countries, does not impinge on the diversity issue at all, although it may have technical and legal implications. Farmers who save seed do so for reasons of cost and have no expectation of modifying the variety, which in any case they could not do.

Seed Sourcing and Seed Flows

De Bruijn *et al.* (1994) note that traditional seed systems are location-specific and also vary greatly within farmers' communities. Overall it is estimated that approximately 80% of all planting material used in the developing world has been saved on-farm from previous seasons. However, this figure masks significant differences between crops, regions and socio-economic groups. For example, the traditional and often major staple crops are generally derived from home-saved material for reasons of cost and convenience. Minor and specialized crops are also sourced locally because there is no alternative. On the other hand, in crops where hybrid varieties are used, that seed will be obtained almost entirely from external sources, although in some cases farmers have discovered that such seed can meet their needs for one or two further multiplications before being replaced.

Regional differences were highlighted in a report by DANAGRO (1989) which showed that, in 1985, 98% of the area sown with maize in Swaziland was from certified seed, compared to only 14% in Tanzania. Seed sourcing has been related to wealth status, with richer farmers maintaining their own seed stock but poorer farmers needing to buy seed. For example, 18% of poor farmers in Rwanda, 36% in Burundi, and 40% in eastern Zaire have to buy all their bean seed from grain markets, i.e. for these farmers no seed is maintained on-farm (Sperling, 1994).

Cromwell *et al.* (1993) give a detailed overview of farmers' seed sources and seed distribution channels. They note that, even though seed sourcing is often relatively complex, with farmers obtaining different crops and varieties from different sources at different times, it is possible to identify four main groups of farmers with regard to seed sourcing behaviour. These are those who:

- are seed secure and can fulfil all their own seed needs;
- source seed off-farm from time to time out of choice;
- source seed off-farm from time to time out of necessity; and
- are seed insecure and consistently need to source seed off-farm.

This has implications for diversity dynamics and maintenance. Seed-secure farmers will tend to maintain their own varieties with limited influx of new varieties, however, they may also manage a wider range of varieties than other farmers. For example, in a study undertaken in the Cuzalapa region of western Mexico, Louette and Smale (1996) showed that those maize farmers who sourced their seed from their own supplies tended to plant more varieties per cultivation cycle than smaller-scale farmers. Curiously, most of the farmers in the survey did not express any particular preference for using their own seed over that of another farmer. This would suggest that 'variety awareness' is not always as well developed in traditional farming communities as visiting researchers would like to believe. It may also reflect the fact that in traditional 'self-contained' seed systems, the same genetic material may be easily available from neighbours, thus reducing the pressure for maintenance by every household.

Seed-insecure farmers may use a wide range of varieties, depending on what is available, but they may not necessarily continue to use them from one year to the next. Typically it is the very smallest scale or tenant farmers who fall into this category (Sperling and Loevinsohn, 1993; Louette and Smale, 1996). Those farmers who source seed off-farm, for whatever reason, are the ones most likely to introduce and subsequently maintain new varieties in their farming system. However, since they may abandon, as well as adopt varieties, the total number of varieties they grow may remain constant although their varietal portfolio alters over time. This is important because it embodies the concept that, under stable conditions, farmers assimilate, evaluate and discard varieties in a reasoned way. This is what would be expected of conservative people, whereas it is often assumed that the availability of a new 'improved' variety causes an abrupt abandonment of traditional materials. Cases where this has happened have been documented for individual farmers (Wright and Tyler, 1994) but these are by no means universal.

Farmers sourcing seed off-farm will often obtain seed from other farmers and, in some communities, these may develop a reputation as being reliable sources of good quality seed (Berg, 1992; Almekinders *et al.,* 1994; Louette and Smale, 1996). The proportion of the farming community involved as seed producers *cum* distributors is usually small. Furthermore, it is often difficult to establish whether these 'local seed suppliers' are making a conscious effort to produce high-quality seed or if they are simply well-endowed farmers who always have surplus grain to sell as 'seed' at the next planting season. Regardless of their precise status and activities, they play an interesting role as they could be the point from which a more organized local supply system could be developed.

In that connection, Sperling and Loevinsohn (1993) found that for newly introduced beans in the African Great Lakes region, 6% of farmers were responsible for almost half the distributions to other farmers and that 55% of the farmers who had used the new varieties for three seasons had not yet begun to pass on their seed. This low figure may reflect the fact that the farmers had not been growing the crop for long enough to have produced a surplus, rather than their unwillingness to pass on seed to others. It is possible, therefore, that the proportion of farmers sharing their seed with others would be higher for better established varieties. It also emphasizes the need to identify and use these key farmers as 'local diffusers' of improved varieties.

Farmers may also become seed sources in response to needs from other areas. For example in Indonesia, to overcome the inherent difficulties of maintaining the quality of soybean seed in a humid climate, some villages in East Java have become specialized in growing soybean on drylands during the rainy season. Although yields are lower at that time of year, this seed forms the basis of the main dry-season crop and so commands a higher price (Linnemann and Siemonsma, 1989).

Vegetables present an interesting case in seed sourcing, with a strong separation between traditional indigenous species/types which are saved/ obtained locally and introduced species which are supplied from the inter-national seed trade. In tropical regions, the introduced vegetables are typically those of temperate regions for which seed production in the tropics is difficult or impossible. In these crops, there are a number of very well-known 'international varieties' which still command a substantial market by reason of their low price but which are now largely superseded by improved forms or hybrids in more developed markets. Examples include the cabbage 'Copenhagen Market', tomato 'Moneymaker' and onion 'Texas Grano'.

It is also possible to find strong local diversity in specific characteristics which are not found in the 'international varieties'. For example the small-(often white) fruited egg-plants of West Africa known as 'Djakhatou', which are appreciated for their bitterness, are the same species as the large-fruited purple varieties of the international markets. Also in West Africa, the local okra varieties are valued for their gelatinous texture when cooked, a charac-teristic not appreciated elsewhere. Therefore, in vegetables in tropical areas one may find very little diversity in the exotic varieties and great diversity in the local ones.

Another interesting feature of vegetable crops is the scope for selection by farmers. Since vegetables are grown for diverse quality attributes such as colour, flavour and texture, as well as yield, there are many opportunities for observant farmers to select and preserve superior forms or interesting variants. However, it is also true that, in traditional farming systems, there may be a reluctance to leave the best plants to seed and indeed there is evidence that in some crops such as cucurbits it is the poor fruit which are used for seed after the good ones have been eaten or sold.

To summarize, there are many factors which can influence farmers' sourcing of seed in traditional agricultural systems. Some of these are delib-erate and purposeful, in that the farmer is pursuing a rational strategy. Others are random and result from instability in the social or natural environment. In the context of managing diversity, several different forces may be at work. Well-endowed farmers may have the luxury of selecting the particular varieties they want to grow and use to meet all their needs and preferences but they may be more likely to pursue the path of commercial production, which may lead to reduced diversity in their fields. On the other hand, poorer farmers in more risky environments may wish to use and conserve varieties which they know and trust, in the cause of risk-aversion, but they may be more vulnerable to sudden changes which cause loss despite their efforts. This has profound implications for developing strategies for *in situ* conservation (see Engels and Wood, Chapter 14, and Smale and Bellon, Chapter 15, this volume).

In developed countries with an established agricultural supply trade, seed sourcing by farmers is essentially a management decision to buy from one supplier or another depending on commercial factors such as price and delivery arrangements. In some cases, proprietary varieties may be available from only one exclusive supplier but generally the farmer has a choice of where to purchase. However, the choice of variety is still limited to those which are currently on the official lists and in the commercial trade. Thus, regardless of the point of purchase, the range of diversity available is determined by bureaucratic and commercial decisions which are taken higher up the supply chain (see Gisselquist, Chapter 16, this volume).

If the farmer decides not to purchase, the alternative is to save seed on the farm. There is a mistaken belief in some quarters that farm-saved seed is now an uncommon practice in developed agriculture, but for the self-pollinated cereals it may still account for a significant share of the total market. Of course, this depends on the crop; seeds of hybrids are re-purchased every year and the more specialized crops such as herbage and oilseeds are also bought in from outside because seed production is managerially complex. One other major incentive for farmers to purchase seed is to obtain chemically treated seed – difficult or impossible for farmers to carry out themselves.

A common problem in sourcing seed of appropriate varieties is that the 'recommendation domains' of varieties – that is, the areas targeted by breeders – may not correspond to farmers' judgement of the value of the variety. A rice variety, Mahsuri, became one of the most popular varieties in India despite having been rejected by researchers for having poor performance (Maurya *et al.*, 1988). Variety release and seed production at state level in India make it difficult for farmers in other states to know about or to obtain seed of a variety, which may be well suited to their needs and conditions (Witcombe *et al.*, 1998).

Divisions within the seed supply system

Much of the discussion in the preceding sections has considered traditional and modern seed systems in isolation. However, in virtually all countries these coexist and supply different parts of the total seed requirement. Development projects implemented over the past 25 years or so have put in place the key components of organized seed supply, such as seed processing plants, stores, vehicles and quality control facilities. The seed industry model adopted by these projects was similar to that which had evolved in countries with a highly developed commercial agriculture. There was an expectation in early seed projects that when new varieties and high-quality seed became available, this would be quickly adopted by farmers and would be a decisive factor in increasing productivity.

With hindsight, these expectations have proved to be optimistic. The organized (or formal) seed sector has made some progress in certain crops, notably with hybrid varieties, but very little in other cases, where the traditional (informal) system remains largely intact. These terms 'formal' and 'informal' to describe the seed sector have come into common usage in recent years but the meanings attached to them vary. We believe the most practical usage is to consider the organized/commercial seed supply as comprising the formal sector while all other supply channels comprise the informal sector. Thus the total supply for any country or crop can be partitioned between these two supply mechanisms. The defining features of the formal sector are:

- planned seed production, separate from grain production;
- use of defined named varieties, usually officially tested and registered;
- use of mechanical processing to improve quality after harvest;
- marketing in sealed, labelled packages, not from a bulk open container; and
- having some declaration or other assurance of quality, often by 'certification'.

In contrast, the informal sector generally lacks these 'seed-specific' procedures and makes a less clear distinction between grain and seed. However, the informal sector may certainly provide seed of modern, improved varieties when seed is saved on-farm. Similarly there is no reason why a traditional or local variety should not be assimilated into the formal supply system when a demand exists.

This partition of the seed supply, in developing countries, is represented diagrammatically in Fig. 13.4. Formal sector suppliers are typically seed companies, either private or public. Informal supplies come mostly from farm-saved seed or from local trading and exchange at the community level. The division between the formal and informal sectors cannot be precise and there may be movement, for example where small local suppliers or traders become more organized and effectively join the formal sector. Furthermore, these two supply mechanisms should not be regarded as independent or competitive; rather they are complementary – each occupying that part of the market in which they have comparative advantage. This conceptual division is useful, for example in considering how best to allocate resources to achieve an overall improvement in seed supply.

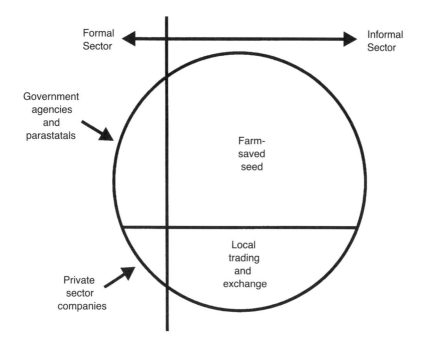

Fig. 13.4. Generalized representation of the national seed supply in many developing countries. (The circle represents the entire seed supply.)

There is interest in bringing these two approaches to seed supply together in a more systematic way; hence the concept of 'integrated seed systems' advanced by Louwaars (1994). There is also much interest by development agencies in 'strengthening the informal sector', but that presents a dilemma because any substantial external intervention may distort the existing system making it less secure and sustainable when the funding is withdrawn.

References

Almekinders, C.J.M., Louwaars, N.P. and de Bruijn, G.H. (1994) Local seed systems and their importance for an improved seed supply in developing countries. *Euphytica* 78, 207–216.

Andrade-Aguilar, J.A. and Hernandez-Xolocotzi, E. (1991) Diversity of common beans (*Phaseolus vulgaris* Fabaceae) and conditions of production in Aguascalientes, Mexico. *Economic Botany* 45, 339–344.

Bandyopadhyay, A.K. and Saha, G.S. (1996) *Survey on Crop Area, Seed Rate, Sources of Seeds and Indigenous Seed Preservation Techniques of Andaman Farmers.* Central Agricultural Research Institute, Port Blair, Andaman and Nicobar Islands, India.

Baumann, M. (1992) Promoting local conservation in Ecuador. In: Cooper, D., Vellvé, R. and Hobbelink, H. (eds) _Growing Diversity: Genetic Resources and Local Food Security._ Intermediate Technology Publications, London, pp. 116–124.

Bellon, M.R. (1996) The dynamics of crop infraspecific diversity: a conceptual framework at the farmer level. _Economic Botany_ 50, 26–39.

Berg, T. (1992) Indigenous knowledge and plant breeding in Tigray, Ethiopia. _Forum for Development Studies_ 1, 13–22.

de Boef, W.S., Berg, T. and Haverkort, B. (1995) _Farmers, Crops and Landraces. Farmers' Roles in the Development and Conservation of Crop Diversity._ CPRO-DLO Centre for Genetic Resources, Wageningen.

Brennan, J.P. and Byerlee, D. (1991) The rate of crop varietal replacement on farms: measures and empirical results for wheat. _Plant Varieties and Seeds_ 4, 99–106.

de Bruijn, G., Almekinders, C. and Keune, L. (1994) Farmers' knowledge and practices in smallholder seed production, with special reference to a case study in Central America. In: Hanson, J. (ed.) _Proceedings of the ILCA/ICARDA Research Planning Workshop._ ILCA, Addis Ababa, Ethiopia, pp. 29–38.

Brush, S.B. (1986) Genetic diversity and conservation in traditional farming systems. _Journal of Ethnobiology_ 6, 151–167.

Brush, S.B. (1995) _In situ_ conservation of landraces in centers of crop diversity. _Crop Science_ 35, 346–354.

Byerlee, D. (1994) _Modern Varieties, Productivity, and Sustainability: Recent Experiences and Emerging Challenges._ CIMMYT, Mexico.

Ceccarelli, S., Grando, S. and van Leur, J.A.G. (1987) Genetic diversity in barley landraces from Syria and Jordan. _Euphytica_ 36, 389–405.

Cleveland, D.A., Soleri, D. and Smith, S.E. (1994) Do folk crop varieties have a role in sustainable agriculture? _BioScience_ 44, 740–751.

Committee on Managing Global Genetic Resources (1993) _Managing Global Genetic Resources: Agricultural Crop Issues and Policies._ Board on Agriculture, National Research Council, National Academy Press, Washington, DC.

Cromwell, E., Wiggins, S. and Wentzel, S. (1993) _Sowing Beyond the State: NGOs and Seed Supply in Developing Countries._ Overseas Development Institute, London.

Cummings, R.C. (1978) Agricultural change in Vietnam's floating rice region. _Human Organisation_ 35, 235–245.

DANAGRO (1989) _SADCC Regional Seed Production and Supply Project: Main Report (Volume 1A)._ DANAGRO Adviser a/s, Glostrup, Denmark.

Fujisaka, S., Guino, R.A., Lubigan, R.T. and Moody, K. (1993) Farmers' rice seed management practices and resulting weed seed contamination in the Philippines. _Seed Science and Technology_ 21, 149–157.

Gore, C.H. (1987) Summary report of ENDA-Zimbabwe's indigenous small-grains programme. In: AGRITEX/GTZ (eds) _Proceedings of a Workshop – Cropping in the Semi-Arid Areas of Zimbabwe._ GTZ-GART, Masvingo, Zimbabwe.

Griffiths, R.G. (1994) Smallholder seed production techniques. In: Hanson, J. (ed.) _Proceedings of the ILCA/ICARDA Research Planning Workshop._ ILCA, Addis Ababa, Ethiopia, pp. 11–13.

Grisley, W. and Shamambo, M. (1993) An analysis of the adoption and diffusion of Carioca beans in Zambia resulting from an experimental distribution of seed. _Experimental Agriculture_ 29, 379–386.

Janssen, W., Luna, C.A. and Duque, M.C. (1992) Small-farmer behaviour towards bean seed: evidence from Colombia. *Journal of Applied Seed Production* 10, 43–51.

Johnson, A.W. (1972) Individuality and experimentation in traditional agriculture. *Human Ecology* 1, 149–159.

Justice, O.L. (1972) Essentials of seed testing. In: Kozlowski, T.T. (ed.) *Seed Biology*, vol. III. Academic Press, New York, pp. 301–370.

van Leur, S., Ceccarelli, S. and Grando, S. (1989) Diversity for disease resistance in barley landraces from Syria and Jordan. *Plant Breeding* 103, 324–335.

Linnemann, A.R. and Siemonsma, J.S. (1989) Variety choice and seed supply by smallholders. *ILEIA Newsletter,* December 1989, pp. 22–23.

Louette, D. and Smale, M. (1996) *Genetic Diversity and Maize Seed Management in a Traditional Mexican Community: Implications for* In Situ *Conservation of Maize.* NRG Paper 96-03, CIMMYT, Mexico.

Louwaars, N.P. (1994) Integrated seed supply, a flexible approach. In: Hanson, J. (ed.) *Proceedings of the ILCA/ICARDA Research Planning Workshop.* ILCA, Addis Ababa, Ethiopia, pp. 39–45.

Maurya, D.M., Bottrall, A. and Farrington, J. (1988) Improved livelihoods, genetic diversity and farmer participation: a strategy for rice breeding in rainfed areas of India. *Experimental Agriculture* 24, 311–320.

Persson, L. (1993) *A Study on Adoption of Improved Maize Among Farmers in Zambia: a Minor Field Study.* Working Paper 224, Swedish University of Agricultural Sciences, Uppsala, Sweden.

Richards, P. (1989) Farmers also experiment: a neglected intellectual resource in African science. *Discovery and Innovation* 1, 19–25.

Sattar, M. and Hossain, S.M.A. (1986) An evaluation of farmers' technology for seed production and post-harvest operations in *aus* rice. *Bangladesh Journal of Extension Education* 1(2), 1–12.

Schneider, J. (1995) Farmer practices and sweetpotato diversity in highland New Guinea. In: Schneider, J. (ed.) *Indigenous Knowledge in Conservation of Crop Genetic Resources. Proceedings of an International Workshop, Bogor, Indonesia.* CIP-ESEAP/CRIFC, Bogor, pp. 63–70.

Smithson, J.B. and Lenné, J.M. (1996) Varietal mixtures: a viable strategy for sustainable productivity in subsistence agriculture. *Annals of Applied Biology* 128, 127–158.

Sperling, L. (1994) *Analysis of Bean Seed Channels in the Great Lakes Region: South Kivu, Zaire, Southern Rwanda and Select Bean Growing Areas of Burundi.* CIAT African Occasional Publications Series, No 13. CIAT/RESAPAC, Butare, Rwanda.

Sperling, L. and Loevinsohn, M.E. (1993) The dynamics of adoption: distribution and mortality of bean varieties among small farmers in Rwanda. *Agricultural Systems* 41, 441–453.

Thomasson, G.C. (1991) Liberia's seeds of knowledge. *Cultural Survival Quarterly*, Summer 1991, 23–28.

Thomson, J.R. (1974) The International Seed Testing Association. 1924–74. *Seed Science and Technol*ogy 2, 267–283.

Thomson, J.R. (1979) *An Introduction to Seed Technology.* Leonard Hill, Glasgow.

Tripp, R. (1996) Biodiversity and modern crop varieties: sharpening the debate. *Agriculture and Human Values* 13, 48–63.

Tripp, R., Miti, F., Mukumbuta, S., Walker, D.J. and Zulu, M. (1998) Farmers' seed management in Zambia – A study of sorghum, cowpea and groundnut seed in Southern and Western Provinces. _NRI Research Report_. Natural Resources Institute, Chatham Maritime, UK.

Tsega, M. (1994) _An Inventory and Investigation of the Optimum Local Seed Storage Methods in Wello and Shewa Administrative Regions_. Seeds of Survival, PO Box 5760, Addis Ababa, Ethiopia.

Voss, J. (1992) Conserving and increasing on-farm genetic diversity: farmer management of varietal bean mixtures in Central Africa. In: Moock, J. and Rhoades, R.E. (eds) _Diversity, Farmer Knowledge, and Sustainability_. Cornell University Press, Ithaca, New York, pp. 34–51.

Widodo, Y. (1995) Sweetpotato cultivation in a rice-based farming system: the dynamics of indigenous knowledge. In: Schneider, J. (ed.) _Indigenous Knowledge in Conservation of Crop Genetic Resources. Proceedings of an International Workshop, Bogor, Indonesia_. CIP-ESEAP/CRIFC, Bogor, pp. 105–114.

Witcombe, J.R., Virk, D.S. and Farrington, J. (eds) (1998) _Seeds of Choice: Making the Most of New Varieties for Small Farmers_. Oxford and IBH Publishing Co, New Delhi.

Wolfe, M.S., Brändel, U., Koller, B., Limpert, E., McDermott, J.M., Müller, K. and Schaffner, D. (1992) Barley mildew in Europe: population biology and resistance. In: Johnson, R. and Jellis, G.J. (eds) _Breeding for Disease Resistance_. Kluwer, Dordrecht, pp. 125–129.

Wood, D. and Lenné, J.M. (1997) The conservation of agrobiodiversity on-farm: questioning the emerging paradigm. _Biodiversity and Conservation_ 6, 109–129.

Wright, M. and Kameswara Rao, N. (1997) _Agricultural Crop Biodiversity in India – the NGO Response to its Decline_. NRI Research Report, Natural Resources Institute, Chatham Maritime, UK.

Wright, M. and Tyler, P. (1994) _Traditional Seed-saving Practices in Northern Ghana and Central Malawi_. NRI Report R2102(S), Natural Resources Institute, Chatham Maritime, UK.

Wright, M., Donaldson, T., Cromwell, E. and New, J. (1994) _The Retention and Care of Seeds by Small Scale Farmers_. NRI Report R2103, Natural Resources Institute, Chatham Maritime, UK.

Wright, M., Delimini, L., Luhanga, J., Mushi, C. and Tsini, H. (1995) _The Quality of Farmer Saved Seed in Ghana, Malawi and Tanzania_. NRI Research Report, Natural Resources Institute, Chatham Maritime, UK.

Zimmerer, K.S. (1991) Labor shortages and crop diversity in the southern Peruvian Sierra. _The Geographical Review_ 81, 414–432.

Zohary, D. (1984) Modes of evolution in plants under domestication. In: Grant, W.F. (ed.) _Plant Biosystematics_. Academic Press, Don Mills, Canada, pp. 579–586.

14

Conservation of Agrobiodiversity

J.M.M. Engels[1] and D. Wood[2]

[1]IPGRI, Via delle Sette Chiese 142, 00145 Rome, Italy; [2]Agrobiodiversity International, 13 Herons Quay, Sandside, Milnthorpe, Cumbria LA7 7HN, UK

Introduction

The purpose of conserving agrobiodiversity is to ensure its availability now and in the future. Thus, there is a specific utilitarian view to conservation that sees agrobiodiversity as a resource. The more potentially useful the germplasm or genes, the more effort will be placed in collecting and conservation, and the more immediately the sample will be used directly in production or indirectly as a genetic resource. Further, there is also a conservationist's view: agrobiodiversity as part of biodiversity. The more threatened the biodiversity, the greater need to conserve it, as otherwise perhaps unique characteristics might be lost.

There is an additional driving force to the conservationist's view, unique to agriculture and forestry: the use of collected samples for agricultural production, directly or after breeding, could lead to the replacement (and therefore loss) of populations in the field. These replaced populations may be a potential agrobiodiversity resource, of crops, domestic animals, and their close or distant wild relatives, in highly managed landscapes or natural vegetation. The level of human influence is another distinctive feature of much agrobiodiversity: crops and domestic animals evolved only because of human management, and such management is needed for continued evolution, or even persistence, of crop varieties and animal breeds (see Wood and Lenné, Chapter 2, this volume).

In addition to the need to conserve domesticates, increasing emphasis is now being given to the conservation of unplanned agrobiodiversity: the useful and harmful microorganisms, insects and pathogens of agroecosystems, and even to transient species such as migratory birds, where the agro-ecosystem may provide an essential seasonal habitat. The technology of conservation will differ markedly within and between different types of plants and animals, and between conservation on-site (*in situ*) and off-site (*ex situ,* often in different countries). Added to complex technical choices there are equally complex policy and institutional issues related to funding, management and control of agrobiodiversity, resulting from the great dependence of nations on access to genetic resources from elsewhere (Kloppenburg, 1988).

We will look first at the historical background to agrobiodiversity conservation, then recent arguments as to why there is a need for conservation, then – briefly – conservation policy and technology. Discussion of the technology required for the conservation of domestic animal breeds is confined to Steane, Chapter 4, this volume.

Evolution of Approaches to Conservation

For most of the history of agriculture, 'conservation' of agrobiodiversity has been a necessary part of food production. Storage of propagating material was all that was needed. In strongly seasonal climates seed was stored for the following season, sometimes separated from grain for consumption, sometimes stored together (see Wright and Turner, Chapter 13, this volume). In more equable climates, vegetative propagating material was planted at once or stored for only short periods. With propagating material under the control of the farmer, and part of the food production cycle, a specific concept of conservation was not needed. This *de facto* conservation of genetic resources by farmers for their own use was, and is, the most important and effective in terms of number of varieties and areas covered. Also for wild relatives of crops, 'non-managed conservation' is still more important than managed conservation.

Crop introduction

At various times in the history of agriculture and horticulture, farmers needed access to the genetic resources of others, perhaps from some distance. This would apply as much to a farmer's trip to market to buy seed as to a naval expedition to obtain breadfruit from the South Pacific. The care invested in conserving the samples would be related to the need to avoid future effort and cost in obtaining the sample again. The common thread was that most early efforts at conservation were associated with plant introduction. The introduction of crop diversity through plant introduction programmes was a boon to colonial expansion. Food-producing species were introduced from their region of origin to homologous climates elsewhere. Grapes were moved from the Mediterranean to South Africa, Chile and California; wheat was grown in Argentina, Canada and Australia; sweet potato diffused after its introduction from South America through the Pacific. The 'Columbian transfer' between the New and Old Worlds was, and still is, very important. The list of introduced crops is extremely wide: most food production in most developing countries is from introduced crops.

Early efforts at introduction had often a narrow genetic base of a very few varieties or diversity within varieties (e.g., limited cattle breeds, a few rubber seeds, limited coffee diversity). Even for staple food species, with dry, easily transported seed, there seems to have been a remarkable lack of within-sample diversity: broom corn (*Sorghum technicum*) was introduced to the USA from a few seeds removed from a brush. As reported by Percival (1925) 'Fife wheat, one of the world's best varieties, which for a long time dominated the vast corn growing regions of Canada and the United States, originated from three ears, apparently the produce of a single plant, grown from a sample of Dantzig wheat sent by a friend from Glasgow about 1842 to David Fife in Ontario.'

Countries with a central responsibility for a great range of agro-ecological conditions were highly active in the introduction of genetic resources. The British system of colonial botanic gardens specialized in the intertropical movement of plantation crops. Most of the world's production of pineapple is based on the clonal descendants of only five plants taken by the French from Cayenne to a glasshouse in Versailles in 1820 and later distributed widely (Purseglove, 1972). The great cost of these early collecting expeditions imposed a correspondingly high value to conserving the collected material as 'back-up' samples in the security of state botanic gardens, if only to avoid the cost of recollecting. A specimen of the first introduction of oil palm (*Elaeis guineensis*) to Java is still growing in the Bogor Botanic Garden after 150 years.

Countries with an expanding agricultural frontier (and an initially limited capacity for plant breeding) needed continued access to a wide range of crops and varieties. Characteristically, the USA and Russia had early and large investments in the introduction and then conservation of genetic resources as the basis for research, trials and improvement. The USA is now the best example of an integrated national germplasm system (Committee on Managing Global Genetic Resources, 1993). The Soviet Union was an early model of scientific plant exploration and introduction through the work of the great agricultural botanist, N.I. Vavilov (Vavilov, 1951).

Breeding

Modern plant breeding, initiated at the end of the 19th century and the beginning of the 20th century, accelerated this process of introduction and evaluation as professional breeders needed access to genetic diversity to use in their programmes (Bauer, 1914; von Tschermak, 1915, for early arguments on the need to conserve and use genetic resources). This diversity was usually maintained in breeders' or working collections of carefully selected germ-plasm, often for immediate use. Use, rather than conservation, was the main priority of these breeding programmes. Many valuable collections were assembled by individuals, and were frequently lost following staff changes or retirement.

After an early narrow introduction through colonial botanic gardens, subsequent collections of a great diversity of major tropical plantation crops were maintained in field collections associated with tropical research institutes (cacao in Trinidad and Costa Rica, cotton in Sudan, coffee in Côte d'Ivoire, and many others). Some of these survived national independence, often as part of regional research programmes; others were lost.

For both humanitarian and political reasons, there was a breeding emphasis in the 1960s on tropical staple food crops such as rice and maize, with notable success by the Ford and Rockefeller foundations. This success was continued by the newly established International Agricultural Research Centres (IARCs) of the Consultative Group on International Agricultural Research (CGIAR). These centres, including the International Rice Research Institute (IRRI) and the Centro Internacional de Mejoramiento de Maíz y Trigo (CIMMYT, for maize and wheat), began to assemble large collections of their mandate crops through planned collecting of landraces and old varieties and by incorporating existing collections of national research/breeding programmes into their collections (Fuccillo *et al.*, 1997). The philosophy was simple: farmers maintained and used their crops in the field (*in situ*) and breeders maintained and used collections of these crops in institutions (*ex situ*). The International Board for Plant Genetic Resources (IBPGR, now IPGRI (International Plant Genetic Resources Institute)) was established in 1974 to coordinate a global effort to systematically collect and conserve threatened crop genetic resources (Pistorius, 1997). We refer to the work of IPGRI as a case study of the changing needs and evolving approaches to conservation (see below).

Genetic erosion

The continuing adaptive change and development of agriculture has always been associated with genetic erosion – the loss of formerly favoured crop varieties or genes and alleles, and domestic animal breeds in the field. This process certainly stretches back to prehistory, as successive waves of first new crops and species of domestic animals, and then varieties and breeds, spread from the centres of origin of agrobiodiversity, replacing earlier varieties and even species.

For Africa alone, crop introduction must have had a great toll on indigenous crops: we do not even know which African crops (if any) were replaced by sugar cane and bananas, first introduced to Africa more than

1000 years ago. Later introductions to Africa included cassava, partly replacing indigenous yams, groundnut replacing (as an oil crop) sesame, maize partly replacing sorghum, and *Phaseolus* beans partly replacing cowpea. Therefore, at the unquantifiable cost of losing some of the variation of its indigenous crops, Africa gained entirely new crops, and there has since been a substantial diversification of these (e.g. maize and *Phaseolus* beans). There are no African crops capable of replacing the productivity of introduced sweet potato, *Phaseolus* beans and bananas in, for example, the fertile Central Highlands of Africa. However, in some regions there was a useful complementarity between introduced crops and local crops, as in the highlands of Ethiopia, where the Near-Eastern crop complement of barley, wheat and chickpea was joined in production by the local domesticates, teff (*Eragrostis tef*), sorghum, noog (*Guizotia abyssinica*), and others.

For domestic animals, pure breeds have, in the past, been conserved worldwide by farmers for their distinctive characteristics. However, ongoing changes in agriculture may threaten minor local breeds (see Steane, Chapter 4, this volume). Even major breeds such as the Sahiwal cattle in Pakistan and India may come under threat from changing agricultural ecologies. Following the rapid spread of canal irrigation in the Punjab, Sahiwal bullocks were found to be inadequate for the new, intensive farm operations (Philipsson, 1992) and began to be replaced on-farm. The dairy qualities of Sahiwals were challenged by buffaloes, producing a higher-fat milk. As with varieties of crops, crosses between animal breeds could produce hybrids more desired than the pure breed. Without the specific conservation of the pure breeds, the ability to produce consistent hybrids would be lost. Conservation of the Sahiwal breed began at Pusa in Bihar in 1912, and has since spread to private, military and state farms in India and Pakistan.

However, the recognition that genetic erosion could be a constraint to future agricultural production is relatively recent. The increasing pace of agricultural development in tropical countries began to impact on the 'centres of origin' and diversity of our crops, and there was alarm that genetic diversity needed for future breeding of crops and domestic animals would be thereby lost. The former exclusive emphasis on 'conservation for convenient access' was supplemented by the idea of 'conservation to safeguard a vanishing resource'. The urgent need was for agricultural genetic resources subject to genetic erosion on-farm to be conserved off-farm (*ex situ*): breeders' working collections became genebanks, and storage technology advanced to meet new needs for long-term conservation.

A key event in the recognition of genetic erosion was the FAO Technical Conference on the Exploration, Utilization and Conservation of Plant Genetic Resources, held in 1967. Commenting on this Conference, Frankel and Bennett (1970) noted concern had been heightened: 'on one hand by the growing awareness of the immense value of representative germplasm collections for plant breeders, geneticists, pathologists, evolutionists and others, and on the other by the growing threat to the continued existence to the gene pools in their areas of diversity.' This period has been reviewed by Frankel (1987) and Pistorius (1997). As a result of the concerns over genetic erosion, there was an explosive growth of national and international genebanks during what FAO terms the 'crisis years' of the 1970s and early 1980s (FAO, 1996a). FAO (1996a) has published a most useful compilation of facts of the 'State of the World' in genetic resources management

Access

A more recent concern has been international access to genetic resources. The Convention on Biological Diversity (CBD), a full legally binding international convention which came into effect in 1993, endorsed national sovereignty over biodiversity – including genetic resources. Existing national and international collections at once became strategic resources: secure conservation was needed against any future restrictions on collecting, access and global exchange of germplasm. The debate continues in the FAO Commission on Genetic Resources on equitable systems of access to agricultural genetic resources, consistent with the provisions of the CBD (and see Gisselquist, Chapter 16, this volume).

The Need for Conservation

Variability for evolution

The overall objective in conserving agrobiodiversity is to ensure the availability of diversity of genetic resources for subsequent utilization (Simmonds, 1962, 1993); however, ultilization in the longer term implies selection between or within varieties – either on-farm, or by breeders (see Thurston *et al.*, Chapter 9, and Witcombe, Chapter 10, this volume). Throughout biology, selection reduces variability. Yet this variability may be useful at a later date, or in other regions. In nature, variability is normally restored over time through mutation, genetic recombination, or gene flow from neighbouring populations, but this may be a slow or uncertain process. The reality of the loss of diversity through selection in agriculture was recognized by Simmonds (1962) who argued that when breeding achieves closer adaptation, it tends to reduce variability and thus reduce long-term adaptability. Simmonds noted that: 'it is a commonplace of genetics that adaptation (fitness to a given environment) and adaptability (capacity for genetic change in adaptation) are antagonistic'. Through conservation we can address this antagonism by maintaining, and rapidly deploying, the variation that is the raw material for the selection and continued evolution of crops and domestic animals.

Decisions on support for conservation will depend on indicators such as:

- expected value of the collections for immediate or future use. A crop of high global value such as wheat (see Cox and Wood, Chapter 3, this volume) will be given more global emphasis than, say, Macadamia nut. A crop amenable to the efforts of breeders, for example, sugar cane (Galloway, 1996) is more liable to be conserved than a 'recalcitrant' crop such as banana. Effort invested in evaluating collections will add value to this indicator.

- Effort invested in obtaining genetic resources (including re-obtaining samples lost during conservation). Rarity in the field would increase this indicator.

- Relative costs of conservation technology. Low-cost technology for cereal seed allows tens of thousands of samples to be stored, while the more expensive *in vitro* or field genebank maintenance limits sample numbers of many of the more tropical crops.

Breeders have been by far the most important immediate clients for conservation programmes. Their access to intraspecific diversity of crops and domestic animals in national and international germplasm collections has been essential to increase food production globally. Vavilov regarded breeding as evolution at the will of man (Frankel and Bennett, 1970, p. 7): agrobiodiversity from conserved samples is now the major source of variation for this human selection for varietal improvement. However, evolution through selection has a cost to diversity: conservation can reduce this cost. With natural selection, genetic traits selected against may be lost, even if they could be of value in other conditions or later. In contrast, with human selection, the diversity of conserved material can be: (i) rapidly deployed through controlled crossing and segregation; (ii) strongly selected for characters of value; and (iii) conserved to meet changing future needs. It is essential, for example, that strong human selection of samples for resistance to one race of pathogen should not destroy within- and between-sample variation encompassing resistance to another race of possible future importance. There are many early examples of breeders assembling, screening for use, and then abandoning collections. The deliberate conservation of diversity, even of 'evolutionary rejects', allows unpredictable future needs to be met.

Diversity for deployment

However, more conservation is not the answer to problems of genetic uniformity in crops. Simmonds (1993) has noted that the genetic vulnerability of crops can be addressed by a better deployment of already available diversity through plant breeding (see also Duvick, 1984, and Witcombe, Chapter 10, this volume). It may also be argued that diversity is needed to ensure the continuing existence of stable (sustainable) ecosystems, but, again, for agroecosystems this is mainly a problem of deploying diversity as needed, rather than conserving each and every agroecosystem. Equally, many arguments for the value of diversity, for example to meet dietary needs, to smooth peaks of labour demand, or to reduce pest pressure (see Thurston *et al.*, Chapter 9, this volume), can be best addressed through better access to, and deployment of, components of diversity, rather than more conservation.

Species and interactions

Other purposes may underlie conservation efforts. The most important purpose of all would be to prevent the extinction of useful species. Crop and domestic animal species depend for their survival to a great extent on human management. If human interest wanes, there is a danger of irreversible species loss. There are many once-important species that are threatened in this way which can be conserved for future needs (National Research Council, 1989). There is a greater danger of the loss of wild relatives of crops – which cannot be 'conserved through use' – and domestic animals: these are of increasing value in breeding as technology for wide crossing and gene transfer is developed (see Witcombe, Chapter 10, this volume). Certainly for crop species, the presence of wild relatives in a conservation area may not be known to wildlife conservationists who, in any case, will be unaware of the specific value of wild relatives to breeders. Wild relatives of domestic animals are receiving specific attention: for example, the gaur (*Bos gaurus*) and other wild cattle of Southeast Asia.

A more general argument for conservation is the 'portfolio value' of the between-species diversity of food species. As a greater and greater proportion of global food is produced from fewer and fewer crops, the risk of catastrophic loss grows (Prescott-Allen and Prescott-Allen, 1990). This was the case with the Irish potato famine of the 1840s. A rapidly rising human population (greater then than now in Ireland) was almost entirely sustained by a single crop, potato, increasingly grown under dangerously marginal conditions where it was highly susceptible to the fungal pathogen, late blight (see Allen *et al.*, Chapter 6, this volume). An overdependence on rice exacerbated the Bengal famine in 1942. Similar reasoning applies to conserving within-species diversity.

The 'option value' of genetic resources provides insurance against changing conditions. The more diverse the range of genetic resources available, the more chance that at least some will flourish under changed conditions.

The 'exploration value' of conserved genetic resources lies in the potential of their as yet unknown characters. This has been given as the reason for maintaining wild ecosystems and traditional farming systems, as plants in these habitats could contain and develop new and valuable genetic characteristics (FAO, 1996b).

The degree and technology of conservation has been determined by the value of individual *components* – genes, varieties or species – of agrobiodiversity. It is these components that are collected, conserved and used. However, in many cases, it is the functional (and therefore evolutionary) *interactions* between these units that are important. Interactions at the genetic level have been conserved within varieties, to be uncovered only during subsequent evaluation by breeders. Interactions between varieties are common – for example, in varietal mixtures of *Phaseolus* beans (see Thurston *et al.*, Chapter 9 and Polaszek *et al.*, Chapter 11, this volume). These interactions may be entirely lost during conservation, with the component varieties being separated for ease of management (Martin and Adams, 1987). For species, intercrops are very common: in the classic Central American intercrop of maize/beans/squash, long-term interactions lead to coadaptation of the components. In the West African intercrop of sorghum and cowpea, the latitude determines the interaction: sorghum varieties flower at different dates (Bunting and Kuckuck, 1970). This relates to the control of competition between the developing cowpea and the sorghum for optimum yield.

In the past, little effort was made to conserve together the potential interactions between these coevolved components at the varietal and species level, although recent proposals for on-farm conservation will address both the value of individual components and also the evolutionary inter-relationships (including the all-important coevolution of domesticates and their pests and pathogens).

Policy and Technology in Conservation

Coordination of conservation efforts

In much of what follows, we will use the role of the IPGRI (formerly IBPGR) as an example of an international approach to conservation.

Conservation of agrobiodiversity started in many instances as an individual breeder's initiative. It was only in the mid-1970s that attempts were made to coordinate global activities, with the establishment of IBPGR in 1974 to coordinate a global effort to collect threatened landraces of the major crops worldwide which were rapidly being replaced by modern varieties. Well-planned and organized collecting missions were launched. Existing genebanks were brought together in a network, and new national genebanks promoted. A so-called 'registry of base collections' consisted of a network of 49 genebanks worldwide that had concluded an agreement with IBPGR to maintain global or regional base collections of one or more crop species (with a total of approximately 250 species or groups of species). Each genebank agreed to follow established technical standards; to allow unrestricted access to samples; to follow appropriate quarantine regulations; to duplicate collections for safety; and to adopt adequate information management (Tao *et al.*, 1989). With the direct or indirect support of IBPGR, well over 200,000 accessions were collected worldwide and securely conserved (Cooper *et al.*, 1994).

Following an agreement with FAO in 1991, it was decided to merge the IBPGR 'registry of base collections' with the International Network of *Ex Situ* Germplasm Collections. The objectives of this Network are to ensure safe conservation and promote unrestricted availability and sustainable use of plant genetic resources for present and future generations, by providing a framework for sharing of benefits and burdens. Subsequently, under a further agreement of October 1994, about 500,000 samples from the CGIAR Centres were placed in the FAO 'Network' (FAO, 1994). According to these agreements, the Centres continue to hold material, designated in annexes to the agreements, in trust 'for the benefit of the international community'; to make it available for 'scientific research, plant breeding or genetic resource conservation'; not to claim ownership over the designated accessions or to apply any form of intellectual property protection; to pass these conditions on to recipients of designated germplasm; to use accepted storage standards; and to accept FAO policy direction over the Network.

Notwithstanding these agreements, national plant genetic resources programmes form the backbone of any regional or global effort. The legal basis of this is sovereignty of each nation over the genetic resources within its borders, a principle clearly endorsed in the CBD. This led to an increased awareness and commitment by countries to conserve their plant genetic resources. Regional 'conservation and use' networks are good examples of this ongoing commitment, as in Southern Africa (SADC) or Europe (the European Cooperative Programme for Crop Genetic Resources Networks, ECP/GR, coordinated by IPGRI). Numerous regional or global crop-specific networks facilitate more efficient conservation; centralized information management; a sharing of genetic diversity; a joint characterization and evaluation; and a collective benefit from breeding efforts. The Latin American Maize Project (LAMP) involved partners from the public and private sectors in developing and developed countries, and an international research centre, CIMMYT. IPGRI plays an active role in a coordinated International Network for the Improvement of Banana and Plantain (through INIBAP); the International Coconut Genetic Resources Network (COGENT); a number of ECP/GR crop networks; networks of underutilized Mediterranean species; etc. Within the Consultative Group on International Agricultural Research (CGIAR), IPGRI is the convening Centre for the System-wide Genetic Resources Programme (SGRP). SGRP has developed a System-wide Information Network for Genetic Resources (SINGER) through which detailed information on accessions can be obtained (SGRP, 1997).

The increased worldwide awareness and recognition of the importance of conserving agrobiodiversity has not been met with increased funding to carry out all the necessary conservation efforts. Difficult policy choices have to be made and efficiency becomes paramount as efforts are made to rationalize conservation. Crop networks play an important role in this by permitting elimination of duplicates, and by setting stricter priorities on what to conserve and where.

Approaches to conservation and utilization are changing rapidly, and feedback and flexibility are always needed. For example, biotechnology offers new tools for the utilization of wild relatives of crops (see Witcombe, Chapter 10, this volume), the value of which was repeatedly emphasized in a recent policy forum on global food supply (Daily *et al.*, 1998). Changing needs for genetic resources – by both farmers and institutional breeders – will influence both what is collected, and the appropriate conservation technology and strategy.

Collecting

What to collect?

Decisions on what to collect will be based on needs that will change over time. Obviously, thorough collecting will reduce the need for further collecting. Also, the targets and intensity of collecting will depend on what will be done with the collected samples. Engels *et al.* (1995) list some objectives of collecting:

- rescue collecting (in case of imminent threat and where *in situ* conservation is not feasible or sufficient);
- collecting for immediate use (ranging from farmers collecting germplasm for local exchange to targeted collecting of specific genotypes by breeders);
- gap-filling for future use (usually as part of a national, regional or global conservation responsibility for one or more species/gene pools);
- collecting for research purposes (to resolve specific problems related to the mating system of a species, taxonomic boundaries, evolutionary relationships among taxa, etc.); and
- collecting for opportunistic reasons.

Planning for collection

A well-defined collecting plan is essential. Bunting and Kuckuck (1970) argue that a plant explorer needs to work 'against the background of knowledge in depth of the ecological, human and agricultural characteristics of the area'. Collecting should be based on an agroecological survey of existing information. Bunting and Kuckuck recommend as a model the classic account of agriculture in the Sudan (Tothill, 1948). This idea has now been extended and refined by Maxted *et al.* (1995) and termed 'ecogeography'. Once the target taxon or group of taxa and the precise geographical area have been defined, the collector can collate information on the distribution of the target taxa and ecological and taxonomic data required to define the best sampling strategy in the field.

As part of a pre-collection survey Maxted *et al.* (1995) recommend: listing of germplasm already conserved; survey of taxonomic, ecological and geographic data sources; collection of ecogeographic data; data verification; and analysis of taxonomic, ecological and geographic data. Such in-depth information is of direct use to collectors (e.g. by defining the best season and localities for collection), and of subsequent use to breeders (as indicated by Bunting and Kuckuck (1970), breeders who have to use the material will be able to work more effectively with it if they know something about the local conditions to which it may be specifically adapted).

Conservation priorities will also become apparent. The planned strategy may have to be further adapted in the field according to the prevailing circumstances. Should insufficient ecogeographic information be available, it may be necessary to launch a survey mission during which the necessary information is collected. This might entail a 'coarse grid sampling' of the target taxa for analysis in the laboratory (Maxted *et al.*, 1997a). However, this is rarely done in practice.

As the objectives and priorities for the conservation of plant genetic resources have changed over time, so have the strategies for collecting. With a change of emphasis from major food crops to wild and weedy relatives and locally important crops, and with increased emphasis on gap-filling and re-collecting for deficient collections, collecting objectives have become more specific (Brown and Marshall, 1995). The evaluation of collections for breeding purposes may uncover geographic 'hot spots' of characters meriting further collecting. For example, valuable resistance to leaf rust in groundnut is restricted to samples from the Taropoto region of Peru. To capture genetic change within varieties, or changes between the varieties grown, collecting could be repeated periodically. Again, this is rarely done.

Field sampling

The fundamental objective of collecting plant genetic resources is to capture the maximum amount of useful genetic variation in the smallest number of samples possible (Marshall and Brown, 1975). The development of efficient sampling strategies is dependent on good information on the kind and amount of genetic variation in target taxa populations and their distribution in the target geographical region (Allard, 1970). The basic parameter for measurement of variation in a given population is allelic richness: the number of distinct alleles at a single locus (see Witcombe, Chapter 10, this volume). This is usually assessed at a large number of marker loci after the sample is taken (Brown and Marshall, 1995). If several populations are to be sampled in a given area, the extent of genetic divergence and the total genetic variation among the populations is important. The latter is reflected in the range and pattern of distribution of numbers of alleles per locus (Schoen and Brown, 1991). The generally accepted benchmark criterion for collecting germplasm is to ensure that at least one copy of 95% of the alleles at a frequency greater than 0.05 should be included in the collected sample

(Marshall and Brown, 1975): a sample of 50 individuals from each population will meet this criterion. Further information on basic sampling strategy (including number and location of sampling sites; number of individual plants sampled at a site; choice of individuals; and number and type of propagules per plant) can be found in Brown and Marshall (1995).

The knowledge and understanding of the genetic structure of many plant populations has increased significantly, providing a more secure base on which sampling techniques can be developed. Different species may be targeted; for example, wild relatives of crops are becoming more valuable to breeders as biotechnology provides tools for wide-crossing. However, wild species differ from crops in many ways: for example, the distribution of the species; local abundance; interpopulation migration; habitat diversity; life-history traits such as duration of the life cycle; population age structure; vegetative reproduction; fecundity; determination of flowering and seed maturation; mating system (outbreeding, self-fertilization or apomixis); pollination mode; and conspicuous polymorphisms. With these differences, sampling strategies for wild species may differ from crops (Brown and Marshall, 1995).

Collecting technology

Collecting techniques and equipment employed will depend on the type of material to be collected, that is, seed, pollen, vegetative propagules or whole plants. To maintain varietal integrity the sample should be of the type used by the farmer. For example, seed should be collected for seed-propagated crops (maize, rice), and vegetative samples for clonally propagated species (e.g. tubers for potatoes and offshoots for bananas; indeed many such crops never produce seed). Where grafting is the propagation technique used by farmers, rootstocks must be grown in advance to allow grafting.

The seed of many crop species, especially tropical fruits with seed of high moisture content, cannot be stored under standard cold dry conditions now used in genebanks. An early recognition of the problem of transporting the collected seed of tropical economic plants led to the development of the 'Wardian case' – in effect, a portable greenhouse – to transport plants as seedlings on deck during long sea voyages from the 1830s until the use of air transport (Hepper, 1989). More recently, zygotic embryos, vegetative tissues such as budwood, shoots or apices have been sampled, treated, transported and subsequently grown under adequate conditions at the base. Withers (1995) discusses in detail the application of *in vitro* techniques to collecting germplasm. Engelmann (1997) specified several situations when *in vitro* collecting can be advantageous, for example for long missions to remote areas when vegetative material or seeds may not survive; when the size and weight of seeds is a problem; when the risk of transferring pests and diseases is higher (when soil particles remain on collected material); and, finally, when there are insufficient seeds for collecting.

For several crops, well-established protocols, procedures and equipment exist which can be adapted to other species. These include the collecting of budwood based on cocoa (Yidana, 1988); the extraction of zygotic embryos based on coconut (Assy Bah *et al.*, 1989); the use of stem nodal cuttings for cotton and related species (Altman *et al.*, 1990); and the use of herbaceous plantlets as explants based on forage grasses (*Digitaria* and *Cynodon* spp.) (Ruredzo, 1989). Collecting of DNA-rich material such as leaves and root nodules can be done with little additional effort when specimens are collected for herbaria or genebanks. The material should be stored with a desiccant immediately after collecting to ensure successful DNA extraction. This provides a simple long-term storage method (Adams, 1997).

Conservation

It is now usual to divide conservation into on-site (*in situ*) and off-site (*ex situ*) conservation, and for simplicity we follow this precedence. However, another possible division is into dynamic and static conservation (Frankel *et al.*, 1995), where evolutionary change in the conserved populations, rather than genetic integrity, is the defining criterion. Approaches to conservation (either *in situ* or *ex situ*) and the techniques used or developed over time, reflect, to a large extent, the thinking, beliefs and modes of a given period, and, always, the changing needs of users. However, whether practised by traditional farmer or institutional breeder, conservation is part of a process to achieve secure food production. In this section the various conservation approaches and methods will be briefly described.

Despite the many differences of scale and user, there are some general features of conservation management. For security, it is always necessary to *duplicate* collections in different localities; *alternative* methods of conservation will also add to security. Adequate computerized documentation is essential, for details of the collecting information, characterization and evaluation of samples, and to aid sample storage or maintenance. Increasingly, genebank documentation is available electronically and globally (Committee on Managing Global Genetic Resources, 1993; SGRP, 1997).

Great care must be taken with the international transport of genetic resources, as: (i) vegetative and seed samples may be contaminated with pests and diseases; and (ii) many species have the potential to turn into noxious weeds following introduction to new areas (see Wood and Lenné, Chapter 2, this volume). At a minimum, international phytosanitary procedures should be followed, but most genebank managers adopt stricter protocols, with seed health testing in dedicated laboratories, and use of virus-freeing techniques for clonal samples (Anon., 1996).

In situ *conservation*
The CBD (UNEP, 1992), covering both wild and domesticated species, uses a complex definition for *in situ* conservation:

the conservation of ecosystems and natural habitats and the maintenance and recovery of viable populations of species in their natural surroundings and, in the case of domesticates or cultivated species, in the surroundings where they have developed their distinctive properties.

This combination of ecosystem and population conservation is usual for wild species. However, there may be substantial differences in approach between the conservation of wild species and domesticates. For example, during wild species conservation, the introgression of alien genes into populations of target species would be strictly avoided. In contrast, for crops, it has been argued that introgression of genes from wild species into crop populations is an evolutionary event and one advantage of *in situ* conservation (Altieri and Merrick, 1987, and many others). The CBD definition is ambiguous on the meaning of 'surroundings' and 'distinctive properties'. Is this on the scale of, for example, maize becoming distinct as a crop in Central America thousands of years ago; or a minor colour variant of a bean originating in the corner of a field recently; or, indeed, a distinctive variety originating in a breeder's selection plot? The implications for management are significantly different in the three cases.

These ambiguities of definition apart, there is limited experience for the *in situ* conservation of domesticated agrobiodiversity, few evaluations of conservation attempts, and, as yet, no agreed protocols for its management *in situ* in compliance with the CBD. However, with the conclusion of the CBD and Agenda 21, and more recently the adoption of the Global Plan of Action (GPA) by the participating countries in the Fourth International Technical Conference on Plant Genetic Resources (FAO, 1996b), a significant impetus has been given to *in situ* conservation. *In situ* approaches may provide a linkage between conservation and development (Hirsch, 1994), but this is a claim which can be disputed. For example, Frankel *et al.* (1995) note that germplasm collections and their users stand to gain much from landraces preserved *in situ*, but caution that 'it is difficult to cast aside the feeling that it is the farmer who foots the bill, whether he knows it or not.'

PROTECTED AREAS Agrobiodiversity represented by the wild relatives of crop and domestic animals may occur beyond the influence of farming, in natural and semi-natural ecosystems. This wild agrobiodiversity can be conserved in the existing system of nature reserves. There have been many proposals for this (Ingram, 1984; Prescott-Allen, 1984; Prescott-Allen and Prescott-Allen, 1984; Wilcox, 1990) but, until recently, little action. There is now a Global Environment Facility (GEF) project on the *in situ* conservation of wild relatives of wheat in Turkey (GEF, 1995).

There are several international systems for designating protected areas. Two of these, the World Heritage Convention and the Ramsar Convention, are full international conventions. Other designations – for example, Biosphere Reserves of the UNESCO Man and the Biosphere Programme

(UNESCO, 1984), and the UN List of National Parks and Protected Areas – while not full conventions, endorse the global value of national protected areas. The role of Biosphere Reserves for genetic resource conservation was specifically surveyed by Ingram (1990). With few exceptions, Ingram found a low incidence of reserves with relatives of staple grain crops (*Triticum, Hordeum, Zea, Oryza* and *Sorghum*) and of other important crops (*Solanum, Ipomoea, Musa*).

Whatever their international or national status, protected areas can be classified under the IUCN Categories of Protected Areas (McNeely, 1996).

At present very few reserves have given emphasis to agrobiodiversity. Those that do include:

- The Sierra de Manatlán Biosphere Reserve in Mexico, which was established to protect a recently discovered and rare species of maize relative, *Zea diploperennis* (Guzman and Iltis, 1991; and see Thurston *et al.*, Chapter 9, this volume): this occurs as an integral component of the traditional maize-based agroecosystem, serving as an important source of fodder.
- The Erebuni Nature Reserve in Armenia, which has a unique richness of the wider *Triticum* gene pool, including *T. urartu*, *T. boeoticum* and *T. araraticum*, and *Aegilops* species. Vavilov himself recommended protection of this site (Damania, 1994).
- The Cuc Phuong National Park in Vietnam is a model, with a full species inventory, classified for their genetic resource value (Thin, 1996). The park contains 292 species of edible plants, including crop relatives such as *Musa itinerans* and *Citrus reticulata*.

A recent review (Murray *et al.*, 1992) showed that only 2.4% of a total of 18,715 protected areas globally had plant inventories, a minimum condition for targeted conservation. With few exceptions, there has been a serious neglect of agrobiodiversity in protected area management. Schlosser (1984) argued the need for 'purpose-oriented' management practices for protected areas, with cooperation between *in situ* and *ex situ* managers, with emphasis on forages, fruit trees, and wild relatives of cultivated plants.

At present the conservation of agrobiodiversity in protected areas is largely unplanned. A feature of this form of conservation is that evolutionary processes continue to operate and that entire populations can be targeted. As 'evolutionary processes' include genetic erosion and even extinction of populations, some form of management may be needed. A disadvantage of protected-area conservation is that the conserved material is not readily available for agricultural use. Also, with limited opportunity for management, little characterization and evaluation can be done on the germplasm, restricting its use as a genetic resource (Maxted *et al.*, 1997b).

CONSERVATION ON-FARM Farmers worldwide have been practising on-farm conservation, or maybe better on-farm management for as long as agriculture has existed, as a necessary part of crop production. For them, the most effective management practices are those which result in the highest yields and the highest food security. Usually, these practices are based on within- and between-species diversity, surviving in areas which are marginal for modern agriculture. In addition to crops, there are wild and weedy or ruderal species associated with farming. There are now suggestions for intervention to boost the effectiveness of this age-old process. Maxted *et al.* (1997b) provide detailed suggestions and procedures for the management of these resources on-farm in the framework of traditional farming systems, thereby continuing the maintenance and evolution of traditional landraces and wild and weedy species which depend on traditional agricultural practices (see Thurston *et al.*, Chapter 9, this volume). Possible advantages of conservation on-farm will need to be weighed against disadvantages, both as a technology for conservation, and from the impact of intervention on farm livelihoods (see Smale and Bellon, Chapter 15, this volume). We return to this in our final paragraphs.

HOME GARDENS For fruits, vegetables and small domestic livestock, home gardens are a reservoir of between-species diversity. Proximity to the home allows detailed selection, for example, of colour variants of most plants and animals, and of the vast morphological variation found in *Brassica oleracea*, the species which includes the cabbage. Several authors, for example Engels (1995), Damania (1996) and Maxted *et al.* (1997a), list the conservation of plant genetic diversity in home gardens as a separate method. As with on-farm conservation, this is dynamic.

A range of gardens may need to be included, as the intraspecific diversity within a garden is often limited and the variation between gardens is often substantial.

There have been many other ideas and proposals for *in situ* conservation of agrobiodiversity, ranging from 'mass reservoirs' (Simmonds, 1962; Frankel and Bennett, 1970; Frankel *et al.*, 1995) to recommendations of ethnobotanists (Brush, 1986; Altieri and Merrick, 1987; Oldfield and Alcorn, 1987; many others reviewed by Wood and Lenné, 1997).

IPGRI action on *in situ conservation*

To investigate the potential of *in situ* conservation, in 1996 the IPGRI initiated a global *in situ* conservation programme in nine selected countries addressing the following pertinent goals through a multidisciplinary research approach:

- to determine and understand the situation in which landraces are maintained by farmers;
- to identify the key factors which affect farmer decisions to maintain landraces and the factors that affect the viability of the maintenance process; and
- to find ways to assist the continued selection of landraces or cultivars that conserve local germplasm.

IPGRI also provides support to national programmes on plant genetic resources to acquire the necessary capacity to develop and sustain *in situ* conservation programmes, in close collaboration with the actual players, the farmers. In addition, IPGRI is prepared to facilitate and foster international collaboration, to contribute to the improvement of strategies and technologies for conservation, and to provide an international information service (Iwanaga, 1995).

Ex situ *conservation*

SEED STORAGE This is the best researched, most widely used, and most convenient method of *ex situ* conservation. Much is known on the optimum treatment of the seed of most of the major food crops. For an early review, see Harrington (1970). Requirements are adequate drying, i.e. seed moisture contents as low as 3% for oily seeds and 5% or more for starchy seeds; appropriate storage temperature (−18°C is recommended for long-term storage, see Plate 16); and careful production of quality seed to ensure the greatest longevity (Rao and Jackson, 1996). Some use is made of liquid nitrogen (at −196°C) for seed storage.

Viability monitoring of stored seeds is an important component of the conservation effort and much research has been carried out on germination and vigour testing. These allow relatively precise predictions when seed viability starts dropping below established thresholds, thus allowing timely regeneration of the accession. Detailed procedures on seed drying, storage and viability monitoring can be found in Roberts (1973). Seed storage is relatively efficient, reproducible and secure in the short-, medium- and long-term. Standards for seed storage in genebanks have been developed and recommended for their international adoption by FAO and IPGRI (Anon., 1994).

However, the seeds of many crop species, especially tropical shrubs and trees, will lose viability if dried (so-called 'recalcitrant' seeds – no better term is available). Seeds of some species can be dried somewhat, but cannot survive low-temperature storage, including coffee (*Coffea arabica*), citrus species, neem (*Azadirachta indica*), rubber (*Hevea brasiliensis*) and others.

There is a recent IPGRI protocol to determine the precise seed storage behaviour of unknown species (Hong and Ellis, 1996) and a compendium of available data on storage behaviour of approximately 7000 species, including references to individual species (Hong *et al.*, 1996).

Most national genebanks now rely on cold stores for seed maintenance. However, these depend on a reliable electricity supply, a possible shortcoming in some countries. To overcome this problem, alternative approaches to low-temperature storage have been developed, including the so-called 'ultra-dry seed' technology. Drying seeds to a moisture content as low as 1% (in the case of oily seeds) or approximately 3% (starchy seeds) and hermetic packing allows storage for long periods at room temperature. Care must be taken to prevent over-drying of the seeds (Walters and Engels, 1998).

Many genebanks which provide a comprehensive service to breeders have problems, not so much with seed storage technology, which is straightforward, but with continually regenerating samples for regular use in evaluation. This is a particular problem with wild species, which tend to be outcrossing, have indeterminate flowering and seeding, may shatter, or may produce little seed (Lyman, 1984).

POLLEN STORAGE This is comparable to seed storage, since pollen can be dried (less than 5% moisture content on a dry weight basis) and stored below 0°C. There is limited experience on the survival and fertilizing capacity of cryopreserved pollen more than 5 years old (Towill, 1985). However, pollen has a relatively short life compared with seeds (although this varies significantly among species), and viability testing may be time-consuming. Pollen has therefore been used to a limited extent in germplasm conservation (Hoekstra, 1995). Other disadvantages of pollen storage are the small amount produced by many species; the lack of transmission of organelle genomes via pollen; the loss of sex-linked genes in dioecious species; and the lack of plant regeneration capacity (although first indications exist that pollen can be regrown into whole plants) (Hoekstra, 1995). Pollen storage may be useful for species which produce recalcitrant seed (IPGRI, 1996). Furthermore, pollen transfer of pests and diseases is rare (excepting some virus diseases), and this allows the safe movement and exchange of germplasm.

FIELD GENEBANKS These are used for the conservation of clonal crops, where seed is recalcitrant, and where seed is rarely produced. The 'rule of thumb' is to use the same propagation techniques as the farmer, for example not disrupting adapted clones through genetic segregation in a seed cycle. Many temperate and tropical fruit trees fall into one or more of these categories, as do many commodity crops such as cocoa, rubber, oil palm, coffee, banana and coconut, as well as most root and tuber crops (Figs 14.1 and 14.2). As an example of the scale of management, oil palm genetic resources in Malaysia are planted at a density of 140 palms to the hectare, and the collection from Nigeria alone occupies 200 ha. However, oil palm seed cannot be stored for more than 2 years, and pollen only for 3 years. This 'living collection', although expensive, is the only practicable method at the moment.

Management may be the same as used during routine farming, so the method can be adapted to local circumstances. Conserved material can be readily characterized and evaluated and then accessed for research and utilization. Some natural selection may take place within and between accessions, but management is designed to avoid this if at all possible. Major constraints to field genebanks include cost and all the natural hazards of farming (pests and diseases, drought, flood, cyclones, etc.).

IN VITRO CONSERVATION When one method of conservation is subject to unavoidable hazards, as with field genebanks, an *alternative* method should also be used. *In vitro* conservation involves the maintenance of explants in a

Fig. 14.1. Cocoa fruit on one of the many varieties of *Theobroma cacao* in the germplasm collections at CATIE (courtesy of Jan Engels, IPGRI).

sterile, pathogen-free environment and is widely used for the conservation of species which produce recalcitrant seeds, no seeds at all, or for material which is propagated vegetatively to maintain particular genotypes (Engelmann, 1997). Although research on *in vitro* techniques only started some 20 years ago, this technique has been applied for multiplication, storage and, more recently, for collecting of germplasm material for more than 1000 species (Bigot, 1987).

　　Different *in vitro* conservation methods are available. For short- and medium-term storage the aim is to increase the intervals between subcultures by reducing growth. This is achieved by modifying the environmental conditions and/or the culture medium, so-called slow-growth conservation. The most widely applied technique is temperature reduction (varying from 0–5°C for cold-tolerant species to 9–18°C for tropical species), which can be combined with a decrease in light intensity or storage in the dark (Engelmann, 1997). Alternatives to standard slow-growth conservation include the modification of the gaseous environment of cultures, desiccation and/or encapsulation of explants. The latter is called synthetic seed: the aim is to use somatic embryos as true seeds. Embryos, encapsulated in alginate gel, can be stored after partial dehydration and sown directly *in vivo* (Janick *et al.*, 1993).

　　For small volumes, long-term storage is practicable through storage of cultures at ultra-low temperature, usually by using liquid nitrogen (–196°C) in cryopreservation. At this temperature all cellular divisions and metabolic processes are stopped and, consequently, plant material can be stored

Fig. 14.2. Fruit of a *Herrania* sp. in the germplasm collection at CATIE (courtesy of Jan Engels, IPGRI).

without alteration or modification for a theoretically unlimited period (Engelmann, 1997). See Steane, Chapter 4, this volume, for cryopreservation applied to domestic animals.

BOTANIC GARDENS AND ARBORETA These have played a historical role in the introduction of crop genetic resources. Usually they maintain only one or a few individuals per species (FAO, 1996a; Engels and Engelmann, 1998) although in recent years there has been a movement towards the establishment of conservation units, including seed-banks (Laliberté, 1997). Unfortunately, most botanic gardens have limited interest or expertise in crop genetic resources, but efforts are being made to change this (Heywood, 1998).

DNA STORAGE This is rapidly increasing in terms of importance. DNA from the nucleus, mitochondrion and chloroplasts is now routinely extracted and immobilized into nitrocellulose sheets where the DNA can be probed with numerous cloned genes. With the development of PCR (polymerase chain reaction) one can now routinely amplify specific oligonucleotides or genes from the entire mixture of genomic DNA. These advances have led to the formation of an international network of DNA repositories for the storage of genomic DNA (Adams, 1997). The advantage of this technique is that it is efficient and simple and overcomes physical limitations or constraints. The disadvantage lies in problems with subsequent gene isolation, cloning and transfer as well as that it does not allow the regeneration of entire plants (Maxted *et al.*, 1997a).

The Complementarity of Conservation Methods

Farming itself is the original method of conservation linked with utilization. But farming is changing and most farmers would not wish to be curators of living museums of agrobiodiversity (as suggested by Wilkes, 1971). Fortunately, the wide spectrum of conservation methods can meet a wide range of needs. With a range of genetic diversity to be included in conservation efforts, security and accessibility can be balanced against feasibility and cost-efficiency. The choice of a single method of conservation alone will not be enough: different and *complementary* methods of conservation will show different advantages and disadvantages. In making such choices it is important to take a holistic view of the intended conservation effort and to place it in a wider context, whenever possible, as part of a development process. This will include local development, national development, and even contributions of agrobiodiversity to global development. This latter is particularly important, as most crop and animal production in most developing countries is from introduced species (Wood, 1988). It is also important to carefully examine the technical and human resources available as well as the administrative and political environment of the conservation effort in order to avoid later constraints.

In choosing alternative or complementary methods of conservation, the most obvious contrast is between *in situ* and *ex situ*. Figure 14.3 presents an overview of how the various conservation methods fit into a continuum from an entire *in situ* approach to the strongest possible *ex situ* approach. The dynamic processes of *in situ* conservation can be combined with the usually more secure and accessible approach of *ex situ*. As a result of disease pressure and natural selection there could be an enhancement of the value of on-farm populations as a source of variability for breeding for disease resistance. However, the rate of this enhancement is unknown, and methods of sampling or evaluation in the field not yet developed. The potential for evolution on-farm was noted by Allard (1990) for disease resistance (of the *Hordeum vulgare*/*Rhyncosporium secalis* pathosystem).

But Allard also argued that there could be sharp changes in gene frequency on-farm. If so, the dynamic evolutionary system on-farm will need the regular back-up of farm-conserved populations by *ex situ* conservation,

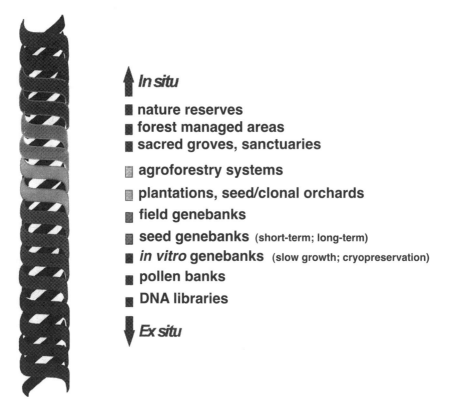

↑ *In situ*

■ **nature reserves**
■ **forest managed areas**
■ **sacred groves, sanctuaries**

▨ **agroforestry systems**

▨ **plantations, seed/clonal orchards**

▨ **field genebanks**

▨ **seed genebanks** (short-term; long-term)

■ ***in vitro* genebanks** (slow growth; cryopreservation)

■ **pollen banks**

■ **DNA libraries**

↓ *Ex situ*

Fig.14.3. Complementary strategies and methods: a continuum of options (courtesy of Hareya Fassil, IPGRI).

to avoid possible loss of genetic diversity (Maxted *et al.*, 1997a). Closer adaptation to the changing and perhaps unmanageable selection pressures on-farm may cause the loss of otherwise valuable genetic diversity, which will need to be conserved *ex situ*.

Even with the advantages of continuing evolution on-farm, and the substantial diversity of material that can be conserved, there will be limited access to those resources, a lack of adequate characterization and evaluation, and the danger that farmers abandon the cultivation of traditional landraces under economic pressures (see Smale and Bellon, Chapter 15, this volume). Careful monitoring will always be needed. Many 'minor' but locally important crops have been neglected by collectors and *ex situ* genebanks. For these and their wild relatives, *in situ* conservation will be indispensable.

'Conservation through use' *in situ* might run the risk of losing specific alleles or genotypes and a 'back-up system' through *ex situ* conservation will be required. This has been emphasized by Hammer *et al.* (1996) who found 96.8% of samples collected in Albania in 1941 to be intact in the Gatersleben genebank in Germany, whereas a survey 50 years later in the same region of Albania showed field erosion of about 50%. They conclude that this 'is an amazing result as the material had to survive the Second World War and two translocations'.

Choice between conservation methods may be dictated by the biology of the species. For instance, if the cultivated species does not form botanical seeds (as for bananas) the choice includes on-farm conservation, maintenance in field genebanks, and *in vitro* slow-growth and/or cryopreservation (Sharrock and Engels, 1997). Table 14.1 contains details of a complementary approach for the coffee gene pool. Cassava provides another example of a gene pool where there has been considerable research to develop *in vitro* techniques and where a broad range of conservation options is available.

There are particular problems with conserving wild species *ex situ* that suggest emphasis on *in situ* conservation. A report from IRRI identified problems associated with dropped seeds and rhizomatous growth habits that made the continuous planting of wild rices impractical (Anon., 1982). Most of the wild rice species were susceptible to the viruses and brown planthoppers prevalent on the IRRI farm.

Agroecosystem conservation and the future

Most of the early work on conservation was for genetic resources of crops and domestic animals – that is, of specific components of agrobiodiversity, whose function was obvious and whose improvement possible with breeding technologies. The practical reasons for this conservation are relatively simple and are outlined in part above. However, more complex approaches to conservation are needed, to address the interrelationships of species in

Table 14.1. Integrating conservation strategies: an example for the coffee gene pool.

Crop	Seed genebank	Field genebank/ In vitro	On farm	Protected areas/reserves
Coffee (*Coffea* spp.)	*Coffea* spp. (intermediate or recalcitrant seed storage behaviour). Seed viability of *C. arabica* (the only self-compatible type) can be maintained up to 1 year (short-term seed conservation useful for germplasm exchange)	*In vitro*: Microcutting/somatic embryogenesis for multiplication/ medium-term conservation; e.g. *in vitro* collection at ORSTOM (France); feasibility of *in vitro* methods depends on species; cryopreservation of seed being investigated. *Field collections:* • Ivory Coast Genetic Resources Centre: > 10 spp. (including wild spp. from Madagascar and W. Africa) comprising> 10,000 genotypes; • Choche and Jima (Ethiopia) • CATIE (Costa Rica)	Some useful new genetic combinations maintained in plantations; *geneflow:* crosses between cultivated types and nearby wild populations have been noted	*In situ* conservation of wild populations in forest reserves: • *C. sessiflora* in East Usambara Mountains (Tanzania); • other endemic wild spp. conserved in Mount Tonkoui Man Research Station (Côte d'Ivoire); • *C. arabica* in Bale Mts National Park (Ethiopa); • endemic spp. conserved in protected areas in Mauritius

Florent Engelmann, personal communication.

systems. There is now considerable interest in the value of dynamic conservation (Prescott-Allen and Prescott-Allen, 1982; Engels, 1995; Wood and Lenné, 1997), where populations are able to respond to evolutionary pressures, including the biotic pressures from other components of agrobiodiversity. This development follows from theoretical and practical advances in wildlife conservation, where a previous emphasis on species is giving way to conservation of ecosystems, where the properties of the whole may be greater than the parts. In agroecosystems the genetic resource component of crops and domestic animals relates to and depends on other agrobiodiversity in the system, for example, soil biota, pests and pollinators. 'Genetic resources' cannot be effectively conserved without 'agrobiodiversity' (see Wood and Lenné, Chapter 2, this volume). Furthermore, under changing conditions, these interactions adapt through evolution. Thus, conserving the agroecosystem will maintain both the component genetic resources and functional species interrelationships, and *also* permit evolution and coevolution, with the expectation of closer adaptation (of the species and also the system) to future conditions. If necessary, management of agroecosystems can both intensify selection pressure (e.g. of drought, pests or competition) and increase diversity as a raw material for selection (see Allen *et al.*, Chapter 6, and Polaszek *et al.*, Chapter 11, this volume). This is a neglected area of conservation, where the needs and technologies have yet to be defined.

But defining needs and technologies for conservation will not be enough. A problem across conservation is that the benefits of conservation may not be apparent to local farmers: breeders' needs may not be the same as farmers' needs. For example, a shepherd in Iran may wish to graze sheep on wild wheat relatives of extreme value to breeders in distant countries. With no knowledge of the value of the wild wheat, and with no national or international mechanism to reward the farmer as a conserver, genetic resource populations can be lost.

Some of the earlier exponents of genetic resource conservation were visionary. Well in advance of our present emphasis on *in situ* conservation, Bunting and Kuckuck (1970) had detailed the information we need for success: a survey of wild vegetation and disturbance, human and historical factors (including anthropological, sociological and linguistic factors, and human migrations), agricultural systems and practices and innovations, biology of crops in the region (and much more). After 30 years, we have yet to reach this level of understanding of any traditional agroecosystem.

References

Adams, R.P. (1997) Conservation of DNA: DNA banking. In: Callow, J.A., Ford-Lloyd, B.V. and Newbury, H.J. (eds) *Biotechnology and Plant Genetic Resources Conservation and Use*. CAB International, Wallingford, pp. 163–174.

Allard, R.W. (1970) Population structure and sampling methods. In: Frankel, O.H. and Bennett, E. (eds) *Genetic Resources in Plants – Their Exploration and Conservation.* IBP Handbook No. 11, Blackwell Scientific, Oxford, pp. 97–107.

Allard, R.W. (1990) The genetics of host–pathogen co-evolution: implications for genetic resource conservation. *Journal of Heredity* 81, 1–6.

Altieri, M.A. and Merrick, L.C. (1987) *In situ* conservation of crop genetic resources through maintenance of traditional farming systems. *Economic Botany* 41, 86–96.

Altman, D.W., Fryxell, P.A., Koch, S.D. and Howell. C.R. (1990) *Gossypium* germplasm conservation augmented by tissue culture techniques for field collecting. *Economic Botany* 44, 106–113.

Anon. (1982) Conservation of the wild rices of Tropical Asia. *Plant Genetic Resources Newsletter* 42, 13–18.

Anon. (1994) *Genebank Standards.* FAO/IPGRI, Rome.

Anon. (1996) Technical guidelines for the safe movement of germplasm. *Plant Genetic Resources Newsletter* 107, 60.

Assy Bah, B., Durand-Gasselin, T., Engelmann, F. and Pannetier, C. (1989) Culture *in vitro* d'embryons zygotiques de cocotier (*Cocos nucifera* L.). Métode, révisée et simplifiée, d'obtention de plants de cocotiers transférables au champ. *Oléagineux* 44, 515–523.

Bauer, E. (1914) Die Bedeutung der primitiven Kulturrassen und der wilden Verwandten unserer Kulturpflanzen für die Pflanzenzüchtung. *Jahrbuch der Deutschen Landwirtschaftsgesellschaft* 29, 104–109.

Bigot, C. (1987) *In vitro* manipulation of higher plants: some achievements, problems and perspectives. In: Boccon-Gibod, J., Benbadis, A. and Shont, K.C. (eds) *Cell Culture Techniques Applied to Plant Production and Plant Breeding. Proceedings of IAPTC French–British Meeting,* 8–9 October, Angers, pp. 5–17.

Brown, A.H.D. and Marshall, D.R. (1995) A basic sampling strategy: theory and practice. In: Guarino, L., Rao, V.R. and Reid, R. (eds) *Collecting Plant Genetic Diversity: Technical Guidelines.* IPGRI/FAO/UNEP/IUCN. CAB International, Wallingford, pp. 75–91.

Brush, S.B. (1986) Genetic diversity and conservation in traditional farming systems. *Journal of Ethnobiology* 6, 151–167.

Bunting, A.H. and Kuckuck, H. (1970) Ecological and agronomic studies related to plant exploration. In: Frankel, O.H. and Bennett, E. (eds) (1970) *Genetic Resources in Plants – Their Exploration and Conservation.* IBP Handbook No. 11, Blackwell Scientific, Oxford, pp. 181–188.

Committee on Managing Global Genetic Resources (1993) *Managing Global Genetic Resources: Agricultural Crop Issues and Policies.* Board on Agriculture, National Research Council, National Academy Press, Washington, DC.

Cooper, D., Engels, J. and Frison, E. (1994) *A Multilateral System for Plant Genetic Resources: Imperatives, Achievements and Challenges.* Issues in Genetic Resources No. 2, May 1994. IPGRI, Rome.

Daily, G., Dasgupta, P., Bolin, B., Crosson, P., du Guerny, J., Ehrlich, P., Folke, C., Jansson, A.M., Jansson, B.-O., Kautsky, N., Kinzig, A., Levin, S., Mäler, K.-G., Pinstrup Andersen, P., Siniscalso, D. and Walker, B. (1998) Food production, population growth and the environment. *Science* 281, 1291–1292.

Damania, A.B. (1994) *In situ* conservation of biodiversity of wild progenitors of cereal crops in the Near East. *Biodiversity Letters* 2, 56–60.

Damania, A.B. (1996) Biodiversity conservation: a review of options complementary to standard *ex situ* methods. *Plant Genetic Resources Newsletter* 107, 1–18.

Duvick, D.N. (1984) Genetic diversity in major farm crops on the farm and in reserve. *Economic Botany* 38, 161–178.

Engelmann, F. (1997) *In vitro* conservation methods. In: Ford-Lloyd, B.V., Newbury, J.H. and Callow, J.A. (eds) *Biotechnology and Plant Genetic Resources: Conservation and Use.* CAB International, Wallingford, pp. 119–162

Engels, J.M.M. (ed.) (1995) *In situ* conservation and sustainable use of plant genetic resources for food and agriculture in developing countries. *Report of a DSE/ ATSAF/IPGRI Workshop*, 2–4 May 1995, Bonn-Röttgen, Germany. IPGRI, Rome, and DSE, Feldafing.

Engels, J.M.M. and Engelmann, F. (1998) Botanic gardens and agricultural gene-banks: building on complementary strengths for more effective global conservation of plant genetic resources. Paper presented at the Fifth International Botanic Gardens Conservation Congress, 14–18 September 1998, Cape Town, South Africa.

Engels, J.M.M., Arora, R.K. and Guarino, L. (1995) An introduction to plant germplasm exploration and collecting planning, methods and procedures, followup. In: Guarino, L., Rao, V.R. and Reid, R. (eds) *Collecting Plant Genetic Diversity: Technical Guidelines. IPGRI/FAO/UNEP/IUCN.* CAB International, Wallingford, pp. 31–64.

FAO (1994) *The International Network of Ex Situ Germplasm Collections.* CPGR-Ex1/94/Inf. 5/Add. 1, FAO, Rome.

FAO (1996a) *State of the World's Plant Genetic Resources for Food and Agriculture.* FAO, Rome.

FAO (1996b) *Global Plan of Action for the Conservation and Sustainable Utilization of Plant Genetic Resources for Food and Agriculture and the Leipzig Declaration* adopted by the International Technical Conference on Plant Genetic Resources, Leipzig, Germany, 17–23 June 1996.

Frankel, O.H. (1987) Genetic resources: the founding years. IV After twenty years. *Diversity* 11, 25–27.

Frankel, O.H. and Bennett, E. (eds) (1970) *Genetic Resources in Plants – Their Exploration and Conservation.* IBP Handbook No. 11, Blackwell Scientific, Oxford.

Frankel, O.H., Brown, A.H.D. and Burdon, J.J. (1995) *The Conservation of Plant Biodiversity.* Cambridge University Press, Cambridge.

Fuccillo, D., Sears, L. and Stapleton, P. (eds) (1997) *Biodiversity in Trust: Conservation and Use of Plant Genetic Resources in CGIAR Centres.* Cambridge University Press, Cambridge,

Galloway, J.H. (1996) Botany in the service of Empire: the Barbados cane-breeding program and the revival of the Caribbean sugar industry, 1880s–1930s. *Annals of the Association of American Geographers* 86, 682–706.

GEF (1995) *Quarterly Operations Report*, April, 1995. GEF Secretariat, Washington, DC.

Guzman, M.R. and Iltis, H.H. (1991) Biosphere reserve established in Mexico to protect rare maize relative. *Laboratory for Information Science in Agriculture* 7, 82–84.

Hammer, K., Knüpfer, H., Xhuveli, L. and Perrino, P. (1996) Estimating genetic erosion in landraces – two case studies. *Genetic Resources and Crop Evolution* 43, 329–336.

Harrington, J.F. (1970) Seed and pollen storage for conservation of plant gene resources. In: Frankel, O.H. and Bennett, E. (eds) *Genetic Resources in Plants – Their Exploration and Conservation*. IBP Handbook No. 11, Blackwell Scientific, Oxford, pp. 501–521.

Hepper, F.N. (1989) *Plant Hunting for Kew*. HMSO, London.

Heywood, V.H. (1999) The role of botanic gardens in *ex situ* conservation of agrobiodiversity. In: Gass, T., Begemann, F. and Frese, L. (eds) *Proceedings of the European Symposium on Plant Genetic Resources for Food and Agriculture*, Braunschweig, Germany, 30 June–4 July 1998. International Plant Genetic Resource Institute, Rome.

Hirsch, L.P. (1994) *Ex-Situ* Conservation of Biodiversity in the Context of Development. *Report of an International Meeting Held at the Smithsonian Institution*, Washington, DC, 16–19 February 1993.

Hoekstra, F.A. (1995) Collecting pollen for genetic resources conservation. In: Guarino L., Rao, V.R. and Reid, R. (eds) *Collecting Plant Genetic Diversity: Technical Guidelines*. IPGRI/FAO/UNEP/IUCN, CAB International, Wallingford, pp. 527–550.

Hong, T.D. and Ellis, R.H. (1996) A protocol to determine seed storage behaviour. In: Engels, J.M.M. and Toll, J. (eds) *IPGRI Technical Bulletin* No. 1, IPGRI, Rome.

Hong, T.D., Linington, S. and Ellis, R.H. (1996) *Seed Storage Behaviour: a Compendium*. Handbooks for Genebanks: No. 4, IPGRI, Rome.

Ingram, G.B. (1984) In Situ *Conservation of the Genetic Resources of Plants: the Scientific and Technical Base*. Forest Resources Division, FAO, Rome.

Ingram, G.B. (1990) The management of biosphere reserves for the conservation and utilization of genetic resources: the social choices. *Impact of Science on Society* 158, 133–141.

IPGRI (1996) *Programme Activities, Germplasm Maintenance and Use. Annual Report*. IPGRI, Rome, pp. 56–65.

Iwanaga, M. (1995) IPGRI strategy for *in situ* conservation of agricultural bio-diversity. In: Engels, J.M.M. (ed.) In Situ *Conservation and Sustainable Use of Plant Genetic Resources for Food and Agriculture in Developing Countries*. Report of a DSE/ATSAF/IPGRI workshop, 2–4 May 1995, Bonn-Röttgen, Germany, pp. 13–26.

Janick, L.V., Kim, Y.H., Kitto, S. and Saranga, Y. (1993) Desiccated synthetic seed. In: Redenbaugh, K. (ed.) *Synseeds, Applications of Synthetic Seeds to Crop Improvement*. CRC Press, Boca Raton, Florida pp. 11–33.

Kloppenburg, J. (ed.) (1988) *Seeds and Sovereignty: the Use and Control of Plant Genetic Resources*. Duke University Press, Durham, North Carolina.

Laliberté, B. (1997) Botanic garden seed banks/genebanks worldwide, their facilities, collections and network. *Botanic Gardens Conservation News* 2(9), 18–23.

Lyman, J.M. (1984) Progress and planning for germplasm conservation of major food crops. *Plant Genetic Resources Newsletter* 60, 3–21.

Marshall, D.R. and Brown, A.H.D. (1975) Optimum sampling strategies in genetic conservation. In: Frankel, O.H. and Hawkes, J.G. (eds) *Crop Genetic Resources for Today and Tomorrow*. Cambridge University Press, Cambridge, pp. 53–80.

Martin, G.B. and Adams, M.W. (1987) Landraces of *Phaseolus vulgaris* (Fabaceae) in Northern Malawi. I. Regional variation. *Economic Botany* 41, 190–203.

Maxted, N., van Slageren, M.W. and Rihan, J.H. (1995) Ecogeographic surveys. In:

Guarino L., Rao, V.R. and Reid, R. (eds) *Collecting Plant Genetic Diversity: Technical Guidelines.* IPGRI/FAO/UNEP/IUCN, CAB International, Wallingford, pp. 255–285.

Maxted, N., Ford-Lloyd, B.V. and Hawkes, J.G. (1997a) Complementary conservation strategies. In: Maxted, N., Ford-Lloyd, B.V. and Hawkes, J.G. (eds) *Plant Genetic Conservation: the* in Situ *Approach.* Chapman & Hall, London, pp.15–39.

Maxted, N., Hawkes, J.G., Ford-Lloyd, B.V. and Williams, J.T. (1997b) A practical model for *in situ* genetic conservation. In: Maxted, N., Ford-Lloyd, B.V. and Hawkes, J.G. (eds) *Plant Genetic Conservation: the* in Situ *Approach.* Chapman & Hall, London, pp. 339–363.

McNeely, J.A. (1996) The role of protected areas for conservation and sustainable use of plant genetic resources for food and agriculture. In: Engels, J.M.M. (ed.) In Situ *Conservation and Sustainable Use of Plant Genetic Resources for Food and Agriculture in Developing Countries.* Report of a DSE/ATSAF/IPGRI workshop, 2–4 May 1995, Bonn-Röttgen, Germany, pp. 27–42.

Murray, M.G., Green, M.J.B. and Walter, K.S. (1992) *Status of Plant and Animal Inventories for Protected Areas in the Tropics: a Contribution to the ODA Strategy Programme for Research on Forestry and Agroforestry.* World Conservation Monitoring Centre, Cambridge.

National Research Council (1989) *Lost Crops of the Incas.* National Academy Press, Washington, DC.

Oldfield, M.L. and Alcorn, J.B. (1987) Conservation of traditional agroecosystems: can age-old farming practices conserve crop genetic resources? *BioScience* 37, 199–208.

Percival, J. (1925) The value of selection work in the improvement of crop plants. In: Brooks, F.T. (ed.) *Report of Proceedings: Imperial Botanical Conference,* London, July 1924, Cambridge University Press, Cambridge, pp. 60–73.

Philipsson, J. (1992) A global review of genetic resources of cattle. In: Hodges, J. (ed.) *The Management of Global Animal Genetic Resources.* FAO Animal Production and Health Paper, No. 104, FAO, Rome.

Pistorius, R. (1997) *Scientists, Plants and Politics: a History of the Plant Genetic Resources Movement.* IPGRI, Rome.

Prescott-Allen, R. (1984) In-Situ *Conservation of Wild Plant Genetic Resources: a Status Review and Action Plan.* Background Document for the First Session of the FAO Commission on Plant Genetic Resources, FAO, Rome.

Prescott-Allen, R. and Prescott-Allen, C. (1982) *In situ* conservation of crop genetic resources. *Nature and Resources* 18, 15–20.

Prescott-Allen, R. and Prescott-Allen, C. (1984) Park your genes: protected areas as *in situ* genebanks for the maintenance of wild genetic resources. In: *National Parks, Conservation, and Development: the Role of Protected Areas in Sustaining Society.* Proceedings of the World Congress on National Parks, Bali, October 1982, Smithsonian Institution Press, pp. 634–638.

Prescott-Allen, R. and Prescott-Allen, C. (1990) How many plants feed the world? *Conservation Biology* 4, 365–374.

Purseglove, J.W. (1972) *Tropical Crops, Monocotyledons.* Longman, London.

Rao, N.K. and Jackson, M.T. (1996) Seed production environment and storage longevity of japonica rices (*Oryza sativa* L.). *Seed Science Research* 6, 17–21.

Roberts, E.H. (1973) Predicting the viability of seeds. *Seed Science and Technology* 1, 499–514.

Ruredzo, T.J. (1989) *Progress Report on IBPGR-ILCA Tissue Culture Project.* IBPGR Report Number 89/11, IBPGR, Rome.

Schlosser, S. (1984) The use of nature reserves for *in situ* conservation. *Plant Genetic Resources Newsletter* 61, 23–25.

Schoen, D.J. and Brown, A.H.D. (1991) Intraspecific variation in population gene diversity and effective population size correlates with the mating system in plants. *Proceedings of the National Academy of Sciences USA* 88, 4494–4497.

SGRP (1997) *Annual Report 1996 of the CGIAR System-wide Genetic Resources Programme.* IPGRI, Rome.

Sharrock, S. and Engels, J.M.M. (1996) *Complementary Conservation.* INIBAP Annual Report 1996, INIBAP, Montpellier, pp. 6–9.

Simmonds, N.W. (1962) Variability in crop plants, its use and conservation. *Biological Reviews* 37, 422–465.

Simmonds, N.W. (1993) Introgression and incorporation: strategies for the use of crop genetic resources. *Biological Reviews* 68, 539–562.

Tao, K.-L., Williams, J.T. and van Sloten, D.H. (1989) Base collections of crop genetic resources: their future importance in man-dominated world. *Environmental Conservation* 16, 311–316.

Thin, Nguyen Nghia (1996) Review of valuable plant genetic resources in the Cuc Phuong National Park. In: *Plant Genetic Resources in Vietnam.* Proceedings of the National Workshop on Strengthening of Plant Genetic Resources Programme in Vietnam, Hanoi, March 1996, Agriculture Publishing House, Hanoi, pp. 87–88.

Tothill, J.D. (ed.) (1948) *Agriculture in the Sudan.* Oxford University Press, London.

Towill, L.E. (1985) Low temperature and freeze-/vacuum-drying preservation of pollen. In: Kartha, K.K. (ed.) *Cryopreservation of Plant Cells and Organs.* CRC Press, Boca Raton, Florida pp. 171–198.

UNEP (1992) *Convention on Biological Diversity.* UNEP, Geneva.

UNESCO (1984) Programme on Man and the Biosphere (MAB). *Action Plan for Biosphere Reserves,* December 1984, UNESCO, Paris.

Vavilov, N.I. (1951) The origin, variation, immunity, and breeding of cultivated plants. *Chronica Botanica* 13, 1–364.

Von Tschermak, E. (1915) Über die Notwendigkeit der Sammlung und Erhaltung unserer bewährten, noch unveredelten Getreidelandrassen. *Wiener Landwirtschafliche Zeitung* No. 104, 750–761.

Walters, C. and Engels, J.M.M. (1998) The effects of storing seeds under extremely dry conditions. *Seed Science Research* 8 (Suppl.), 3–8.

Wilcox, B.A. (1990) *Requirements for the Establishment of a Global Network of in Situ Conservation Areas for Plants and Animals.* Draft prepared for FAO Forest Resources Development Branch.

Wilkes, H.G. (1971) Too little gene exchange. *Science* 171, 955.

Withers, L.A. (1995) Collecting *in vitro* for genetic resources conservation. In: Guarino, L., Rao, V.R. and Reid, R. (eds) *Collecting Plant Genetic Diversity.* IPGRI/FAO/UNEP/IUCN, CAB International, Wallingford, pp. 511–525.

Wood, D. (1988) Introduced crops in developing countries: a sustainable agriculture? *Food Policy* 3, 167–172.

Wood, D. and Lenné. J. (1997) The conservation of agrobiodiversity on-farm: questioning the emerging paradigm. *Biodiversity and Conservation* 6, 109–129.

Yidana, J.A. (1988) The development of *in vitro* collecting and isozyme characterization of cocoa germplasm. PhD thesis, University of Nottingham.

A Conceptual Framework for Valuing On-farm Genetic Resources

15

M. Smale* and M.R. Bellon

CIMMYT, Apartado Postal 6-641, 06600, Mexico DF, Mexico

Introduction

Biological scientists have long been aware of the importance of diversity in agroecosystems. Diversity among species provides multiple benefits to agriculture. For example, beneficial insects and soil organisms can contribute to crop health and reduce the need for agricultural chemicals. Plant breeders and other scientists rely for crop improvement on the genetic diversity they find in their own working collections, *ex situ* collections, and among the varieties maintained and selected by farmers.

To many biological scientists, therefore, the economic value of diversity within and among species is self-evident; assigning it a monetary value is objectionable on both methodological and ethical grounds. Economists tend to approach conservation issues with 'the arrogance of humanism' (Ehrenfeld, 1988). Although questions of species and subspecies existence are inherently moral judgements, economics is based on utilitarianism: its focus is human society and how to choose the best means of achieving a predetermined social goal (Randall, 1986).

Economists argue that important decisions are already being made about the re-allocation of scarce resources from some research activities, such as crop improvement, to the protection and conservation of genetic resources and their diversity. They generally believe that the costs of conservation should be viewed seriously, and the benefits should be quantified to the greatest extent possible; 'if we can't save all species, we need a ranking based on one or more criteria, from which we select the highest ranked for conservation' (Brown, 1990). Without such information, we are forced to make arbitrary decisions.

*See Contributors list for current address.

© 1999 CAB International. *Agrobiodiversity*
(D. Wood and J.M. Lenné)

In this chapter, while recognizing the ethical difficulties involved in assigning values to biological diversity, we propose a general approach for identifying which crop populations to conserve on-farm (*in situ*) and *ex situ*. The approach relates the choice of conservation strategy (*ex situ*, on-farm) to the biological characteristics of crop populations in a reference area and their use-value as recognized by the farmers who cultivate them (see also Engels and Wood, Chapter 14, this volume). Given a well-defined reference area within a crop centre of diversity where genetic resources have been targeted for conservation, can we identify the least-cost conservation strategy?

For decades, in recognition of the critical importance of crop genetic resources to global society, conservationists have collected and stored the seeds of traditional varieties and crop relatives in *ex situ* genebanks. Seed storage is a safe and efficient way of conserving orthodox seeds[1], and has the advantage of making the germplasm readily available for use by breeders and study by other researchers (Ford-Lloyd and Jackson, 1986; see Wright and Turner, Chapter 13, this volume). *Ex situ* conservation is static, aiming to preserve as closely as possible the genetic structure of the original population (Guldager, 1975).

On-farm conservation of crop genetic resources is increasingly viewed as being complementary to *ex situ* conservation (Altieri and Merrick, 1987; IPGRI, 1993; Brush, 1995; Bellon *et al.*, 1997a). On-farm conservation is the continued cultivation and management by farmers of a diverse set of crop populations in the agroecosystem where the crop has evolved or in secondary centres of diversity (Bellon *et al.*, 1997a). A dynamic form of conservation, it allows crop populations potentially to continue their evolution in response to natural and human selection (Jackson, 1995; Pham *et al.*, 1996; see Engels and Wood, Chapter 14, this volume).

The role of farmers in on-farm conservation is fundamental because crop populations are the result of human selection and management as well as natural factors. Conservation is not the moral obligation of farming communities; farmers transform or abandon particular varieties to suit their own needs. For crop genetic resources to be conserved on-farm, maintaining crop populations and varieties must be advantageous to farmers. Farmers in any reference area must have an economic or cultural incentive to continue growing the crop populations that are identified as key genetic resources.

Not all crop populations will be conserved, and the criteria on which they are chosen are subject to debate. In the past, collectors selected their samples based on a fairly *ad hoc* visual inspection of plants and their own hypotheses about the selective pressures affecting the crop populations grown in a given environment. The conceptual framework we propose below has been designed to support strategic decisions about which crop populations in a given reference area are suitable candidates for *ex situ* and on-farm conservation initiatives. We assume that the reference area, such as a crop centre of diversity, has already been selected based on prior beliefs about the likelihood of encountering potentially valuable genotypes. The framework is both preliminary and conceptual, and it should be seen as a heuristic tool to facilitate discussion. To implement it fully would require a much more extensive development than is possible within the confines of this chapter.

Farmer Management of Diversity

Components of farmer management of diversity

Farmers' varieties are crop populations that they identify and name as units, including landraces, modern varieties, and modern varieties that farmers have selected or mixed with their own landraces. The latter are sometimes referred to as 'creolized', 'rusticated', or 'locally adapted' modern varieties[2]. To design on-farm conservation strategies, we need to understand and be able to influence the process by which farmers select and manage their seed and their repertoire of crop varieties. Bellon *et al.* (1997a) have identified three components of farmers' management of diversity:

1. *Seed flows* refer to the process by which farmers obtain the physical unit of seed for a given variety. The seed a farmer plants may have been selected from his or her own crop in the preceding season, exchanged or purchased from other farmers or institutions, or mixed from a combination of sources.
2. *Variety choice* is the process by which farmers decide which crop varieties to plant.
3. *Seed selection and management* is the process by which a farmer who keeps seed from his or her own crop: (i) selects the specific part of the plant material to be used as seed, and (ii) handles the seed from harvest to planting (Fig. 15.1).

The three components can be conceptualized as the dependent or behavioural variables that we want to explain and eventually to influence in a study of the bases of on-farm conservation (Fig. 15.2). Their systematic variation is caused by variation in a number of independent variables, such as the environmental, socio-economic and cultural factors that influence the farmers' decision-making. These factors also operate at different scales. Some are related to the characteristics of individual farmers, such as farmer age, gender, education and knowledge. Others are household characteristics that are fixed in the short term and may be interrelated, such as wealth or socio-economic status, or access to land, labour and capital. Other factors describe the institutions of the farmer's community, including patterns of labour exchange, land tenure, social obligations and ethnic identity. Some are regional or national features, such as the availability of physical infrastructure (irrigation, roads, telecommunications), the degree of market development, and government policies. Finally, the community as a whole is situated within a biophysical environment that affects crop performance, defined by rainfall conditions, temperature, soils and topography.

Fig. 15.1. Farmers in Oaxaca, Mexico, have an intimate familiarity with the maize varieties they grow (courtesy of Mike Listman, CIMMYT).

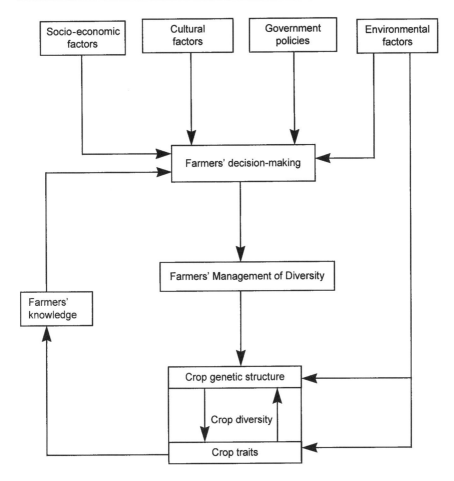

Fig. 15.2. A conceptual model of the factors that influence the farmers' management of diversity. Source: Bellon *et al.* (1997a).

Differences in management of diversity among farmers should translate into different genetic structures and levels of diversity, with consequences for the genetic diversity of the crop in a reference area. Although farmers cannot directly observe the genetic structure of their crop, they gain knowledge with experience of the yield, functional and morphological characteristics of the varieties they grow. This knowledge is in turn used in the decision-making process that defines their management of diversity. Crop genetic structure is also shaped by environmental factors through natural selection. The genetic structure of the crop populations in a reference area evolves over time, but in which ways and within what time period is unknown and cannot be predicted with certainty (see Maxted *et al.*, 1997).

Farmers' varieties as bundles of characteristics

Researchers have documented that small-scale farmers in areas of crop diversity often plant several crop varieties in one season (Brush *et al.*, 1981; Richards, 1986; Dennis, 1987). Small-scale farmers usually have multiple interests or concerns and confront numerous problems in attempting to fulfil them. Since it is unlikely that one variety has all of the traits demanded by the farm household, and since desirable traits may be associated with undesirable ones, the choice of varieties can be seen as a process by which farmers assemble various bundles of traits to fulfil specific production conditions, consumption preferences, or marketing requirements (Bellon, 1996). There are trade-offs in the selection of varieties, and the farmer can change the bundle of traits by changing the allocation of crop area among varieties.

Economic framework for analysing farmers' incentives to maintain diversity

We can incorporate some of these concepts into an economic framework that depicts farmer incentives to grow varieties that are identified as key genetic resources. The approach outlined here draws on aspects of characteristics models (Lancaster, 1966; Ladd and Suvannunt, 1976; Adesina and Zinnah, 1993) and also draws on the theory of impure public goods (Cornes and Sandler, 1986).

Characteristics models state that farmers or farm households maximize the utility from the multiple attributes of the crop produced by their choice of varieties, rather than the varieties themselves or only a single trait, such as grain output or yield. Farmers choose varieties based on the bundle of observable characteristics that each seed type embodies and produces. Each variety supplies or 'yields' different amounts of each characteristic. Seed is unique as a commodity in that it has characteristics that are private as well as those that are public (Morris *et al.*, 1998). The private characteristics of seed are those that cannot be consumed by two farm households at once, such as the grain, fodder and the amounts of characteristics that are produced in each farm household's crop. The public characteristics of the seed are those related to its genetic attributes, including its contribution to genetic diversity. Each choice of seed amount and variety combination jointly produces or 'yields' characteristics of use to the farmer as well as a characteristic of public interest – a contribution to the diversity of genetic resources in the reference area. While farmers can sell seed as a production input, the full value of the genetic resources embodied in the seed is not realized in the seed price.

We can view the decision more formally and generally as a model of decision-making in agricultural households, using maize farming in Mexico as an example. In each season, the household chooses a set of n seed lots for varieties ($\mathbf{x}$) to combine with non-seed inputs (represented by an index Y)

that maximizes the expected utility from a set of consumption attributes (**q**). The household also chooses the amount it will consume (**Q**) from the maize outputs (**X**) produced by the inputs:

Max $EU(\text{q}|\Omega,Z)$

$$\text{x,Q} \tag{1}$$

The set of parameters Ω represents socio-economic, agroecological, and other exogenous factors which condition farmers' decisions.

Consumption characteristics q = $(q_1, \ldots q_j)$ include ease of hand processing, or suitability for particular dishes. The q_j are the total quantities of the *j*th characteristic of maize output, and q_{ij} is the quantity of the *j*th output characteristic produced by one unit of the *i*th of *n* maize types. The set of q_{ij}, which express input–output relationships, are technical coefficients which are fixed in the short term to the farmer but are amenable to change through crop improvement[3]. There is no commercial market for characteristics, although the market for maize may value them indirectly. Farm households may sell or buy any or all of the maize varieties, so that (**X** – **Q**) is negative for a net consumer of maize and positive for those selling more than they consume, at price **p**.

The household faces the constraint that its expenditure on seed at price or cost **w** and on non-seed inputs cannot exceed its exogenous income *I* (such as income from off-farm labour, other crops or migration that is earned before planting) and its expected returns from sales of its maize varieties:

$$Y + \mathbf{w'x} = I + \mathbf{p'}(\mathbf{X} - \mathbf{Q}) \tag{2}$$

The decision of the farm household is also constrained by the technology for jointly produced varieties:

$$F(\mathbf{X}, z^{\text{h}}|\mathbf{x}, \mathbf{r}, Y) = 0 \tag{3}$$

X is a vector of maize outputs for the *i* varieties grown[4], which is in turn a function of their production characteristics $\mathbf{r} = (\rho_1, \ldots \rho_k)$, the amount of seed planted, and non-seed inputs. Production characteristics include tolerance of abiotic and biotic stresses, and performance on a specific soil type. As defined above in the case of consumption attributes, ρ_k represents the total quantities of the *k*th production characteristic, and ρ_{ik} is the quantity of the *k*th production characteristic produced by one unit of the *i*th of *n* maize types. Features of the variety yield distribution might also be expressed in these characteristics.

A seed lot is the physical unit of grain that is planted (Louette *et al.*, 1997), and the household may obtain it from its own crop, other farmers' crops, or the formal seed system. Choice of any set **x** generates not only maize outputs (**X**), but the farm household's individual contribution (z^b) to a public good, genetic diversity in the region (*Z*). If genetic diversity is defined over characteristics that are not observed by farmers (such as allele frequencies), the household's individual contribution to diversity may not be observable and we would not expect the utility function to be defined over z^h. Utility may be defined over Z when genetic diversity is observable as morpho-phenological variation, however, Z could then be interpreted as the supply of characteristics in the farmers' reference area.

As in the Bellon *et al.* (1997a) definition of farmer management of diversity, the diversity in the maize grown by a farmer can be expressed as:

$$z^h = z^h(\mathbf{x}, \theta, \beta) \tag{4}$$

where θ and β are parameters related to seed flows and seed management practices. The choice of varieties also affects diversity at the farm level through a technical relationship x_{iz}. Further, since 'diversity' is a public characteristic, it is affected by the decisions of all farmers in the region of reference, although not linearly, and probably interactively:

$$Z = Z(z^1, \ldots z^h, \ldots z^m) \text{ for all farmers h} = 1, \ldots m \tag{5}$$

One analytical result of this type of model is that farmers as a group may choose seed amounts and variety combinations that are less or more than socially optimal, because they do not take into account the interaction of their choices with the choices of other farmers. For example, if many farmers assume that others are growing a variety they have ceased to plant, the variety may be lost from the reference area[5]. The interaction may create a social cost, as in the case where a large number of farmers choose to grow one variety because it provides certain desirable characteristics, but a high degree of genetic uniformity in resistance genes increases vulnerability to disease (Heisey *et al.*, 1997). Whether the 'diversity' that is generated by any seed choice is observable to farmers and affects their decisions is also a testable hypothesis.

Finally, farmer demand for a variety is determined by the characteristics it embodies, the importance of the characteristic in the goals of the farm household, the importance to the farmer of the supply of traits in the locality (alternatively, 'what everybody else does', or Z), real prices and costs of production ($\mathbf{P}$, $\mathbf{w}$, Y), and the exogenous factors Ω which condition their production choices:

$$x_i = x_i(\mathbf{P}, w_i, Y, I; q_{i1}, \ldots q_{if}, \rho_{i1}, \ldots \rho_{ik} \mid \Omega, Z) \tag{6}$$

Note that in the case of a commercial producer, utility is maximized over expected profits and a homogeneous output that is sold in its entirety. Consumption characteristics will not matter, but production characteristics may still affect decisions.

On-farm conservation requires that farmers have an incentive to continue growing the varieties identified as contributing to the diversity of genetic resources in their reference area. How can the application of this framework assist us in designing a strategy for on-farm conservation? Firstly, it may guide us in the choice of which varieties to conserve, on-farm or *ex situ*. Secondly, estimation of equation (6) provides information about the technical incentives that can be provided by breeding interventions (**q,r**), as well as the potential effects of policies that influence the conditions under which farmers operate (Ω). Similarly, estimation of equation (4) may inform us about how technical interventions in seed flows, seed selection and management practices may affect the crop's genetic diversity. The following sections describe these points in greater detail.

Strategic Decisions in Collection and Conservation

Choosing populations to conserve

The framework can guide us in the choice of crop populations to conserve, given that a reference area has already been identified based on prior expectations of its relative importance among other candidate conservation areas. The set of crop populations in the reference area can be analysed by classifying them along two axes: (i) the probability that farmers will maintain the population; and (ii) the contribution of the population to the overall genetic diversity in the area (Fig. 15.3).

The probability that farmers will maintain any variety or crop popu-

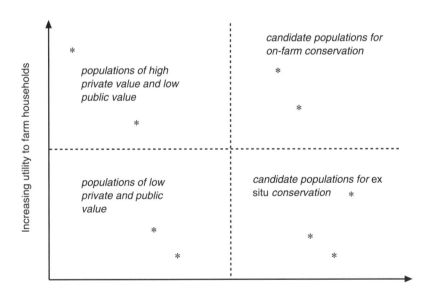

Fig. 15.3. Framework for choosing crop populations to conserve on-farm and *ex situ*, in a given reference area.

lation is clearly a function of its value to them. In our framework, the probability that a variety will be maintained reflects the number of production and consumption characteristics for which a variety ranks highly, and the relative importance of these characteristics to farmers in meeting their objectives. Farmers' demand for a variety is higher the greater its usefulness. All varieties or populations in a given reference area can be ranked according to their capacity to supply the characteristics demanded by farmers. When a variety's value is only partially revealed in market prices, its utility rank represents its relative value in current use.

Similarly, each population can be ranked according to its contribution to the genetic diversity of the crop populations in the reference area. Let us define the collection of populations of interest in the reference area as a meta-population, that is, a set of populations interconnected through migration[6] (Olivieri and Gouyon, 1990; David, 1992). Within a metapopulation, some populations are more similar than others in terms of their alleles, allele frequencies and agromorphological characteristics. Clearly, two populations that are very similar contribute less to the overall genetic diversity of the metapopulation than two that are different. The contribution of any population is relative, since it depends on the other constituent populations of the metapopulation in a reference area.

Combined, the two indicators can be used to choose which population to conserve on-farm and *ex situ*. Those populations located to the northeast of the graph in Fig. 15.3 have both high utility and a high contribution to genetic diversity. They are prime candidates for on-farm conservation since they have both a high private value to farmers and a high public value as genetic resources. These are the 'least cost' to conserve on-farm, since farmers already value them highly. A population located to the southeast is a prime candidate for *ex situ* conservation; although it has great value as a contributor of genetic diversity, farmers themselves value it relatively little. Such a population would 'cost' a lot to conserve on-farm in order to encourage farmers to grow it. Populations to the west of the quadrant can in essence be 'ignored'; those highly valued by farmers will be maintained by them, and those valued less may be discarded – but without major consequences for farmers or the society in the reference area. Given a fixed conservation budget, spending money to conserve these by any means would seem relatively wasteful.

Measuring the usefulness of varieties

The mathematical assumptions underlying the full development of the economic model we describe above can be restrictive, and there is considerable theoretical debate over the pros and cons of various methods for eliciting preferences. In practice, some fairly simple methods for eliciting rankings and compiling utility indices for varieties have been developed (see, for example, Chambers, 1988; Sall *et al.*, 1997). When a farmer ranks the varieties he or she grows with respect to its characteristics and can state the relative importance of the characteristics in his or her farming objectives, a preference ordering or utility index is 'revealed'[7].

Farmers usually have a good knowledge of the performance of the varieties they grow, and can rank them accordingly, for a set of characteristics (Figs 15.1 and 15.4). The ranking of varieties with respect to their characteristics, combined with the relative importance of the characteristics to target-farmers, can be used to predict which varieties are likely to be maintained and which are likely to be dropped. For example, this approach predicts several varieties are maintained if each one of them ranks highest for a different subset of characteristics of importance. Specifically, for characteristics of comparable importance: (i) one variety displaces another if it ranks higher on at least one characteristic and equal on all others; and (ii) a variety is included in the set but does not displace another variety if it ranks higher than the others for a subset of characteristics, but ranks lower on the remainder (Bellon, 1996). If a farmer grows only one variety but has access to many or has tried others, we may conclude that the variety dominates all other available varieties with respect to the attributes that matter to that farmer. A variety with a high ranking for many relatively important characteristics or a high ranking for a single very important characteristic may be

Fig. 15.4. Evaluating the characteristics of varieties requires close interaction between farmers, genetic resource specialists and social scientists (courtesy of Mauricio Bellon, CIMMYT).

more likely to be maintained than one that ranks high for only one, and low for the remainder. Superior performance for a single high-ranking concern may compensate, however, for inferior rank among other concerns of lesser importance.

Measuring the contribution of a variety to genetic diversity in a reference area

The aim of on-farm conservation is to maintain the evolutionary processes that have shaped crop populations for generations. Its underlying principle is that when crop populations (varieties) are managed by farmers in their own biophysical environments, they will continue to evolve in the future, as in the past, adapting to new and changing circumstances. Therefore the value of any variety to on-farm conservation lies on its evolutionary potential. Two possible indicators of this potential are genetic polymorphism and breadth of adaptation.

Polymorphism is defined as 'the simultaneous and regular occurrence in the same population of two or more discontinuous variants or genotypes in frequencies which cannot be accounted by recurrent mutation' (Ford, 1940). Selection pressures, both environmental as well as human, act upon these populations by discarding less adapted genotypes and maintaining the more adapted ones. However, at what rate this process takes place is open to question. Selection affects the allele frequencies across varieties. Three additional factors may affect allele frequencies: (i) rare mutation events, (ii) migration from other varieties, and (iii) genetic drift, or the random process of allele loss that depends particularly on population size.

The first two of these factors may act to increase polymorphism although, in general, the importance of mutations is very small compared to the migration factor within a few cropping seasons. The third factor exerts pressures in the opposite direction. The methods for measuring genetic diversity that are relevant for on-farm conservation are those that focus on assessing polymorphism within and among varieties, and their genetic structure in a given reference area or potential site.

Currently, a number of procedures are used to estimate genetic diversity at the biochemical and molecular (DNA) levels. Electrophoresis, which is the separation and visualization of allozymes, has been the most frequently used because it is widely applicable, cost-effective, and relatively rapid (Hamrick and Godt, 1997). Recent advances in biotechnology are providing more sophisticated molecular methods for detecting genetic variation at the DNA level, such as restriction fragment length polymorphism (RFLP) markers, and markers based on polymerase chain reactions (PCR) (Szmidt, 1995). Although such methods are increasingly popular, allozymes are still commonly used. Over time, a great deal of knowledge has accumulated regarding their use. Table 15.1 shows several genetic parameters that are used

Table 15.1. Common parameters used to assess genetic diversity.

Parameter	Definition
Per cent polymorphic loci	No. of polymorphic loci/ no. of studied loci (a population is polymorphic at the locus *n* if the most frequent allele is less than 95%, but other limits may be defined)
Mean number of alleles per polymorphic locus (mean allelic richness)	$$\frac{1}{L} \sum_{i=1}^{L} n_i$$ n_i = number of polymorphic loci observed L = total no. of studied loci
Genetic diversity at a single locus (*H* = expected heterozygosity on basis of Hardy–Weinberg)	$$H = 1 - \sum_{i=1}^{i=n} (p_i)^2$$ n = number of alleles at the locus p_i = frequency of the allele *i*
Mean diversity within populations	$$H = \sum (H_i)/N$$ N = number of loci H_i = Hardy–Weinberg expected heterozygosity at locus *i*
Proportion of genetic diversity residing between populations	$$G_{st} = (H_t - H_s)/H_t$$ H_s = mean (for all varieties) of the set of the within-variety genetic diversity H_t = genetic diversity (total) of the set of varieties considered as a unique population
Nei's genetic distance (*D*)	$$D = -\log(I)$$ where $$I = J_{xy}/(J_{xx}\,J_{yy})^{1/2}$$ $$J_{xy} = \%p_i\,q_i$$ I = normalized identity *I* for the locus J_{xy} probability that two alleles are identical when one allele is chosen from population *x* and the other from population *y* p_i and q_i are frequencies of the allele *i* in the populations *x* and *y*, respectively

with allozyme markers to assess genetic diversity and its structure. Good examples of how they have been used to estimate genetic variation are found in Hamrick and Godt (1997) for different crops, and Doebley *et al.* (1985) for maize in Mexico.

Indices constructed as defined in Table 15.1 capture different types of genetic variation. The genetic variation in a single variety or a group of varieties, and particularly in landraces, can occur as a large number of alleles at a single locus, as variation in the frequency of alleles among the populations of the same or distinct varieties, or as a high level of heterozygosity. The way diversity is structured within and between populations depends on the mating system of the crop (Hamrick and Godt, 1997). Even where diversity of individual varieties is low in terms of allele numbers or heterozygosity, the alleles they contain may be unique to the different varieties (Hodgkin *et al.*, 1993).

It should be pointed out that the existence of extensive polymorphism at the DNA or isozyme level does not necessarily relate to any kind of adaptive or evolutionary potential (see Witcombe, Chapter 10, this volume). Nor is the absence of polymorphism at the molecular level due to a lack of adaptive or evolutionary potential. However, the assumption is that the degree of polymorphism is an indicator, albeit imperfect, of this potential. In any case, the use of biochemical or molecular markers can determine whether two populations are very similar or not, and therefore it can indicate the contribution each of the constituent populations to the metapopulation overall genetic diversity.

The breadth of adaptation[8] of a variety is the another possible indicator of its evolutionary potential[9]. Although there is not necessarily a correlation between molecular polymorphism and breadth of adaptation, it is likely that given similar levels of polymorphism, a population than is broadly adapted may have higher capacity to evolve under new conditions that one that is narrowly adapted. However, even a narrowly adapted population, with a more limited capacity to evolve, may be of interest for on-farm conservation if it contributes unique alleles to the metapopulation in the reference area.

If, on scientific grounds, one accepts that genetic diversity is better assessed at the biochemical and molecular (DNA) levels than at the morphological level, the fact remains that while farmers can observe morphological traits and their variation, they cannot detect polymorphism at a molecular level (see also Cox and Wood, Chapter 3, and Witcombe, Chapter 10, this volume). As a consequence, what may be of interest to those who seek to promote on-farm conservation may be of no concern to farmers. The challenge of on-farm conservation is to identify the varieties that are both attractive to the farmers and contribute most to future flexibility of the genetic resource system, as represented by the polymorphism in the crop populations of the reference area.

Farmers' responses to economic change

Incentives are necessary to enable on-farm conservation because of economic change and farmers' adaptation to change. Through applying the framework, we can investigate how changes in the exogenous technological, socio-economic and cultural parameters that condition farmers' decisions (Ω) influence the repertoire of varieties they maintain, providing a more dynamic view of farmer management of diversity. Some of the factors included in Ω, as well as **P** and **w**, can be influenced by policies. The relative importance of factors of different scale can also be tested, such as variety characteristics, household characteristics, characteristics of the environment in which the crop is produced, or the extent of market development. The framework may assist us in the development of policy recommendations to support on-farm conservation.

In a more dynamic view of farmer management of diversity, we are most concerned about how changes in exogenous factors modify: (i) the set of farmers' concerns, (ii) the distribution of traits among available varieties, and (iii) the relative performance of a particular trait in a variety.

We can think of changes in the number or types of concerns as mapping into the characteristics space over which farm families define their goals and objectives, changing its dimension. As the relative importance of characteristics changes, so do the trade-offs among them. Market integration and the availability of technologies provide new ways of solving old problems, create new goals and new problems, and cause other problems to become irrelevant. For example, the opportunity to work off-farm generates income that is not correlated with agricultural income, providing a means other than crop diversity of managing risk. On the other hand, when marketing infrastructure is improved and costs of marketing decrease, subsistence farmers may begin to market their crop, choosing to grow higher-yielding varieties alongside the lower-yielding local varieties they prefer for home consumption.

The distribution of traits among varieties changes when a variety that 'packs' more traits is introduced (not necessarily a modern variety), leading to a reduction in the farmer's repertoire because fewer varieties are needed to fulfil the same farming objectives. Since modern plant breeding can 'pack' many traits in a single variety, it is not surprising that the introduction of modern varieties into a farming system often has been associated with the loss of other varieties (however, not necessarily with the loss of diversity of traits of interest to the farmer) (see Witcombe, Chapter 10, this volume). The third type of change is the simplest and is most frequently (but perhaps inaccurately) used to explain the loss of crop diversity: a variety is introduced by farmers or formal seed systems that performs better with respect to one characteristic, e.g. yield (Smale, 1997).

The three types of changes are not mutually exclusive and may occur simultaneously. Most likely the processes of crop diversity loss that have occurred worldwide result from a combination of them; in general, the range of characteristics that matter to farmers has contracted while new varieties have been introduced that farmers rank higher with respect to the remaining traits. It is important to remember, however, that the forces of economic change have limits. Market imperfections are common throughout the developing world (de Janvry *et al.*, 1991), constraining farmers' ability to substitute diversity through market transactions (Brush *et al.*, 1992). Plant breeders cannot always break the linkages among certain desirable and undesirable traits, or may not be paying attention to some of the concerns that farmers value.

Economic changes that affect the private incentives for growing a repertoire of varieties also influence the public good: crop genetic diversity in the reference region. The level of genetic diversity in a region can drop if farmers choose to specialize in a few varieties that are genetically similar. If many farmers assume that somebody else is keeping the varieties they are discarding, there is the danger that none will keep them and they will be lost from the reference area (see also Thurston *et al.*, Chapter 9, this volume).

As Brush and Meng (1998) have suggested based on their work in Turkey, as long as some farmers continue to select landraces for cultivation, conservation occurs *de facto*, and a formal programme may not be necessary. We may incorrectly deduce from Fig. 15.3 that no interventions are required to encourage farmers to continue growing the crop populations found in the northeast portion of the graph. Changes in the exogenous parameters Ω may alter the usefulness of these crop populations to farmers, and hence the probability that they will be maintained – shifting them to the southeast of the graph. One option may be to counteract this process by enhancing their desirable traits under the new conditions.

For example, in the Philippines, Bellon *et al.* (1997b) used multidisciplinary methods to identify a cluster of rice varieties that were both genetically diverse and highly valued by farmers for their consumption characteristics and tolerance to biotic and abiotic stresses (Fig. 15.4). Although still cultivated in the rainfed system, they had been 'discarded' in the irrigated system. In the irrigated system, varieties with shorter duration could be grown and two crops produced. High opportunity costs were associated with growing the older, favoured varieties with longer growing cycles. A breeding intervention aimed at reducing cycle length might enhance the desirability of the old varieties under the new system.

Participatory plant breeding, or any breeding system which enhances what farmers want, may be a way of providing incentives for farmers to maintain crop populations that are identified as key genetic resources (see Witcombe, Chapter 10, this volume). In terms of our model, breeding offers the opportunity to enhance the output of a variety per unit of seed in terms of an observable characteristic of value to farmers. Farmers' demand for the variety will increase. If the variety also contributes to the genetic diversity of the reference population, both private and public goals will be addressed.

Further Issues

Identifying which crop population is worth conserving in a given reference area is only part of the operational problem of on-farm conservation. Finding the appropriate social mechanisms through which to work with farmers and encourage the diffusion between them of new techniques, information and seed is challenging (Ashby *et al.*, 1996; Sperling *et al.*, 1996).

The approach we have presented here also leads to the hypothesis that the management of maize diversity (in terms of its components of varietal choice, seed flows, seed management) is not homogenous within a region, and there may be distinct types of management associated with different kinds of farmers. Using different research methods, in several locations in Mexico, researchers have found that some farmers retain their seed for successive generations and some replace their seed frequently for both traditional and modern varieties, while others deliberately mix their seed lots for the same varieties (Aguirre, 1997; Louette *et al.*, 1997; Rice *et al.*, 1997). Each management practice requires a unique approach to on-farm conservation.

In addition to identifying the crop populations most likely to be maintained by farmers, we need to identify the farmers who are most likely to maintain them. Farmers facing more heterogeneous agroecological conditions, who are located in more isolated areas, or who have culturally based preferences for certain products may be more interested in maintaining a large repertoire of varieties than others. Aguirre (1997) found that households located in isolated areas maintained higher levels of varietal diversity than those located in more market-integrated areas. Even within a community there may be differences. Among Turkish wheat farmers, Meng *et al.* (1997) found that the probability of planting a landrace in a plot was influenced by household characteristics such as availability of family labour, education and ownership of livestock. She uses information on household characteristics and the probability that a household will grow landraces to identify potential participants in an on-farm conservation programme (Brush and Meng, 1998).

The application of this framework in a specific region may require a baseline study and periodic follow-up studies to monitor whether the targeted populations are maintained or not, as well as the evolutionary changes that may be happening in them and in the overall metapopulation in the reference area.

This framework also has some implications for *ex situ* conservation (see Engels and Wood, Chapter 14, this volume). The populations classified in the four quadrants may be suitable for different genebank management strategies. For those located in the northwest quadrant of Fig. 15.3, it may be useful to evaluate samples more thoroughly and to isolate specific traits of value that could then be placed in other materials. For those located in the southwest, small samples may be stored in long-term deep cold storage and may not require any grow outs. For those in the northeast quadrant, samples in the genebank may be used as a back-up. For those in the southeast, the samples stored may be part of the active collection (G. Wilkes, CIMMYT, 1997, personal communication).

Conclusions

Increasingly, on-farm conservation is viewed as complementary to *ex situ* conservation because it is by nature dynamic, allowing crop populations to continue to evolve in response to natural and human selection pressures. Farmers' roles in on-farm conservation are fundamental since much of the cost of on-farm conservation strategies will be borne by them. One role of socio-economic research in this field is to assist us in understanding, analysing, and predicting the ways in which farmers manage their genetic resources. A second is to enable us to identify least-cost methods or strategies for enhancing resource management and conservation. Costs, as they are defined in economics, refer to opportunities forgone rather than monetary values.

Seed has both private and public characteristics. The feasibility of on-farm conservation depends on whether farmers are able to benefit from it at the same time that genetic diversity is enhanced in the reference area (see Engels and Wood, Chapter 14, this volume). Our framework enables us to begin answering questions about the effects of breeding interventions and other policy incentives on variety choice, and the effects of variety choice, seed flows, and seed selection and genetic diversity. We can also apply the framework to the choice of conservation strategies in a reference region, using a utility index to represent value in current use, and genetic distance as a proxy for evolutionary potential.

The application of the framework can be directly interpreted in terms of the relative costs of alternative conservation strategies. Crop populations that are highly valued by farmers will cost less to conserve on-farm than those with low private value, for both farmers and society as a whole. Crop populations with high public value and low private value are best conserved in genebanks, because inducing farmers to maintain them would be more costly. Society should not invest in crop populations that have both low public value and low private value.

Although our framework may guide us in identifying the crop populations to conserve on-farm and *ex situ*, we should recognize that, in some cases, changes in exogenous environmental and socio-economic factors may be so overwhelming that farmers' incentives to maintain varieties will disappear despite substantial efforts to make them attractive. Put simply, the opportunity cost of growing the varieties may be too high. We should not, therefore, overestimate our ability to promote on-farm conservation.

Acknowledgements

The authors thank Douglas Gollin, Daniel Grimanelli, Jean-Louis Pham and Garrison Wilkes for helpful comments.

Notes

[1] Orthodox seeds can be dried to a relatively low moisture content (±6%), and stored at sub-zero temperatures. Under these conditions, their viability can be assured for many years, certainly decades if not considerably longer.

[2] A modern variety has been selected or bred for certain traits (such as high yield, short stature, or good response to fertilizers) using scientific methods. Landraces are populations or races that have become adapted to farmers' conditions through natural and artificial selection. In open-pollinated crops such as maize, creolized varieties are improved varieties that have mixed with landraces in farmers' fields for at least several years (see Thurston *et al.*, Chapter 9, this volume).

[3] Here, the marginal products of each seed variety with respect to each trait are assumed to be constant and equal to average products. Their expression in any given year is clearly affected by farmer management and climatic variation, so that Q is in reality stochastic.

[4] Here, maize output is defined as grain amounts produced from each seed lot. Output could also be defined as fodder, or grain and fodder.

[5] The loss need not be absolute, and it may be related to the fact that a variety may be planted in such small populations that it is genetically swamped by surrounding varieties, i.e. lost genetically.

[6] The fact that we focus on metapopulations indicates that this framework may only apply to crops where migration and recombination between differentially adapted populations allow allele interchange, such as open-pollinated crops (maize, sorghum). It may be of more limited value for self-pollinated crops (wheat, rice). However, even in these crops a small outcrossing rate can allow important allele interchange over time. This may not be the case for clonally propagated crops (potato).

[7] For this ordering to be consistent with economic principles, two minimum conditions must be satisfied: (i) farmers know how to rank the varieties in terms of the desirability of their characteristics, and (ii) if farmers rank A over B and B over C, they also rank A over C. The first principle is called 'completeness' and implies that individuals are not 'paralysed by indecision' – either A is preferred to B, B to A, or A = B. The second principle is called 'transitivity' and implies that individuals do not contradict themselves (Deaton and Muellbauer, 1980). Neither condition seems very restrictive, although they imply that researchers should be careful and thorough in eliciting rankings. Utility is an ordinal rather than a cardinal concept.

[8] Breadth of adaptation of a variety refers to its performance under different environments or management conditions. A broadly adapted variety is one that performs well in many different environments, while a narrowly adapted variety is one that performs well only in one or a few different environments.

[9] The breadth of adaptation of a variety also has implications for the willingness of farmers to maintain it. A broadly adapted variety may be of interest to more farmers than one that is narrowly adapted, increasing the chances that it may be maintained.

However, a population with broad adaptation to environmental conditions or management patterns does not automatically imply that it will be of interest to farmers, if, for example other characteristics such as taste or storage are not appropriate to them. A narrowly adapted variety may be more difficult to maintain since the conditions under which it is attractive to farmers are more limited.

References

Adesina, A.A. and Zinnah, M.M. (1993) Technology characteristics, farmers' perceptions and adoption decisions: a Tobit model application in Sierra Leone. *Agricultural Economics* 9, 297–311.

Aguirre Gómez, J.A. (1997) Analisis regional de la diversidad del maiz en el sureste de Guanajuato. PhD thesis, Universidad Nacional Autonoma de Mexico, Facultad de Ciencias, Mexico, DF.

Altieri, M.A. and Merrick, L.C. (1987) *In situ* conservation of crop genetic resources through maintenance of traditional farming systems. *Economic Botany* 41, 86–96.

Ashby, J.A., Gracia, T., del Pilar Guerrero, M., Quirós, C.A., Roa, J.I. and Beltrán, J.A. (1996) Innovation in the organization of participatory plant breeding. In: Eyzaguirre, P. and Iwanaga, M. (eds) *Participatory Plant Breeding: Proceedings of a Workshop on Participatory Plant Breeding, 26–29 July 1995, Wageningen.* IPGRI, Rome, pp. 77–97.

Bellon, M.R. (1996) The dynamics of crop infraspecific diversity: a conceptual framework at the farmer level. *Economic Botany* 50, 26–39.

Bellon, M.R., Pham, J.L. and Jackson, M.T. (1997a) Genetic conservation: a role for rice farmers. In: Maxted, N., Ford-Lloyd, B.V. and Hawkes, J.G. (eds) *Plant Conservation: the* in situ *Approach*. Chapman & Hall, London, pp. 263–289.

Bellon, M.R., Pham, J.L., Sebastian, L.S., Francisco, S.R., Loresto, G., Erasga, D., Sanchez, P., Calibo, M., Abrigo, G. and Quilloy, S. (1997b) Farmers' perceptions and variety selection: implications for on-farm conservation of rice. Presented paper, International Workshop on Building the Basis for Economic Analysis of Genetic Resources in Crop Plants, CIMMYT and Stanford University, 17–19 August 1997, Palo Alto, California.

Brown, G.M., Jr (1990) Valuation of genetic resources. In: Orians, G.H., Brown, G.M., Jr, Kunin W.E. and Swierzbinski, J.E. (eds) *The Conservation and Valuation of Biological Resources*. University of Washington Press, Seattle, pp. 203–228.

Brush, S.B. (1995) *In situ* conservation of landraces in centers of crop diversity. *Crop Science* 35, 346–354.

Brush, S.B. and Meng, E. (1998) Farmers' valuation and conservation of crop genetic resources. *Genetic Resources and Crop Evolution* 45, 139–150.

Brush, S.B., Carney, H.J. and Huaman, Z. (1981) Dynamics of Andean potato agriculture. *Economic Botany* 35, 70–88.

Brush, S.B., Taylor, J.E. and Bellon, M.R. (1992) Biological diversity and technology adoption in Andean potato agriculture. *Journal of Development Economics* 39, 365–387.

Chambers, R. (1988) *An Interim Note on Ranking Methods*. Institute of Development Studies, University of Sussex, Brighton.

Cornes, R. and Sandler, T. (1986) *The Theory of Externalities, Public Goods, and Club Goods*. Cambridge University Press, Cambridge.

David, J. (1992) Approche Methodologique d'une Gestion Dynamique des Ressources Genetiques chez le Blé Tendre (*Triticum aestivum* L.). These de Doctorat de l'Institut National Agronomique Paris-Grignon.

Deaton, A. and Muellbauer, J. (1980) *Economics and Consumer Behavior*. Cambridge University Press, Cambridge.

Dennis, J.V. (1987) *Farmer Management of Rice Variety Diversity in Northern Thailand*. PhD thesis, Cornell University, Ithaca, New York.

Doebley, J.F., Goodman, M.M. and Stauber, C.W. (1985) Isozyme variation in the races of maize from Mexico. *American Journal of Botany* 72, 629–639.

Ehrenfeld, D. (1988) Why put a value on biodiversity? In: Wilson, E.O. and Peter, F.M. (eds) *Biodiversity*. National Academy Press, Washington, DC, pp. 212–216.

Ford, E.B. (1940) Polymorphism and taxonomy. In: Huxley J.S. (ed.) *New Systematics*. Clarendon Press, Oxford, pp. 493–513.

Ford-Lloyd, B.V. and Jackson, M.T. (1986) *Plant Genetic Resources – an Introduction to their Conservation and Use*. Edward Arnold, London.

Guldager, P. (1975) *Ex situ* conservation stands in the tropics. In: Roche, L. (ed.) *Methodology of Conservation of Forest Genetic Resources*. FAO, Rome, pp. 85–92.

Hamrick, J.L. and Godt, M.J.W. (1997) Allozyme diversity in cultivated crops. *Crop Science* 37, 26–30.

Heisey, P., Smale, M., Byerlee, D. and Souza, E. (1997) Wheat rusts and the costs of genetic diversity in the Punjab of Pakistan. *American Journal of Agricultural Economics* 79, 726–737.

Hodgkin, T., Rao, V. Ramanatha and Riley, K. (1993) Current issues in conserving crop landraces *in situ*. Presented paper, On-Farm Conservation Workshop, Bogor, Indonesia, 6–8 December 1992.

IPGRI (1993) *Diversity for Development: the Strategy of the International Plant Genetic Resources Institute*. International Plant Genetic Resources Institute, Rome.

Jackson, M.T. (1995) Protecting the heritage of rice biodiversity. *Geojournal* 35, 267–274.

de Janvry, A., Fafchamps, M. and Sadoulet, E. (1991) Peasant household behavior with missing markets: some paradoxes explained. *Economic Journal* 101, 1400–1417.

Ladd, G. and Suvannunt, V. (1976) A model of consumer goods characteristics. *American Journal of Agricultural Economics* 58, 504–510.

Lancaster, K.J. (1966) A new approach to consumer theory. *Journal of Political Economics* 74, 132–157.

Louette, D., Charrier, A. and Berthaud, J. (1997) *In situ* conservation of maize in Mexico: genetic diversity and maize seed management in a traditional community. *Economic Botany* 51, 20–38.

Maxted, N., Ford-Lloyd, B. and Hawkes, J.G. (eds) (1997) *Plant Conservation: the in situ Approach*. Chapman & Hall, London.

Meng, E., Taylor, J.E. and Brush, S. (1997) Household varietal choice decisions and policy implications for the conservation of wheat landraces in Turkey. Presented paper, International Workshop on Building the Basis for Economic Analysis of Genetic Resources in Crop Plants, CIMMYT and Stanford

University, 17–19 August 1997, Palo Alto, California.

Morris, M., Rusike, J. and Smale, M. (1998) Maize seed industries: a conceptual framework. In: Morris, M. (ed.) *Maize Seed Industries in Developing Countries: Technical, Economics, and Policy Issues.* Lynne Rienner, Boulder, Colorado, pp. 35–54.

Olivieri, I. and Gouyon, P.H. (1990) The genetics of transient populations: research at the metapopulation level. *Trends in Ecology and Evolution* 5, 207–210.

Pham, J.L., Bellon, M.R. and Jackson, M.T. (1996) A research program for on-farm conservation of rice genetic resources. *International Rice Research Notes* 21, 10–11.

Randall, A. (1986) Human preferences, economics, and the preservation of species. In: Norton, B.G. (ed.) *The Preservation of Species: the Value of Biological Diversity.* Princeton University Press, Princeton, New Jersey, pp. 79–109.

Rice, E., Smale, M. and Blanco, J.-L. (1997) *Farmers' Use of Improved Seed Selection Practices in Mexican Maize: Issues and Evidence from the Sierra Santa Marta.* Economics Working Paper 97–04. CIMMYT.

Richards, P. (1986) *Coping with Hunger: Hazard and Experiment in an African Rice-Farming System.* Allen & Unwin, London.

Sall, S., Norman, D. and Featherstone, A. (1997) Adoption of improved rice varieties in the Casamance, Senegal: Farmers' preferences. Poster, XXIII International Association of Agricultural Economists Conference, Sacramento, California, 10–16 August 1997.

Smale, M. (1997) The Green Revolution and wheat genetic diversity: some unfounded assumptions. *World Development* 25, 1257–1269.

Sperling, L., Scheidegger, U. and Buruchara, R. (1996) *Designing Seed Systems with Small Farmers: Principles Derived from Bean Research in the Great Lakes Region of Africa.* Agricultural Research and Extension Network Paper No. 60. Overseas Development Institute, London.

Szmidt, A. (1995) Molecular population genetics and evolution: two missing elements in studies of biodiversity. In: Boyle, T.J.B. and Boontawee, B. (eds) *Measuring and Monitoring Biodiversity in Tropical and Temperate Forests.* Center for International Forestry Research, Bogor, Indonesia, pp. 177–193.

Regulatory Issues

16

D. Gisselquist

International Economics Department, The World Bank, 1818 H Street NW, Washington, DC 20433, USA

Among the many ways in which governments can and do influence agrobiodiversity, government regulations of inputs trade and use constitute an important set of instruments that can promote or undermine agrobiodiversity. The purposes of this chapter are: (i) to describe the standard practices that governments use to regulate inputs; (ii) to identify points where input regulations impact on agrobiodiversity; and (iii) to note changes in common regulatory practices that may be considered to more effectively promote or protect agrobiodiversity.

Since regulatory practices and issues vary according to the input under consideration, much of this chapter is organized around inputs. The first section deals with pesticides, the second with seed regulations. The next section discusses several additional topics, including introduction of exotic non-plant species and fragile environments. The final section summarizes conclusions and suggestions.

Regulating Pesticides

In regulating inputs, possibly the biggest single thing that governments can do to protect and promote agrobiodiversity – and also general biodiversity – is to regulate pesticides in a way that creates pressures and incentives for companies and farmers to shift from conventional pesticides – broad-spectrum poisons – to low and zero risk pesticides and other non-chemical methods for pest management. Broad-spectrum poisons work through organic processes that are common to a broad range of organisms, so that in normal use they damage and kill not only target pests but also beneficial insects, birds, mammals, plants, beneficial soil bacteria and fungi, etc. (see Polaszek *et al.*, Chapter 11, this volume). Most governments have banned or restricted use of many of the more dangerous and long-lasting broad-spectrum poisons such as DDT and other organochlorines. However, dependence on broad-spectrum poisons continues at high levels in developed countries and is expanding in developing countries. In ex-socialist countries in transition, farmers from the late 1980s have drastically reduced use of broad-spectrum poisons, but that is an anomaly due more to economic collapse than to any new commitment to environmental protection, and can be expected to reverse as economies recover.

Essentially all countries register – list – allowed active ingredients and pesticide products. Governments also control import, production, handling, residues in food, etc. Many aspects of pesticide regulation are important to protect public health as well as agrobiodiversity and the environment. However, pesticide registration – preparing lists of allowed pesticides – is arguably the aspect of pesticide regulation where the design of rules and procedures has the most impact on agrobiodiversity.

An active ingredient (ai) is the chemical – or microorganism – that does the work. Companies mix ai's into formulations for sale. A pesticide product is a formulation with a brand name. Some pesticide products are proprietary or research-based products, for which the company owns one or more patents that blocks others from producing the same formulation. For such formulations, a company that is able to register its products has a monopoly. Other pesticides are commodity products, using ai's and other inputs that are old and out-of-patent, so that a company that registers a product may or may not hold a monopoly, depending on whether or not the government has allowed any other competing company to register another product with the same formulation.

Governments register new pesticide ai's and products after reviewing results from laboratory and field tests designed to show environmental and public health impact and also efficacy. Since the biggest markets for pesticides are in the major OECD (Organization for Economic Cooperation and Development) countries, companies focus on those markets in developing new products. Companies also face the biggest challenges to get their products registered in those countries. OECD countries have coordinated their pesticide regulations to the extent that they generally ask for the same tests, which holds down registration costs for the companies. However, the review procedures and decisions to approve or reject may be different, so that an ai or pesticide may be registered in the USA and Spain, for example, but not in Sweden (General Accounting Office, 1993). This situation remains after long debate. Countries have different priorities and sensitivities about pesticide risks and also different climates and crops, so that the perceived and real risks and benefits from the same pesticide are not the same everywhere. Developing countries generally copy registration processes in developed countries.

In most countries where they want to sell, pesticide companies are able to satisfy demands for information on environmental and public health risks from tests already performed for registration in one of the major developed countries. However, most countries also ask for data on efficacy based on in-country tests. In the USA, the federal government's Environmental Protection Agency (EPA) does not ask for efficacy tests for agricultural pesticides, but states often do. The EU forces member countries to accept results from efficacy tests from another country with comparable agro-ecological conditions, but does not force them to accept the same pesticides.

Over the past several decades, concerns about environmental and public health damage from broad-spectrum poisons – along with evidence of pest resistance, secondary pests, and failure to control economic losses – has stimulated efforts to limit use of broad-spectrum poisons and at the same time to develop and disseminate alternative pest control technologies (Gisselquist and Benbrook, 1996; also see Polaszek *et al.*, Chapter 11, this volume). A broad range of alternative technologies are available, including, for example:

- microbiological pesticides, e.g. viruses and bacteria that attack insect pests, or soil fungi that control the spread of other harmful fungi;
- agronomic practices, including plant rotation, planting crops that repel or attract pests, destroying crop residues, etc.;
- insect growth regulators, such as chitin inhibitors that prevent maturation of insects by blocking production of a new exoskeleton;
- pheromones, such as sex attractants, that are used to confuse or trap insects;
- inoculants that can be applied to seeds to stimulate the plant's immune system to resist fungi or other diseases;
- plant growth regulators that can be used to accelerate ripening to avoid pest attack, etc.; and
- biological control through rearing and releasing large numbers of an indigenous insect predator, such as ladybugs (ladybirds) in North America, as well as classic biological control that involves introduction of an exotic predator to provide permanent pest control (often for an introduced pest).

Standard rules and procedures for pesticide registration which have been designed to deal with broad-spectrum poisons create problems for low and zero risk pesticides. For example, compared with broad-spectrum poisons, biopesticides such as bacteria, pheromones and plant growth regulators are more sensitive to management. They must be applied when pests are in a particular growth phase. Also, they may not kill pests outright but rather prevent maturation or reproduction, so that effectiveness is not so easy to see as with broad-spectrum poisons. Hence, results from efficacy tests may be less conclusive and impressive than those obtained with broad-spectrum poisons. Also, many of the risk tests are unnecessary. Finally, low and zero risk products inherently have smaller markets, since they are effective against a more narrow range of pests, and sometimes against only one species, whereas broad-spectrum poisons kill many pests and have large markets. Hence, companies with low risk products may not be able to afford to put them through an expensive registration process considering relatively low expected sales.

In short, with low and zero risk pesticides, governments do not need so much information to determine that products are safe, while at the same time, markets are smaller so that companies cannot afford to develop so much information. When governments put low and zero risk products through the same registration procedures as broad-spectrum poisons, many safer products will not be registered, and farmers are left with broad-spectrum poisons.

Many countries have already begun to adjust their regulatory processes to favour low and zero risk pesticides. A survey prepared for the OECD Pesticide Forum (1995, p. 9) found that:

- Many countries have reduced their data requirements for biopesticides, or established entirely new data requirements, to make registration faster and easier and therefore to have the products on the market more quickly.
- Sweden and Norway have created a product substitution programme that allows safer new products to replace older products, registered for the same purpose.

In 1994, the US EPA established the Biopesticides and Pollution Prevention Division, assigning it responsibility to register biopesticides, including microbial pesticides, biochemical pesticides (with non-toxic mode of action and natural occurrence), and plant pesticides (pesticide substances produced by transgenic plants, such as *B.t.* toxins in transgenic maize).

After revisions in regulations to favour low risk products, the cost and time to register a new biopesticide in the USA in the mid-1990s has been estimated at $1 million and 1 year, which can be compared to $50 million and 3–4 years to register a broad-spectrum poison. Adjustments in the US registration system create huge incentives for companies to shift research attention to low risk products. This is having results. In the financial year 1995, for example, half of the 40 new active ingredients registered in the USA were biopesticides.

Incentives favouring low risk products through registration systems could be further enhanced if governments would also begin to tax pesticides according to risk. These incentives can be further reinforced by progressive banning of broad-spectrum poisons, which can be easier over time as companies and farmers come to rely more on alternative low risk pesticides and pest management technologies.

Regulating Seeds

Essentially all governments have some legal and institutional framework in place for regulating seed trade. The one aspect of seed regulation that is most relevant for agrobiodiversity is variety registration. Other seed regulations that impact on agrobiodiversity include controls on introduction of exotic plant species and transgenic cultivars and phytosanitary controls on seed trade. Although intellectual property rights (IPRs) have little to do with agrobiodiversity, many people think otherwise. Hence, this section also discusses IPRs for seed technology (see also Wright and Turner, Chapter 13, this volume).

Compulsory or voluntary variety registration

Variety registration means that some government body accepts that plants grown from a specific lot of seeds are distinct and uniform, and these characteristics are stable over at least 2 years (i.e. plants meet DUS criteria). Many governments also ask that a variety demonstrates value in cultivation and use (i.e. meets VCU or performance criteria) before granting registration, or before allowing seed sale.

Governments around the world follow two common patterns in registering varieties.

- In the USA, India and some other countries, variety registration is available but voluntary, which means that governments allow companies to sell seeds of cultivars that governments have not registered. In these countries, there are no lists of allowed cultivars, but governments may on the other hand list banned varieties (i.e. weed seeds not allowed in seed packages).
- In the EU and also in many ex-socialist and developing countries, governments enforce compulsory variety registration, so that companies are not allowed to sell seeds until some government agency has registered the variety, usually after several years of official DUS and VCU tests. With rare exceptions, each EU country automatically accepts varieties registered in any other EU country, which provides multiple 'doors' for varieties to enter. Also, the large EU market provides strong incentives for companies to go through the registration process, so that farmers enjoy a steady supply of new commercial varieties despite regulatory barriers. For major crops, EU Common Catalogues list all cultivars registered in at least one EU country and approved for sale throughout the EU. On the other hand, many ex-socialist and developing countries maintain their own single-country lists of allowed cultivars, so that registration becomes a serious barrier to introduction of new commercial cultivars. For small countries, non-hybrid (low cost)

seeds, and minor crops, farmers may have access to few if any private cultivars.

Countries may enforce compulsory registration for some crops, and allow voluntary registration for others. For example, Bangladesh does not allow seed sale for unregistered varieties of rice, wheat, sugar cane, jute, and potatoes, but allows seed sale for unregistered varieties of all other crops. Similarly, Malawi enforces compulsory variety registration for two crops only, hybrid maize and tobacco.

In countries with compulsory variety registration, seed sale for old cultivars – including traditional varieties as well as out-of-date commercial cultivars – is often illegal. In the EU, for example, companies must pay for a cultivar to be tested and registered, then also pay an annual fee for as long as they want the cultivar to be listed for sale. The company that pays those fees is known as the 'maintainer' (because it maintains breeders' seeds for the variety). The maintainer may not actually own the variety, since it may be a public variety from another country (e.g. the USA), but the EU allows the maintainer to act more or less as the owner: he or she has exclusive right to produce the variety, and can collect a fee from anyone else who produces or imports the variety (though this has been successfully challenged in the several cases where someone imported seeds of public varieties from the USA). When a maintainer stops paying the annual fee to maintain registration, the variety is dropped from the list of varieties allowed for sale, and subsequent seed sale is illegal. The EU has grandfathered in some exceptions to these rules, so that seed sale is allowed for a short-list of old varieties approved before 1972. For example, the EU allows seed sale for 'bere' barley from northern Scotland, a variety that can be traced to the 8th century, which is probably the oldest cereal variety still grown in Europe and possibly the world (Jarman, 1996).

In countries with compulsory variety registration it is hard to see how organic farmers will get seeds, since they will normally not want to grow the same cultivars – bred for high input conditions – that conventional farmers plant. Many of the cultivars suitable for organic farming may not be grown on a large scale, so that no seed company will be interested in registering them. If so, seeds will not be available. Hence, in countries with compulsory variety registration, lack of market access to seeds of suitable cultivars can undermine efforts by NGOs, private farmers, extension agents, universities and other organizations to promote organic agriculture.

In countries that enforce compulsory variety registration – and for crops for which compulsory registration is applied – it is illegal to sell seeds of landraces. Landraces cannot be registered as varieties since they do not meet DUS criteria (plants are not uniform). If any company wanted to sell seeds of a landrace, the best it could do would be to breed and select a genetically uniform strain out of the landrace that would meet DUS criteria for registration. In other words, the company would have to deliberately reduce the genetic diversity of the landrace before seed sale would be legal. However, the incentive to do so would be limited by the market, since expected revenues would have to cover breeding and registration costs before a company would go ahead, and a landrace – or selection from a landrace – would probably have a small market of organic farmers and hobby gardeners only.

There are some ways to get around compulsory variety registration to allow seeds of landraces and old varieties to circulate, but they are not suitable for general commercial practices. In the UK, for example, farmers who want to grow old or antique varieties meet to exchange seeds (selling would be illegal); farmers who take seed of an antique variety contract to return a quantity of multiplied seed the next year. Also, some countries allow informal (farmer-to-farmer or non-commercial) seed sale for unregistered varieties.

Governments that enforce compulsory variety registration use state power to diminish agrobiodiversity – to hold or drive older varieties, minor cultivars and landraces out of the market. What is the trade-off for this loss? The most common defence of compulsory variety registration is to protect farmers from buying seeds of inferior varieties. One expert argues that 'the farmer should be given some protection against those who might try to market an unsatisfactory variety simply to recoup breeding costs' (Kelly, 1989, p. 105). However, evidence from countries and crops without compulsory variety registration demonstrates that farmers are able to assess variety performance and do not need such protection. Also, companies are motivated to offer consistently good varieties to protect their brand name. On the other hand, compulsory variety registration inflicts large economic costs in the form of forgone gains, that farmers may not realize because they do not see the superior varieties that are not allowed. Several studies show large losses (forgone gains) for cotton growers in California and wheat growers in Canada from government restrictions on introduction of new varieties (Ulrich et al., 1987; Constantine et al., 1994) and large gains to maize growers in Turkey from easing such barriers (Gisselquist and Pray, 1997).

Promoting agrobiodiversity by limiting farmer access to modern cultivars

Casual observation suggests that when countries shift from traditional to modern agriculture, at least some measures of agrobiodiversity fall as farmers adopt modern cultivars and specialize for the market. Some experts discourage the introduction of modern private cultivars so that farmers in developing countries will continue with traditional cultivars. Whether or not giving farmers access to modern private cultivars leads to erosion of agrobiodiversity, it is arguably unethical to deliberately block farmers in developing countries from having access to the same range of choices that is available to farmers in developed countries (see Witcombe, Chapter 10, this volume).

Recommendations in this chapter – to give farmers more choices – are based on the vision that agrobiodiversity can be consistent with high incomes and modern farming. With this vision, the most developed OECD countries are important nurseries for developing the technological, policy and organizational innovations that can promote and preserve agrobiodiversity. In other words, strategies for agrobiodiversity can be seen as improvements for farmers in the most developed agricultural countries, and not as obstructions that block farmers in poor countries from choosing high-income options. This is a 'one-world' strategy, that endorses the principle of equality and hence comparable technology and systems for rich and poor countries, though it may take time to realize equality.

Controlling introduction of new plant species

Most countries put inputs of planting material on the import control or restricted list, so that some government agency (normally part of the ministry of agriculture) is empowered to decide whether or not to allow deliberate introduction of a new plant species (e.g. kiwi fruit from New Zealand into Chile). Decisions about whether or not to allow an exotic plant species are more relevant for biodiversity than for agrobiodiversity. The biggest risk is that an introduced plant species may spread into the wild, pushing indigenous wild species out of ecological niches (see Wood and Lenné, Chapter 2, this volume). While governments have an accepted role to limit negative externalities, if these are not serious, then strong arguments favour introduction to give farmers more choices. Most agricultural production today comes from species that are grown outside of their Vavilovian centres. Further introductions of exotic plants could lead to more or less agrobiodiversity, depending on how farmers adjust cropping patterns with new choices. However, as proposed above, policies for agrobiodiversity should be consistent with OECD policies and patterns, with high productivity, incomes, and farmer choice.

Controlling introduction of transgenic cultivars

There is as yet no scientific or political consensus on the risks associated with introduction of transgenic cultivars. Aside from the health concerns about transgenic foods, environmental concerns focus on risks that genes could jump to wild relatives of agricultural species and somehow diminish the wild gene pool (e.g. for wild maize in Mexico or potatoes in Peru) or to related weed species, creating super weeds. On the other hand, transgenic cultivars could have higher resistance to pests and diseases, allowing farmers to reduce pesticide use, which would improve agrobiodiversity and biodiversity (see Polaszek *et al.*, Chapter 11, this volume).

Phytosanitary controls on seed imports

International seed trade – as well as trade in grain and other organic products – threatens agricultural production and biodiversity, but also carries some threat to agrobiodiversity. The main concerns are to stop imports of weeds, seedborne insect pests, and diseases that could spread across agricultural fields (cutting yields and income) and also into the wild. To limit those threats, essentially all countries enforce phytosanitary import controls, which normally include: banning seeds of particular species from countries with dangerous pests and diseases; insisting on particular tests or treatment for seeds according to species and origin; and post-entry quarantine, including plant grow-out tests, depending on species and origin (see Polaszek *et al.*, Chapter 11, this volume). However, import regulations are often not focused on realistic phytosanitary threats, so that protection is weak, even though regulations may be so restrictive that they severely disrupt trade.

Potential damage to crop agrobiodiversity from imported plant pests and diseases flows from threats to agricultural production, which are also often controllable. For example, an imported pest could so severely reduce incomes from a crop that farmers abandon it. If the crop remains viable anywhere in the world, scientists can usually address any such set-back for a particular region with breeding to introduce resistance, classical biological control to control the introduced pest, etc. Hence, the focus of phytosanitary controls is reasonably on concerns other than agrobiodiversity.

Phytosanitary controls are subject to abuse, and are often controversial. Controls are often used to block imports to protect local producers, without any logic in terms of pest or disease threats. On the other hand, many countries do not address realistic pest and disease threats, do not effectively monitor seed imports, and do not have scientific resources at border posts and quarantine stations to test for pests and diseases, so that there is little real protection, even though official import controls that are nominally based on phytosanitary concerns may be so tight that they block legitimate seed and other trade.

Intellectual property rights (IPR)

Most OECD governments and a growing list of ex-socialist and developing country governments have laws that provide mechanisms for breeders to register ownership of varieties (plant varietal protection (PVP)). The new World Trade Organization insists that members accept the Trade-related Intellectual Property Rights (TRIPS) Agreement, which entails, among other obligations, that countries establish some kind of PVP system for plant varieties and also patents for microorganisms and microbiological and non-biological processes for producing plants and animals. So far, essentially all countries that establish a PVP system have become members of UPOV (Union for the Protection of New Plant Varieties), an international treaty organization that accepts members with PVP laws meeting UPOV guidelines. However, the TRIPS Agreement and WTO allow countries to design and enforce PVP legislation that is not acceptable to UPOV.

Fears that big seed companies would somehow be able to gain unfair advantages in seed trade have been associated with the spread of IPRs and PVP legislation. However, these fears are confused. Standard IPR legislation allows inventors to register newly developed or discovered technology only, not technology that has been in use or is traditional and undeveloped, and UPOV's guidelines for PVP legislation fit this design. Under legislation fitting UPOV guidelines, companies cannot register ownership of landraces (since they are not varieties, and furthermore they are not novel or new), and cannot register ownership of a variety that has been in general use for several years. These materials are public goods, that can be multiplied and sold by anyone.

However, there is a potential problem if a country enforces compulsory variety registration. Then, sale of seed for old and traditional cultivars and landraces is illegal due to compulsory variety registration. In such circumstances, companies that register ownership of new cultivars (through PVP legislation) may end up with the only seeds that are legal for sale. However, this situation comes from compulsory variety registration rather than from PVP.

Recognizing ownership of wild relatives, landraces and traditional varieties

The Convention on Biological Diversity recognized that countries own germplasm within their borders. This recognition is on the one hand obvious, since countries can regulate economic activities within their borders. However, working out whether and how to exercise ownership rights is not so easy, and proposed schemes to exercise ownership rights may threaten rather than protect agrobiodiversity.

If the concern is to protect agrobiodiversity, then it is by no means clear that a government improves things by trying to make money – or by helping some group of citizens to make money – out of current agrobiodiversity. If, for example, government establishes some meaningful system of ownership for landraces or traditional varieties (see LaSalle, Chapter 7, this volume for biological control agents), that means that government identifies some people or groups as owners, and empowers them to take a cut from sales of the seeds. If an owner is not sufficiently motivated to sell his or her variety or landrace, he or she could presumably pull it out of the market. Hence, the notion of extending ownership over landraces and traditional varieties does not necessarily protect them, but can rather limit their use.

Some development experts encourage governments to assign ownership rights and subsequent revenues to indigenous or tribal people, to pay them for protecting and developing landraces over centuries. However, no one has – to my knowledge – proposed any realistic way to collect or to distribute such money. Even if those problems could be solved, it is not clear that payoffs for actions long past will encourage recipients or others to continue to preserve agrobiodiversity.

In some scenarios, a country may try to collect royalties from companies that sell seeds in foreign markets based on germplasm from within its borders. This can seem like a good idea for a country like India as long as it expects to tax foreign companies that take Indian germplasm to other countries. However, as soon as Indian companies get into breeding and selling overseas, royalties to the Indian government for using Indian germplasm become an excess tax burden that puts Indian-origin cultivars and Indian seed companies at a disadvantage in foreign markets against companies taking their germplasm from other sources. Benefits for agrobiodiversity appear to be nil, and there do not seem to be other benefits except for tax revenues, which would be modest relative to the damage done to Indian breeders and seed companies.

Some governments have been able to collect fees from companies for gene prospecting (collecting wild plants and animals). Such payments do provide clear incentives for governments to preserve undeveloped areas, but they may be too small to have much of an impact on decisions about whether or not to protect or to develop an area. Since the current market value of biodiversity in any particular region (e.g. a national park) is limited by current knowledge, payments for gene prospecting may understate the long-term value of that biodiversity.

As for patents on genes and processes related to transgenic crops, such patents do not affect farmer access to inputs for pre-existing agrobiodiversity. As above, efforts on the part of some governments to collect royalties from undeveloped germplasm that goes into transgenic crops may have a bigger effect on seed company profits and investments than on government incentives to preserve agrobiodiversity. On the other hand, gene-modified organisms do present some direct risks to agrobiodiversity and biodiversity, as discussed in the next section (see also Polaszek *et al.*, Chapter 11, this volume).

Other Inputs Regulations Related to Agrobiodiversity

Reproductive material for exotic non-crop species

Governments normally put all such material on import control lists, so that prospective importers must gain special permission before proceeding. There are a variety of reasons that someone might want to import a new non-crop species. In some cases, farm interests might want to import a new species for commercial agriculture, such as emus into the USA, or a new species of carp into Bangladesh for fish-farming. Pest management specialists sometimes import a new species for biological control; for example, experts from the International Institute for Tropical Agriculture, in collaboration with experts from many African countries, introduced a parasitic wasp from South America to control the cassava mealybug (Neuenschwander and Herren, 1988), which was an earlier inadvertent introduction from South America, possibly through import of infected reproductive material for cassava (see LaSalle, Chapter 7, this volume). Hobbyists may want to introduce exotic fish or other species.

Governments' decisions about whether to allow import of an exotic species are assigned to particular agencies, and this is normally indicated in import regulations. For example, import of animals, birds and reptiles into India is allowed on the recommendation of the Chief Wildlife Warden of a state government.

New exotic species may enhance agrobiodiversity by increasing the number of agricultural species in a country. On the other hand, introduced species may drive other species out of use. Also, many errors have been made over the years with both accidental and deliberate introductions of exotic agricultural species that have become pests, and have damaged agriculture and biodiversity; threats to agrobiodiversity are probably minor in comparison. For example, snails introduced for production in Vietnam have become a pest in rice paddies, damaging farm income, but farmers continue to grow multiple rice varieties.

Fragile environments

Governments deal with fragile environments through special controls on agricultural activities, including limits on use of agricultural inputs and controls on livestock production. Other controls (e.g. on settlement and industry) may also be necessary to protect these regions, but are outside the scope of this chapter. For example:

- a karst region in Lithuania and Latvia has highly porous subsoil conditions, so that agrochemicals enter groundwater and move long distances without any filtering. In response to this risk, the government of Lithuania promotes organic agriculture in the area and has also reviewed large-scale livestock production and sewage treatment practices.
- Much of the Netherlands is below sea level, so that drainage is constrained. The Government of the Netherlands implements a tight programme to control fertilizer applications on each plot to what is necessary to replace nutrients taken off the plot in the form of agricultural products.

- Migratory birds pass through some sites in large numbers. Governments working with and through an international treaty organization have identified many of these flyway stopovers as Ramsar Sites for special attention and protection. For example, Lithuania's Nemuno Deltos Regional Park in Kursiu Lagoon on the Baltic Sea is a Ramsar Site. While much of the land in the park continues to be used for pasture and field crops, government bans all pesticide and fertilizer use.

Government initiatives to protect biodiversity in fragile environments often – inadvertently – promote agrobiodiversity by forcing farmers in the area to use less inputs, adopt practices that are different from those of farmers in other regions of the country, etc. Fragile environments can also serve as 'laboratories' for testing and observing policies for agrobiodiversity – for example pesticide restrictions – that are not yet accepted in other parts of the country.

Conclusions

One of the important questions facing advocates for agrobiodiversity may be the relative importance of agrobiodiversity in fields versus agrobiodiversity in wilderness areas of Vavilovian centres (i.e. biodiversity in wilderness areas for crops and other species important to agriculture). If, for example, it is determined that 90% of agrobiodiversity that is important for the security of our food supply is in wilderness areas, an optimal strategy for agrobiodiversity may be to maximize production from existing fields, sacrificing in-field agrobiodiversity, so as to reduce pressures to bring more wilderness areas under cultivation. The following recommendations about inputs regulations are made with the assumption that both fields and wilderness areas are important for agrobiodiversity, but that some wilderness areas are essential, so that the remaining wilderness area becomes more important as and if it shrinks over time.

The most important single aspect of inputs regulation for agrobiodiversity is what is done to turn pest management away from broad-spectrum poisons and towards low and zero risk pesticides and pest management strategies (see Polaszek *et al.*, Chapter 11, and Lenné and Wood, Chapter 18, this volume). This chapter recommends that governments introduce a sliding scale of regulatory barriers, with low or no barriers for low and zero risk pesticides at one end, increasing tests and registration expenses and also risk-based sales taxes for more dangerous products, and many more banned pesticides for which risks are no longer acceptable. Ideally, incentives will generate research and adjustments in farming practices that will allow governments to progressively ban more broad-spectrum poisons over time. Banning of conventional pesticides already occurs in sensitive environments, and these can be used as laboratories to study how best to extend general pesticide bans.

Secondly, agrobiodiversity and organic agriculture would be easier to promote if governments would adopt voluntary variety registration, allowing companies, NGOs and others to sell seeds of not only new cultivars but also traditional varieties, minor cultivars and landraces without explicit government review and approval.

Thirdly, to reduce risks that deliberately or accidentally introduced species will disrupt agriculture, agrobiodiversity or biodiversity, governments can focus on realistic threats and rationalize import controls. A number of international agencies provide guidance to governments to identify risks. These include the International Plant Protection Convention, which advises import controls to limit plant pests and diseases; the Office International des Epizooties, which advises import controls to limit animal diseases; and the International Council for the Exploration of the Sea, which advises on the introduction of new aquatic species. Governments that have extensive land borders may not be able to block introductions at their own borders, but must rather work together with neighbouring countries to protect an entire region within natural boundaries such as oceans and deserts. Also, reducing incentives for smuggling – cutting duties – draws more seed imports through formal channels, which allows the government to address realistic import threats.

Fourthly, we do not know enough about benefits and risks from deliberate imports of exotic species and genetically modified organisms, nor about how to control problems when they do arise. Much more scientific work can be considered in this area (see Polaszek et al., Chapter 11, this volume). One question of some importance is to reconsider the benefits and risks from classic biological control, which has currently fallen into disfavour but may deserve much more support (see LaSalle, Chapter 7, this volume).

Input regulations are standard features of government operations. Regulations can often be adjusted in ways that not only protect agrobiodiversity but at the same time favour higher yields and incomes. For example, countries that move to voluntary variety registration and refocus phytosanitary import controls on realistic pest and disease threats not only promote agrobiodiversity but also ease farmer access to inputs. However, there are also some trade-offs. Strong farm and industry lobbies favour broad-spectrum poisons that damage agrobiodiversity. Without prior introduction of alternate low risk pest management inputs and strategies, immediate bans on broad-spectrum poisons could have serious impacts on farm production.

Promoting and protecting agrobiodiveristy is an activity that requires widespread participation and support from farmers and private environmental groups to succeed. Government cannot do it alone, even with all of the best regulations and controls. Hence, input regulations can be designed to allow farmers and others to take part. Reducing regulatory barriers to introduction of low and zero risk pesticides and removing government controls on introduction of new cultivars empowers interested and committed people outside government to maximize their contributions to promoting and protecting agrobiodiversity.

References

Constantine, J.H., Alston, J.M. and Smith, V.H. (1994) Economic impacts of the California One-variety Cotton Law. *Journal of Political Economy (US)* 102, 951–974.

General Accounting Office (1993) *Pesticides: a Comparative Study of Industrialized Nations' Regulatory Systems.* US General Accounting Office, Washington, DC.

Gisselquist, D. and Benbrook, C. (1996) Technology transfer, competition, and trade liberalization for low-risk pest management. *UNDP–World Bank Trade Expansion Program Occasional Paper* 11. World Bank, Washington, DC.

Gisselquist, D. and Pray, C. (1997) The impact of Turkey's 1980s Seed Regulatory Reforms. In: Gisselquist, D. and Srivastava, J. (eds) *Easing Barriers to Movement of Plant Varieties for Agricultural Development.* World Bank, Washington, DC.

Jarman, R.J. (1996) Bere barley – a living link with the 8th century? *Plant Varieties and Seeds* 9, 191–196.

Kelly, A.F. (1989) *Seed Planning and Policy for Agricultural Production: the Roles of Government and Private Enterprise in Supply and Distribution.* Belhaven, London.

Mooney, P.R. (1979) *Seeds of the Earth: a Private or Public Resource?* Canadian Council for International Cooperation, Ottawa.

Neuenschwander, P. and Herren, H.R. (1988) Biological control of the cassava mealybug, *Phenacoccus manihoti*, by the exotic parasitoid *Epidinocarsis lopezi* in Africa. *Philosophical Transactions of the Royal Society, Series B*, 318, 319–333.

OECD Pesticide Forum (1995) *Activities to Reduce Pesticide Risks in OECD and Selected FAO Member Countries: Results of a Survey.* Paper prepared for the workshop, Pesticide Risk Reduction, Uppsala, Sweden, 16–18 October 1995.

Ulrich, A., Furtan, W.H. and Schmitz, A. (1987) The cost of a licensing system regulation: an example from Canadian prairie agriculture. *Journal of Political Economy* 95, 160–178.

17

Agrobiodiversity and Natural Biodiversity: Some Parallels

D. Wood and J.M. Lenné

Agrobiodiversity International, 13 Herons Quay, Sandside, Milnthorpe, Cumbria LA7 7HN, UK

Hordeum spontaneum, Triticum boeoticum and *T. dicoccoides* are common plants in their distribution centres. Frequently they build extensive, massive stands . . . *Hordeum spontaneum, T. dicoccoides* and *Avena sterilis* are spread in masses. On uncultivated slopes, natural fields of these wild cereals extend over many kilometres. In their growth and total mass, these wild fields of wheat, barley and oats are not inferior to their cultivated counterparts.

(Zohary, 1969, pp. 55–56)

Introduction

In contrast with the diversity of nature, our common perception of agriculture is of uniformity. The unmanaged diversity of nature assembles itself into functioning ecosystems, capturing energy, regulating nutrient flow and maintaining a natural, stable and productive balance between organisms. Furthermore, evolution, particularly the coevolution of interdependent species, continues. Agriculture is seen as unnatural: overall productivity is low, nutrients leak from cropping systems, and the genetic uniformity of crops and domestic animals causes instability through disease susceptibility and lack of an adaptive fit to the environment. The problems of unnatural uniformity in agriculture then demand the unnatural disturbances of chemical inputs and mechanical tillage to maintain production. In turn, these unnatural disturbances stress wild biodiversity and the wider environment. With this polarized perception of agriculture and nature, agriculture can only become sustainable if in some way it mimics nature, replacing external inputs with endogenous biological processes, all dependent on biodiversity which should always be nurtured and allowed to evolve in agricultural systems.

Such is the power of our image of the naturalness of nature that we have accepted both the polarity between nature and agriculture and the consequent argument to move agricultural management closer to nature to resolve the apparent conflict. But our starting point for deriving agriculture from nature has always been the species-diversity and structural complexity of nature: we assume a complex origin in nature and a progressive, and damaging, simplification in agriculture. Our 'natural' model for sustainable agriculture then can be no other than diversity and complexity (Harris, 1969). Indeed, many present-day prescriptions for agricultural management and research are mainly based on the need to mimic the 'diversity–complexity' model supposedly found in nature (Altieri, 1987; Conway, 1997).

However, through the concept of 'agrobiodiversity' we can step back and reexamine the interface between agriculture and nature. With a greater knowledge of the complexity of crop-associated biodiversity we can see its diverse origins from nature, and its present functioning often as an extension of nature. This viewpoint may perhaps allow a reassessment of natural models currently proposed for agriculture.

To illustrate this positive view of agrobiodiversity as a legitimate extension of nature, this chapter will briefly review ecological concepts of disturbance, succession, community structure and the evolving debate on biodiversity and stability. Within each section we relate these concepts – developed for natural ecosystems – to agriculture and then try to identify agricultural homologues of natural systems. Emphasis will be given to natural systems which are simple – at least in the above-ground parts. There is some complementarity of this approach with Chapter 2, where we demonstrated the continuities between hunter–gathering and agriculture.

Edwards *et al.* in Chapter 8 (this volume), have noted the differences between agroecosystems and natural ecosystems – often depending on human management: this chapter will stress the similarities. It should be read as a speculative attempt to generate new (and question old) hypotheses: it does not have the experimental base of the sound science of other chapters. To some extent, we are playing 'the Devil's advocate': questioning whether more agrobiodiversity is always needed in all systems. We hope that our arguments can reduce polarity between nature and agriculture, emphasize continuities, and point to a new focus for the management of agricultural ecosystems.

Disturbance

In the first century AD, the Roman Columella (*c.* AD 65) wrote '. . . *at sine agri cultoribus nec consistere mortales nec ali posse manifestum est* [. . . without tillers of the fields it is obvious that mankind can neither subsist nor be fed]'. By definition and long usage, agriculture is disturbance: the tilling of fields. In contrast, our perception of nature may be of undisturbed vegetation, often the climax vegetation of forest. However, in nature, too, there is considerable disturbance, which may be overlooked by ecologists. For example, Johns (1986) claims that such disturbance is characteristic, even for tropical forest ecosystems, and that disturbance events have had little impact on the thinking of rainforest ecologists. Harper (1977, p. 628) argues yet more strongly that disasters are the main determinants of community structure. Rindos (1984, p. 268) suggests that crops typically are 'weedy heliophytes' – sun-loving colonizers of disturbed habitats: this functional similarity between crops and weeds has been noted often (see Edwards *et al.*, Chapter 8, this volume).

There are many types of natural disturbance, related to climate, mountain-building (with subsequent erosion and sedimentation), and to fire. Seasonally dry or wet climates allow burning and flooding which kills or stresses existing vegetation. Longer climatic cycles such as 'El Niño' events and localized cyclones disturb existing vegetation. Even longer-term climate changes such as glacial cycles provide open areas for plant colonization on the retreat of glaciers.

Over long time scales the geological cycles of mountain-building, major erosion and the deposition of eroded sediment and lava and volcanic ash control land form and in turn, the stability of global vegetation. Over shorter periods landslips, river meanders and coastal sedimentation provide new substrates for plant growth. Continuing erosion of volcanoes and uplifting mountains provides not just new substrates, but an increment of nutrients in flood water. In seasonally wet, particularly monsoon, climates this enrichment will be annual and regular, encouraging a natural annual cycle of disturbance and rapid regrowth of vegetation in nutrient-rich conditions.

As argued in Chapter 2, this volume, fire was a feature of the environment well before agriculture (Goldhammer, 1993; Bird and Cali, 1998). Huston (1994) notes the ubiquity and significance of fire in virtually all terrestrial ecosystems. It has been claimed that fire plays an important role in the maintaining of stable and vigorous savannah and grassland communities throughout the world (Stocks and Trollope, 1993). A combination of elephant grazing and fire has been credited with the maintenance of 'multiple stable states' in African woodlands (Dublin *et al.*, 1990).

In regard to the regularity of disturbance, Harper (1977, p. 770) argues that 'somewhere along a scale of frequency and intensity of hazards must come a stage at which the hazard is sufficiently part of the "order or system of things" for the responses to it to be thought of as contributions to fitness and for the hazards to become part of natural selection.' Harper (1997, p. 629) gives as examples species of pine and *Eucalyptus* where fire is needed for the release and germination of seed. As a more general adaptation, weeds globally have evolved to survive disturbance and repeated natural hazards. In nature, it is reasonable to argue that regular disturbance is a stabilizing factor for vegetation, a factor to which component species can and do adapt.

Indeed, removing the regular disturbance will cause instability of the vegetation: species adapted to disturbance will be replaced by species adapted to
less disturbance. Conceptually, as with Stocks and Trollope (1993) and
Dublin *et al.* (1990) in the above paragraph, we must accept that a repeatedly
disturbed environment can result in a stable plant community. As argued by
Rindos (1984, p. 169), stability in this sense 'is compatible with dynamic
change: a system that cycles through a set series of changes may be as stable
as one that does not change at all – although the type of stability is different'.
A humbler 'cycle' analogy is the 'dynamic stability' of riding a bicycle: stop
pedalling and you fall off. A biological analogy is mowing a grass tennis
court: stop regular mowing and weed invasion begins the process of change
and instability leading to perhaps another stable state (but one on which you
will be unable to play tennis).

Parallels with agriculture

So too with agriculture – regular disturbance allows productivity and dynamic
stability. For example, exploiting the annual disturbance of flood has been a
common feature of agriculture. More than 30 years ago, Allan (1965)
suggested that:

> Systematic agriculture . . . may have begun in the flood plains of the great
> rivers – first by utilising the natural floods and then by controlled flooding or
> irrigation, for the step from one to the other is natural and not very
> difficult. . . . This is not at all surprising, for these soils are the most
> persistently fertile in the world: they have an almost inexhaustible supply of
> available plant nutrients brought down from the upper lands drained by the
> rivers.

A feature of this early agriculture would have been annual cropping in
nutrient-rich fields: an exact parallel of the annual weeds that grew in such
areas before agriculture (see Wood and Lenné, Chapter 2, this volume). In
developing his 'rubbish-heap' theory of the origin of agriculture through
weeds, Anderson (1954, p. 130) used an agricultural analogy in noting that
big rivers: 'plough up the mantle of vegetation and leave raw scars in it.
Rivers are weed breeders; so is man, and many of the plants which follow us
about have the look of belonging originally on gravel bars and mud-banks'.
Hawkes (1969) noted that there is little basic difference between cultivated
plants and weeds and argues that the ancestors of our cultivated plants were
restricted to disturbed soil 'along river banks, on gravel, rocks, screes,
landslide areas . . . '.

The disturbance of annual natural flooding seems to have been a feature
of some areas of origin and early development of agriculture. Neolithic
settlements in the Fertile Crescent and along the edges of the Indo-Gangetic
Plain were close to gravel fans and alluvium where annual rivers debouched
from nearby mountains into desert. 'Décrue' agriculture – sowing and

planting as floods subsided annually – was and is a feature of the river agriculture in West Africa – notably the inland delta of the Niger (Harlan, 1995). The meaning of 'Mesopotamia' is the land between the rivers, the annual flooding of which stimulated early agriculture. Even at some distance from mountains, the annually disturbed flood plains of major rivers such as the Mekong and (until recently) the Nile were rich agricultural areas, supported by the erosion of the Himalayas and the Ethiopian highlands respectively (probably half the world's rice is grown on Himalayan silt). In another region of seasonal floods, during the Chinese Neolithic, rice agriculture was developed in low-lying regions of abundant marshes and lakes along the Yangzi valley (Pearson and Underhill, 1987). The need for greater control over flooding – while maintaining the advantages both of nutrient input and the disturbance of an alternating wet and dry regime – led to centrally controlled irrigation systems (Wittfogel, 1957).

Agriculture has continually mimicked the role of fires in nature for clearing vegetation – particularly in the suppression of forest and the opening of productive grassland. Savannahs worldwide are burned annually to provide fresh grazing for cattle. Vast areas of forest throughout the tropics are lopped and burned each dry season to clear fields for the widespread practice of shifting cultivation. We estimate that in 1 hour with a machete, a farmer can sever all lianes in a hectare of forest, causing perhaps 50% of the canopy leaves to fall rapidly. A further 1 week will suffice to ring-bark and kill all trees and shrubs, with the loss of all remaining leaves. Advantages of this fire-associated disturbance for agriculture include greatly reduced competition from canopy species for the subsequent crop, heat destruction of soil seed-bank of weeds, and added inorganic nutrients to what are otherwise nutrient-poor forest soils.

Some examples of the use of fire in African agriculture show parallels with natural fire regimes. In Zambia, a fire-based system called 'citemene', from a Bemba word meaning 'to cut', is used for finger millet cultivation. In 'large-circle' citemene, brushwood is lopped from trees over a wider area, piled in a circle, and, after drying, burnt (Allan, 1965). This concentration of nutrients from a wider area allowed reliable cropping and high yields from land otherwise marginal for agriculture. There was a further advantage: it was found that burning of vegetation on the surface soil will enhance the amount of nitrate produced when the rains come (Allan, 1965, p. 74). Burning may not just add nutrients: in experiments at Gezira, in the Sudan, firing of soil was found to increase water permeability, and the subsequent activity of nitrogen-fixing microorganisms (Crowther, 1948). Burning, both for land clearing and for shifting cultivation, was a feature of agriculture in Europe from the end of Roman times until recently. The 'temporary cultivation of ground burnt over' was noted for Corsica and the Ardenne forest in modern times (Parain, 1941).

Many of our important crops, including legumes, palms and grasses are resistant to burning. The legumes which provide much of our vegetable protein evolved from ancestors with hard seeds, resistant to vegetation fires. Palms (including oil, date and coconut palms) have their vascular tissue distributed through the trunk (noted by Sauer, 1969), unlike the superficial and more readily damaged vascular tissue of dicotyledonous woody plants. Adult palms can only be killed with persistence by fire, and palm fruits may be fire resistant (the date genus *Phoenix* is named from the Arabian bird of legend, arising from the ashes).

It is perhaps the grasses that show the most pronounced adaptations to disturbance in the form of fire. Clayton and Renvoize (1986, p. 17) are of the opinion that:

> the success of the grasses lies primarily in the evolution of a versatile lifestyle adapted to unstable or fluctuating environments, particularly those associated with strongly seasonal rainfall regimes or the early stages of succession following disturbance. This life-form then proved readily adaptable to a partnership with fire and herbivores, creating the highly competitive grassland ecosystem. Finally their propensity for exploiting instability has made them partner to the revolutionary changes in landscapes induced by man.

While grass seeds are not fire resistant, they have remarkable mechanisms to remove the seed from danger. As noted by Zohary (1969, p. 58), 'The spikelet in wild wheats and the triplet in wild barley are, in fact, specialized arrow-shaped seed dissemination units which very effectively insert the mature fruiting units into the soil.' Sorghum and its ancestors have fruits with geniculate awns: these are twisted above the fruit, then bent at almost a right-angle. When they fall, diurnal changes in moisture cause the awn to twist and bury the fruit in the ground. Such species are characteristic of fire-maintained savannahs where natural fruit-burying would be a selective advantage against fire damage. These features of wild ancestors of some of our important crops indicate that regular disturbance by fire was a natural part of the environment of early farming.

As crops and crop-associated biodiversity became adapted to regular disturbances, the resulting agroecosystem became more stable. Indeed, ecosystem instability would now ensue from removing the disturbances characteristic of agricultural management.

Succession

Disturbance creates open areas for plant colonization. Colonization begins the process of succession (Clements, 1916). Succession 'is the sequential process of replacement of early colonizing species by later arriving or developing ones. Typically the process is characterized by a general increase in the structural complexity of communities, a progressive change from annual to perennial habit and, in very many cases, an increase in species diversity' (as defined by Frankel *et al.*, 1995). Succession, rather than fixity, seems to be part of the natural order: most plant communities are successional and in successions 'each species is doomed to local extinction' (Harper, 1977, p. 82).

It is widely accepted that species characteristic of early succession are fundamentally different in their competitive, reproductive and eco-physiological strategies from species later in the succession (Grime, 1977; del Moral and Bliss, 1993, p. 4). Dover and Talbot (1987, p. 27) define two types of plant strategy: plants of open habitats with a 'productivity' strategy; and plants of 'saturated' environments, with an efficiency strategy. The 'product-ivity' strategy is a feature of plants in the r-phase of succession: Harper (1977, p. 760) argues that there is nothing in r-selection to favour length of life. Rather, r-selection favours fecundity and precocity – the allocation of a large fraction of plant resources to seed production: during this phase there is no premium on attributes that confer competitive ability (see Edwards *et al.*, Chapter 12, this volume).

Ecologists have long used the concept of population growth under r- and then K-conditions to illuminate the adaptation of species to succession. The concept was first developed for human demography (Pearl and Reed, 1920) and was extensively applied to plant populations by Harper (1977). Following Harper, the concept is best explained by considering r as the 'intrinsic rate of natural increase' of a population: the rate at which it will expand provided there is no shortage of resources or inhibiting con-sequences of population growth. For example, the r-phase in the life of a population is the rapid increase after colonization of an unexploited environment. In contrast, K is the upper asymptote or equilibrium popu-lation size prevailing as more and more species and individuals attempt to colonize and compete (Odum, 1969). A K-strategist channels all available assimilates in early life into establishing canopy height and achieving competitive dominance, with 'selfish and conservative growth strategies' based on plant height and a perennial canopy – best exemplified by invest-ment of the plant resource in a woody trunk (Harper, 1977). K-strategy is a feature of density-stressed, resource-limited communities. By definition, early stages of succession favour species which have evolved an r-strategy. In later stages of succession, when inter-plant competition becomes important, K-strategies predominate. As noted by Harper (1977, p. 771) 'the properties that favour leaving descendants during an r-phase are quite different from those in a K-phase.'

This concept has been extended to associated biodiversity. For example, Bongers (1990) noted that nematodes which have short life-cycles, high fecundity, are often viviparous, and live in ephemeral habitats could be considered as 'colonizers' or r-strategists. In contrast, nematodes with low fecundity, long life-cycles, low colonization ability, which live in habitats with long durational stability could be considered as 'persisters' or K-strategists. Similarly, for fungal pathogens, Burdon (1992) suggested that rust pathogens could be described as r-strategists: obligate biotrophs, with poor off-season survival, high fecundity, short generation time, efficient dispersal, a narrow host range and pathogenic on annuals and ephemeral plant parts; conversely, soil pathogens like *Rhizoctonia* and *Pythium* species could be considered *K*-strategists: necrotrophs or hemibiotrophs, with good off-season survival, low fecundity, long generation time, inefficient dispersal, a wide host range and commonly pathogenic on perennial hosts (see Allen *et al.*, Chapter 6, this volume). The possibility of coevolution between plants and associated biodiversity in relation to colonization strategies merits further study.

Parallels with agriculture

Bazzaz (1987, p. 248) argued that 'Disturbances generate early successional habitats and select for plants with short life span, fast growth, and high reproductive output.' Although Bazzaz was describing natural successions, the plant features given are exactly those of our annual staple food crops, and those characteristic of the *r*-phase of succession. In contrast, tree crops follow a *K*-strategy: the build-up of a canopy and trunk biomass.

Odum (1969, and see commentary by Drury and Nisbet, 1973) recognized the implications for agriculture in the different stages of ecological succession: early stages had a high net yield, with linear food chains (in nature, mainly grazing); later stages a low yield and weblike, predominantly detritus, food chains. Odum then argued that man was preoccupied with obtaining as much 'production' from the landscape as possible by developing and maintaining early types of ecosystems, usually monocultures.

Harper (1977) presented two simple strategies which are open to successional species faced with local extinction in nature – to 'wait until the right habitat reappears', or to 'escape to somewhere else'. The parallel between natural succession and farm practice is close: agriculture continually opens new habitats for colonization and efficiently places seed of crops in this 'right habitat'. In addition, competition is very strictly controlled in agriculture by adjusting planting density, thinning seedlings, and – usually – considerable weeding (apart from sleeping, most human time is spent weeding). The annual cropping that provides most of our food seems to be a close homologue of early-successional, *r*-phase, vegetation, and it is perhaps from this type of vegetation in nature that models for staple crop production are best sought.

Succession is related to the degree of persistence of species in the ecosystem: there may be useful parallels between nature and agriculture in host–pathogen systems. The evolutionary dynamics of host–pathogen interactions in natural ecosystems has been the subject of considerable research during the past 25 years (see Burdon, 1987 for an extensive coverage). Of particular interest, the dogma that race specific resistance is an artefact of agriculture is being convincingly challenged (Burdon *et al.*, 1996). Two different types of resistances are found in natural plant populations – race specific and race non-specific resistance – which coexist as complementary mechanisms whereby any pathogen may be firstly delayed then reduced. Burdon *et al.* (1996) suggest that the evolution of gene-for-gene systems may be most likely to develop where susceptible host tissues are ephemeral, such as in annual crops or where deciduous leaves die at the end of the season, for example poplar rust (*Melampsora occidentalis*) under the dry continental climate east of the Cascade mountains in the Pacific Northwest of the USA. In contrast, the moist maritime climate west of the Cascade mountains provides a very favourable environment for poplar rust development throughout the year on more persistent host tissue, and resistance to rust is based on race non-specific resistance (Hsiang and Chastagner, 1993). Understanding of the processes – whether due to environmental or other factors – which determine the genetic basis of resistance to pathogens in natural populations can have important applications to agriculture. In 1977, Harper noted that the stage was ripe for bringing the ideas and techniques of plant pathology in managed crops and forests to the exploration of natural vegetation. Now after 25 years of intensive effort and tremendous advances in the knowledge of host–pathogen associations in natural ecosystems, one could argue that the time is ripe for applying this knowledge in turn to management of pathogens in agroecosystems.

The finest example of farmers' understanding of succession is compressed in one cropping system – that of pan-tropical shifting cultivation. The natural climax of trees is suppressed by ring-barking and lopping, and debris burnt after drying. The 'agro-succession' normally begins with seed crops (rice, maize, millets); continues with root crops – cassava, yams and sweet potato – and bananas, and matures with residual (and preserved) economic trees (fruits and palms). After a long period of fallow to re-establish the 'climax' tree cover, and suppress weeds, agro-succession is again induced. In one field the ecological strategies of 'seed-culture', 'vegeculture', and 'arboriculture' are successively followed (amplified in 'Conclusions' below).

Pure Stands in Nature

In a single-minded preoccupation with complex late-successional models for all agriculture, agroecologists have exclusively promoted mixed stands as a model for agriculture – as a consequence, 'monoculture' has become a term of abuse. For example, a recent policy statement from the World Bank (Johnson, 1998) directly equated 'agriculturally sustainable development' with 'increasing the productivity of complex (as opposed to monoculture) farming systems.' The current paradigm of what agriculture 'should' be like always includes species diversity as a requirement: as an example, the view of Ferrar (1992) that 'the safest way to ensure an unhindered supply of food may be mixed cropping in species-rich environments – systems that approximate as closely as possible to natural ecosystems while still being able to produce food, with minimum amounts of human perturbation.' There are many hundreds of similar prescriptions for agriculture in nature's image, based on within and between species diversity, and sometimes a deterministic view of ecosystem development (as in Odum, 1969). Pervading these views is the idea that species-diverse and complex stands are 'stable' (despite a considerable literature questioning this view, some of which will be reviewed in the section on 'Diversity in Nature', below).

These prescriptions for 'agriculture in nature's image' ignore the existence, indeed commonness, of pure stands in nature. Such pure stands could provide a functional model for sustainable agriculture in addition to the now-pervasive models based on species-diversity and structural complexity.

The two sections above, on disturbance and succession, are necessarily linked: disturbance begins succession. Pure stands of plants – natural monocultures – occur throughout successional sequences, and many seem not to depend on disturbance for their persistence. However, the pure stands which are found in early natural successions, as a result of regular natural disturbance, are of relevance to agriculture, indeed, could be a useful model for agriculture. Early succession is often characterized by low species diversity (Loucks, 1970). Bazzaz (1987, p. 249) predicted that early successional communities 'are likely to have low species diversity as each species occupies a large portion of total available resources (niche space). Low diversity has been shown for a number of early successional sequences . . . '.

Perhaps the earliest of early successional vegetation types is that of open water – progressing zonally to mud-flats, seasonally flooded land, and then dry land. Pure stands are common in open water, including *Papyrus*, *Eichhornia*, *Salvinia* and many others. Pure stands of mud-flats and seasonally flooded land include *Spartina* (as noted by May, 1975) and *Phragmites*. Haslam (1971) considered monodominant *Phragmites communis* to be the commonest community of British reedswamps and wet marshes, with popuations that can be self-perpetuating for hundreds of years in stable conditions. Marine pure stands include many mangroves – especially *Rhizophora* – and sea-grasses such as *Zostera* (but see Watt, 1947, for his ideas on stability and decline in *Zostera* and other pure stands).

On dry land, fire is responsible for many pure stands (including those of crop relatives, to which we return later). Goldhammer (1993) listed pines, *Melaleuca quinquenervia* in Florida (as both an exotic and a fire-resistant plant), and *Imperata* grasslands that are maintained by short fire-return intervals.

Harper (1977) notes bracken (*Pteridium aquilinum* – which is allelopathic) as a persistent monodominant stand, and one of the 'five most common plants on the face of the earth, though there may not be a great many real individuals', and that other clone-forming species may also produce large areas of 'specific and genetic monotony' (examples: *Erica*, *Festuca rubra* and *Holcus mollis*).

Introduced species may form pure stands, indeed, this is a major problem of invasive aliens (for example *Rhododendron ponticum* in Britain). Darwin noted that 'Several of the plants now most numerous over the plains of La Plata covering square leagues of surface almost to the exclusion of other plants have been introduced from Europe' (quoted by Harper, 1977). Daehler (1998), in a review of the taxonomic affinities of invasive species, noted that the grass family was spectacularly over-represented among weeds and natural area invaders and suggested that the grass growth habit favours monocultures. He also noted that another family with important crops – the *Leguminosae* – had a propensity to form nearly monospecific stands.

Insular floras have characteristic single-dominant species, such as *Metrosideros polymorpha* in the Hawaiian Islands and *Scalesia pedunculata* in the Galapagos Islands (Vitousek *et al.*, 1996). Invasive species on islands have such a potential for forming pure stands and replacing native species that they are the nightmare of conservationists (as with *Leucaena leucocephala* on Pacific islands).

Late-successional vegetation may also consist of pure stands (Connell and Lowman, 1989; Hart, 1990). For forest trees, Schulze *et al.* (1996) note that in all regions of the deciduous forests there are stands composed of only one, or a few, species, giving as examples *Fagus sylvatica*, which dominates deciduous forests in Europe, *F. orientalis* forms nearly pure stands in the montane regions of the Caucasaus, and *F. crenata* forming pure stands under high rainfall in Japan. Further, *Betula* and *Populus* may form pure stands in North America, Siberia and Japan, and *Larix gmelinii*, the Siberian larch, forms monotypic stands over huge areas of Siberia.

Tropical examples of pure stands of trees include mopane (*Colophospermum mopane*) in Malawi and *Gilbertodendron dewevrei* monodominant in the Congo rainforest. The tree legume, *Peltogyne gracilipes*, forms monodominant stands in Amazonian forest, and recovers from intense damage by insects (Nascimento and Proctor, 1994). Southern temperate forests in New Zealand have pure stands of *Notofagus solandri* var. *cliffortioides* (Wiser *et al.*, 1998).

The natural homologues most relevant to agriculture will be of stands of wild species ancestral to crops. In the grass family, the evidence of genetic and ecological continuity between natural ecosystems and agroecosystems seems to be strong.

There are many cases of grasses growing in pure stands – for example, Cox and Morton (1986) describe almost pure stands of *Sporobolus wrightii* in seasonally burnt, semi-arid areas, of the southwestern USA. Also, indigenous annual *Sorghum* species dominate certain landscapes at the end of the growing season in tropical northwest Australia (Andrew and Mott, 1983). Dry season burning has little effect on their persistence, and plant densities reach 9–220 plants m^{-2}. The wild *Sorghum bicolor* subsp. *arundinaceum* is closely related to domesticated sorghum and still harvested for human food. The races *aethiopicum* and *verticilliflorum* of *S. bicolor* are often dominant grasses in the northern savannah of Africa (de Wet and Shechter, 1977).

Perhaps the most significant findings for agricultural management are those of Zohary on natural stands of wheat and barley, and Harlan on natural stands of wheat and sorghum. Zohary (1969, p. 55) emphasizes the fact that the direct ancestors of our cereals *Hordeum spontaneum* (for barley), *Triticum boeoticum* (for einkorn wheat) and *T. dicoccoides* (for emmer wheat) are common wild plants in the Near East, frequently building extensive, massive stands in primary habitats, where they are dominant annuals. For example, *H. spontaneum* 'is massively and continuously spread over primary habits' in its distribution centre in the Fertile Crescent (Zohary, 1969, p. 53). In the 'park-forests' (open *Quercus* woodland) there are extensive 'natural fields' of wild cereals (Zohary, 1969, his quotes). It is from these natural ecosystems that our more important cereals were domesticated.

Harlan (1981, p. 11) also uses the field analogy for dense natural stands in Near Eastern woodlands: 'Even now, stands of wild cereals develop as dense as sown fields when protected from livestock.' Harlan (1992) described natural stands of wild grass species from Africa, where 'Massive stands of truly wild races of sorghum can be found widely distributed over the savanna zones', and the Near East, where 'massive stands of wild wheats cover many square kilometers'; and natural stands of wild cereals 'are as dense as a cultivated field'.

Perhaps it is to these same 'ancestral ecosystems' that we should look for ecological models for cereal crop management. Indeed, Zohary (1969) claims that we already possess ample information on the distribution, ecology and biology of the wild cereals in the Near East. Harlan's 1981 paper was specifically on the ecological settings for the emergence of agriculture. Harlan (1981, p. 16) considered that the most productive plant formations in terms of agricultural development were the Near Eastern woodlands and the tropical savannahs and dry forests. It was from these same wild cereals, in the exact locations described by Zohary and Harlan, that early farmer-ecologists developed the more productive ancestors of today's high-yielding varieties. Ecological knowledge from the 'ancestral ecosystems' of our cereal crops will be of far greater relevance to agricultural management than models based on late-successional species-diverse ecosystems, as often proposed by agroecologists.

Agriculture is perhaps the most important example of applied ecology: ecologists need to elucidate the parallels with agriculture of natural 'mono-cultures', especially in 'ancestral ecosystems' where they may be common. From the above examples, it seems that a range of factors may contribute to the persistence of pure stands, including regular nutrient input (from silt and burning), water availability (from seasonal flooding), availability of open habitats through disturbance (and mechanisms of resistance to fire, flood and grazing), absence of competition, ability to recover from pest invasions, invasibility, breeding system and plant type. Many of these have become features of the human management of agriculture – we venture to suggest not by chance.

Diversity in Nature

With great reluctance we turn to the confused debate on the possible connection between diversity and stability in nature. To be of any guidance for agriculture, this debate needs to demonstrate two propositions: that species-diverse, complex communities are stable; and that species-poor, simple communities are not stable. Neither proposition seems to be generally valid.

Firstly, there may be instability in natural species-rich communities. Swift and Anderson (1994) recognize that many natural communities have cycles of collapse and recovery which are a response to major perturbations in their productive capacity. Indeed, the most damaging plant disease out-break of recent times has been dieback in species-diverse natural Australian forests. This disease, caused by the fungus *Phytophthora cinnamomi*, has destroyed whole plant communities: for example, 59 indigenous species in 34 genera and 13 botanical families have been eliminated from infested jarrah (*Eucalyptus marginata*) forest (Weste and Marks, 1987). On theoretical grounds, May (1975) argued that complex ecosystems are in general dynamically fragile (with the example of tropical forest). Nilsson and Grelsson (1995) noted that 'little empirical progress has been made in answering the question of whether diverse systems resist disturbance, i.e. are less fragile, than simple ones.'

In the parallel debate on species diversity and ecosystem function, Grime (1997) noted that, so far, neither evolutionary theory nor empirical studies had presented convincing evidence that species diversity and eco-system function were consistently and causally connected. Huston (1997; and see also Aarssen, 1997; Doak *et al.*, 1998; and Wardle *et al.*, Chapter 5, this volume) strongly queried experiments purporting to show the link between species diversity and ecosystem functioning, arguing that results which supported the diversity–productivity and diversity–sustainability hy-potheses and also that 'the loss of species threatens ecosystem functioning and sustainability' were not supported by the data (see Lenné and Wood, Chapter 18, this volume). Huston (1997) further argued that the Earth's most productive systems generally had low plant diversity while high plant diversity is found under much less productive systems. He held that this demonstrated that the number of plant species had relatively little effect on productivity. With reference to agriculture, Swift *et al.* (1996) accepted that for the generality of agroecosystems it was not really clear what the ecological function of biodiversity might be. Wardle *et al.* (Chapter 5, this volume) note that the experimental evidence was ambiguous (and see Lenné and Wood, Chapter 18, this volume).

Secondly, natural monocultures may be stable (we give many examples of apparently stable pure stands above). May (1975) noted that there was no reason to expect simple natural monocultures to be unstable, giving examples of bracken (*P. aquilinum*) and the tidal mud-flat colonizer, cord grass (*Spartina*). For tropical rain forests, the ultimate example of plant diversity, Janzen (1974) noted that species-poor forests could exist, and that the 'existence of such forests . . . falsifies the dogma that diversity is mandatory for ecosystem stability in highly equitable climates.' Grime (1997) noted 'It would be naïve to assume that species-poor ecosystems are always malfunctional; some of the world's most extensive and ancient ecosystems – boreal forests, bogs, and heathlands – contain few species'.

Rindos (1984, pp. 166–172) reviews the application of ideas of complexity and stability to agriculture, concluding that the correlation, although intuitively sound, remains unproven. Rindos (1984) quotes Colinvaux (1978): 'The claim that complex communities are more stable than simple communities . . . is invalid. . . . It has done mischief by distracting people from real problems.' Dover and Talbot (1987) noted that, if true, the diversity–stability hypothesis:

> provides a strong practical argument for conservation – for maintaining diversity in ecosystems whenever possible as a natural buffer against perturbation. But if the theory is flawed, basing policies on it could have unexpected and undesirable environmental consequences. For instance, if diversity causes stability, the most species-rich communities – such as tropical rain forests and coral reefs – should be able to withstand the greatest disruption at human hands. In fact, these communities are among the most fragile. . . . experimental evidence and theoretical analysis reveal the notion that diversity causes stability as oversimplified at best, if not dead wrong.

Goodman (1975, p. 261) was yet more critical:

> The diversity–stability hypothesis may have caught the lay conservationists' fancy, not for the allure of its scientific embellishments, but for the more basic appeal of its underlying metaphor. It is the sort of thing that people like, and want, to believe.

The very usage of the term 'stability' has been questioned. Orians *et al.* (1996) recognized that discussions of the degree and causes of stability were frequently hampered by vagueness and inconsistency about what is meant by stability, and that the causes and consequences of species diversity could be different. As argued by Grimm and Wissel (1997): 'The ambiguity of the term "stability" is one of the main sources of confusion in the ecological stability discussions' . . . "stability" alone is so ambiguous as to be unusable as a concept.'

The concept of agroecosystems mimicking the supposed stability of nature may therefore be confused and inappropriate. Consider the analogy of an English 'landscape' garden. There will be rhododendrons from the wet eastern Himalayas, roses from eastern Asia, alpines from the screes and cliffs of the higher altitudes of the world's mountains, pond plants from temperate wetlands worldwide, conifers from the Pacific Northwest and much besides. The 'biological stability' of such a garden is almost irrelevant: if at all, stability depends to only a very small extent on the biological interaction between the ornamental species in the garden ecosystem. In contrast, the stability of this garden system depends predominantly on management. This management reflects, as far as possible, the ecosystems of origin of the species. Alpines are planted in a rockery, pond plants in perennial water, rhododendrons on acid soil, less hardy plants on south-facing sheltered slopes. There is reason to believe that this 'management based on ecosystem properties' analogy is already practised in agriculture, with the various crop, animal and landscape elements being matched to natural homologues. Its value has perhaps been needlessly obscured by the debate over diversity and stability.

We have provided evidence of ecosystem disturbance in nature, as in agriculture. We have suggested that the *r*-strategy characteristic of early successions in nature is exactly mimicked by crop species. We have shown that there are species-poor 'monocultures' in nature homologous with the monocultures of cropping systems. We have argued that there may be a direct genetic connection – for example, in the origin of sorghum as a crop – between stands of wild relatives in nature and the crop in agriculture. All this provides substantial evidence for a conceptual and practical link between nature and agriculture, based on ecosystem homology. For staple grain crops, ecosystem homologues seem to be simple natural systems – the annual 'fields of wild grain' that so impressed Zohary (1969) and Harlan (1992). Yet 'natural' models, based on the supposed – but unproven – link between diversity and stability, and supposed – but mainly inappropriate – homologies with complex tree-based ecosystems, are repeatedly proposed for agriculture, and are now firmly part of the development paradigm. This is a misreading of nature and perhaps even a misunderstanding of the motives of farmers as they 'mix and match' diversity: mixed to meet their own diverse needs, matched by their diverse capacities and farming environments.

Conclusions

We argue that different types of crops have arisen by different ecological trajectories from nature. This is not a new idea: Harris (1976) uses the terms 'vegeculture' for roots and tubers and banana cropping systems, and 'seed-culture' (sorghum and millet in his example from West Africa). We could add 'arboriculture' for tree crops, from Yen (1974). Despite this range of evolutionary trajectories, originating from different natural ecosystems, great effort has been devoted to deriving agricultural models only from the *K*-strategy of complex late-successional vegetation. Annual grain crops provide 72% of our food production (Harlan, 1995). For these crops, application of models derived from late-successional woody vegetation may be conceptually and practically sterile – a certain breach with the lessons of nature and the perceptive actions of farmers. We do not differ from the conclusions of Soule and Piper (1991) that 'agroecosystems should mimic the vegetation structure of natural plant communities', but argue that a far wider range of natural communities than hitherto considered – including natural pure stands – can provide models for productive agriculture – including crop monocultures.

As an example, the seed-culture trajectory seems firmly based on simple natural models that, if there is any meaning to the word, are stable. The close linking of seed-culture with, for example, natural savannah ecosystems, can establish an 'evolutionary pedigree' for agriculture – a genetic and ecological continuity with the wild. Indeed, both an 'evolutionary' and an 'ecological' continuum between ancestral species and crops has already been suggested by Frankel *et al.* (1995, p. 39). This continuum or pedigree is of more than theoretical interest: May's concern over agricultural monocultures was exactly that they 'lack an evolutionary pedigree', and would therefore not be as stable as natural pure stands (May, 1975). Demonstrating such a pedigree has implications for the management of crop monocultures, not least in their relation to pests and diseases.

Under this view, agriculture and agrobiodiversity, rather than representing a breach with nature, can be linked conceptually and biologically with all nature, rather than, as hitherto, only the more complex parts of nature. However, as repeatedly noted by Edwards *et al.* (Chapter 8, this volume), recent and relatively rapid changes in agriculture may disrupt the long continuity of agricultural practice that allows agroecosystems and their constituent species to adjust.

The challenge for agrobiodiversity management will be to understand the past linkages, and to recognize present disruptions. Certainly (see Thurston *et al.*, Chapter 9, this volume), a greater knowledge of traditional farming can be a key to understanding these linkages and disruptions. Farmers have sensibly mixed and merged crops of different ecological origins in a great variety of cropping systems, each one producing the range of food, plant products and environmental services required by each farmer. But farmers have also done as nature does – employed a range of separate and different systems to meet different conditions and requirements. The common combination of field and garden within a farming system is an example of this.

There is a need to investigate the natural homologues of important cropping systems to provide management models for crops. For example, how do pure stands of wild wheat and sorghum maintain their status in nature? There seems to have been few such studies. One was the 'Ammiad' project in eastern Galilee (Anikster and Noy-Meir, 1991), on the wild tetraploid wheat, *Triticum turgidum* var. *dicoccoides*, the progenitor of most cultivated tetraploid and hexaploid wheats (including bread wheat). The lessons of Ammiad could be extended to other natural ecosystems which are 'ancestral homologues' of cereal agriculture, with an emphasis on the factors controlling the persistence of wild progenitors – disturbance, disease, microclimate, competition and many more.

Another valuable study, by Haggar and Ewel (1997), constructed experimental models of tropical ecosystems; noted that monocultures 'are capable of effective resource use and high productivity'; and proposed that 'ecologically designed mixtures' could be of use. Further such studies could provide evidence for management of crops based on the direct properties of ecosystems, rather than our present confused, and derivative, concepts of diversity and stability. In parentheses, but significantly, Ewel's experience with this detailed field research on diversity led him to the conclusion that crop monocultures were essential for the conservation of wild biodiversity: without monocultures 'vastly greater quantities of land would have to be devoted to agriculture and forestry if we are to make any reasonable attempt to feed and shelter the populace of this overpopulated world' (Ewel, 1991). We return to this argument in Chapter 18.

There is another prospect: the use of information from the massive experimental base of agriculture – unclouded by preconceptions – to refine our approaches to the management of wild biodiversity. The practical needs have been suggested in general by Mooney *et al.* (1996) – to explore more fully the role of genetic, species and landscape diversity in agroecosystems vs. natural ecosystems. Following such studies, a productive middle ground could develop between nature and agriculture – perhaps recognizing, with Adams (1997): 'Gone, therefore, are the days when conservationists could conceive of "nature" in equilibrium and hence portray human-induced changes in those ecosystems as somehow "unnatural"'.

References

Aarssen, L.W. (1997) High productivity in grassland ecosystems: effected by species diversity or productive species. *Oikos* 80, 183–184.

Adams, W.M. (1997) Rationalization and conservation: ecology and the management of nature in the United Kingdom. *Transactions of the Institute of British Geographers* (New Series) 22, 277–291.

Allan, W. (1965) *The African Husbandman.* Oliver & Boyd, Edinburgh.

Altieri, M.A. (1987) *Agroecology: the Scientific Basis of Alternative Agriculture.* Westview Press, Boulder, Colorado.

Anderson, E. (1954) *Plants, Man and Life.* Melrose, London.

Andrew, M.H. and Mott, J.J. (1983) Annuals with transient seed banks: the population biology of indigenous *Sorghum* species of tropical north-west Australia. *Australian Journal of Ecology* 8, 265–276.

Anikster, Y. and Noy-Meir, I. (1991) The wild-wheat field laboratory at Ammiad. *Israel Journal of Botany* 40, 351–362.

Bazzaz, F.A. (1987) Experimental studies on the evolution of niche in successional plant populations. In: Gray, A.J., Crawley, M.J. and Edwards, P.J. (eds) *Colonization, Succession and Stability.* Blackwell Scientific, Oxford, pp. 245–272.

Bird, M.I. and Cali, J.A. (1998) A million-year record of fire in sub-Saharan Africa. *Nature* 394, 767–769.

Bongers, T. (1990) The maturity index: an ecological measure of environmental disturbance based on nematode species composition. *Oecologia* 83, 14–19.

Burdon, J.J. (1987) *Diseases and Plant Population Biology*. Cambridge University Press, Cambridge.

Burdon, J.J. (1992) The growth and regulation of pathogenic fungal populations. In: Carroll, G.C. and Wicklow, D.T. (eds) *The Fungal Community: its Organisation and Role in the Ecosystem*. Marcel Dekker, New York, pp. 173–181.

Burdon, J.J., Wennstrom, A., Elmquist, T. and Kirkby, G.C. (1996) The role of race specific resistance in natural plant populations. *Oikos* 76, 411–417.

Clayton, W.D. and Renvoize, S.A. (1986) *Genera* Graminum: *Grasses of the World*. HMSO, London.

Clements, F.E. (1916) Plant succession: an analysis of the development of vegetation. *Carnegie Institute of Washington Publication* 242, 1–512.

Colinvaux, P. (1978) *Why Big Fierce Animals are Scarce*. Princeton University Press, Princeton, New Jersey.

Columella, L.J.M. (*c.* AD 65) *De Re Rustica*. Liber I. Praefatio. Loeb Classical Library edn (1941), Heinemann, London.

Connell, J.H. and Lowman, M.D. (1989) Low-diversity tropical rain forests: some possible mechanisms for their existence. *American Naturalist* 134, 88–119.

Conway, G. (1997) *The Doubly Green Revolution: Food for All in the Twenty-First Century*. Penguin Books, London.

Cox, J.R. and Morton, H.L. (1986) The effects of seasonal fire on live biomass and standing crop in a big sacaton (*Sporobolus wrightii*) grassland. In: Joss, P.J., Lynch, P.W. and Williams, O.B. (eds) *Rangelands Under Seige*. Proceedings of the 2nd International Rangeland Congress, Adelaide, May, 1985, Cambridge University Press, Cambridge, pp. 596–597.

Crowther, F. (1948) A review of experimental work. In: Tothill, J.D. (ed.) *Agriculture in the Sudan*. Oxford University Press, Oxford, pp. 439–593.

Daehler, C.C. (1998) The taxonomic distribution of invasive angiosperm plants: ecological insights and comparison to agricultural weeds. *Biological Conservation* 84, 167–180.

Doak, D.F., Bigger, D., Harding, E.K., Marvier, M.E., O'Malley, R.E. and Thomson, D. (1998) The statistical inevitability of stability–diversity relationships in community ecology. *American Naturalist* 151, 264–276.

Dover, M. and Talbot, L.M. (1987) *To Feed the Earth: Agro-ecology for Sustainable Development*. World Resources Institute, Washington, DC.

Drury, W.H. and Nisbet, I.C.T. (1973) Succession. *Journal of the Arnold Arboretum* 54, 331–368.

Dublin, H.T., Sinclair, A.R.E. and McGlade, J. (1990) Elephants and fire as causes of multiple stable states in the Serengeti-Mara woodlands. *Journal of Animal Ecology* 59, 1147–1164.

Ewel, J.J. (1991) Yes, we got some bananas. *Conservation Biology* 5, 423–425.

Ferrar, P. (1992) Classical biological control and social equity: a reply to Altieri. *Bulletin of Entomological Research* 82, 297–298.

Frankel, O.H., Brown, A.H.D. and Burdon, J.J. (1995) *The Conservation of Plant Biodiversity*. Cambridge University Press, Cambridge.

Goldhammer, J.G. (1993) Historical biogeography of fire: tropical and subtropical. In: Crutzen, P.J. and Goldhammer, J.G. (eds) *Fire in the Environment: the Ecological, Atmospheric, and Climatic Importance of Vegetation Fires*. John Wiley & Sons, Chichester, pp. 297–314.

Goodman, D. (1975) The theory of diversity–stability relationships in ecology.

Quarterly Review of Biology 50, 237–266.

Grime, J.P. (1977) Evidence for the existence of three primary strategies in plants and its relevance to ecological and evolutionary theory. *American Naturalist* 111, 1169–1194.

Grime, J.P. (1997) Biodiversity and ecosystem function: the debate deepens. *Science* 277, 1260–1261.

Grimm, V. and Wissel, C. (1997) Babel, or the ecological stability discussions: an inventory and analysis of terminology and a guide for avoiding confusion. *Oecologia* 109, 323–334.

Haggar, J.P. and Ewel, J.J. (1997) Primary productivity and resource partitioning in model tropical ecosystems. *Ecology* 78, 1211–1221.

Harlan, J.R. (1981) Ecological settings for the emergence of agriculture. In: Thresh, J.M. (ed.) *Pests, Pathogens and Vegetation.* Pitman, London.

Harlan, J.R. (1992) *Crops and Man,* 2nd edn. American Society of Agronomy, Madison, Wisconsin.

Harlan, J.R. (1995) *The Living Fields: our Agricultural Heritage.* Cambridge University Press, Cambridge.

Harper, J.L. (1977) *Population Biology of Plants.* Academic Press, London.

Harris, D.R. (1969) Agricultural systems, ecosystems and the origins of agriculture. In: Ucko, P.J. and Dimbleby, G.W. (eds) *The Domestication and Exploitation of Plants and Animals.* Duckworth, London, pp. 3–15.

Harris, D.R. (1976) Traditional systems of plant food production and the origins of agriculture in West Africa. In: Harlan, J.R., de Wet, J.M.J. and Stempler, A.B.L. (eds) *Origins of African Plant Domestication.* Mouton, The Hague, pp. 311–356.

Hart, T.B. (1990) Monospecific dominance in tropical rain forests. *Trends in Ecology and Evolution* 5, 6–11.

Haslam, S.M. (1971) Community regulation in *Phragmites communis* Trin. I. Monodominant stands. *Journal of Ecology* 59, 65–88.

Hawkes, J.G. (1969) The ecological background of plant domestication. In: Ucko, P.J. and Dimbleby, G.W. (eds) *The Domestication and Exploitation of Plants and Animals.* Duckworth, London, pp. 17–29.

Hsiang, T. and Chastagner, G.A. (1993) Variation in *Melampsora occidentalis* rust on poplars in the Pacific Northwest. *Canadian Journal of Plant Pathology* 15, 175–181.

Huston, M.A. (1994) *Biological Diversity: the Coexistence of Species on Changing Landscapes.* Cambridge University Press, Cambridge.

Huston, M.A. (1997) Hidden treatments in ecological experiments: evaluating the ecosystem function of biodiversity. *Oecologia* 110, 449–460.

Janzen, D. (1974) Tropical blackwater rivers, animals, and mast fruiting in the Dipterocarpaceae. *Biotropica* 6, 69–103.

Johns, R.J. (1986) The instability of the tropical ecosystem in New Guinea. *Blumea* 31, 341–371.

Johnson, I. (1998) Letter from the new Vice President Environmentally and Socially Sustainable Development. *Environment Matters* Annual Review, The World Bank, Autumn, 1998, p. 3.

Loucks, O.L. (1970) Evolution of diversity, efficiency, and community stability. *American Zoologist* 10, 17–25.

May, R.H. (1975) Stability in ecosystems: some comments. In: van Dobben, W.H.

and Lowe-McConnell, R.H. (eds) *Unifying Concepts in Ecology.* W. Junk, B.V., The Hague, pp. 161–168.

Mooney, H.A., Cushman, J.H., Medina, E., Sala, O.E. and Schulze, E.D. (1996) What have we learned about the ecosystem functioning of biodiversity? In: Mooney, H.A., Cushman, J.H., Medina, E., Sala, O.E. and Schulze, E.D. (eds) *Functional Roles of Biodiversity: a Global Perspective.* John Wiley & Sons, Chichester, pp. 475–484.

del Moral, R. and Bliss, L.C. (1993) Mechanisms of primary succession: insights resulting from the eruption of Mount St Helens. *Advances in Ecological Research* 24, 1–66.

Nascimento, M.T. and Proctor, J. (1994) Insect defoliation of a monodominant Amazonian rainforest. *Journal of Tropical Ecology* 10, 633–636.

Nilsson, C. and Grelsson, G. (1995) The fragility of ecosystems: a review. *Journal of Applied Ecology* 32, 677–692.

Odum, E.P. (1969) The strategy of ecosystem development. *Science* 164, 262–270.

Orians, G.H., Dirzo, R. and Cushman, J.H. (1996) Impact of biodiversity of tropical forest ecosystem processes. In: Mooney, H.A., Cushman, J.H., Medina, E., Sala, O.E. and Schulze, E.D. (eds) *Functional Roles of Biodiversity: a Global Perspective.* John Wiley & Sons, Chichester, pp. 213–244.

Parain, C. (1941) The evolution of agricultural technique. In: Clapham, J.H. and Power, E. (eds) *The Cambridge Economic History of Europe from the Decline of the Roman Empire,* Vol. 1, *The Agrarian Life of the Middle Ages.* Cambridge University Press, Cambridge, pp. 118–168.

Pearl, R. and Reed, L.J. (1920) On the rate of growth of the population of the United States since 1790 and its mathematical representation. *Proceedings of the National Academy of Sciences USA* 6, 275–288.

Pearson, R. and Underhill, A. (1987) The Chinese Neolithic: recent trends in research. *American Anthropologist* 89, 807–822.

Rindos, D. (1984) *The Origins of Agriculture: an Evolutionary Perspective.* Academic Press, London.

Sauer, C.O. (1969) *Seeds, Spades, Hearths, and Herds: the Domestication of Animals and Foodstuffs,* 2nd edn. MIT Press, Cambridge, Massachusetts.

Schulze, E.D., Bazzaz, F.A., Nadelhoffner, K.J., Koike, T. and Takatsuki, S. (1996) Biodiversity and ecosystem function of temperate deciduous broad-leaved forest. In: Mooney, H.A., Cushman, J.H., Medina, E., Sala, O.E. and Schulze, E.D. (eds) *Functional Roles of Biodiversity: a Global Perspective.* John Wiley & Sons, Chichester, pp. 71–98.

Soule, J.D. and Piper, J.K. (1991) *Farming in Nature's Image: an Ecological Approach to Agriculture.* Island Press, Washington, DC.

Stocks, B.J. and Trollope, W.S.W. (1993) Fire management: principles and options in the forested and savanna regions of the world. In: Crutzen, P.J. and Goldhammer, J.G. (eds) *Fire in the Environment: the Ecological, Atmospheric, and Climatic Importance of Vegetation Fires.* John Wiley & Sons, Chichester, pp. 315–326.

Swift, M.J. and Anderson, J.M. (1994) Biodiversity and ecosystem functioning in agricultural systems. In: Schulze, E.D. and Mooney, H.A. (eds) *Biodiversity and Ecosystem Function.* Springer-Verlag, Berlin, pp. 15–41.

Swift, M.J., Vandermeer, J., Ramakrishnan, P.S., Anderson, J.M., Ong, C.K. and Hawkins, B.A. (1996) Biodiversity and agroecosystem function. In: Mooney,

H.A., Cushman, J.H., Medina, E., Sala, O.E. and Schulze, E.D. (eds) *Functional Roles of Biodiversity: a Global Perspective*. John Wiley & Sons, Chichester, pp. 261–298.

Vitousek, P.M., Loupe, L.L., Andersen, H. and d'Antonio, C.M. (1996) Island ecosystems: do they represent 'natural experiments' in biological diversity and ecosystem function. In: Mooney, H.A., Cushman, J.H., Medina, E., Sala, O.E. and Schulze, E.D. (eds) *Functional Roles of Biodiversity: a Global Perspective*. John Wiley & Sons, Chichester, pp. 245–259.

Watt, A.S. (1947) Pattern and process in the plant community. *Journal of Ecology* 35, 1–22.

Weste, G. and Marks, G.C. (1987) The biology of *Phytophthora cinnamomi* in Australian forests. *Annual Review of Phytopathology* 25, 207–229.

de Wet, J.M.J. and Shechter, Y. (1977) Evolutionary dynamics of sorghum domestication. In: Seigler, D.S. (ed.) *Crop Resources*. Academic Press, New York, pp. 179–191.

Wiser, S.K., Allen, R.B., Clinton, P.W. and Platt, K.H. (1998) Community structure and forest invasion by an exotic herb over 23 years. *Ecology* 79, 2071–2081.

Wittfogel, K. (1957) *Oriental Despotism: a Comparative Study of Total Power*. Yale University Press, New Haven, Connecticut.

Yen, D.E. (1974) Arboriculture in the subsistence of Santa Cruz, Solomon Islands. *Economic Botany* 28, 247–284.

Zohary, D. (1969) The progenitors of wheat and barley in relation to domestication and agricultural dispersal in the Old World. In: Ucko, P.J. and Dimbleby, G.W. (eds) *The Domestication and Exploitation of Plants and Animals*. Duckworth, London, pp. 47–66.

Optimizing Biodiversity for Productive Agriculture

18

J.M. Lenné and D. Wood

Agrobiodiversity International, 13 Herons Quay, Sandside, Milnthorpe, Cumbria LA7 7HN, UK

Agricultural intensification on the land best suited for crop production is the best way to help preserve global soil and water resources, forests and rangelands, and wildlife habitats. The hallmark of 20th Century agriculture has been the adoption of yield-increasing, land-saving technology, which has permitted us to keep pace with rapid population growth.

(Borlaug, 1998)

Introduction

The most immediately valuable part of global biodiversity is the agro-biodiversity on which farming and, in turn, global food security, depends. Agrobiodiversity includes not only crops and livestock and their wild relatives but also interacting species of pollinators, symbionts, pests, parasites, predators and competitors (Qualset *et al.*, 1997). Agriculture is the largest global user of biodiversity; it is the most important land use type on earth; and the source of most of our food.

Agrobiodiversity depends on people, and people must be continually – and knowledgeably – involved in its management. This is especially so in most developing countries where over 60% of the economically active population is involved in utilizing and managing biodiversity in agriculture. Agriculture has conserved biodiversity on the hoof and as seed and planting materials over 10,000 years. Agriculture extracts value from biodiversity at each harvest or cull, but nurtures the productive and renewable base. Our future depends on optimizing existing biodiversity for productive agriculture while reducing degradation of the natural resource base and the environment.

This chapter focuses on the lessons learnt from past management and the challenges facing future management of agrobiodiversity. Ways of optimizing agrobiodiversity for productive agriculture are highlighted. We recognize that study, increased understanding, and the sustainable management of agrobiodiversity as presented in this book may well be critical not just to agricultural production, but also to the future of biodiversity globally.

Lessons from the Past

The dynamics of domesticated biodiversity

Frankel *et al.* (1995) have argued that an evolutionary continuum links wild predomesticates with present-day crop varieties, through natural selection over millions of years, domestication, selection and often wide dispersal over 10,000 years, followed by 100 years of modern plant and animal breeding. Frankel *et al.* (1995) also identify an ecological continuum: field type has also been selected from the landscape and then managed at various levels of intensity (see Wood and Lenné, Chapter 17, this volume). During these evolutionary processes, there have been progressive and often major changes in agrobiodiversity.

Millions of years of natural selection of wild relatives of crops and livestock has occurred in response to both abiotic and biotic influences. For example, the widespread occurrence of plant defence compounds and batteries of resistance genes in wild relatives of crops are evidence of powerful selective forces (Harper, 1990). We do not know how much diversity was generated or lost along the way and, to a great extent, our ignorance of this process continues to this day.

Domestication was a key event for crop (Frankel *et al.*, 1995) and animal biodiversity. Early farmers selected from a limited range of plant families and animal genera in nuclear areas of domestication (Lenné and Wood, 1991; Wood and Lenné, 1993; see Wood and Lenné, Chapter 2, and Steane, Chapter 4, this volume). Domestication was therefore strongly selective, emphasizing the quality rather than the quantity of diversity, as noted by Odum (1969). Crop varieties and animal breeds are, after all, the functional units of food production. Much of the diversity which was excluded from the crop through selection was not needed by the 'crop version' of the species (e.g. toxins, shattering heads, dormancy, perenniality, etc.) (Harlan, 1975; Simmonds, 1979). However, the fact that we often return to the wild for useful characters (Lenné and Wood, 1991) is an inevitable consequence of 'founder effects' (Ladizinsky, 1985; Pickersgill, 1989) and ongoing natural and human selection.

For example, in wheat, the single domestication event which resulted from a cross between *Triticum turgidum* and *Triticum tauschii* transformed a rather ordinary cereal into the most widely grown crop on earth (see Cox and Wood, Chapter 3, this volume). But it also produced a crop depauperate in some useful characters. Breeders return to the wild gene pool of *T. tauschii* for resistances to rusts and other diseases. Similarly, *Brassica oleracea*, which represents the C genome of the *Brassica* complex, has produced many popular vegetables, e.g. cabbage, broccoli, cauliflower, Brussels sprouts (Smartt and Simmonds, 1996). *Brassica rapa* (chinese cabbage, turnip), *B. napus* (oilseed rape, swede) and *B. juncea* (brown mustard), which represent the A and B genomes, are much richer sources of resistances to bacterial and viral diseases. In contrast, during domestication sorghum and its wild relatives continued to grow in close proximity throughout much of Africa (Doggett, 1988; Aldrich *et al.*, 1992; see Cox and Wood, Chapter 3, this volume). Virtually the entire gene pool of *Sorghum bicolor* has been available to farmers and breeders to develop varieties. These examples demonstrate the critical role played by domestication in setting the genetic boundaries for crop diversity.

Domestication was followed by up to 10,000 years of natural selection through exposure to a diversity of climates, pests, pathogens and weeds (Frankel *et al.*, 1995); human selection for specific plant and animal traits and dietary and market needs; and wide dispersal. Agriculture allowed crops and livestock to spread far beyond the range of their wild ancestors, especially in the last 500 years, exposing them to a great diversity of environments. In a classic paper, Bennett (1970) (p. 125) argued that: 'The environments which crops occupy are many times more numerous and diverse than the environments of the relatively few centres in which they originate, in spite of the general uniformity of environments under modern conditions.' Crops and domestic animals are now found well beyond the limits of ecological tolerance of their immediate wild relatives. For example, from an origin in the Near East, bread wheat is now grown very successfully in North and South America, Africa, Asia and Australia. This combination of natural and human selection accounts for the remarkable variability found among and within crop landraces and animal breeds and their extraordinary ranges of adaptation (Peel and Tribe, 1983; Wood and Lenné, 1993; see Chapters 2, 3, 4 and 9, this volume).

A tremendous diversity of crop varieties and livestock breeds have been deployed in farmers' fields since domestication. Selection by farmers was mostly based on morphological markers, often controlled by single genes (such as seed and animal colour variants). Most varieties and breeds have names, but the number of different names representing the cultivars and breeds in a given field, farm, or region is not usually a good indicator of genetic diversity (Morris and Heisey, 1997; Wood and Lenné, 1997; see Cox and Wood, Chapter 3 and Steane, Chapter 4, this volume). For example, bean seed colour is controlled by very few genes, and different colours are not a measure of genetic distinctness (Briand *et al.*, 1998). Just as high morphological diversity may be no indication of high genetic diversity, morphological uniformity cannot be equated with a lack of genetic diversity. The pedigrees of modern wheat and rice varieties may include over 50 parents including many landraces (Hargrove *et al.*, 1988; Frankel *et al.*., 1995; see Witcombe, Chapter 10, this volume). A mythology may have arisen through an over-emphasis on the value of traditional 'micro-varieties' based on morphological diversity (Wood and Lenné, 1997).

It is difficult to generalize about the effects of evolutionary processes and selection pressures on diversity for useful characters during this significant period in crop and livestock development. The human ability to identify, promote, partition, discard and enhance genetic diversity through selection was driven by the need to produce food as well as economic, social, cultural and aesthetic factors: 'So long as it produces variants which man finds pleasing and amusing, it may survive, though it be less lusty, less capable of caring for itself in the ordinary field and woodland struggle for existence' (Anderson, 1954). Intelligent decisions by the earliest farmers at the time of domestication promoted the most important plant characters which underpin food production and rejected many detrimental characters. The development of agriculture has involved a continuous process of generation and selection of new superior varieties and rejection and loss of old inferior ones. The socio-economic reasons for growing and maintaining diversity are often far more compelling than the biological reasons (Clawson, 1985; Wood and Lenné, 1997; see Smale and Bellon, Chapter 15, this volume).

There is also strong evidence that diversity for some useful characters was lacking in early crop varieties in some locations or was present at too low a frequency for farmers to select for (Wood and Lenné, 1997; see Cox and Wood, Chapter 3, this volume). In addition, farmer selection for quality characters may have resulted in selection against or random loss of useful characters such as disease resistance. Indeed, the immense variety of crop management practices developed by farmers since the beginnings of agriculture (see Thurston, 1992; Thurston *et al.*, Chapter 9 and Polaszek *et al.*, Chapter 11, this volume) strongly suggests that landraces of major staple food crops were inherently poor in the 'evenness' of some useful characters such as disease resistance (Lenné, 1998). Landraces have a population genetic structure which accommodates an indefinite range of genetic diversity; much of it may be of limited immediate value but more a store for opportunity and adversity (Frankel *et al.*, 1995).

The development of modern plant and animal breeding techniques has greatly facilitated wider use of a wealth of diversity from many sources including landraces and traditional animal breeds, and especially, has allowed food production to keep pace with population growth. However, the lack of a generally usable classification system for landraces and traditional animal breeds is a serious constraint to utilizing, managing and conserving this important component of agrobiodiversity (Wood and Lenné, 1997; see Steane, Chapter 4, this volume).

The era of modern plant and animal breeding

The development of modern plant breeding in the late 19th century initiated a process of plant introduction, evaluation and assemblage of collections of crop diversity for future use never seen before. Modern plant breeding allowed the recombination of diversity from widely different backgrounds, countries, climates and cultures in an infinite number of combinations and applied intense selection pressure to remove unwanted characters (see Cox and Wood, Chapter 3 and Witcombe, Chapter 10, this volume). It protected farmers from the unpredictable effects of environmental selection and therefore crop loss.

The high yielding varieties (HYVs) which heralded the Green Revolution were productive and profitable and billions of farmers adopted them (Tripp, 1996; also see Witcombe *et al.*, 1998). This may have led to a decrease in 'crop and varietal diversity' on farm but as many landraces and old varieties were collected and conserved in genebanks for future use, especially during the last half of this century, the actual loss of varietal, and more so, genetic diversity was probably small during this period. It was also probably substantially less than has occurred during the previous 10,000 years of crop development when farmers continuously discarded inferior varieties when superior varieties were identified. It is probable that discarded varieties possessed useful cryptic traits which could not be detected by farmers because of their low frequency in the crop population. The diversity located in genebanks has been extensively tapped for breeding programmes during the past 40 years and will continue to be used as needs arise. This is considered as 'diversity in reserve' (Duvick, 1984; see Cox and Wood, Chapter 3 and Witcombe, Chapter 10, this volume).

Modern plant breeding is supported by gene pools of currently unused cultivars, experimental lines, old varieties, ancestral taxa, and wild relatives – any genotype which can be crossed to produce new cultivars, either in the short term (within the primary gene pool) or the long term (within the secondary and tertiary gene pools – see Witcombe, Chapter 10, this volume). Both farmers and breeders can maintain diversity in reserve in diverse cropping systems, advanced lines, and genebanks to provide future useful varieties and characters to cope with changing conditions (see Chapters 3, 10 and 14, this volume).

Modern plant breeding therefore greatly increases the potential for broadening the diversity for useful traits in crops locally, regionally and globally. The genetic diversity of landraces is the most immediately useful part of crop biodiversity. However, wild relatives have also contributed many useful traits to some crops and in the process, increased the diversity in the crop gene pool. A major impediment to further progress in utilizing valuable biodiversity is the lack of evaluation of wild relatives. Ongoing methodological improvements, such as molecular marker technologies, for accessing useful characters from exotic sources (including unrelated plants and microorganisms) will continue to improve the potential and efficiency for using a wider spectrum of diversity for useful characters in most staple crops in future (Tanksley and McCouch, 1997; Staskawicz, 1998; see Witcombe, Chapter 10, this volume). This utilization of more and more sources of traits in crop improvement programmes has resulted in an overall increase in the crop biodiversity for use by farmers. Farmers in developing countries also continue to grow varietal mixtures as well as HYVs together with traditional varieties for sound socio-economic reasons (Brush, 1981; Smithson and Lenné, 1996; Wood and Lenné, 1997; Witcombe et al., 1998; see Thurston et al., Chapter 9, this volume), further increasing diversity on-farm. In fact, recent studies have shown that the farmers' distinction between traditional and modern varieties is often unclear (see Thurston et al., Chapter 9, this volume).

Domesticated animal diversity is needed to ensure that variation is accessible for future improvement (see Steane, Chapter 4, this volume). Changes in feed availability, disease, climate and increasing demands for live-stock products exert pressures to achieve the most efficient production system possible. Traditional animal breeds are now considered to be sources of useful genetic variation (Frankham, 1994) and are being increasingly used in breeding programmes. This will optimize the diversity for useful traits in important livestock breeds in support of productive agriculture.

The realization by plant breeders in the 1950s that the productivity of staple food crops was limited genetically stimulated considerable effort in increasing the genetic diversity of useful traits, especially in wheat and rice. Emphasis was placed on improving plant type and increasing resistance to important pests and diseases to increase yields and reduce losses. For example, stem and leaf rusts have been the most intractable diseases of wheat since domestication (Heisey et al., 1997) and have caused catastrophic losses of food and resulting famine for millennia (Carefoot and Sprott, 1969). During the 1960s–1980s, a major effort was made in developing countries to improve the resistances of wheat and rice to important diseases and pests. Improved wheat varieties have resistances to rusts, leaf blotches, scab, bacterial leaf streak, barley yellow dwarf virus, tan spot and karnal bunt (Rajaram et al., 1988; Heisey et al., 1997; Smale et al., 1998) while improved rice varieties have resistances to leafhoppers, stemborers, gall midge, blast, bacterial leaf blight and several viruses (Evenson et al., 1996; Pingali et al., 1997).

The increased diversity for pest and disease resistances and widespread availability of improved HYVs of staple cereals (a key component of Green Revolution technologies) have made a major contribution to increased genetic diversity for useful characters on-farm and in reserve; food production; decreased food prices; and increased profitability (Lipton and Longhurst, 1989; Greenland, 1997; Harrington, 1997; Pingali et al., 1997; Smale et al., 1998). There has been a substantial increase in the diversity of useful traits in widely grown wheat and rice varieties during the past 30 years. For example, the number of landraces in the background of popular wheat varieties increased from five to 50 during 1960–1990 (Frankel *et al.*, 1995). Similarly, the number of landraces in the backgrounds of rice varieties released from 1966–1994 increased from four to 46 (see Witcombe, Chapter 10, this volume). Many recently released HYVs have considerable diversity through extensive pedigrees and multiple stress resistances from multiple sources (Greenland, 1997). This has also resulted in greater yield stability (Hazell, 1989; Wood, 1996; see also section on future challenges).

The importance of crop-associated biodiversity

Crop-associated biodiversity – the organisms associated with the crop in the agroecosystem – includes harmful (pests, pathogens and weeds) and beneficial (pollinators, symbionts, natural enemies, etc.) components, both above- and below-ground. To date, the critical role of associated biodiversity as part of agrobiodiversity has not been sufficiently appreciated and this book analyses this in some detail in Chapters 5, 6, 7, 11 and 12. Although all of our major food crops (including wheat, rice, maize, potato, barley, cassava, sugarcane, sweet potato, sorghum and soybean) are either self- or wind-pollinated or clonally propagated, nutritionally valuable crops such as fruits and some grain and pasture legumes rely on insect pollinators for food production (Allen, 1974; see LaSalle, Chapter 7, this volume).

The many and complex interactions between the crop and the harmful and beneficial components in modern and traditional cropping systems are major determinants of whether there will be more or less food, feed and fuel. Such interactions may be relatively simple, such as foliar pests and pathogens directly damaging crops but are often quite complex, such as endophytes within pasture grasses reducing damage from foliar insects and, at the same time, suppressing or stimulating root infection by mycorrhizal fungi that may confer resistance to root-feeders or hyper-parasites parasitizing natural enemies of foliar insect pests which cause damage to crops (see Chapters 5, 7 and 11, this volume). Such associations are sensitive to environmental conditions and agricultural management practices and it is unfortunate that not enough is known about their evolutionary history. In addition to co-evolved associations, crop introduction, especially the intercontinental movement of major food crops, has placed crops in association with exotic associated biodiversity (Buddenhagen, 1977; see also Chapters 6, 7 and 11, this volume).

Interactions which occur between coevolved plants and pathogens in both natural ecosystems and agroecosystems range from minor to epidemic (Burdon, 1987, 1993; Dobson and Crawley, 1994). Wild and weedy relatives can be important sources of pathogen and pest inoculum for crops (Lenné and Wood, 1991) and there are numerous examples of pathogens and pests from wild ecosystems moving to crops (see van Emden and Williams, 1974;

Thresh, 1981). There is no doubt that the evolution of disease and pathogen variability has played an important role in fostering diversity in ecosystems (Gillett, 1962; Burdon, 1987; see Polaszek *et al.*, Chapter 11, this volume). Much of the diversity for disease resistance present in crops today represents the 'ghosts of diseases past' (Harper, 1990). There is growing evidence that at least some resistances may be derived from interactions between ancestral plant–pathogen associations (Gale, 1994; Allen *et al.*, 1998).

With knowledge generated by research, associated biodiversity can be managed: promoted if useful, controlled if harmful. Perhaps the best examples of using beneficial associated biodiversity to manage harmful biodiversity are of biological control of crop pests. The most dramatic example of successful application of classical biological control as a pest management strategy is the control of the cassava mealybug in sub-Saharan Africa. Biocontrol of the mealybug resulted in enormous savings in the production of cassava, with expected benefits over a 40-year period of £6 billion from a total expenditure of £40 million, a cost : benefit ratio of 1500:1 (Herren and Neuenschwander, 1991; see LaSalle, Chapter 7, this volume). Reestablishment of effective natural biological control of the rice brown planthopper in Southeast Asia, especially Indonesia, in the late 1980s through substantial reduction in the use of pesticides in irrigated rice systems is a good example of using beneficial associated natural biodiversity to manage harmful biodiversity (Perfect, 1998).

It is probable that the management of fields, to increase or stabilize crop production, has had more profound impacts on crop-associated agrobiodiversity than the diversity of the actual crop itself. Settle *et al.* (1996) give an example of the indirect effect of rice production on the insect agrobiodiversity of the irrigated rice field. As the rice crop develops, detritus from previous crops in the flooded field allows the build up of detritus feeders; these are then preyed upon by predators, whose populations build up to a level that permits the early and effective control of rice pests (see also Way and Heong, 1994 and Edwards *et al.*, Chapter 12, this volume, for more on the agrobiodiversity of rice systems).

In most agroecosystems, the majority of the biodiversity present occurs below the soil surface, not above it (Beard, 1991; Giller *et al.*, 1997). Below-ground organisms are critical in determining the functioning of agroecosystems, influencing plant growth and the productivity of the agroecosystem in the long term (Ingham *et al.*, 1985; Lavelle, 1994; see Wardle *et al.*, Chapter 5, this volume). For example, root herbivores in many terrestrial ecosystems may consume two to ten times more than the total above-ground mass although the effects of below-ground herbivory remain largely unseen unless productivity is severely reduced (French *et al.*, 1979; Masters *et al.*, 1990).

Agricultural practices such as tillage (see Edwards *et al.*, Chapter 12, this volume) can induce shifts in the composition of the soil biota, however there is little evidence that agricultural intensification has had predictable, harmful consequences for the diversity of organisms in the soil (see Wardle *et al.*, Chapter 5, this volume). Indeed, soil organisms are also often subjected to repeated wetting and drying cycles, variations in substrate inputs and in crop growth, which are sometimes more significant than agricultural practices (Domsch *et al.*, 1983; Wardle and Parkinson, 1991; MLURI, 1997). The extent to which soils can be abused, and yet still continue to produce yields indicates the robust nature of associated biodiversity below ground (see Wardle *et al.*, Chapter 5, this volume).

The manipulation of beneficial associated biodiversity to manage harmful associated biodiversity provides *additional* approaches to proven genetic approaches to optimize biodiversity for productive agriculture. The interaction between the crop, its associated biodiversity and environmental and management factors has no doubt resulted in evolutionary changes in both. Understanding the functions of different components and the interactions between them (for example, the interaction between host plant resistance and biological control of insect pests – see Thomas and Waage, 1996) will be important both for the wise management of associated biodiversity for productive agriculture and sound management of the environment.

Partitioning of diversity in agroecosystems

Diversity among species in agroecosystems is important because of the complementary contributions that crop and animal species make to the human diet, as well as to the functioning and productivity of agroecosystems of which they are a part. Niche expansion will have a direct impact in enhancing the productive agrobiodiversity of farms, allowing crops from many different ecogeographical origins to be brought together and to add diversity to any one agroecosystem. The productivity of cropping systems can also be increased by optimally partitioning diversity in space and time through the use of multiple cropping and intercropping (Francis, 1986; see Cox and Wood, Chapter 3, and Thurston *et al.*, Chapter 9, this volume), and there is often a positive effect of genetic diversity for pest resistance in mixtures and multilines in buffering the spread and severity of pathogens and pests as well as increased yield stability (Marshall, 1977; Wolfe, 1985; Andow, 1991; Smithson and Lenné, 1996; see also Chapters 6, 7, 9 and 11, this volume) but it is often associated with only minor increases in productivity. The degree of the relationship between diversity and increased food (vs. biomass) production merits much more study in agroecosystems (Allen, 1974; Marshall, 1977; Frankel *et al.*, 1995; Smithson and Lenné, 1996). Indeed, the majority of studies on diseases, for example, have targeted specialized cereal systems in temperate regions. We suspect that the same principles will apply to subsistence agriculture in tropical regions where diverse cropping systems are extensively grown. However, the gains in production from temporal and spatial diversity within fields may be countered by the extended and overlapping seasons and the close proximity of neighbours' fields which will exacerbate diseases. Clearly this is an area where further research is needed.

Much of the perception of stability in diverse agroecosystems is illusory (Janzen, 1973) and 'the predisposition to expect greater stability in complex ecosystems is probably a combined legacy of eighteenth century theories of political economics . . . and perhaps religiously motivated attraction to the belief that the wondrous variety of nature must have some purpose in an orderly world . . .' (Goodman, 1975; see also Wood and Lenné, Chapter 17, this volume). As noted by Way and Heong (1994), there is an uncritical acceptance that decreased diversity, as in crop monocultures, increases instability by encouraging pest outbreaks and crop losses. In fact, increased vegetational diversity in some agroecosystems may increase rather than decrease pest and disease problems (Way, 1966, 1977; Southwood and Way, 1970; van Emden and Williams, 1974; see Polaszek *et al.*, Chapter 11, this volume) and simple communities may not be unstable (May, 1973; Goodman, 1975; Murdoch, 1975; Pimm, 1984; Way and Heong, 1994; see Chapters 6, 7, 11 and 17, this volume). *Phytophthora cinnamomi* is among the world's most destructive pathogens of diverse native vegetation, destroying almost 60 species in 13 botanical families in southern Australia's forests (Newhook and Podger, 1972; Weste and Marks, 1987). The expectations of the diversity–stability hypothesis, including for pest management strategies, are borne out neither by experiments, by observation, nor by models, and its theoretical formulations have no necessary connection with secure scientific law (Goodman, 1975).

Most of the debate on diversity and stability relates to plant diversity and, in some cases, its interaction with pests. The effects of soil biodiversity on ecosystem stability remain entirely unresolved and results from recent studies are ambiguous. Further, key questions such as the extent to which the composition and diversity of the soil biota may affect such things as the stability of nutrient cycles, consistency of crop production across years, and crop resistance and resilience to disturbances (e.g. drought) remain to be investigated (see Wardle *et al.*, Chapter 5, this volume).

Different agroecosystems support widely differing levels of diversity above and below ground and in the crop and associated biodiversity components. It is probable that agroecosystems which promote a greater heterogeneity of plant residues or organic matter inputs contribute to maintaining a greater diversity of soil organisms (Wardle *et al.*, Chapter 5, this volume). But we may well ask: is it necessary for diversity to be equally partitioned among all components of the agroecosystem for productive agriculture? The irrigated rice ecosystem is perhaps the best understood example where uniform cropping conditions and stable water supply allow manipulation of relatively few manageable components of diversity (specifically natural enemies) so that pests are controlled and high production is assured (Way and Heong, 1994). Optimizing agrobiodiversity for productive agriculture should be based on developing similar understanding of other agroecosystems, especially those in which most of our staple crops are grown. Clearly this is another area where further research is needed.

Conservation of agrobiodiversity

Crop and livestock biodiversity is a resource for present and future food production. Crop plants can be conserved *ex situ*, usually as seed or field genebanks, or *in situ* on farm (see Engels and Wood, Chapter 14, this volume) while animal breeds are perhaps best conserved *in situ* (see Steane, Chapter 4, this volume). The strategy used to prioritize what to conserve should be based on the evolutionary history of the crop: for example, whereas conservation of landraces may be a high priority in sorghum, wheat's future may depend more on maintenance of the Middle Eastern ecosystems that harbour its wild relatives (see Cox and Wood, Chapter 3, this volume).

Although *ex situ* conservation is regarded as 'static' conservation, each sample represents many thousands of years of evolution and past interactions with diverse biotic and abiotic influences. On-farm conservation is better described as *de facto in situ* conservation. Saving seed for the next season is essential for crop production by farmers who either have no resources to purchase seed or no access to formal seed systems (see Wright and Turner, Chapter 13, this volume). *In situ* conservation is considered to be 'dynamic', but the rate or extent of evolution for useful characters on farm is unknown and perhaps no obvious changes may be observable within 100 years or more (Maxted *et al.*, 1997). Genetic diversity is more likely to be lost through *in situ* rather than *ex situ* conservation strategies (see pp. 112–113 in Maxted *et al.*, 1997).

Effective integration of *ex situ* and *in situ* conservation is needed to allow valuable diversity identified *in situ* to be conserved *ex situ* for future use. Various possibilities for optimizing conservation of crop and animal biodiversity involve different levels of intervention by farmers and scientists (Brown, 1991; Witcombe *et al.*, 1998; see Chapters 4, 10 and 14, this volume). For example, for crops, mass reservoirs (populations derived from diverse parental lines or landraces) could be developed to accelerate coevolution with the prevailing stresses (Allard, 1990). To support coevolution between plants and pathogens, Browning (1974) suggested the establishment of 'pathogen parks'. Scientists could work directly with farmers to characterize diversity for useful characters on-farm and in wild populations (backed up by controlled evaluation). Farmers would benefit from superior lines selected from their populations, increasing the probability of their continued maintenance of diversity, and global food production would benefit from increased knowledge of diversity for useful characters.

Whatever option is chosen, care must be taken that the strategy to conserve crop associations with harmful biodiversity does not lead to increased crop losses (Ingram, 1998; Lenné, 1998). Human population pressures and the need to produce more food will restrict the scope for conventional *in situ* on-farm conservation. If growing specific landraces on-farm is no longer seen as profitable, small farmers will change to more productive varieties (Tripp, 1996; see Smale and Bellon, Chapter 15, this volume). Conventional *in situ* conservation is, in most cases, probably not sustainable and it is imperative that it is supported by other methods.

Shortcomings of current biodiversity initiatives

Agriculture – the scientific basis, the knowledge base and the practice, both traditional and formal – has been seriously neglected by ecologists and conservation biologists. This neglect has even extended to topics on which agricultural knowledge could provide essential guidance to the biodiversity agenda, including the behaviour of introduced species (most crop and animal production is from such species); genetic mapping; mating systems; plant and animal physiology and stress tolerance; plant and animal health; pest management strategies (including quarantine); the rhizosphere; *ex situ* conservation; and not least, indigenous knowledge as a key to environmental management. This neglect needs to be clearly articulated by the global scientific community.

The 1992 Convention on Biological Diversity (CBD) recognized for the first time the importance of agrobiodiversity and the need to ensure its sustainable use and conservation in meeting future food needs. The associated Agenda 21 document also gave extensive coverage to agriculture. Despite paying some attention to agrobiodiversity, the global biodiversity agenda is still focused on the conservation of wild species. Indeed, despite the potential for the CBD to redress the neglect of agrobiodiversity, the former tensions between agricultural production and biodiversity conservation reemerged during a meeting of the Conference of the Parties (COP) in Buenos Aires in 1996. A position paper prepared for the meeting (UNEP, 1996) noted that 'modern intensive agriculture has made it possible for the ever-increasing human population to be fed without the extensive destruction of habitats to provide the needed food' (para. 21). However, para. 22 noted 'modern agriculture . . . has been responsible for considerable damage to biological diversity'.

In May, 1998, the COP-IV called for 'efforts to identify and promote sustainable agricultural practices, integrated landscape management of mosaics of agricultural and natural areas as well as appropriate farming systems that will reduce possible negative effects of agricultural practices on biological diversity and enhance the ecological functions provided by biological diversity to agriculture'.

Despite the high international profile given to agriculture in the CBD process and its continued emphasis through four COP events, new funding for research to improve the understanding and enhance the sustainable utilization and conservation of agrobiodiversity has largely been lacking (as noted by Wood and Lenné, Chapter 1, this volume). The Global Environment Facility (GEF) of the World Bank/UNEP/UNDP has an 'Initiative on Agrobiodiversity' which has earmarked US$160 million specifically for projects on agrobiodiversity in four ecosystems: semi-arid and arid; forest; mountain; and coastal and wetland systems. Few projects have been funded as yet. Unfortunately, there appears to be limited interest in some agro-ecosystems which are more important for food production: monsoon rainfed, irrigated valley bottoms, and subhumid savannahs, indeed, the many high potential areas which probably contain as valuable an agrobiodiversity as arid, forest, mountain and wetland systems.

The changing approach to desertification in international environmental policy shows that a rigorous research agenda should be applied as early as possible if time and funding are not to be needlessly wasted on unfounded concepts. This caution will apply strongly to agrobiodiversity.

Challenges for the Future

This book is especially timely in the light of the challenges facing global food production during the next 20–30 years, especially the need to produce more food from less land using technologies which have minimal adverse environmental impacts. The joint objectives of increasing food, feed and fibre production and optimally using and conserving agrobiodiversity will require a range of different approaches. The objectives of growing diversity may differ sharply between different farmers and different agroecologies; there may be conflicts of interests between local, national and global needs for diversity; and there may be a cost to maintaining diversity whose future role in crop evolution is uncertain. It is likely that trade-offs will be necessary to support the development of technologies which favour increased food production in high potential production systems, at the expense of local biodiversity, but to the benefit of regional and global biodiversity. Such choices – the success of which will depend on acceptance by farmers – should be based on sound science and knowledge and not on speculation.

Increased food production

Food production has tripled in the past 40 years, outpacing population growth (Lipton and Longhurst, 1989; Conway, 1997). History records no increase in food production that was remotely comparable in scale, speed, spread and duration. Much of the increase in food production globally during the past 40 years can be attributed to research, with internal rates of return of 30–50% (Pinstrup-Andersen *et al.*, 1997). Agricultural research has played a key role in overall economic development of many countries.

Agricultural research is 'the best hope – the only hope – of winning the race between population and food' (Sen, 1975). This view is supported by millions of small, poor farmers who have adopted improved HYVs of staple food crops with a diversity of useful characters (Tripp, 1996).

The development and continuing deployment of HYVs, in preference to landraces, is driven by the need to produce more food. World population will grow substantially over the next 20 years and 95% of this growth will occur in developing countries. The world's farmers will have to rapidly increase food production as in the next two generations the world will consume as much food as has been consumed in the entire history of human-kind. By 2015, a 70% increase in global rice production (about 350 Mt more rice) will be needed to feed the additional mouths and raise nutritional standards to a satisfactory level in major rice-consuming countries (Pingali *et al.*, 1997; Pinstrup-Andersen *et al.*, 1997). It is estimated that the demand for meat and milk will rise by 300% and 155%, respectively, by 2020 (Rosengrant *et al.*, 1995; Pinstrup-Andersen *et al.*, 1997; see Steane, Chapter 4, this volume).

However, available arable land per person is declining rapidly (from 0.5 to 0.25 ha per person during the last 40 years) and will be reduced to 0.1 ha per person during the next 50 years (Krattiger, 1998). Cropping intensity has increased in many developing countries, especially where there is access to irrigation. For example in Bangladesh, the cropping intensity is over 180% and Nigeria, Kenya and Ethiopia now surpass major Asian countries in intensity of land-use (Byerlee and Eicher, 1997). Cropping intensity will need to increase on the best land as the pressure to produce more food from the same amount of land increases rapidly. This presents an immense challenge to the world's food production capability. Future optimal utilization of agrobiodiversity must respond to this overriding need to produce more food and to reduce pressure on natural ecosystems with useful biodiversity (e.g. wild relatives of crops and domestic animals). Technologies which optimize crop, animal and associated biodiversity have an increasingly im-portant role to play in the future.

Integrated crop management

Characterization, utilization and management of agrobiodiversity is ultimately about developing integrated crop management strategies which optimally use genetic diversity and encourage parasitoids, predators, symbionts and antag-onists to increase yields and manage harmful pests and pathogens. Success will be dependent on multidisciplinary approaches operationalized through integrated activities for diagnosis, assessment and monitoring of agro-biodiversity; understanding functions and interrelationships between and among different components of agrobiodiversity; and developing management options which optimize biodiversity for productive agriculture.

Enhanced genetic resistance and integrated pest management (IPM) technologies will greatly contribute to increased and stabilized food production without increased use of pesticides, which harm the environment. For example, improved crop management methodologies as well as advances in the development and wide-scale application of IPM technologies in rice have resulted in insecticide use falling substantially (Pingali and Gerpachio, 1997) to the benefit of associated predators and parasites. The agricultural productivity increases needed to lift the populations of low-income developing countries out of poverty and food insecurity without doing irreparable damage to the environment and natural resources will be possible only, however, if the appropriate government policies are pursued and continued investment in agricultural research is supported.

Transgenic crops

From 1996 to 1998, there has been a 15-fold increase in the global area (in industrial and developing countries) of transgenic crops to 27.8 Mha (James, 1998). Adoption rates for transgenic crops are some of the highest for any new technology in agriculture and reflect farmer satisfaction with the significant and multiple benefits of higher productivity and reduced pesticide use. Expected wider availability of transgenic crop varieties with insect and disease resistance and herbicide tolerance in the near future will further reduce pesticide use. Today's insect resistance technology could already substitute for £2.3 billion worth of the £7 billion market in insecticides used on major crops as well as increasing yields by 5–10% (Krattiger, 1998) and with substantial environmental and human benefits. Herbicide technology could reduce the degrading drudgery of weeding for the disadvantaged parts of the labour force (women) who, in sub-Saharan Africa, may spend up to 80% of their working week pulling weeds (see Gressel, 1996). In addition, the potential for biotechnologies to access more distant sources of useful traits will broaden the diversity for such traits in the crop gene pool. It is anticipated that the greatest contribution from transgenic crops will be in improving yield stability through resistances to pests and tolerance to abiotic stresses (see Polaszek *et al.*, Chapter 11, this volume).

More stable, less risk-prone agriculture

Sound crop management practices which optimize the use of agrobiodiversity can contribute to more stable, less risk-prone food production to the benefit of all. The coefficient of variation (cv) for global rice production and global rice yields has declined from 1960 onwards (Hazell, 1989). In six of eight countries surveyed in South and Southeast Asia (India, Thailand, Bangladesh, Pakistan, Vietnam and the Philippines), yields have become significantly more stable, partly due to improved crop management practices, including increased genetic diversity for useful characters. Hazell (1989) specifically drew attention to the yield and production stability in rice and wheat, the two crops which are the flag-bearers of the Green Revolution. The higher and more stable yield potential and profitability of Green Revolution wheat and rice varieties permits farmers to invest in inputs for producing even more food (Wood, 1996). For wheat and rice, much of this stability is related to the widespread use of varieties with durable and multiple resistances to major pests and pathogens as well as the deployment of efficient IPM practices (Rejesus *et al.*, 1996; Greenland, 1997; Pingali *et al.*, 1997).

Protection of the environment and the natural resource base

The contribution of agricultural research, including the sound use of agro-biodiversity, to reducing pressure to cultivate biologically diverse, fragile, marginal or forested areas and protect the natural resource base has not been widely recognized (Harrington, 1997). The diffusion of improved rice and wheat technologies has resulted in sustained economic growth, especially in Asia, and the saving of many millions of hectares from the plough and axe. Without improved agricultural technologies, another 350 Mha would have had to be cropped to produce enough food for the world's population – an area about the size of India (Krattiger, 1998) – which would have resulted in an environmental disaster of immense proportion (Thompson, 1998). In India alone, another 40 Mha would have been needed to meet the demand for rice and wheat (Harrington, 1997). In China also, the cultivated cereal area would have had to increase threefold (Borlaug, 1998). Farming would have expanded on to highly erodible soils and forests and other natural vegetation would have been destroyed and with it much biodiversity. There is repeated emphasis to this danger in the agricultural literature, for example, Abdulai and Hazell (1995) argue that a move to extensive farming systems would lead to mismanagement of resources and damage environmentally fragile areas.

Specialization and intensification in agriculture on high quality land could be the best hope for maintaining increased productivity and reducing damage to marginal ecosystems (Borlaug, 1998). The hallmark of 20th century agriculture has been the widespread adoption – by both poor and rich alike – of yield-increasing, land-saving technologies, which have permitted agricultural research to keep pace with rapid population growth. In addition, intensification has taken pressure off the useful biodiversity of natural vegetation, which greatly enhances livelihood options for rural people (e.g. gathering fuel; collecting wild foods such as honey and herbs to improve diet; and harvesting natural products for crafts, such as basket making, for income generation).

Severe food shortages, increased poverty, livelihood insecurity, political and social instability and natural resources and environmental degradation could occur in many developing countries early in the next century unless more food is produced and prices are kept low through continued support for agricultural research and optimal use of agrobiodiversity. This will need considerably more research to understand and exploit the potential of agrobiodiversity without damaging nature and, as part of nature, agrobiodiversity will remain a sensitive indicator of the health of agroecosystems. There is no alternative for feeding the world's population except intensification (Krattiger, 1998; Thompson, 1998). This is now widely recognized by researchers but not yet by many policymakers.

The need to inform policy decisions

Our greatest concern is that the rich knowledge base for agrobiodiversity demonstrated in this book will not be translated into effective agricultural research and development policy. We are now awash with technically dubious environmental and biodiversity projects, all taking scarce funding from agriculture. This is not a problem of lack of information, but of presentation by scientists – and indeed farmers, whose activities are also misrepresented (Chibnik, 1981). International bodies and funding agencies are lobbied daily with persuasive and often inaccurate or scientifically naive information. With regard to the similar debate over forests of a decade ago, the tree geneticist Namkoong (1991) chastised forest scientists:

> We might ask if we have been forced into this [weak] position by vociferous advocates of the extremes or if we have simply abdicated the position by our silence. . . . By hesitating to enter the debate, we only accede the field to the biologically naive and find ourselves able to serve only as peripherally significant technicians in pursuit of the objectives of the uninformed.

As an example of policy needs, pesticides remain a concern in agricultural policy (see Gisselquist, Chapter 16, this volume). Pressing forward with the right policies on pesticides could accelerate crop productivity and ensure sustainable management of agrobiodiversity. The wrong policies will deny poor farmers access to the right inputs, such as pest-resistant crops, and allow them subsidized access to damaging inputs, such as Category I pesticides. The indiscriminate and injudicious use of pesticides in developing countries, especially in the 1980s, was a direct result of inadequate pesticide policies (Pingali and Gerpachio, 1997). Such policies will lead to further environmental degradation, reduced crop productivity, and food insecurity. Enlightened policies, together with the introduction and development of pest-resistant varieties, have generally resulted in reduced pesticide inputs, use of safer chemicals and improved farmer safety.

Over pesticides and many other issues, agriculture is now a battleground of lobbyists: but much of this lobbying is based on the impacts of agricultural practices on the immediate agricultural environment – that is, on agrobiodiversity. Farmers and agricultural scientists have a broader and very practical responsibility to the environment: an environment damaged will in turn damage future food production. Agrobiodiversity is a sensitive indicator of sustainable agricultural and environmental policies. A knowledge of agrobiodiversity itself, and of its links with nature, is therefore an essential tool first to quantify and then address legitimate concerns over agricultural practices, and then to develop environmentally sound agricultural practices. There is an urgent need to reduce uncertainty and to increase the efficiency of decision-making over agriculture. Scientists characteristically draw back from policy issues, but the credibility of agricultural science and its continued funding depend on an honest perception and vigorous presentation of policy advice.

Valuable time in international fora is increasingly being diverted to single-issue campaigns, often with a distinct anti-technology slant. Much of this lobbying is well-meaning, but ill-informed; some lobbying is misrepresentation; very little contributes anything of value to agricultural policy. Further sound research on agrobiodiversity, and more importantly, its presentation and promotion, is urgently needed as a basis for policy decisions, especially where trade-offs are inevitable. As an example of research needs to support policy, Tinker (1988) contrasted two scenarios for European crop production: one with all high potential areas used for 'efficient intensive farming . . . avoiding the grossest forms of environmental damage' with 'low-productivity areas supported by Government for mainly environmental objectives'; the second where all crops were grown under 'less-intensive agriculture'. A greater knowledge of agrobiodiversity is a key to efficient decision-making between and within these extreme scenarios. This is increasingly relevant to developing countries, as public research declines and agricultural inputs cannot always be ensured.

We should remember the message of Schultz (1964). Farmers are 'poor but efficient'. It will not be enough to exhort farmers to recycle components of their own indigenous technology and detailed knowledge of agrobiodiversity: they have done this over time and space to the limit of possibilities. Agricultural development will depend on the introduction of new factors of production, to take advantage of the 'dynamic responsiveness of farmers' (Ball and Pounder, 1996). A synthesis of information on agrobiodiversity will allow the development and introduction of appropriate technology to farming systems in a manner that can benefit farmers. This can increasingly be a precision approach, rather than the present blunt-instruments of development policy – one of which is the questionable attempt to 'maximize biodiversity' in farming. An increasing range of 'agrobiodiversity options' will allow farmers to continue to make rational decisions.

It is inevitable with continued population growth that further agricultural intensification on the best land will be needed to produce sufficient food. The only alternative in the short term is to expand the agricultural frontier into often biodiversity-rich marginal land. This is not only *not* likely to produce sufficient food but is also likely to cause substantial loss of biodiversity and irreparable damage to the environment. A knowledge of agrobiodiversity in a wide range of agroecosystems is essential to ensure that increased food production can be sustained. This book aims to redress the past neglect of agrobiodiversity and demonstrate that the long and productive history of the human management of agrobiodiversity can provide a corpus of knowledge and practice which is both of supreme value in its own right for productive agriculture, and also of the greatest value as a model for wider biodiversity conservation and utilization.

Acknowledgements

Parts of this chapter are based on unpublished reports prepared for the DFID Crop Protection Programme and CABI Bioscience. The first author is grateful for their support.

References

Abdulai, A. and Hazell, P. (1995) The role of agriculture in sustainable economic development in Africa. *Journal of Sustainable Agriculture* 7, 101–119.

Aldrich, P.R., Doebley, J., Schertz, K.F. and Stec, A. (1992) Patterns of allozyme variation in cultivated and wild *Sorghum bicolor. Theoretical and Applied Genetics* 85, 451–460.

Allard, R.W. (1990) The genetics of host–pathogen coevolution: implications for genetic resource conservation. *Journal of Heredity* 81, 1–6.

Allen, R. (1974) Does diversity grow cabbages? *New Scientist* 29 August, 528–529.

Allen, D.J., Lenné, J.M. and Wood, D. (1998) New encounter diseases and allopatric resistance. In: *Seventh International Congress of Plant Pathology*, Edinburgh, Abstract 4.1.5.

Anderson, E. (1954) *Plants, Man and Life*. Melrose, London.

Andow, D.A. (1991) Vegetational diversity and arthropod population response. *Annual Review of Entomology*, 36, 561–586.

Ball, D. and Pounder, L. (1996) 'Efficient but poor' revisited. *Economic Development and Cultural Change* 44, 736–760.

Beard, J. (1991) Woodland soil yields a multitude of insects. *New Scientist* 131(1784), 14.

Bennett, E. (1970) Tactics in plant exploration. In: Frankel, O.H. and Bennett, E. (eds) *Genetic Resources in Plants – their Exploration and Conservation*. IBP Handbook No. 11, Blackwell Scientific Publications, Oxford, pp. 157–179.

Borlaug, N.E. (1998) Food security, plant pathology and quarantine. Public Forum Discussion on the Role of Plant Pathology in Global Food Security, 7th International Congress of Plant Pathology, Edinburgh, August, 1998.

Briand, L., Brown, A.E., Lenné, J.M. and Teverson, D.M. (1998) Random amplified polymorphic DNA variation within and among bean landrace mixtures (*Phaseolus vulgaris* L.) from Tanzania. *Euphytica* 102, 371–377.

Brown, A.D.H. (1991) Population divergence in wild crop relatives. *Israel Journal of Botany* 40, 512.

Browning, J.A. (1974) Relevance of knowledge about natural ecosystems to development of pest management programs for agro-ecosystems. *Proceedings of the American Phytopathological Society* 1, 191–199.

Brush, S.B., Carney, H.J. and Huaman, Z. (1981) Dynamics of Andean potato culture. *Economic Botany* 35, 70–88.

Buddenhagen, I.W. (1977) Resistance and vulnerability of tropical crops in relation to their evolution and breeding. *Annals of the New York Academy of Sciences* 287, 309–326.

Burdon, J.J. (1987) *Diseases and Plant Population Biology*. Cambridge University Press, Cambridge.

Burdon, J.J. (1993) The structure of pathogen populations in natural plant communities. *Annual Review of Phytopathology* 31, 305–323.

Byerlee, D. and Eicher, C.K. (1997) *Africa's Emerging Maize Revolution*. Lynne Rienner Publishers, London.

Carefoot, G.L. and Sprott, E.R. (1969) *Famine on the Wind: Plant Diseases and Human History*. Angus & Robertson, London.

Chibnik, M. (1981) Small farmer risk aversion: peasant reality or policymakers' rationalization? *Culture and Agriculture* 10, 1–5.

Clawson, D.L. (1985) Harvest security and intraspecific diversity in traditional tropical agriculture. *Economic Botany* 39, 56–67.

Conway, G. (1997) *The Doubly Green Revolution: Food for All in the 21st Century*. Penguin, London.

Dobson, A. and Crawley, M. (1994) Pathogens and the structure of plant communities. *Trends in Ecology and Evolution* 9, 393–398.

Doggett, H. (1988) *Sorghum*. John Wiley & Sons, New York.

Domsch, K.H., Jagnow, G. and Anderson, T.-H. (1983) An ecological concept for the assessment of side-effects of agrochemicals on microorganisms. *Residue Reviews* 86, 65–105.

Duvick, D.N. (1984) Genetic diversity in major farm crops on the farm and in reserve. *Economic Botany* 38, 161–178.

van Emden, H.F. and Williams, G.C. (1974) Insect stability and diversity in agro-ecosystems. *Annual Review of Entomology* 19, 455–475.

Evenson, R.E., Herdt, R.W. and Hossain, M. (1996) *Rice Research in Asia: Progress and Priorities*. CAB International, Wallingford, and IRRI, the Philippines.

Francis, C.A. (ed.) (1986) *Multiple Cropping Systems*. Macmillan, New York.

Frankel, O.H., Brown, A.H.D. and Burdon, J.J. (1995) *The Conservation of Plant Biodiversity*. Cambridge University Press, Cambridge.

Frankham, R. (1994) Conservation of Genetic Diversity for Animal Improvement. *Proceedings of the 5th World Congress on Genetics Applied to Livestock Production*, 21, 385–392.

French, N.R., Steinhorst, R.K. and Swift, D.M. (1979) Perspectives in grassland ecology. In: French, N.R. (ed.) *Perspectives in Grassland Ecology*. Springer-Verlag, New York, pp. 59–87.

Gale, M. (1994) Genetics, markers, maps and wheat breeding. *Journal of the Royal Agricultural Society of England* 155, 162–176.

Giller, K.E., Beare, M.H., Lavelle, P., Izac, A.M.-N. and Swift, M.J. (1997) Agricultural intensification, soil biodiversity and agroecosystem function. *Applied*

Soil Ecology 6, 3–16.

Gillett, J.B. (1962) Pest pressure, an underestimated factor in evolution. *Systematics Association Publications* 4, 37–46.

Goodman, D. (1975) The theory of diversity–stability relationships in ecology. *Quarterly Review of Biology* 50, 237–266.

Greenland, D.J. (1997) *The Sustainability of Rice Farming.* CAB International, Wallingford, and IRRI, the Philippines.

Gressel, T. (1996) Plant biotechnology can quickly offer solutions to hunger in Africa. *The Scientist* 19, 10.

Hargrove, T.R., Cabanilla, V.L. and Coffman, W.R. (1988) Twenty years of rice breeding. *BioScience* 38, 675–681.

Harlan, J.R. (1975) *Crops and Man.* American Society of Agronomy, Madison, Wisconsin.

Harper, J.L. (1990) Pest, pathogens and plant communities: an introduction. In: Burdon, J.J. and Leather, S.R. (eds) *Pest, Pathogens and Plant Communities.* Blackwell Scientific Publications, Oxford, pp. 3–14.

Harrington, L. (1997) Diversity by design. *CGIAR News* 4, 5–8.

Hazell, P.B.K. (1989) Changing patterns of variability in world cereal production. In: Anderson, J.R. and Hazell, P.R. (eds) *Variability in Grain Yields: Implications for Agricultural Research and Policy in Developing Countries.* Johns Hopkins University, Baltimore, Maryland.

Heisey, P.W., Smale, M., Byerlee, D. and Souza, E. (1997) Wheat rusts and the costs of genetic diversity in the Punjab of Pakistan. *American Journal of Agricultural Economics* 79, 726–737.

Herren, H.R. and Neuenschwander, P. (1991) Biological control of cassava pests in Africa. *Annual Review of Entomology* 36, 257–283.

Ingham, R.E., Trofymow, J.A., Ingham, E.R. and Coleman, D.C. (1985) Interactions of bacteria, fungi and their nematode grazers on nutrient cycling and plant growth. *Ecological Monographs* 55, 119–140.

Ingram, D.S. (1998) Everything in the garden's lovely. *New Scientist* 2154, 44.

James, C. (1998) Global Review of Commercialized Transgenic Crops: 1998. *ISAAA Briefs* 8–1998.

Janzen, D. (1973) Tropical Agroecosystems. *Science* 182, 1212–1219.

Krattiger, A. (1998) The importance of ag-biotech to global prosperity. *ISAAA Briefs* 6–1998.

Ladizinsky, G. (1985) Founder effect in crop plant evolution. *Economic Botany* 39, 191–199.

Lavelle, P. (1994) Faunal activities and soil processes: adaptive strategies that determine ecosystem function. In: *XV ISSS Congress Proceedings,* Vol. 1: *Introductory Conferences.* Acapulco, Mexico, pp. 189–220.

Lenné, J.M. (1998) The biodiversity and conservation of crops for disease resistance. Symposium presentation. In: *7th International Congress of Plant Pathology,* Edinburgh, August, 1998, Abstract 4.1.1S.

Lenné, J.M. and Wood, D. (1991) Plant diseases and the use of wild germplasm. *Annual Review of Phytopathology* 29, 35–68.

Lipton, M. and Longhurst, R. (1989) *New Seeds and Poor People.* Unwin Hyman, London.

Marshall, D.R. (1977) The advantages and hazards of genetic homogeneity. *Annals of the New York Academy of Sciences* 287, 1–20.

Masters, G.J., Brown, V.K. and Gange, A.C. (1990) Plant mediated interactions between above- and below-ground insect herbivores. *Oikos* 66, 148–151.

Maxted, N., Ford-Lloyd, B.V. and Hawkes, J.G. (1997) *Plant Genetic Conservation: the* in situ *Approach.* Chapman & Hall, London.

May, R.M. (1973) *Stability and Complexity in Model Ecosystems.* Princeton University Press, Princeton, New Jersey.

MLURI (1997) *Annual Report of the Macaulay Land Use Research Institute.* Craigiebuckler, Aberdeen, UK, p. 78.

Morris, M.L. and Heisey, P.W. (1997) Achieving desirable levels of crop genetic diversity in farmers' fields: factors affecting the production and use of improved seed. In: *Proceedings of an International Conference on Building the Basis for the Economic Analysis of Genetic Resources of Crop Plants.* CIMMYT and Stanford University, Palo Alto, California.

Murdoch, W.W. (1975) Diversity, complexity, stability and pest control. *Journal of Applied Ecology* 12, 795–807.

Namkoong, G. (1991) Biodiversity-issues in genetics, forestry and ethics. *The Forestry Chronicle* 68, 438–443.

Newhook, F.J. and Podger, F.D. (1972) The role of *Phytophthora cinnamomi* in Australian and New Zealand forests. *Annual Review of Phytopathology* 10, 299–326.

Odum, E.P. (1969) The strategy of ecosystem development. *Science* 164, 262–270.

Peel, L. and Tribe, D.E. (1983) *World Animal Science, Domestication, Conservation and Use of Animal Resources.* Elsevier Science, Amsterdam.

Perfect, T.J. (1998) *Brown Planthopper: Threat to the Green Revolution.* Inaugural Lecture, The University of Greenwich and the Natural Resources Institute, Chatham Maritime, Kent, 27 May 1998, 17 pp.

Pickersgill, B. (1989) Cytological and genetic evidence on the domestication and diffusion of crops within the Americas. In: Harris, D.R. and Hillman, G.C. (eds) *Foraging and Farming: the Evolution of Plant Exploitation.* Unwin Hyman, London, pp. 426–439.

Pimm, S.L. (1984) The complexity and stability of ecosystems. *Nature* 307, 321–326.

Pingali, P.L. and Gerpachio, R.V. (1997) *Towards Reduced Pesticide Use for Cereal Crops in Asia.* Economics Working Paper 97–04, CIMMYT, Mexico.

Pingali, P.L., Hossain, M. and Gerpacio, R.V. (1997) *Asian Rice Bowls: the Returning Crisis?* CAB International, Wallingford, and IRRI, the Philippines.

Pinstrup-Andersen, P., Pandya-Lorch, R. and Rosengrant, M.W. (1997) The world food situation: recent developments, emerging issues, and long-term trends. *2020 Vision, Food Policy Report,* Washington, DC.

Qualset, C.O., Damania, A.B., Zanatta, A.C.A. and Brush, S.B. (1997) Locally based crop plant conservation. In: Maxted, N., Ford-Lloyd, B.V. and Hawkes, J.G. (eds) *Plant Genetic Conservation: the* in Situ *Approach.* Chapman & Hall, London, pp. 160–175.

Rajaram, S., Singh, R.P. and Torres, E. (1988) Current CIMMYT approaches in breeding wheat for rust resistance. In: Simmonds, N.W. and Rajaram, S. (eds) *Breeding Strategies for Rusts of Wheat.* CIMMYT, Mexico, pp. 101–118.

Rejesus, R.M., Smale, M. and van Ginkel, M. (1996) Wheat breeders' perspectives on genetic diversity and germplasm use: findings from an international survey. *Plant Varieties and Seeds* 9, 129–147.

Rosengrant, M.W., Agcaoili-Sombilla and Perez, N.D. (1995) Global Food Projections to 2020. Implications for Investment. *Food, Agriculture and Environment Discussion Paper* 5, IFPRI, Washington, DC.

Schultz, T.W. (1964) *Transforming Traditional Agriculture.* Yale University Press, Newhaven, Connecticut.

Sen, S. (1975) *Reaping the Green Revolution: Food and Jobs for All.* Tata McGraw-Hill, New Delhi.

Settle, W.H., Ariawan, H., Astuti, E.T., Cahayana, W., Hakim, A.L., Hindayana, D., Sri Lestari, A. and Pajarningsih (1996) Managing tropical rice pests through conservation of generalist natural enemies and alternative prey. *Ecology* 77, 1975–1988.

Simmonds, N.W. (1979) *Principles of Crop Improvement.* Longman, Essex.

Smale, M., Hartell, J., Heisey, P.W. and Senauer, B. (1998) The contribution of genetic resources and diversity to wheat production in the Punjab of Pakistan. *American Journal of Agricultural Economics* 80, 482–493.

Smartt, J. and Simmonds, N.W. (1995) *Evolution of Crop Plants.* Longman, London.

Smithson, J.B. and Lenné, J.M. (1996) Varietal mixtures: a viable strategy for sustainable productivity in subsistence agriculture. *Annals of Applied Biology* 128, 127–158.

Southwood, T.R.E. and Way, M.J. (1970) Ecological background to pest management. In: Rabb, R.L. and Guthrie, F.E. (eds) *Concepts of Pest Management*, North Carolina State University, Raleigh.

Staskawicz, B.J. (1998) Molecular biology of disease resistance: from genetic understanding to control. *Seventh International Congress of Plant Pathology*, Edinburgh, 1998, Abstract 3.2.PL.

Tanksley, S.D. and McCouch, S.R. (1997) Seed banks and molecular maps: unlocking genetic potential from the wild. *Science* 277, 1063–1066.

Thomas, M. and Waage, J. (1996) *Integration of Biological Control and Host Plant Resistance Breeding: Scientific and Literature Review.* Technical Centre for Agricultural and Rural Cooperation (ACP-EEC), CTA, Wageningen.

Thompson, R.L. (1998) Viewpoint: Public policy for sustainable agriculture and rural equity. *Food Policy* 23, 1–7.

Thresh, J.M. (1981) *Pests, Pathogens and Vegetation.* The Pitman Press, Bath.

Thurston, H.D. (1992) *Sustainable Practices for Pest Management in Traditional Farming Systems.* Westview Press, Boulder, Colorado.

Tinker, P.B. (1988) Efficiency of agricultural industry in relation to the environment In: Park, J.R. (ed.) *Environmental Management in Agriculture: European Perspectives.* Belhaven Press, London, pp. 7–20.

Tripp, R. (1996) Biodiversity and modern crop varieties: sharpening the debate. *Agriculture and Human Values* 13, 48–63.

UNEP (1996) *Consideration of Agricultural Biodiversity under the Convention on Biological Diversity.* CBD Secretariat Paper UNEP/CBD/COP/3/14. UNEP, Nairobi.

Wardle, D.A. and Parkinson, D. (1991) Relative importance of the effects of 2,4-D, glyphosate and environmental variables on the soil microbial biomass. *Plant and Soil* 122, 29–37.

Wardle, D.A., Bonner, K.I. and Nicholson, K.S. (1997a) Biodiversity and plant litter: experimental evidence which does not support the view that enhanced species richness improves ecosystem function. *Oikos* 79, 247–258.

Wardle, D.A., Zackrisson, O., Hörnberg, G. and Gallet, C. (1997b) The influence of island area on ecosystem properties. *Science* 277, 1296–1299.

Way, M.J. (1966) The natural environment and integrated methods of pest control. *Journal of Applied Ecology* 3 (Suppl.), 29–32.

Way, M.J. (1977) Pest and disease status in mixed stands vs monoculture: the relevance of ecosystem stability. In: Cherrett, J.M. and Sagar, G.R. (eds) *Origins of Pest, Parasite, Disease and Weed Problems.* Blackwell Scientific Publications, Oxford, pp. 127–138.

Way, M.J. and Heong, K.L. (1994) The role of biodiversity in the dynamics and management of insect pests of tropical irrigated rice – a review. *Bulletin of Entomological Research* 84, 567–587.

Weste, G. and Marks, G.C. (1987) The biology of *Phytophthora cinnamomi* in Australasian forests. *Annual Review of Phytopathology* 25, 207–273.

Witcombe, J.R., Virk, D.S. and Farrington, J. (1998) *Seeds of Choice. Making the Most of New Varieties for Small Farmers.* Oxford and IBH Publishing Co., New Delhi.

Wolfe, M.S. (1985) The current status and prospects of multiline cultivars and varietal mixtures for disease resistance. *Annual Review of Phytopathology* 23, 251–273.

Wood, D. (1996) The benign effect of some agricultural specialization on the environment. *Ecological Economics* 19, 107–111.

Wood, D. and Lenné, J.M. (1993) Dynamic management of domesticated biodiversity by farming communities. In: *Proceedings of the Norway/UNEP Expert Conference on Biodiversity, Trondheim, May, 1993*, pp. 84–98.

Wood, D. and Lenné, J.M. (1997) The conservation of agrobiodiversity on-farm: questioning the emerging paradigm. *Biodiversity and Conservation* 6, 109–129.

Index

Note: page numbers in *italics* refer to figures and tables